The Ecor
Tou

D1428732

This new edition of *The Economics of Tourism* reflects the tremendous changes that have occurred in the tourism sector in the last twelve years. It recognises that the nature of tourism demand and supply is being transformed by innovations in information communication technologies, market liberalisation and climate change. Paralleling this there is much greater interest in the study of tourism by both students and researchers in mainstream economics.

The text is now in four parts covering: demand; supply; national, regional and international matters; and environmental issues. The concluding chapter appraises the state of the economic research into tourism. The increased interest in tourism has engendered the development of new methods of analysis and the refinement of established ones. Accordingly, the book has been extensively restructured, revised and expanded with two new chapters: chapter six of the first edition is now broken down into two and a new chapter has been added on environmental issues to take account of new developments, critically review the associated literature and consider future trends in tourism economics research. The reader friendliness of the book has also been enhanced in various ways, such as the extensive chapter cross-referencing to refresh the reader's memory and the inclusion of a detailed list of abbreviations.

The Economics of Tourism will continue to make accessible for the non-specialist the application and relevance of economics to tourism. Extensively revised and updated to include research and case studies the textbook will be an indispensable resource for both students and researchers.

Mike J. Stabler is Senior Research Fellow at the University of Reading; his principal research has been on the environmental impacts of economic activity focusing on the tourism sector, published in journals and books. He was joint author of the original edition of this book.

Andreas Papatheodorou is Assistant Professor at the University of the Aegean, Greece, and Honorary Research Fellow at the Nottingham University Business School, UK. His tourism research focuses on competition, pricing and corporate strategy in air transport, cruising and travel distribution and has been published in international academic journals.

Prior to her death in 2006, **Professor M. Thea Sinclair** was Director of the Tourism and Travel Research Institute (TTRI) in the Business School of the University of Nottingham.

The Economics of Tourism

Second edition

Mike J. Stabler,
Andreas Papatheodorou and
M. Thea Sinclair

Routledge
Taylor & Francis Group

LONDON AND NEW YORK

First edition published 1997
Second edition 2010
by Routledge
2 Park Square, Milton Park, Abingdon, Oxon, OX14 4RN

Simultaneously published in the USA and Canada
by Routledge
270 Madison Avenue, New York, NY 10016

*Routledge is an imprint of the Taylor & Francis Group,
an informa business*

© 1997, 2010 Mike J. Stabler, Andreas Papatheodorou, M. Thea Sinclair

Typeset in Times New Roman and Franklin Gothic
by Keyword Group Ltd
Printed and bound in Great Britain by CPI Antony Rowe,
Chippenham, Wiltshire

British Library Cataloguing in Publication Data
A catalogue record for this book is available from the British Library

Library of Congress Cataloging in Publication Data
Stabler, Mike.
Economics of tourism / Mike J. Stabler, Andreas Papatheodorou,
M. Thea Sinclair. – 2nd ed.
p. cm.
Previous edition has main entry under M. Thea Sinclair.
Includes bibliographical references and index.
1. Tourism–Great Britain–Marketing. 2. Tourism–Economic aspects.
3. Leisure–Economic aspects. I. Papatheodorou, Andreas, 1974– II.
Sinclair, M. Thea. III. Title.
G155.G7S73 2010
338.4′791–dc22
2009023195

ISBN 13: 978-0-415-45938-9 (hbk)
ISBN 13: 978-0-415-45939-6 (pbk)
ISBN 13: 978-0-203-86427-2 (ebk)

ISBN 10: 0-415-45938-9 (hbk)
ISBN 10: 0-415-45939-7 (pbk)
ISBN 10: 0-203-86427-1 (ebk)

To Betty, Thea's mother, and
Laurie, Thea's partner

In memory of Thea Sinclair,
16 April 1950 to 26 September
2006

CONTENTS

FIGURES AND TABLES

Figures

Tables

AUTHOR PROFILES

The late M. Thea Sinclair

Thea largely took responsibility for Chapters 2, 3 and 6 (which reflected her research interests outlined below) of the first edition of this book, published in 1997. Such was the quality of her contribution that, apart from adding developments in analysis that have occurred since 1997 and updating the case studies and references, her material remains as the core of the same Chapters, except that Chapter 6 has been split into two to form the current Chapters 6 and 7 by distinguishing between regional and international aspects of tourism.

It should also be acknowledged that Thea had a strong influence on all the content and structure of the original edition that still pervades this second one. Therefore it seems only fitting, with the agreement of her family, that she continues to be named as an author, on an equal footing with the current authors.

Thea studied at the University of Reading in the late 1960s and obtained a BA in economics and sociology and MA in European studies. She registered for a MPhil. at the same university as a preliminary step in gaining her doctorate after being appointed as a lecturer and subsequently senior lecturer in Keynes College at the University of Kent. While at Kent she later became Director of the Tourism Research Centre. She was appointed as the first Director, with the title of Professor, of the Tourism and Travel Research Institute (TTRI) in the Business School at the University of Nottingham in 1998, a position held until her death in September 2006.

Her specialization in the economics of tourism throughout her academic career grew out of a study of the application of the Keynesian multiplier analysis to tourism in Spain. She widened her research by investigating tourist demand and pricing, collaborating with others in both the UK and Spain, and also considered international issues, especially with respect to those of developing countries using tourism as a means of securing economic growth and development; she took a particular interest in gender roles and inequities. At TTRI, in conjunction with colleagues, she expanded her study of the impact of tourism through the application of economic models such as CGE and TSA. She also undertook research into tourism policy and forecasting, crisis management, productivity and poverty relief in the sector.

Mike J. Stabler

Mike was a late entrant into academia, having previously been in local government and commercial management. He gained a BSc at the University of Southampton in the social sciences, specializing in economics, sociology and statistics. On graduating he was appointed as a Lecturer and eventually Senior Lecturer in the Centre for Spatial and Real Estate Economics in the Faculty of Urban and Regional Studies at the University of Reading. He took early retirement in 1997 and was made an Honorary Senior Research Fellow in the Centre, a position which he still holds.

Mike's research interests originated in the leisure field, in which he published a number of papers and gave presentations at many conferences in the UK and internationally. He was a member of the Leisure Studies Association and served on the editorial board of the journal *Leisure Studies*, becoming editor for two years. He formed and chaired the Leisure Economics Group for a number of years. He moved into tourism research, concentrating on its environmental effects, and became Joint Director of the Tourism Research and Policy Unit at Reading until his retirement. He was a member of the editorial boards of *Tourism Management* and *Tourism and Hospitality Research* for several years. Mike is the author of books or chapters in books and several articles in academic journals on rural and urban environmental issues related to tourism; he also has edited or co-edited books on the same topic areas.

E-mail: M.J.Stabler@reading.ac.uk and mike.j.stabler@btinternet.com

Andreas Papatheodorou

Currently Andreas is an Assistant Professor in industrial economics with an emphasis on tourism at the School of Business Administration, University of the Aegean, Greece. He is also an Honorary Research Fellow at the University of Nottingham Business School. He gained an MPhil. in economics and a DPhil in geography at the University of Oxford and commenced his academic career at the University of Surrey. He is a fellow of the UK Tourism Society, a board member of the Hellenic Aviation Society; he is also a Partner in the Air Consulting Group; and he presided over the executive committee of the 2008 Air Transport Research Society world conference, held in Athens, Greece. In 2009 he was recognized as an Emerging Scholar of Distinction by the International Academy for the Study of Tourism.

His principal research interest is tourism, focusing on issues related to competition, pricing and corporate strategy in air transport and travel distribution, mostly in the Mediterranean region. Andreas has had articles published in many international academic journals and edited or co-edited three books. He is an advisor to the Greek government on tourism policy, education and development; he also runs air transport and tourism executive courses in Africa and the Middle East.

E-Mail: academia@trioptron.org and a.papatheodorou@aegean.gr

Personal Website: www.trioptron.org

PREFACE TO THE FIRST EDITION

Tourism is one of the world's most important activities, involving millions of people, vast sums of money and generating employment in developing and industrialized countries. Yet many aspects of tourism have been ignored.

This book makes a key contribution from an economic standpoint to the understanding of tourism. Examining such issues as demand for tourism, how tourism firms operate in national and global contexts and the effects of tourism on destination areas, the authors explain how economic concepts and techniques can be applied to the subject.

Particular attention is paid to the importance of market failure with reference to identifying the environmental implications of tourism and the means of pursuing the sustainability of both tourism and the resources on which it depends.

The Economics of Tourism presents new insights into the intricacies of tourism demand, firms and markets, their global interrelations and the fundamental contribution of the environment to tourism activities, to offer an accessible, interdisciplinary analysis of the interwoven fields of tourism and economics.

PREFACE TO THE SECOND EDITION

It is over twelve years since the preparation of the first edition of this book was initiated, publication being in late 1997. The publishers started pressing for a new edition in 2002, but it was not until 2006 that serious negotiations took place as both authors were engaged fully in other academic activities. The tragic death of Thea Sinclair in September 2006 was the trigger for the remaining author to be committed to a new edition to be dedicated to her memory. Thea had already identified Andreas Papatheodorou as a person who might be involved in contributing to revisions and additions to the book and he readily agreed to become co-author of this edition.

In the introduction to the original edition it was argued that tourism was an important economic activity that should command serious study, but that it was a subject that at times did not fully engage the interest of mainstream economists. Except for a few areas, such as analysis of demand and the derivation of forecasting models and the impact on income and employment in tourism destinations that were rigorously researched, mostly published in economic journals or texts directed at an economic readership, the literature tended to be angled towards tourism development and management. There were only a handful of books that took a specifically economic perspective and these were essentially introductory texts containing elementary analysis and applications of the subject's principles to tourism.

Over the last ten years, until the advent of the recent global recession, the sector continued to grow in size and importance. Education in and analysis of tourism have increased and developed markedly. More colleges and universities are offering courses in the subject where the relevance of economics has been acknowledged. This century, the expansion of western universities into Asia, particularly China, where the importance of tourism to economic development and growth has been recognized, has in part also encouraged students there to attend courses in western countries and subsequently fostered the study of the industry. Consequently, Asian scholars are making an increasingly valuable contribution to research and the literature. The rise in the study of tourism is exemplified by the incredible explosion in the number of academic journals related to it, now nearing 100. The economic study of the subject has also burgeoned, becoming much more quantitative, increasingly reflecting the long established and accepted approaches in its conventional proposals.

During the same period there has been a shift of focus as now greater emphasis is given to tourism's impact on sociocultural structures and physical environments, in addition to its economic effects. The influence of the almost exponential growth of information technology has transformed demand and supply transactions of tourism goods and services, revolutionizing the structure of the intermediaries, especially travel agency and tour operators, spreading also into the transport and accommodation sectors. Two global issues have become more dominant. One has been the growth of international trade coupled with agreements on and regulation of it. Institutions like the World Trade Organization (WTO), General Agreement on Trade in Services (GATS) and the periodic meetings and actions of the G8 countries hold implications for tourism as well as trade, particularly for developing countries. The other issue is that of the environmental impact of human activity, for which there is now firm evidence of the detrimental effects. It is only recently that politicians have accepted the need for action, which scientists and conservationists have advocated for many years. Tourism both suffers from and contributes to environmental problems.

This new edition retains the exposition of basic concepts, principles and analysis as an essential prerequisite to demonstrating the relevance of economics and its application to tourism. It also maintains its objectives of acting as a source book for the economic tourism literature; likewise its original intention to identify and review trends in the sector and assess their significance is reinforced. Furthermore, it incorporates and examines the introduction of new analytical methodologies and techniques and appraises their explanatory and predictive efficacy.

The revisions and additions to the text have been extensive. The introductory chapter that traces the economic study of tourism has been shortened in order to indicate more strongly the relevance of economics to it. Chapter 2 has been updated to include more recent research on demand. The Gorman–Lancaster characteristics theory is examined because of its relevance to the Hedonic Pricing Method (HPM), which has a number of applications to tourism, as indicated in Chapters 3 and 10, while the section on the social context takes account of changing patterns of behaviour with more examples. Chapter 3 also incorporates developments in demand modelling, with revisions to both the single and system of equations models. The section on demand forecasting has been expanded with a more detailed examination of the techniques. The chapter also revisits the characteristics theory, leading on to choice modelling and error correction techniques, hardly introduced at the time of the publication of the first edition of this book.

Chapters 4 and 5 have been restructured. The theoretical explanation of the various market competitive structures in Chapter 4 has been retained. The examination of the structure-conduct-performance (SCP) model and game theory scenarios applied to tourism, formerly in Chapter 5, has been moved into Chapter 4. Chapter 5 now focuses on the market structures of the tourism supply sectors of transport, accommodation, the intermediaries and attractions.

The original Chapter 6 on the international context of tourism has been split into two. The new Chapter 6 considers the national and regional context of tourism, covering initially the size of the sector, its economic impact, linkages, leakages and displacement effects. The section on multiplier analysis and input–output modelling has been expanded and computable general equilibrium (CGE) introduced. The sections on the instability of earnings, tourism growth models, tourism structural issues and gender roles in employment in the original Chapter 6 have been retained. The New Economic Geography (NEG) approach to regional aspects of tourism and the implications for tourism policy are new.

An additional chapter has been added to the two chapters on the environmental issues associated with tourism. A new Chapter 8 acts as a preface to former Chapters 7 and 8 by putting them in the context of environmental problems and sustainability at an international level, including the likely impact of globalization. The chapter reflects current approaches and concerns in the environmental economics field and relates them to tourism in host countries. International agreements and policies to counter the impacts of environmental degradation are reviewed, as well as the impact of tourism on human-made and natural environments. The structure of Chapters 9 and 10 is retained, but they have been revised to take account of new methods of analysis. The case studies have been updated.

The final Chapter (11) has been recast to give a more forward look of trends in tourism and research into it.

FOREWORD

As a service sector activity, tourism is challenging to economists, as it does not fit the traditional mould of an industry or economic sector. Yet, it is this very challenge that has led to the long pedigree of distinguished economists applying themselves to tourism. At the same time, with the maturing of tourism as a subject area, authors and publishers are beginning to deliver themed reviews of the tourism literature. The first edition of *The Economics of Tourism* was published in 1997 and was one of the early examples of these reviews.

The first edition of *The Economics of Tourism* has already become a classic. This second edition is more poignant following the tragic death of Thea Sinclair, one of the drivers of the first edition, and much of Thea's thinking and work remains in the second edition; indeed the book is dedicated to her memory.

Two leading tourism economists, Mike J. Stabler and Andreas Papatheodorou, have written the second edition. The book is impressive because not only does it cover extensively the classic economic subject areas of tourism demand, the economics of tourism supply and issues of competitiveness and the international context, but it also addresses contemporary issues such as environmental economics, sustainability and tourism futures.

The book begins with a wide-ranging and insightful chapter on conceptualizing tourism and the economics of tourism. This chapter sets the scene for a book that impresses with both its scholarship and thoroughness. In this second edition, the authors have delivered a volume that is readable, international in scope and extensive in coverage of the field – offering insights into how economists view the activity that is tourism.

Each of the chapters represents a major critical review of the literature and thinking in the key areas of tourism economics, drawing upon the literature from mainstream economics as well as tourism economics and other disciplines. Each chapter builds upon basic concepts and theories before demonstrating their relevance and application to tourism. Here the authors have ensured that the chapters are well researched, thoughtful and will become classics of the field.

This second edition has been extensively revised and updated to make it not only an exceptional work of scholarship and a 'must have' book for all tourism researchers, but also one that is accessible to non-economists – such as myself. This is a book that belongs on the shelves of all tourism researchers and teachers, as well as in the libraries of the world.

Chris Cooper
Dean of School,
Oxford Brookes University Business School.
July 2009

ACKNOWLEDGEMENTS

We wish to thank a number of people for their encouragement and assistance in bringing to fruition the project to prepare a second edition of this book dedicated to the memory of Thea Sinclair. Many people in academia who knew and loved Thea were very supportive and willingly gave their time in answering our many questions about their research and understanding of tourism.

First, we are grateful to the three anonymous reviewers who unanimously stated that a second edition of the book was certainly warranted and who were very complimentary about a number of features of the original publication, urging us to retain its general content and rigorous approach. They were constructive in their observations on the weaknesses and omissions they found and brought to our attention the developments in the study of tourism that should be included in the new edition, directing us also to what they considered to be seminal publications in the field.

We are indebted to Steve Wanhill, who most graciously and generously gave us permission to draw freely on his keynote address to the Tourism and Travel Research Institute's annual conference in December 2007, in which he analysed the submissions to the journal *Tourism Economics*. It was Steve's interpretation in his presentation of the trends in tourism study, reflecting what economic scholars perceived as important research areas over the years since the inauguration of the journal, which informed the authors that their general approach to the selection of the content and emphasis of the new edition was basically sound and gave them confidence to proceed with the structure decided on.

Many, many thanks are due to Erlet Cater, with an admirable competency in economic analysis, in the Department of Geography at the University of Reading, for the insights she gave us on the direction being taken by her own subject's investigation of tourism. Her help and support are much appreciated in offering the run of her extensive collection of publications, and her willingness to read and give advice on the drafts of chapters submitted to her for an opinion.

We also wish to thank Elliott Kefalas, Editorial Advisor of *Air Transport News*, for providing us with the photograph that graces the front cover of the book. Furthermore, the authors wish to acknowledge the support of Irini Dimou for allowing us to use material from her research on transaction costs and agency theories. We are grateful to Pavlos Arvanitis, Manos Dardoufas, Panagiotis Panagopoulos, Konstantinos Polychroniadis and Petros Zenelis for contributing to the preparation of the manuscript. We would also like to thank Adam Blake and Neelu Seetaram for their constructive comments on passages within their areas of expertise of tourism analysis. Abi Swinburn, past secretary to CSpree, the Business School, the University of Reading, took on the onerous task of compiling the complicated Figures 9.1 and 9.2 from somewhat scrappy hand-prepared copies. She is thanked most sincerely for her perseverance; the resulting figures are excellent.

We thank Chris Cooper, Dean of School, Oxford Brookes University Business School, for his ready willingness to provide a foreword. Also, we are most grateful for the endorsments of the book by Geoffrey Lipman, Chair, Advisory Board, De Haan Institute, University of Nottingham, Assistant Secretary General of UNWTO and Haiyan Song, Chair Professor of Tourism, School of Hotel and Tourism Management, The Hong Kong Polytechnic University.

Michael P. Jones, at Routledge, was very supportive and displayed great patience in the face of many, and often rather naïve and obvious, questions put to him over many months. Moreover, we appreciate the extension granted by him for the submission of the manuscript, when we discovered the original deadline was unrealistic, given the magnitude of the task of updating and revising the book.

Finally, we are grateful to our respective families for their encouragement and support during the prolonged and stressful process of producing the second edition of this book.

Mike J. Stabler
Andreas Papatheodorou

ABBREVIATIONS, ACRONYMS AND GLOSSARY OF TERMS

ABTA	Association of British Travel Agents
ACI	Airports Council International
ADB	Asian Development Bank
ADLM	Autoregressive Distributed Lag Model
AERE	Association of Environmental and Resource Economists
AGE	Applied General Equilibrium
AIC	Akaike Information Criterion
AIDS	Almost Ideal Demand System
AIT	Air Inclusive Tour
APC	ATOL Protection Contribution
AR	Average Revenue
ARIMA	Autoregressive Integrated Moving Average
ASTA	American Society of Travel Agents
ATOL	Air Travel Organizers' Licensing
AuER	*Australian Economic Review*
BA	British Airways
BAA	British Airports Autority
BAT	Best Available Technology
BATNIEC	Best Available Technology Not Involving Excessive Cost
BPEO	Best Practicable Environmental Option
BPM	Best Practical Means
BWIA	British West Indian Airways
CAA	Civil Aviation Authority
CAMPFIRE	Communal Area Management Programme for Indigenous Resources
CBA	Cost–Benefit Analysis
CCA	Conservation Corporation Africa

CEA	Cost–Effectiveness Analysis
CFCs	Chloro Fluoro Carbons
CGE	Computable General Equilibrium (analysis)
CIA	Community Impact Analysis
CIE	Community Impact Evaluation
CITES	Convention on International Trade in Endangered Species
CM	Choice Modelling
CMEV	Commercial Marginal Economic Value
CPI	Consumer Price Index
CRS	Computer Reservation Systems
CVM	Contingent Valuation Method
DA	Damage Assessment
DFID	Department for International Development (in the UK)
DIY	Do It Yourself
DMIS	Destination Management Information Systems
DTLR	Department for Transport, Local Government and the Regions (in the UK)
EC	European Commission
ECAA	European Common Aviation Area
EC–LAIDS	Error Correction–Linear Almost Ideal Demand System
ECM	Error Correction Model (technique)
EE	*Ecological Economics*
EEA	European Economic Area
EIA	Environmental Impact Analysis
EKC	Environmental Kuznets Curve
ERR	External Rate of Return
ESA	Environmentally Sensitive Area
EU	European Union
FDI	Foreign Direct Investment
FFP	Frequent Flyer Programme
FOE	Friends of the Earth
FSC	Forestry Stewardship Council
FTO	Federation of Tour Operators
GAMS	General Algebraic Modelling System
GATS	General Agreement on Trade in Services
GATT	General Agreement on Tariffs and Trade
GDP	Gross Domestic Product
GDS	Global Distribution Systems
GEF	Global Environmental Facility
GEMPACK	General Equilibrium Modelling Package
GMEF	Global Ministerial Environmental Forum
GNP	Gross National Product

GPI	Genuine Progress Indicator
HDI	Human Development Index
HERMES	Harnessing Employment Regional Mobility Entrepreneurship in South East Europe
HFCs	Hydro Fluoro Carbons
HKIA	Hong Kong International Airport
H–O	Heckscher–Ohlin (theorem)
HP/HPM	Hedonic Pricing Method
HPF	Household Production Function
HPTCM	Hedonic Pricing Travel Cost Method
IATA	International Air Transport Association
IATE	International Association for Tourism Economics
ICAO	International Civil Aviation Organization
ICOR	Incremental Capital–Output Ratio
ICT	Information Communication Technologies
IISD	International Institute for Sustainable Development
ILO	International Labour Organization
I–O	Input–Output (analysis)
IRR	Internal Rate of Return
ISEE	International Society for Ecological Economics
ISEW	Index of Sustainable Economic Welfare
ISIC	International Standard Industrial Classification
ITTO	International Tropical Timber Organization
IUCN	International Union for the Conservation of Nature
JAL	Japan Air Lines
JEEM	*Journal of Environmental Economics and Management*
JOST	*Journal of Sustainable Tourism*
Kh	Human Capital
Km	Human-Made Capital
Kn	Natural Capital
Kt	Total Capital (stock)
LAIDS	Linear Almost Ideal Demand System
LCC	Low-Cost Carriers
MARPOL	International Regulations on Maritime Pollution
MC	Marginal Cost
MCA	Multiple Criteria Analysis
MCDA	Multiple Criteria Decision Analysis
MPB	Marginal Private Benefit
MPC	Marginal Private Cost
MR	Marginal Revenue
MSB	Marginal Social Benefit
MSC	Marginal Social Cost
MSY	Maximum Sustainable Yield

NAFTA	North American Free Trade Agreement
NAIRU	Non-Accelerating Inflation Rate of Unemployment
NDP	Net Domestic Product
NEF	New Economics Foundation
NEG	New Economic Geography
NERA	National Economic Research Associates
NGO	Non-Governmental Organization
NMEV	Net Marginal Economic Value
NNP	Net National Product
NOAA	National Oceanic Atmospheric Administration
NPV	Net Present Value
NRR	Non-Renewable Resource(s)
NSW	New South Wales
ODI	Overseas Development Institute (in the UK)
OECD	Organization for Economic Cooperation and Development
OLI	Ownership–Location–Internalization (Framework) Localization
PBSA	Planning Balance Sheet Analysis
PCB	Polychlorinated Biphenyls
PFC	Perfluorinated Compounds
PFM	Production Function Method
PSO	Public Service Obligation
PTCM	Pooled Travel Cost Method
PTE	Passenger Transport Executive
PV	Present Value
REEP	*Review of Environmental Economics and Policy*
RIA	Resources Impact Analysis
RoA	Rest of Australia
ROR	Rate of Return
RR	Renewable Resource(s)
RSPB	Royal Society for the Protection of Birds (in the UK)
RSS	Really Simple Syndication
SABRE	Semi-Automated Business Research Environment
SARS	Severe Acute Respiratory Syndrome
SBC	Schwarz–Bayesian Criterion
SCP	Structure–Conduct–Performance (model)
SD	Sustainable Development
SDIC	Sensitive Dependence on Initial Conditions
SEA	Strategic Environmental Assessment
SEM	Structural Equation Modelling
SIC	Standard Industrial Classification
SMS	Safe Minimum Standards

SNA	System of National Accounts
SSNIP	Small but Significant Non-Transitory Increase in Price
ST	Sustainable Tourism
STD	Sustainable Tourism Development
TALC	Tourism Area Life Cycle
TCM	Travel Cost Method
TEV	Total Economic Value
TGDP	Tourism Gross Domestic Product
TGF	Trip Generation Function
TMEV	Tourism Marginal Economic Value
TNC	Transnational Corporation (or Company)
TOPS	Thomson Online Program System
TSA	Tourism Satellite Accounts
TTRI	Tourism and Travel Research Institute (in Nottingham University UK)
TVA	Tourism Value Added
TVP-EC-LAIDS	Time Varying Parameter–Error Correction–Linear Almost Ideal Demand System
UCARIMA	Unobserved Components Autoregressive Integrated Moving Average
UK	United Kingdom
UN	United Nations
UNCSD	United Nations Commission on Sustainable Development
UNDP	United Nations Development Programme
UNEP	United Nations Environmental Programme
UNFCCC	United Nations Framework Convention on Climate Change
UNWTO	United Nations World Tourism Organization
US	United States
USTOA	United States Tour Operators Association
VAR	Vector Autoregression
VAT	Value Added Tax
VATI	Value Added of Individual Tourism Industries
VFR	Visiting Friends and Relatives
WCED	World Commission on Environmental Development
WHO	World Health Organization
WTA	Willingness to Accept
WTO	World Trade Organization
WTP	Willingness to Pay
WTTC	World Travel and Tourism Council
WWF	World Wildlife Fund

PART I
Introduction and demand theory in tourism

1 THE SCOPE AND CONTENT OF THE ECONOMICS OF TOURISM

1.1 Introduction

The aim of this book is to demonstrate the extent to which tourism is susceptible to economic analysis and to provide a review and critical evaluation of the tourism literature emanating from a number of disciplines, also to indicate possible directions for future research. The book is not conceived as simply a text aiming to introduce the application of economics to tourism to non-specialist readers. It attempts to go further by demonstrating the subject's ability to strengthen the theoretical foundations of the more descriptive, diffuse and pragmatic approaches to the analysis of tourism and by showing the potential of economics to explain and predict tourism phenomena and inform government and business operational policy in the sector. By examining tourism, using concepts, principles and methodologies from mainstream and more specialized fields of economic analysis, it also contributes new perspectives on an activity of major and increasing economic importance which until the 1990s was largely neglected in the economics literature. As identified in the preface to this new edition, the book introduces more recent and advanced applications of economic analytical methods of a quantitative nature, so going beyond a simple exposition of principles. It also raises issues of global importance that reflect rapid changes in the economic, ecological, environmental, political and social systems that tourism, an activity of major importance, both has an impact on and is affected by.

An important objective of the book has been to make it as accessible as possible, both to an economic audience and to those to whom economic concepts and methods are less familiar. To this end, theories and analytical approaches have been discussed in a way that avoids, as far as is possible, the excessive use of technical economic terms and the subject's econometric methods. Given the vast expansion in the economics of tourism in recent years, the topics and references to studies of them covered are necessarily selective. Two factors that determined the selection are, first, the desirability of including the main areas in which economics is relevant to a comprehensive analysis of tourism, and second, the perception of key issues in tourism.

The topics covered include demand, supply, market structure, pricing, output, growth and the environment, in a domestic and international context. The emphasis is on examining the role of quantifiable variables (the very essence of economics) in analyzing tourism phenomena, as this is where the contribution of economics to tourism research is likely to be most significant. It is worth repeating the point made in the previous paragraph that the ultimate focus of economic analysis is on its implications for tourism policies. In particular economists consider that the underlying rationale for their investigations of tourism is to indicate the impact, often leading to unintended outcomes, of its operation and actions to support its development and growth.

It should be noted that, to an extent, the analysis of tourism within the book proceeds from the simple to the more complex in the early chapters in that economic concepts are initially used to examine tourism as though it were a non-composite commodity, similar to other goods, purchased by a relatively homogeneous set of consumers and likewise supplied by businesses in an identifiable industrial sector. That in reality this is not the case is discussed below in examining the debate about conceptualization of tourism, both in the wider and economics literature. Further complexities are gradually introduced throughout the book as attention is paid to the different components of tourism, initially in a closed economy and subsequently in an open or international one. Given the increasing concern over both global and local environmental challenges, a significant proportion of the book is devoted to the interrelationship of tourism with natural and human-made resources, including the use of those that are not traded in markets, necessitating the exploration of methods for estimating their value. Problems associated with the operation of tourism concerning externalities, such as pollution and waste, and inequalities in income and wealth are investigated and the policy instruments, both regulatory and priced-based, that can be applied to mitigate their detrimental effects are analysed. The conclusion of the book evaluates the progress made in the development of the economic analysis of tourism and suggests its explanatory and predictive efficacy. The possible directions required of future research and analysis are indicated.

1.2 The state of analysis of tourism

1.2.1 Background

It is acknowledged that tourism has characteristics that set it apart from other economic activities and therefore pose particular analytical problems. Even as a service sector, it is distinct in that it is usually purchased without inspection, consisting of a range of goods and services which are purchased and/or used, often in sequence, such as reservation agencies, financial services, acquisition of specialized clothing and equipment, transportation, accommodation, food and human-made and natural attractions. Because the last two components are a substantial proportion of total inputs, it is the set of markets giving rise to the non-priced features, referred to above, that create analytical difficulties. It is, thus, the unusual composite nature of tourism, frequently taking place internationally and therefore involving generating and destination countries, which requires specific analysis relating to spatial and temporal factors, very much within the purview of geography, but to which spatial economics, concerned with urban and regional issues, is able to contribute.

This composite nature of the sector and the many-faceted components that constitute what can be referred to as the 'tourism product' supplied through many markets have given rise to a long-standing and occasionally acrimonious debate in the wider literature, echoed in economics, on the nature of tourism. Those studying the phenomenon need to agree on this in order that it is certain that research approaches are consistent, the definition of it is unequivocal and its boundaries are clearly delineated. However, before the current position on its conceptualization can be established, it is instructive to examine briefly the complex nature of tourism studies and how it is bedevilled by controversy because of the many approaches that exist.

With very few exceptions, notably Gray (1966), there was little analysis of tourism until the 1970s and even then much research tended to be uncritically descriptive, employing inadequately defined goals and an agreed approach. No attempt is made here to conduct a comprehensive review of the development of studies since the 1970s as this has been carried out elsewhere in the tourism literature (for example, Sheldon, 1990; Eadington and Redman, 1991; Sinclair, 1991a; Van Doren et al., 1994; Wahab and Cooper, 2001; Shaw and Williams, 2002; Lew and Hall, 2004; Hall and Page, 2006), some of which has offered views on its content and scope. There is, however, a measure of agreement with Pearce and Butler (1993) when they state that the many disciplines that have studied tourism bring with them all their conceptual and methodological baggage.

A considerable number of disciplines have made contributions to the subject, the principal ones being anthropology, ecology, economics, environmental studies, ethics, geography, political economy, politics, social psychology and sociology. The main themes which have been studied are the cultural, economic, environmental and social effects of tourism, travel patterns and modes between origins and destinations, tourism's relationship with economic development, tourists' motivations and behaviour, forecasting trends and practical aspects of planning, management and marketing. More recently attention has been paid to tourism and sustainability, the balance between economic, environmental and social goals (what is known as the triple bottom line), social intra-generational inequity and the relationship of tourism with poverty in destinations, and responsible tourism that is sensitive to the conservation of natural and fragile environments.

As pointed out by those who have long worked in the wider field of tourism studies, a number of aspects of the subject suffer from a poor or even the absence of a theoretical framework through a lack of appropriate research (Sessa, 1984). However, the lack of a theory of tourism is understandable, given its characteristics, identified above. Similarly, concern has been expressed that much writing on the subject has no firm sense of direction and is methodologically unsophisticated (Pearce and Butler, 1993), which can have detrimental consequences for the operation of the sector and policies concerning its development and growth. Nevertheless, there has been a steady flow of contributions over the last two decades to the debate on the need for a theory, emphasizing the difficulties of deriving one, and a definition of tourism and its conceptualization see, for example, Ritchie et al. (2005). It is the definition of tourism that has become the focus of attention recently because of the perception that it is necessary to define it, in order to establish what the scope and content of tourism activity is, to facilitate the pursuit of sustainable tourism operations and development. The link between sustainability and the need to conceptualize tourism is examined in Chapter 8, therefore it is not considered at this point.

Unlike the study of tourism in other disciplines, the economic study of it is set within an established academic one, possessing a formal structure in educational establishments with vehicles for the dissemination of research in a firm theoretical foundation, with clear principles and well-tested methodologies. Its approach to the study and analysis of the sector is described below. This notwithstanding, the subject has similar concerns as to the exact nature and structure of tourism that has identified conceptualization issues, most notably whether or not it is an industry or a system or merely a collection of markets. Thus, an outline of the current state of the attempts to agree on exactly what tourism consists of and the construction of a theoretical framework is now examined.

1.2.2 The conceptualization of tourism in the wider literature

In the wider academic study of tourism there have been many publications on the reasons why it is necessary to define it as a sector and its concomitant conceptualization. The issues can be encapsulated by outlining what Lew et al. (2004), from a largely geographical perspective, assert is the need to define it, and why, by arguing that it is essential, in an empirical context, to do so to identify the nature of its impacts, especially economic, and the associated policy implications. They consider that the supply and demand sides should be distinguished in what they perceive is a system, rather than an industry. The supply side is deemed to be an aggregation of all the businesses that directly provide goods and services involving distribution, commodification and a lifestyle in an interaction with the demand side, consisting of the leisure, recreational and tourism pursuits of those away from home; Lew et al. (2004) recognized that the relationship between tourists and suppliers gives rise to economic, environmental, political and socio-cultural effects. They also acknowledged that the mobility, space and time dimensions of tourism are important determinants of demand, very often underpinning constraints such as cultural and social structures, institutions, occupation, work patterns, wealth, income, gender, age and class. Overall, their conclusions are that, rather than a single disciplinary approach, analysis should include the perspectives of many disciplines, or involve cooperation between them. However, they express concern that an obstacle to this development is that currently there appears to be increased disciplinary specialization within academic journals, notwithstanding the incredible growth in their number.

In the wider literature, the study of tourism is mainly through case studies, emanating from the newer universities and colleges rather than the more prestigious older universities, that concentrate on quite a narrow range of topics that are supported by government departments, non-governmental organizations, tourism bodies and businesses. The social sciences, particularly sociology, predominate in research that focuses on issues such as cultural identities and threats to them, ecotourism, ethnicity, heritage, history, gender roles, post-modernity, poverty, sustainability, tourism development and growth and its impact, especially on natural, human-made and human environments (issues considered from an economic perspective are identified in the next section).

However, the crucial issue of the lack of a theory of tourism, which is perceived as vital for progress to be made in overcoming the shortcomings of research into it, identified earlier in this section on the state of analysis, is that there is an increasing imperative to conceptualize the subject. Related to this is the assertion that the study of tourism should transcend disciplinary boundaries. Most of the key references on the debate have appeared in journal articles. The position is confusing, as there is no clear consensus either on what the disciplinary framework to be adopted should be, or its conceptualization.

Before these two issues are discussed, it is instructive to examine what has been stated about the nature of tourism that, it is argued, should inform the way in which it is analysed and perceived. Parry and Drost (1995), in a relatively early paper on management related to marketing, and Faulkner and Russell (1997) strongly argued for an alternative approach to the study of tourism founded in complexity and chaos theory, which Lorenz (1963) first propounded as a result of his dissatisfaction with the models then used in meteorological forecasting. His observations of uniformity, emerging from what appeared to be ostensibly random occurrences from the initial condition, did not attract attention until several years later when other scientists rediscovered his findings and rapidly saw the relevance of his work to their own. Lorenz's theory has been widely applied in mathematics, statistics, physics and the natural sciences, particularly ecology. Complexity and chaos theory is outlined in the next section in general terms to indicate why it has appealed to those working in tourism studies who perceive that the sector parallels the characteristics of natural systems. The theory is also very relevant to economics both in its mainstream context and as applied to tourism. Reference to its relevance to the subject is justified below and elsewhere in this book.

1.2.3 Complexity and chaos theory

Complexity and chaos theory, hereafter referred to simply as chaos theory, recognizes that many human systems are characterized by openness, complexity, non-linearity, the operations of which are non-deterministic, dynamic and unpredictable. The theory posits that such systems, which appear to be disordered, unstable and subject to seemingly random and entirely unconnected, often exogenous events, do have a discernible underlying pattern. The inherent and periodic instability of systems facilitates change as their states are transformed in an evolutionary and adaptive manner. It is argued that the 'butterfly effect' of chaotic systems, also first suggested by Lorenz, means that relatively minute changes in the early evolutionary stage lead to much larger and often unintended outcomes as they develop in a ripple effect, both over time and spatially. McKercher (1999) interpreted the butterfly effect and the subsequent evolution stages as the sensitive dependence on initial conditions (SDIC). Another aspect of their development is for systems to enjoy increasing returns of survival, where successful changes incorporated into them persist, resulting in a degree of stability until another disruptive change occurs; it is argued that such characteristics are a fundamental feature of natural systems. Furthermore, it is asserted that the application of the theory should be on a 'bottom-up' basis, as opposed to the more traditional top-down positivist method. The disadvantage of positivist analysis is that it tends to over-simplify the representation of the real world, being reductionist and often static in breaking systems down into their constituent parts and considering them in isolation. In contrast a bottom-up approach opens out the analysis by identifying the complexities of systems where stability and dynamic forces influencing them are juxtaposed. Also, the method gives the opportunity to conduct the analysis at a higher level of scale. It is an approach especially relevant to human interaction, through society's institutions and structures and its economic and political systems. Thus chaos theory holds two implications for all forms and fields of research. The first is that the positivist method should be abandoned in favour of a more inductive approach that is wider ranging. The second is that research should not be confined within the boundaries of a single discipline. Without necessarily asserting that tourism is a system, the nature of which examined below, it is apparent that it possesses features that suggest it can be analysed within a chaos theoretical framework.

1.2.4 Complexity and chaos theory in the context of tourism

The interest in chaos theory in relation to crisis management was firmly established in the 1980s (see for example Fink, 1986), but it was not until the 1990s that it was extensively considered in the context of tourism, with many publications on the subject that Ritchie (2004) refers to in a review. Ritchie shows there is widespread consensus that tourism is sensitive to exogenous disaster events and over which those involved in it have no control, but this kind of occurrence should be distinguished from a crisis, which is attributable to internal issues, for example failure to recognize and adapt to change or poor management practices. However, the development of crisis management and its application to tourism is not the principal interest at this point; it is considered in Chapters 4 and 5 on the economics of tourism supply. The focus at this juncture is on tourism's characteristics and the approach to analysing it.

Though not the first to recognize the complex nature of tourism, Russell and Faulkner (1997), in a well-received article, were in the forefront of the academic interest in chaos theory at a theoretical level and its possible application to the sector. They considered tourism to be essentially about human behaviour, consisting of many diverse elements and subject to much uncertainty, and that analysis of it does not fit easily into the linear, deterministic and reductionist positivist paradigm. They argued that as it is now accepted in the natural sciences that this paradigmatic framework no longer applies, it is equally invalid in the social sciences in which tourism has been studied. In reality they consider that chaos theory offers a feasible explanation, that all tourism activity takes place in economic, political and social systems that are inherently complex, disordered, dynamic and subject to continuous disequilibrium, being only periodically stable. Russell and Faulkner also reinforce the argument for breaking out of the straitjacket of the traditional single disciplinary analysis of tourism. In common with many other researchers, such as Gunn (1987), Lieper (1990), Przeclawski (1993), Tribe (1997), Hardy (2002), Farrell and Twining Ward (2004) and Bramwell (2007), they acknowledge that, though, clearly, there is a multidisciplinary approach to its study, that is from the perspective of many subjects, as noted above, this does not mean that it is truly inter- or trans-disciplinary. As the term suggests, interdisciplinary research is that where approaches by a number of scholars from different subjects are combined to investigate a particular topic. Trans-disciplinar approaches essentially consist of joint problem-solving collaborative projects between academic researchers and those in practice. The former offer their expertise in their field and the theoretical framework, while the latter, often stakeholders, provide the empirical content based on their experiences; see, for example, expositions and illustrative case studies of the approach in Dickens (2003) and Nowotny (2003). Russell and Faulkner (1997) go on to relate chaos theory to tourism examples at a macro and micro level and liken both to the theory and experiences of the sector and to Butler's (1980) destination life cycle model. McKercher (1999) is highly critical of thinking on the lack of progress in defining tourism and its functioning, arguing that it is locked in a time warp of its own making. He echoes the line of reasoning of Russell and Faulkner (1997) on the unsatisfactory state of tourism studies in relation to reality and reinforces the merits of chaos theory as a framework that reflects the characteristics of the sector. He identifies eight components or factors, both internal and external, that interrelate with each other to determine the outcomes, in his derivation of a multidimensional chaos model of tourism, linking tourist destination communities via tourism agencies, competing destinations, and economic, political and social systems. Interestingly, McKercher coins the term 'rogues' for individuals, firms or organizations that trigger revolutions in travel and tourism; for example Thomas Cook in the nineteenth century, Freddie Laker, who was the first to introduce cheap trans-Atlantic air travel, and Stelios Haji-Ioannou, who founded easyJet, a low cost airline, and has extended his business into cafes, cinemas, cruise ships and most recently buses.

Following his article with Russell in (1997), Faulkner (2001) indicates the possible application of chaos theory to crisis or disaster management by setting out the stages of its cycle, therefore developing its similarities to Butler's (1980) resort life cycle hypothesis, which he referred to in the earlier paper. The application of chaos theory to crisis management has also been examined in Blake et al. (2003) in a paper concerning the destruction of the World Trade Center in 2001. Ritchie (2004) gives a number of examples related to tourism. What is emerging is that researchers into tourism are increasingly accepting that chaos theory, in addition to contributing to an understanding of the complexity of tourism and its analysis, is also capable of explaining its variability, the uncertainty of intended outcomes, unexpected shocks and its evolutionary nature.

1.2.5 The economic conceptualization of tourism

To an extent economics has not debated the conceptualization of tourism for as long a period as it has been debated in the wider academic context, and accordingly there is not a very large literature on the topic within the discipline. The subject has generally been comfortable within its well-established paradigm stemming from the nineteenth century. The emphasis in economics is more on concerns as to how tourism's structure and operation within the discipline ought to be studied, largely in the field of industrial economics. Wilson (1998), in a review of the relevant literature, alights on an important aspect of economic empirical research in a discussion as to whether tourism is an industry or a market. He considers that it is crucial to clarify the distinction between the two terms in economic analysis, citing the paper by Leiper (1979) and his attempts to define tourism, describing it as a system. The argument Wilson puts forward for a clear definition is to avoid confusion on the methodology to be used for its investigation; should it be one employed in investigating an industry or that applied to markets, or elements of both? He asserts that the conceptualization of tourism is required to settle the issue, also suggesting that there are grounds for accepting that the controversy is not simply one of semantics and therefore is not an arid and fruitless exercise.

Wilson's examination of the various definitions of tourism offered in the literature reveals that they are based on identifying the characteristics of tourists and tourism supply. These are not rehearsed here, because their interest from an economic perspective is that of Lancaster's (1971) characteristics approach, which is discussed in Chapters 2 and 3. Wilson cites Nightingale's (1978) paper on the definition of an industry and market, in drawing out the distinction between them. Wilson concludes that a market is a vehicle for firms to sell goods or services with similar characteristics, whereas an industry is a collection of businesses that use similar processes to produce basically uniform goods and services. Tourism is therefore not an industry, although there are many industries that are engaged in supplying it as a product, where a significant proportion of the supply of its elements arises from businesses solely engaged within it as a phenomenon.

The debate is inconclusive: after ten years, no agreed definition and conceptualization of tourism have emerged. However, the distinction between an industry and a market remains important in interpreting, for analytical purposes, the sector's identification and its contribution to economies, using official statistics based on standard industrial classifications (SIC) that are derived from industry-based data. This has implications for economic models concerned with the impact of tourism, such as Keynesian multiplier, input–output, computable general equilibrium and tourism satellite accounting analyses, which are considered in Chapter 6.

The prescription for economists in these areas of the investigation of tourism is that it cannot be viewed as a single industry. Therefore, the methodologies normally adopted in an industrial economics context can be used as long as the research results are qualified estimates of the sector's role in economies and the assumptions on which they are based are fully spelled out.

Treating tourism as a market, or more correctly as a set of markets, is more problematic. Tourism transactions, like those for any other commodities, involve consumers or tourists, the goods and services they demand and the businesses and industries supplying them. The complex nature of the product, as already indicated above, necessitates several transactions between buyers and sellers to provide all the elements of a holiday. Analysing this, relying on secondary data, has greater implications than an industry-based approach as it is much more difficult, also carrying the danger that measuring tourism's significance as an economic activity is merely indicative, therefore suffering from wide margins of error.

The immediate foregoing discussion, on the distinction between industries and markets in the economic study of tourism, raises the issue as to whether the subject's analysis has reached a threshold where research needs to adopt a broader approach to reflect the nature of the phenomenon identified earlier. This has been acknowledged in a fairly recent publication by Matias et al. (2007) devoted entirely to the economic perspective on tourism research. In the first chapter on new analysis frameworks in tourism economics, Jennings (2007) presents her perception of the theoretical paradigms that should be employed to secure advances in economics tourism research. Jennings' review echoes the debate in the wider tourism literature above concerning the continued unwarranted adherence to positivist, quantitative oriented orthodoxy in the face of tourism's complexity, rapidly changing characteristics and instability, so different from its nature in the twentieth century. She posits that a more holistic approach by economics is required, suggesting seven paradigms that might be selected to inform the diversity of the discipline's research. Other analytical frameworks, models and operational tools are examined in Matias (2007) that extend the range of approaches that economics might adopt. Another direction which economic tourism research can take has been posited by A.M. Williams (2004), who takes a political economy stance. He argues that theoretical developments in the approach have relevance to issues in tourism and illustrates this by considering commodification in the sector's markets, its labour structures and processes and its regulation. Williams' advocacy of a political economy perspective has an affinity with new economic geography, which is examined in Chapter 6 with regard to growth theory.

The broadening of economic approaches to the study of tourism outlined here is instructive in that it indicates the likely directions that it might take in the future. Of particular note is economics' links to other disciplines and the possibilities, and indeed necessity, of participating in enlarged paradigms in cooperative research. The reader should appreciate that notwithstanding the rigour given to the study of tourism by economics, working within its single disciplinary framework, the findings of the subject's analyses must be relevant to the operation of the sector in an empirical context.

The identification here of the fundamental issues confronting the approach to and current state of tourism studies in its broadest context gives an indication of where economics stands in relation to other disciplines.

1.2.6 Concluding observations on the definition and conceptualization of tourism

The inference to be drawn form the foregoing sections, in which a number of opinions on the definition and conceptualization of tourism in both the wider context and within the discipline of economics were identified, is that there are fundamental and unresolved issues concerning research into the sector. Irrespective of the fact that the debate is ongoing, and therefore seemingly has not resulted in a consensus as to how to articulate a theory of tourism that clearly defines and conceptualizes it to inform the future direction of its study, the examination of it has revealed implications for the economic analysis of the sector and its consideration here. The review of the literature above has shown that there is a measure of agreement that it is inherently different from other forms of human and economic activity and appears to breach single disciplinary boundaries, and also requires a specific methodological framework. However, currently, while acknowledging the need for new approaches, the study of tourism largely continues to proceed within traditional disciplinary paradigms, in which, as indicated, there are a variety of definitions and conceptions of it.

This poses a problem as to how to refer to tourism throughout the remainder of this book. Is it a sphere of activity, a system, a number of systems, a sector, a series of sectors, an industry, a collection of industries, a market or set of markets, a loose interrelationship of businesses concerned with the supply of tourism, a circuit even, or simply a phenomenon? There is no generic term that adequately describes it and facilitates its analysis. Agreeing with Wilson (1998) that it cannot be considered as a single industry, the authors of this book are minded to subscribe to the view that it is best described as a set of markets, or a sector of the economic system. However, these are perhaps unduly clumsy ways of describing it. Possibly ducking the issue of the terminology to be adopted, it has been decided that, in discussing the literature, the terms employed by the researchers referred to will be respected and thus adhered to in quoting their work. The upshot is that the various terms for tourism tend to be used interchangeably in the book, although most often tourism is referred to as a sector, notwithstanding the possible confusion that in economics they relate to the classification of industries.

This discussion on the definition and conceptualization of, and the approach to the analysis of, tourism understandably begs the question as to why so much attention should be devoted to what might seem to be simply semantics regarding what it is and how research into it should be conducted. Prima facie, most disciplines have little difficulty in working within their respective methodological frameworks to yield meaningful results. For example, to assist businesses involved with practical matters, such as marketing and management, or to advise organizations and governments on tourism policies and their implementation at a more strategic level.

The justification for examining the controversies is that despite nearly a decade of debate since the turn of the century, as shown above, a discernible convergence of the perspectives of many disciplines has emerged. This holds out the prospect in the wider field of tourism of giving a more cohesive and tighter analytical framework of analysis that should meet the criticisms that much research is largely descriptive, fragmented and lacks theoretical rigour. However, the intention here is not to expand on the impact of the convergent trend in the more general context of tourism study, but to consider the implications for the economics of the phenomenon.

The section above on the definition and conceptualization of tourism within economics indicated that it is important to establish both, for there are implications for empirical research using secondary data. It was shown that the distinction as to whether it is an industry or market is crucial when the data has been collected on an industrial sector basis, but attempts are made to analyse and interpret it to reflect the characteristics and structure of tourism. In effect there is a lack of correspondence between the data and reality. Furthermore, it was implied that traditional methodological approaches embodied in industrial economics are inappropriate; this issue is discussed in Chapters 4 and 5. There are more fundamental arguments for re-conceptualizing economics and its methods when considering the interrelationship between it and the wider biosphere, covered in Chapter 9. The need to extend and develop the forms of economic analysis is examined in Chapters 8–10, which consider the linkage between tourism and the environment, especially moving towards the ultimate goal of sustainability. Finally, in the last chapter, the examination above of the debate on possible redefinition and re-conceptualization of tourism economics is taken up in a discussion that assesses its development over the recent past and its likely future directions. A fundamental issue examined is whether the relevance of the economic analysis of tourism at an operational level is likely to be impaired if, as currently, it largely remains within a single disciplinary framework.

1.3 The content and scope of the economics of tourism

The reviewers of the first edition of this book, prior to the decision to proceed with a new one, gave valuable guidance on what content should be expanded and updated and the nature of the new material that ought to be incorporated, also directing the authors to seminal publications. Mindful of the developments in the study of tourism since the original publication of this book, the literature search for the preparation of this new edition concentrated on developments in the study of tourism over the last decade or so. Initially attention was centred on the wider academic fields of tourism research to identify contributions that took an economic approach. Then the literature more specifically located in economic publications was examined.

Prior to adopting the position identified in the next paragraph, which is considered to be representative of the current nature of economics research into tourism, it is acknowledged that the literature reflecting its approach to the sector takes a variety of forms and comes from a number of sources. There are papers, mainly by academics, emanating from: university schools and departments (under the auspices of a wide range of disciplines); conferences; commissioned research by governments, tourism bodies and research organizations. Many such papers are subsequently submitted for publication as chapters in books, or as whole texts, devoted to specific aspects of tourism. Alternatively, research material from the same sources are submitted directly to book publishers (usually those specializing in tourism studies), either as proposals by the authors or by invitation from commissioning editors. Likewise, papers are most often submitted directly to journals. A significant number of papers of relevance to tourism have been published in mainstream economics journals such as the *American Economic Review, Australian Economic Review, Journal of Economic Surveys, Econometrica, Economica, Economic Journal, Journal of Political Economy, Quarterly Journal of Economics* and *Review of Economic Studies*. There have also been articles in the more applied fields of economics, for example in *Applied Economics, Industrial Economics, Journal of Law and Economics, Journal of Urban Economics* and *Regional Studies*. More recently the newer *Ecological Economics, Journal of Environmental Economics and Management* and *Review of Environmental Economics and Policy* have published papers that have a bearing on the operation of the tourism sector. In the wider context, where there are now a large number of tourism related journals, publications such as the *Journal of Travel and Tourism Research, Pacific Tourism Review* and *Tourism Analysis* have included articles with an economic perspective. Additionally, the relatively new *Journal of Ecotourism* and *Journal of Sustainable Tourism* contain articles that have environmental economic implications. However, the long-standing and well-regarded *Annals of Tourism Research and Tourism Management* are pre-eminent with regard to the inclusion of articles which focus on the economics of the sector.

Notwithstanding the acknowledgement in the previous paragraph that studies of tourism from an economic perspective have been published in a wide range of books and journals, both in those considered as within the conventional context of the subject and in the tourism field, it was the advent of the journal *Tourism Economics* in 1995 that began to delineate more clearly the economic interest in the phenomenon and thus the content and scope of research. From a paper delivered at the first IATE (International Association for Tourism Economics) and the TTRI (Tourism and Travel Research Institute) conferences by the journal's editor, Wanhill (2007), who subsequently generously gave permission to quote from it extensively, it is possible to gauge from submissions to the journal, as opposed to published articles, the tourism issues that researchers consider are important. A content analysis of papers published in the journal throughout 2008 (the manuscript for this book was completed for publication early in 2009), was resisted as an update to Wanhill's review of submissions. It was assumed that most articles published in the 2008 volume would have already been considered by the end of 2007.

The broad distribution of seven categories of subject areas identified by Wanhill were:

- Tourism industry
- Macroeconomic assessment
- Demand modelling
- Tourism governance
- Supply issues
- Data analysis
- Statistical theory

Ten special issues between 1999 and 2008 covered economic impact (three), tourism modelling and policy, competitiveness, safety and security, seasonality, issues in Australian and island tourism, and sustainability. Of these special publications, the third and fourth involved the late Thea Sinclair, one of the authors of the first edition of this book.

Over a quarter of submissions related to what Wanhill describes as the tourism industry. Macroeconomic assessment, demand modelling and tourism governance each accounted for just under a fifth. Supply issues constituted a tenth and the remaining two, data analysis and statistical theory, were 5 per cent and markets 4 per cent of total submissions.

Looking at Wanhill's categories in more detail reveals that researchers' perceptions of the fundamental issues in the tourism industry are very much concerned with facilities and services, where just over a third of submissions related to heritage/cultural attractions and events. Of the remainder, transport (largely air travel) accounted for 16 per cent, finance (embracing management accounting, leverage, yield, portfolio analysis and risk assessment) 15 per cent, accommodation (mostly hotels) 12 per cent, recreational activities 12 per cent, travel intermediaries, including information technology, 7 per cent and restaurants 4 per cent. The tourism sector, examined in Chapters 4 and 5, covers the principal features identified here.

Macroeconomic assessment in submissions to *Tourism Economics* consists of five topics, of which destinations and economic impacts each comprise 31 per cent, growth and development 26 per cent, while sustainability (7 per cent) and international trade (6 per cent) are of relatively minor interest. This is surprising and somewhat inexplicable, given their prominence in the wider economic, geographical and sociological academic fields and the importance accorded these two topics in this book in Chapters 6, 8, 9 and 10, being considered currently as crucial issues. Subsumed under destinations are the research areas of management, marketing and partnership formation for a wide range of locations. Linked to papers on destinations are those on tourism development and growth that are increasingly being grounded in mainstream and spatial economic theories. Papers arising from research on economic impact are beginning to apply more macroeconomic models, in particular computer general equilibrium (CGE) analysis, thus widening the economic investigation of the effect of tourism from the long established multiplier and input–output (I–O) analyses. Blake and Sinclair (2003) in the UK and Dwyer, Forsyth and Spurr (2004) in Australia have made a huge contribution to the development of macroeconomic modelling, which is reviewed in Chapter 6. It is shown that the advantage of the CGE approach is that it is more closely related to real-life situations, being set firmly in an empirical context relating to goods and services flows, employment, consumption, production and investment behaviour, also showing the contribution tourism makes to national, regional and local economies.

The subject of sustainability is bedevilled by a lack of secondary data on which research can be based. The methods being developed by ecological and environmental economists entail the collection of data from primary sources, which is expensive, and its analysis is complex and protracted. The subject also suffers from a lack of clear conceptualization and definition. The issues identified here are extensively discussed and evaluated in Chapter 8 at the global level, especially with regard to the economic perspective on the conditions and indicators of sustainability. The divergence between prescriptions by environmental economists and those in the wider tourism literature and practice, operationally, are assessed in the same chapter. Evidence on tourism's performance, especially with respect to ecotourism, as a responsible form, is provided in Chapter 9. The affect on the pursuit of sustainability of market failure, such issues as the public goods nature of many tourism resources, externalities that degrade natural and human-made environments and intra- and intergenerational factors, are also assessed in the same chapter. In Chapter 10, the appraisal of methods of the valuation of non-market, open-access resources, touched on in Chapter 8, that constitute a significant proportion of tourism's product is explained.

The very small percentage of submissions to *Tourism Economics* on the international trade aspects of tourism reflects its status as a rather neglected area of research in the literature. Yet two issues are important. First, how international trade, both in goods and services, coupled with globalization, affects the structure, development and growth of destinations and consequently their natural, human-made and human environments, very often detrimentally. An indication of the nature of this environmental impact is given in Chapter 8. The second issue in the context of the tourism sector is concerned with how far the economic theory of international trade is applicable to such a service phenomenon. In Chapter 7 the Ricardian and Ohlin–Heckler theories applied in mainstream economics are examined and it is shown that they do not fully accord with what is required to analyse tourism. In particular, there are problems in relating them to how trade influences infrastructural investment and strategies, drawing on examples concerning branding, niche and segmentation marketing, considered in Chapters 4 and 5. The theories by Linder (1961) and Porter (1998) that concern market structures, emphasizing the relevance of inter-industry trade that is a feature of tourism, are cited as being apposite in Chapter 7. Also, the chapter notes the need to apply empirical measures that consider tourism's specific characteristics.

Demand modelling is a long-established area of economic research in tourism and continues to be so. Nearly two-thirds of submissions to *Tourism Economics* are econometric studies on the topic, with a further quarter consisting of time series approaches. This undoubtedly has occurred because of the availability of appropriate data. There are, however, problems where the studies of predicted demand are more complicated, with the result that estimations are unreliable. Error correction models, considered in Chapters 2 and 3, have been used to improve forecasting. Seasonal and spatial aspects are of minor research interest, as are studies on the typology of demand models, notwithstanding the fact that spatial ones are significant in the estimation of travel expenditures to attractions, particularly natural ones, and the valuation of non-priced resources investigated in Chapter 10.

More than half the submissions on tourism governance were on the two related topics of resources management and evaluation and public sector economics, the latter actually being focused on the access to and valuation of public goods. These two areas of study are investigated in Chapters 8 and 9, regarding global and destination environmental issues, respectively. Of particular importance is the valuation of non-market resources vis-à-vis those that are market-based, with the emphasis on the methods employed, explained and evaluated in Chapter 10. Policy and taxation constitute a third of papers submitted under tourism governance to the journal. The policy implications of the economic analysis of tourism should be the foundation of the application of the subject's concepts, principles and methods, drawing on a number of fields ranging over ecological, environmental, ethical, industrial, spatial and welfare economics. The issues that policy articles cover include the impact and assessment of central and local government intervention with respect to tourism's contribution to economic development and growth nationally, regionally and locally, investigated in Chapter 6. Tourism policy papers also include matters relating to intervention, such as those concerned with competitive market structures, concentration, scale, entry/exit costs and infant industry arguments which are discussed in Chapters 4 and 5. Increasingly, tourism has had to contend with unexpected events having a huge impact on the sector, such as the aircraft attack on the World Trade Center in New York in 2001, the bombing in Bali in 2002, the tsunami in Asia in 2006 and currently the global financial crisis and economic recession. The management of events such as these is a fairly recent occurrence in academic studies, stemming from suppositions in the literature that complexity and chaos characterize tourism. Therefore, theories on these, outlined above in the section on the conceptualization of the sector, are applicable. However, few possible strategies to deal with them have been derived.

Issues concerning tourism supply, representing a tenth of submissions to the journal, cover economic efficiency (mainly relating to productivity), employment, industrial structure, entrepreneurship and management and information communication technologies (ICT). Employment as a topic has largely been concerned with structural changes and their impact, levels of income and its distribution. Papers on the tourism sector's structure echo issues in a wider industrial economics context in relation to the structure, content and performance (SCP) of tourism, including the intermediaries; SCP is examined at some length in Chapter 4. The study of ICT in tourism supply is obviously of growing importance, particularly in its effect on the hospitality, tour operation, travel agency and transportation (especially airline) sectors. It is a little surprising that the impact of ICT has not generated more studies.

The interest in market analysis is very low as less than a twentieth of submissions to *Tourism Economics* were in this field. The two main topics covered are competition and price. A special issue on competition, jointly initiated by the late Thea Sinclair, was published in 2005. It is an area of analysis where competition in destinations, which embraces management issues on strategies and marketing, ought to command more attention. Chapters 4 and 5 contain extensive coverage of these aspects. The studies on price have developed into a focus on hedonic pricing (HP) modelling that is closely related to Lancaster's (1966; 1971) characteristics theory, explained in Chapters 2 and 3; HP modelling applied to the valuation of non-priced resources is considered in Chapter 10.

The somewhat modest interest in data analysis, sampling and statistical theory, together accounting for less than a tenth of submissions, is puzzling for what is a burgeoning interest in studies on the measurement of the contribution of tourism to economic development and growth through the analysis of national income accounts. Some statistics are collected by tourism bodies and in research projects on specific aspects of the sector, such as the sampling of household leisure behaviour and visitors to attractions, which have yielded primary data. Nevertheless, tourism has suffered, with other service sectors of economies, in the overwhelming concentration on the analysis of the productive industries. Official statistics on tourism are merely a subordinate outcome of data on transport, employment and the balance of payments.

However, the encouragement to derive tourism satellite accounts (TSA), initiated by the World Tourism Organization (UNWTO), has begun to gather pace with the recommendation to produce statistics on tourism consumption, production, demand and supply, employment, capital investment and the identification of indicators. Going into greater detail, to facilitate comparisons, involves obtaining data conforming to standard industrial codes that creates difficulties for tourism if it has to be collected by survey.

1.4 The structure of the book

Wanhill's (2007) analysis of submissions to *Tourism Economics* has partly informed the preparation of the new edition of this book. However, more importantly, it has given the authors a measure of confidence that Wanhill's findings largely confirm the impression gained by the review of the literature search for the book that the principal issues and trends in tourism studies, to which economics contributes, have been appropriately identified and covered.

The book is in four parts. Following this introductory chapter, Part I, containing Chapters 2 and 3, covers the theory of demand for tourism and its empirical context, respectively. Similarly, Part II, Chapters 4 and 5, considers the theory of supply, covering the structure, construct and performance of supply and the various market competitive structures and the application of game theory, before examining the practice of the supply of tourism relating to transport, accommodation, the intermediaries and attractions. Part III examines the macroeconomic impact of tourism in a spatial context, in which Chapter 6 analyses it at the regional and local level. Chapter 7 investigates tourism's economic effect as a service sector at an international level based on trade theories. Part IV, comprised of Chapters 8, 9 and 10, concentrates on environmental issues, in which due regard has been paid to book and journal publications in the fields of ecological and environmental economics that are applicable to tourism. Chapter 8 looks at the significance of globalization and growth in international tourism world-wide and its implications for the pursuit of sustainability. Chapters 9 and 10 consider the environmental impact of tourism at the destination level, concentrating on such issues as the assessment of the economic benefits and costs it generates, its impact on the allocation of resources, especially natural ones, and policy instruments to mitigate its adverse environmental effects. As alluded to above in the concluding observations on the conceptualization of tourism, the final chapter revisits the issues, particularly as they bear on the disciplinary boundaries and research performance of the economic study of the phenomenon. Given the continuing debate on the nature of tourism and how it should be analysed, an evaluation of where economics stands in relation to calls for inter- and trans-disciplinary approaches in order that research is more effective and valid empirically is necessary.

2 MICROECONOMIC FOUNDATIONS OF TOURISM DEMAND

2.1 Introduction

The relative and absolute importance of tourism in people's expenditure budgets has risen dramatically, with consequences not only for the welfare of tourists themselves but also for the residents of the areas they visit. The large numbers of tourists and the scale of their expenditure have considerable effects on the income, employment, government revenue, balance of payments, environments and culture of destination areas. A fall in demand can bring about decreases in living standards and rises in unemployment, while increased demand can result in higher employment, income, output and/or inflation and may threaten environmental quality and sustainability. Furthermore, tourism firms are confronted by changing revenue and profits and governments experience changing tax revenue and expenditure. Thus, tourism demand affects all sectors of an economy – individuals and households, private businesses and the public sector.

The significant level and repercussions of tourism demand provide a strong case for better understanding of the nature of tourists' decision-making process. A further reason for doing so is that policies which are formulated in relation to tourism demand ultimately depend upon the relevance of the theories which have been used to explain and estimate it. The incorporation of an inappropriate theoretical framework in empirical studies of demand can result in incorrect specification of the equations which are used to estimate tourism demand and biased measures of the responsiveness of demand to changes in its determinants. Any policy decisions which are based on such measures are also likely to be misguided.

Accordingly, this chapter will examine the economic theories which underlie tourist decision-making at the microeconomic level, while Chapter 3 will concentrate on explaining and evaluating the models which have been used to estimate and forecast demand. The discussion in the two chapters aims at providing a theoretical framework which can be used to evaluate the many empirical studies which have been undertaken. It should also provide the basis for the formulation of appropriate models for estimating tourism demand and, hence, for more accurate results and more appropriate policy implications. The discussion will introduce recent developments in the economic literature relating to consumer behaviour, particularly the micro-economic basis of more aggregate relationships at the national level and inter-temporal preferences.

The emphasis in the theoretical analysis in this chapter is on variables which can be measured quantitatively, including the effective demand for tourism, which is the amount that consumers are willing and able to spend rather than the notional demand which they would like to exercise, but which is not backed by the ability to pay. This does not imply that non-measurable, qualitative variables are unimportant. Indeed, research in economic psychology takes some cognitive variables into account and mainstream economics acknowledges the important role which expectations can play. However, measurable 'material' variables are the main focus of attention since this is the area where economics has most to contribute. Thus, the approach which is taken in the two chapters complements the non-economic analyses of tourism demand which have been undertaken to date.

The second section of this chapter will examine the economic theory which explains tourism demand. Initially, the relationship between employment, income, demand for consumer goods and unpaid time is investigated. This provides an explanation of a person's demand for all consumer goods and services, including tourism. The choice of how much tourism to purchase relative to other goods and services is then discussed and the analysis is subsequently extended to consider choices between different types of tourism, including the cases where one type of tourism complements or substitutes other types. The effects of changes in income and relative prices on tourism demand are examined and some determinants of the timing of spending are then explained. In addition to the neoclassical approach regarding tourism consumption, the Gorman–Lancaster characteristics framework is also discussed (Gorman, 1980; Lancaster, 1966; 1971). Although the theoretical explanation in the second section of the chapter may appear somewhat technical, it is crucial in providing a rigorous basis for applied work. Empirical studies which are undertaken without an explicit theoretical underpinning may produce biased results with misleading policy implications for the area concerned. Section three of the chapter departs from the earlier emphasis on the individual by considering the social context of tourism decision-making, showing how insights from other disciplines can contribute to the economic analysis of tourism. The relationship between the theory of tourism demand and empirical studies will be discussed in Chapter 3.

2.2 Optimal choice in tourism demand

2.2.1 Consumption, paid work and unpaid time

Both people's preferences and their expenditure budgets are key determinants of the demand for tourism. A person who is considering whether to spend a holiday away from home has an amount of money, or budget, which is available for expenditure on tourism and other goods and services. The size of the budget depends upon the number of hours that he or she spends in paid work per time period (labour supply), on the income per hour and on the rate of taxation on income which yields the disposable income available for purchasing goods and services. People trade off paid work against unpaid time; some people prefer more income, resulting from more paid work, while others prefer to have more unpaid time for leisure or household activities and therefore spend less time in paid work. If they undertake more paid work and have less unpaid time, their level of income rises but leisure and household work are forgone; conversely, taking more leisure reduces income. There is, however, a tension as income is often required to undertake leisure pursuits (including tourism) so that the latter have an imputed 'price' or opportunity cost. This is often associated with the so-called 'leisure paradox' (Cooper et al., 2008), which depicts the negative relationship between time and discretionary income in an individual's life cycle: for example, students have substantial leisure time but low income resources, whereas urban professional people have a relative affluence at the expense of limited free time. Each combination of paid work and unpaid time provides a different amount of earnings, or budget, which may be spent on goods and services. The highest ratio of paid work to unpaid time usually provides the largest budget, corresponding to the largest potential consumption value, and vice versa.

The different combinations of consumption and unpaid time which a person may have are illustrated by the line CBU in Figure 2.1. The vertical axis measures the value of consumption and the horizontal axis measures increases in unpaid time when read from left to right (or increases in paid time when read from right to left). The point OC shows the maximum consumption which the person can achieve, resulting from spending the maximum possible time in paid work. Someone who is not in paid work has a combination of consumption and unpaid time shown by B, with OC* being the value of consumption which the person obtains while unemployed, for example by unemployment benefits. Positions between C and B show intermediate combinations. The line CBU is known as the budget line, the slope of which indicates the rate of remuneration; so if, for example, the wage rate rises, it becomes steeper.

People also receive satisfaction, or utility, from consumer goods and from unpaid time. For example, a person may receive the same utility from

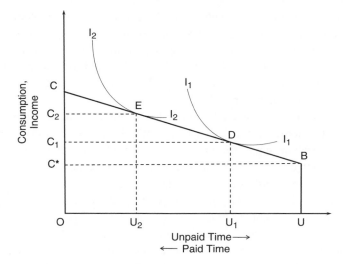

Figure 2.1 Consumption, paid work and unpaid time

a large amount of consumer goods and small amount of unpaid time as from a small amount of goods and high amount of unpaid time, or from intermediate combinations of the two. The different combinations of consumption and unpaid time which provide a given level of satisfaction are depicted by the curves I_1I_1 and I_2I_2 in Figure 2.1. The curves are known as indifference curves because the person is indifferent between alternative positions on a given curve, since he or she receives the same satisfaction from every choice of quantities of goods and services depicted by it. Curves further away from the origin of the graph would correspond to higher combinations of consumption and unpaid time, and hence to higher satisfaction, and vice versa.

Economists assume that people want the maximum satisfaction possible and obtain this by choosing the combination of paid work and unpaid time which provides them with their preferred, attainable combination of consumer goods and unpaid time. One person's preferred position might be at D in Figure 2.1, which is determined by the point of tangency between the indifference curve I_1I_1 (determined by the person's preferences) and the budget line and shows the person's optimal combination of consumption, OC_1, and unpaid time, OU_1 (with paid work time UU_1). An individual with different preferences would have an alternative combination of consumption and unpaid time. For example, the individual might obtain the same level of satisfaction from higher ratios of consumption and paid work to unpaid time, depicted by the curve I_2I_2. Thus, the individual's optimal position, with given preferences and income per hour, would be at E, corresponding to consumption of OC_2 and a lower value of unpaid time, OU_2 (and higher paid time UU_2). There are, of course, circumstances when a person is unable to choose his/her preferred combination of consumption and unpaid time and is constrained by structured hours of work to a specific sub-optimal position, for example because of having to work a 38-hour working week with no possibilities of part-time work or overtime. Moreover, he/she might be forced to take additional unpaid time because of a downturn in the economy or might even be made redundant, with unpaid time OU and consumption OC*.

The amount which is available for spending on tourism and other goods is, therefore, the income or budget which results from the person's paid work (labour supply) and from his or her preferences between consumption (permitted by paid work) and unpaid time. Hence, consumption and labour supply are jointly determined and should be considered simultaneously. Changes in remuneration for work bring about changes in people's consumption and unpaid time. For example, an increase in the wage rate or a decrease in income tax results in higher income and greater consumption and more, less or the same amount of unpaid time. This is because an increase in effective remuneration per hour encourages a person to substitute higher paid work and higher consumption for unpaid time, i.e. the substitution effect. Conversely, the person can use higher earnings from a given amount of paid work time to purchase more goods while simultaneously taking more unpaid time, the income effect. The net effect is that depending on personal preferences, he/she can increase consumption while having more, less or the same unpaid time. The application of the concepts of substitution and income effects to tourism will be considered following discussion of the extent to which people demand tourism vis-à-vis other goods and services.

2.2.2 Demand for tourism relative to other goods and services

The demand for tourism depends upon the total budget which is available for spending (resulting from the person's labour supply or unemployment benefits, as discussed above) and on preferences for tourism relative to other goods and services. At one extreme, the person could allocate all of his/her budget to tourism and, at the other, none of it to tourism and all of it to other goods. Between these two extremes, a range of combinations of tourism and other goods are feasible. All possible combinations are given by the budget line, the slope of which indicates the relative prices of goods and services and which is depicted by TG in Figure 2.2. The point OT is the amount of tourism which would be consumed if a person spends all of his/her budget on tourism and OG is the amount of other goods which would be consumed if there were no expenditure on tourism, with the line TG showing intermediate combinations. The amounts of tourism and of other goods which it is possible to consume depend upon the relative prices of tourism and other goods so that lower tourism prices would permit more tourism

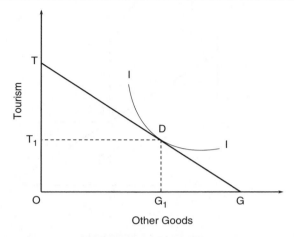

Figure 2.2 Consumption of tourism and other goods

and vice versa. The effect of changes in relative prices, which are depicted by changes in the slope of the budget line, will be discussed below.

The combination of tourism and other goods which the person decides to purchase depends upon his/her preferences. Alternative combinations of tourism and other goods can provide the consumer with the same level of satisfaction, so that, for example, low tourism consumption and high consumption of other goods provides the same satisfaction as high tourism consumption and low consumption of other goods, as illustrated by the indifference curve II in Figure 2.2. The person allocates his/her budget between tourism and other goods by choosing the combination which maximizes satisfaction. This is the point D, where the indifference curve is tangential to the budget line, resulting in OT_1 tourism and OG_1 consumption of other goods. An individual with a stronger preference for tourism would consume a combination to the left of point D, whereas someone keener on consuming other goods would have an indifference curve tangential to TG to the right of D. From a destination management perspective, it is important to hold information on the allocation of consumer's discretionary income. For example, individuals who must repay loans for durable goods purchased in previous time periods are unlikely to be an effective target market for tourism advertising purposes at present (Dolnicar et al., 2008).

People have to decide not only on their preferred combination of tourism relative to other goods, but also on their preferred combination of different types of tourism. This is essentially related to the separability concept (Deaton and Muellbauer, 1980b), according to which consumers first allocate a specific part of their budget to tourism activities and subsequently make a choice among the available tourist products: budget allocation at the former stage is assumed to be independent of the latter. For example, a tourist

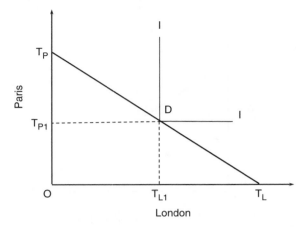

Figure 2.3 Tourist destinations as complements

could spend all of his/her tourism budget on visits to friends and relatives or all of it on holidays in new locations abroad, or could choose some combination of the two. The optimal position again depends upon the person's budget and preferences and it is again assumed that the budget is allocated between different types of tourism so as to maximize satisfaction. The optimal combination of visits to friends and relatives and holidays abroad can be illustrated using a graph similar to Figure 2.2 but with the different types of tourism measured on the axes and is shown in Figure 2.3. In reality, of course, there may be more than two combinations; these could be shown mathematically but not diagrammatically, being constrained by the existence of only two dimensions.

In the case of different types of tourism, a person may choose a combination of tourism products. However, this is not the only outcome which may occur as one type of tourism may be a substitute for or complement to another. For example, some American tourists who travel to Europe regard destinations in different European countries as a complementary part of the tourist experience, rather than substitutes, so that, for example, London and Paris may be regarded as complements and fixed proportions of expenditure are allocated to each. This may be related to an implicit need of the consumer to generate economies of scale by spreading the monetary and time costs of the transatlantic journey over a potentially large number of European destinations. This case is depicted in Figure 2.3, where the budget line T_PT_L shows how different combinations of tourism expenditure could be allocated to the two destinations, but the L-shaped indifference curve II shows that the person wishes to allocate set proportions of the budget to each.

The alternative case of tourist destinations as substitutes might apply to holidays in Sydney and New York, as illustrated in Figure 2.4. The budget line, T_ST_{NY}, indicating the relative prices of the two holiday destinations,

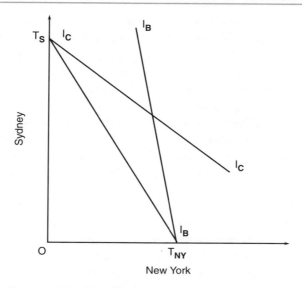

Figure 2.4 Tourist destinations as substitutes

again shows that different proportions of the budget might be allocated to tourism in each destination. However, the indifference curve I_BI_B shows that person B regards the two destinations as substitutes and selects New York as the preferred destination. Another individual, C, also regards the two destinations as substitutes but has different preferences, illustrated by the indifference curve I_CI_C, and chooses Sydney rather than New York. Knowledge of the extent to which different types of tourism or tourist destinations are substitutable or complementary is particularly useful for tourism planning and marketing.

In any case, temporal issues should also be explicitly considered. By definition, a tourist cannot be simultaneously present in more than one tourist destination, i.e. place consumption is essentially discrete at a given point in time. As a result, complementarity of tourism destinations implicitly assumes that a prospective tourist has sufficient time resources to allocate among different places given his or her travel preferences. Lew and McKercher (2006) model tourist movements using a local destination analysis to highlight spatial patterns in tourism behaviour. On the other hand, discreteness in choice is more easily understood in the context of substitute tourist areas.

2.2.3 The effects of changes in income and prices on tourism demand

Economists posit that tourism demand is affected principally by income and prices and information about the extent to which changes in demand result from each of these variables is also important for both tourism suppliers

and policy-makers. It is helpful, initially, to examine the effects of each of these variables separately. In the case of a rise in income with constant relative prices, the effect on most types of tourism and most tourist destinations is likely to be positive. Thus, an increase in income results in a rise in tourism purchases, similar to the effect of increasing income on the demand for most other goods and services; i.e. it is a normal good because demand for it is positively related to income. However, it is possible for a rise in income to bring about a fall in demand, such as for tourism in mass market destinations, implying that this form of tourism is an inferior good. This might be the case where a beach holiday in the Caribbean is substituted for one in the Costa Brava.

The two effects are illustrated in Figure 2.5. The vertical axis measures tourism and the horizontal axis measures other goods. The lines TG and T′G′ are the budget lines before and after the rise in income respectively and are parallel because of the assumption of constant relative prices for tourism and other goods. Indifference curves are included to illustrate the person's preferences. If tourism is a normal good, preferences may be illustrated by the indifference curve I_2I_2 so that demand rises from OT_1 to OT_2 at E. If it is an inferior good, indicated by the indifference curve I_3I_3, an increase in income brings about a decrease in tourism from OT_1 to OT_3 at F. If demand is positively related to income and rises more than proportionately, the good is known as a luxury and if demand rises less than proportionately, it is known as a necessity. In terms of the concept of elasticity, the demand for a luxury good is said to be elastic (greater than one) with respect to changes in income while that for a necessity is inelastic (less than one).

The second case to be considered concerns the effect on tourism demand of a change in relative prices with income held constant. Demand and prices

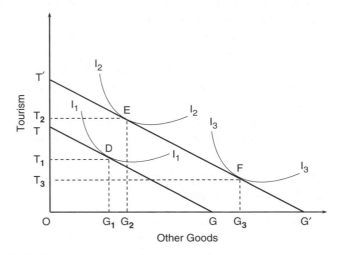

Figure 2.5 Effects of a rise in income on tourism consumption

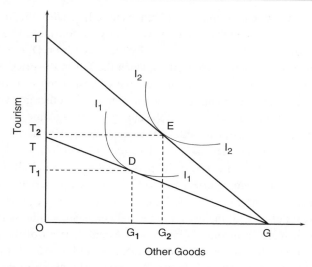

Figure 2.6 Effects of a fall in price on tourism consumption

are usually negatively related, so that a fall in price is normally associated with a rise in demand and vice versa. The effect of a decrease in the price of tourism is depicted in Figure 2.6. Since tourism is now cheaper, the person's budget can now purchase a maximum of OT′ tourism instead of OT, while the maximum amount of other goods which can be purchased remains constant at OG since their prices are assumed to remain constant. The combinations of tourism and other goods which can be purchased after the price decrease are given by the line T′G. The original and subsequent optimal combinations of tourism and other goods are respectively points D and E in Figure 2.6, so that a fall in the price of tourism results in an increase in quantity demanded and satisfaction as the person purchases OT_2 tourism and OG_2 other goods as opposed to OT_1 and OG_1 before the fall in price. It is also possible to consider the choice between two similar forms of tourism where the price of one changes relative to the other. Thus, for example, a resident of the UK may be contemplating one of two resort holidays in the Mediterranean, one in France and the other in Tunisia, but should the euro appreciate in value against sterling while the dinar remained unchanged, the Tunisian resort holiday would be chosen.

It is possible to depict both the income and price effects on tourism demand in the same diagram, as illustrated in Figure 2.7, which effectively combines Figures 2.5 and 2.6. Suppose, for example, that there is a change in the price of tourism so that tourism becomes cheaper relative to other goods and the person's budget line shifts from TG to T′G. The optimal point of consumption was originally at D. The effect of the change in relative prices is demonstrated by drawing the broken line, PP, with the same slope as the new budget line T′G, and hence with the new relative prices, tangential to the original indifference curve I_1I_1. Since the line is tangential to the original indifference curve, utility is constant. The effect of the change in relative prices is given by the move from D to S. This effect is known as the substitution effect, since the fall in the price of tourism has caused the person to substitute relatively cheaper tourism for other goods, so that tourism demand rises and the demand for other goods decreases.

The second effect is that of the change in real income; as tourism is now cheaper, the person is better off in real terms. He/she has the option of spending all the increase in real income on tourism or all of it on other goods or some on each. If the person chose to spend all the increase on tourism, the income effect would be illustrated by a move from S to E in Figure 2.7 where OT_2 tourism and OG_2 other goods are purchased (since PP is parallel to $T'G$, the price ratio is held constant so that only the effect of the increase in income is taken into account). If all of the increase were spent on other goods, the income effect would be from S to F where OT_3 and OG_3 are purchased, respectively. If he/she chose to demand more of both tourism and other goods, the optimal point would be somewhere between E and F. The net effect of the change in relative prices is the move from D to E, from D to F or from D to a point between E and F respectively. Hence, tourism demand rises while the demand for other goods falls, remains constant or rises, depending upon the person's preferences.

Before moving on from the above discussion, it is worth noting that the substitution effect resulting from the change in relative prices, as demonstrated in Figure 2.7, was defined with utility held constant (Hicksian effect).

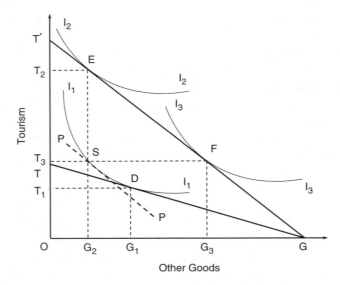

Figure 2.7 Effects of a fall in price and a rise in income on tourism consumption

An alternative definition of the substitution effect takes account of the change in demand resulting from a change in relative prices, when income rather than utility is held constant, so that the original bundle is still attainable (Slutsky effect). In the former case, the demand for tourism is known as the compensated (rather than uncompensated) demand because the person is said to have been 'compensated' for the change in relative prices by allowing his/her level of satisfaction to remain the same (Varian, 2006).

2.2.4 Tourism demand over time

People's choices concerning the timing of consumption are known as inter-temporal choice and have attracted increasing attention within economics (for example, Heckman, 1974; Obstfeld, 1990; Deaton, 1992). If, for example, two time periods are considered, a person might choose to spend all the income which he/she receives in time period 1 within that time period and none of it in the future (period 2) in order to maximize initial consumption. Alternatively, the person could elect to spend and consume less in the first period in order to increase future spending and consumption. These possibilities are illustrated in Figure 2.8.

The person has an income of Y_1 in time period 1 and Y_2 in time period 2 and could choose to spend all of Y_1 on tourism and other goods in period 1 and all of Y_2 on tourism and other goods in period 2. In this case, the optimal consumption point would be at D. On the other hand, the person might decide to consume less in the first period in order to consume more in the second period. In the extreme case of an individual who chose to consume nothing

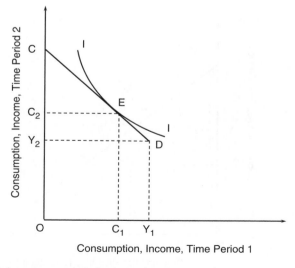

Figure 2.8 Inter-temporal choice in tourism consumption

in the first period so as to maximize consumption subsequently, the optimal consumption point would be at C. The range of consumption possibilities which are available from spending less than the full amount of income initially and more subsequently are given by the line CD. The combination of consumption in the two time periods which is chosen depends upon the person's preferences. These may be illustrated by an indifference curve which, in this case, shows the combinations of consumption in each period which provide the same level of satisfaction. In the example in Figure 2.8 the person chose to consume at point E, consuming less tourism and other goods, OC_1, in the first period in order to consume more, OC_2, in the ensuing period.

This case assumes a zero interest rate (hence the slope of the budget line is −1 and $Y_2C = Y_2D$) and ignores the possibility that people can borrow in order to increase their current consumption or may lend in order to consume more in the future. Introducing borrowing and lending increases the range of inter-temporal consumption combinations which are available, as is illustrated by the budget constraint CDC^* in Figure 2.9. The line CD is steeper than the equivalent line in Figure 2.8 since a person could receive a higher future income, and hence higher future consumption, via the interest earned from lending out current income. He/she could also borrow to increase consumption in the first period and the line DC^* shows the combinations of (higher) first and (lower) second period income and consumption which could be attained by borrowing. The line CDC^* shows all the first and second period consumption possibilities which the person could achieve by lending or borrowing, with OC^* depicting the limiting case of maximum first period consumption and zero second period consumption, OC showing the opposite limiting case and CC^* the intermediate combinations. The indifference

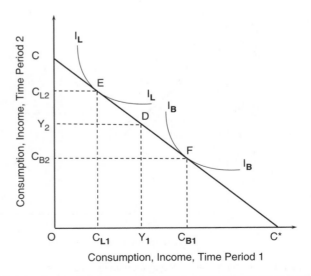

Figure 2.9 Inter-temporal tourism consumption with borrowing and lending

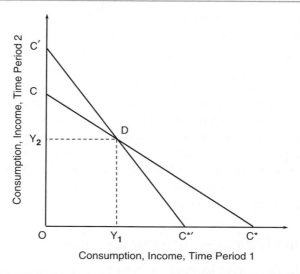

Figure 2.10 Effect of a rise in interest rates on the consumer's budget line

curve I_BI_B illustrates a person who chooses to borrow in order to consume more initially, while I_LI_L depicts an individual who prefers to lend in order to increase his/her income and consumption in the second period.

There are some complications to this analysis because changes in interest rates and inflation alter people's inter-temporal consumption possibilities and provide an incentive to consume less now and more later or vice versa. For example, in Figure 2.10, the initial budget line for tourism and other goods is CC^*. An increase in interest rates increases the possibilities of future consumption by permitting a rise in future income and decreases the possibilities of present consumption by increasing the cost of borrowing. The new budget line is $C'C^{*'}$ in the figure. The effect of the rise in interest rates may be disaggregated into substitution and income effects. The former is always negative, i.e. there is an inverse relationship, since an increase in interest rates encourages people to substitute (cheaper) future consumption for (more expensive) current consumption. The income effect is positive since the higher income permitted by higher interest rates has a positive, overall effect on current consumption, i.e. the relationship is positive. For a borrower, the net effect of the rise in interest rates on current consumption is negative but for a lender, the net effect could be positive or negative, depending on the relative sizes and therefore net effect of the substitution and income effects.

The credit crunch and the rise of interest rates in the Eurozone and Britain in 2007–8, as a result of inflationary pressures emerging from high commodity prices, had a negative effect on tourism consumption given the unprecedented borrowing experienced by European households in the previous years. With the subsequent burst of the commodity price bubble, fears about inflation eased substantially; hence, to revitalize the economy and curb recession both the European Central Bank and the Bank of England followed an aggressive expansionary monetary policy in 2008–9 by reducing interest rates to almost zero levels. Pessimism about the future, however, discouraged consumers from engaging in a spending-spree which would expose them into significant financial risk; consequently, the application of monetary policy per se proved partially ineffective. It is also interesting to note that a fall in inflation has similar effects on current consumption to those brought about by a rise in interest rates, since lower inflation results in a rise in real interest rates. The outcome is more complicated if changes in the aggregate rate of inflation are accompanied by a change in the relative prices of tourism and other goods, as additional substitution and income effects have to be taken into account.

2.2.5 The Gorman–Lancaster characteristics framework

So far the analysis has focused on the theoretical pillars of the neoclassical microeconomic analysis, which assumes that individuals derive utility by consuming goods per se. An alternative theory of demand has been proposed by Lancaster (1966, 1971) and Gorman (1980) based on the so-called characteristics approach. According to this theory, what really matters for the individual is the consumption of characteristics associated with goods, i.e. the latter are important only because of the specific features they possess. The theory has been applied in tourism by various researchers, with notable contributions made by Rugg (1973), Morley (1992) and Papatheodorou (2001b). The essence of the theory is now discussed in Figure 2.11.

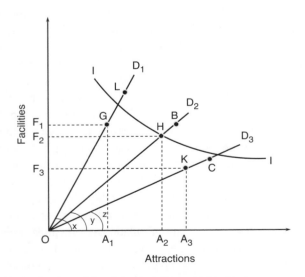

Figure 2.11 Application of the characteristics model in tourism

Two bundles of tourist characteristics are assumed, namely attractions (natural and built – measured on the x axis) and facilities (such as hotels, airports and ancillary services – measured on the y axis). By visiting destinations D_1, D_2 and D_3, tourists accumulate quantities of characteristics in the ratios of x, y and z as depicted by rays OD_1, OD_2 and OD_3 respectively. These rays encapsulate the information provided by the consumption technology matrix, which refers to the transformation of vacation days spent in tourist destinations into quantities of meaningful characteristics (attractions and facilities). Case D_1 is an example of a tourist destination emphasizing the provision of facilities, D_3 is a resort with a higher weight on attractions, whereas D_2 is in an interim position. The maximum feasible consumption of characteristics is determined by a budget (as in the case of traditional microeconomics) but also of a time constraint. It should be noted, however, that the current framework assumes that consumers choose destinations in a discrete manner, i.e. tourists spend all their vacation time and expenditure either in D_1, D_2 or D_3; linear combinations are excluded from the analysis. Hence, the budget and time constraints refer to specific points (i.e. combinations of characteristics called vertices) rather than to continuous lines connecting these points. In particular, by spending all his or her monetary resources on vacation days at destination D_1, the tourist may enjoy a characteristics combination (called vertex) given by point L; alternatively, a vacation in D_2 generates vertex B and a holiday in D_3 vertex K. Vertex K yields a smaller quantity of facility characteristics compared to L or B (i.e. F_3) but offers a larger quantity of attraction characteristics (i.e. A_3); hence, K is not dominated by either L or B and D_3 cannot be excluded from the tourist market. With respect to the time constraint (which jointly considers the duration of the journey and the sojourn in situ), a tourist may at most consume vertex G by visiting D_1, H by going to D_2 and C by choosing D_3. As the two constraints notionally intersect both should be considered to derive the effective (i.e. most binding) constraint for the tourist. In fact, this is given by points G for D_1 (corresponding to attractions level A_1 and facilities quantity F_1), H for D_2 (related to $A_2 > A_1$ and $F_2 < F_1$) and K for D_3 (with $A_3 > A_2$ and $F_3 < F_2$). If tourist preferences are given by the indifference curve II, then the individual should optimally choose to visit D_2 and consume vertex H.

The characteristics approach provides an excellent framework to study comparative statistics on a number of issues. For example, if any of the three destinations became more expensive (or conversely cheaper), then its maximum attainable combination (shown by L, B or K) would shift southwest (northeast) along the relevant characteristics ray. Likewise, if the journey time to a destination became shorter (longer), then its vertex (given by G, H or C) would move to the right (left) along OD_1, OD_2 or OD_3. Moreover, a destination which decided to strategically invest in tourist facilities (attractions) would see its characteristic ray becoming steeper (flatter). Similarly, improvement of quality at a destination level means that a tourist can potentially consume a larger quantity of both facilities and attractions on a daily basis (related to the time constraint) – hence, any improvement of quality will lead to a rightward shift of the respective time vertex. On the other hand, and with respect to the budget constraint, it is important to know whether the improvement in quality means that the destination has now become more expensive or not. If the cost of the holiday does not change, then the budget vertex will shift to the right as the product becomes better value for money (i.e. the consumer can receive more characteristics for the same price). If the cost of the holiday rises to the same proportion as quality, then value for money (i.e. the quality/price ratio) will remain the same, hence the budget vertex will not move. Finally, if the cost of the holiday rises more than proportionally compared to the improvement in quality, then the value for money deteriorates and the budget vertex shifts to the left.

If a tourist becomes richer, then he or she can afford to stay longer and consume more characteristics in all destinations. Hence, and similarly to the classical microeconomic theory, the budget constraint will move to the right in a parallel manner. As for the time constraint, the result of an increase in consumer affluence is ambiguous: according to the 'leisure paradox' (discussed earlier), the time constraint should move to the left if affluence is achieved at the expense of available leisure time. If this is not the case, then the time constraint could remain unchanged or even move to the right if the consumer can now afford to use faster transport technology than previously. For example, a tourist may now decide to fly by aeroplane instead of going somewhere by boat or use his or her private car instead of a coach. 'Outcome' driven tourists who are interested in minimizing the duration of the journey so as to maximize the sojourn in a tourism destination (Lew and McKercher, 2006) are likely to take advantage of this transport mode substitution if possible.

Destination advertising can also be incorporated in the framework: by generating an increase in the perceived quantities of consumed characteristics in a particular destination on a daily basis, advertising results in a shift of the respective vertices (L and G for D_1, B and H for D_2, K and C for D_3) to the right and/or in a change in the slope of the characteristics ray of the destination in question. In essence, advertising raises the profile of a specific tourism destination and reduces cognitive distance (Papatheodorou, 2001b): for example, although Albania is closer to Germany than Greece, most Germans are likely to be much more familiar with the characteristics of Greek tourism destinations compared to the rather isolated and only now-emerging resorts on the Albanian Adriatic coast. On the other hand, the opposite results seem to emerge in the case of negative publicity due, for example, to a terrorist attack (such as 9/11) or an epidemic like SARS or swine flu; fortunately, however, the adverse impact is often likely to last for a limited period of time only (Graham, 2008a).

Changes in consumer preferences are depicted by a change in the slope of the indifference curve, which if substantial enough may lead to the optimal choice of another destination. On these grounds, the simultaneous existence of more than one tourism destination may be attributed to heterogeneity of tourism preferences. For example, and to pre-empt the discussion in the following section, Plog's (1973) allocentric tourists would be primarily interested in attractions choosing D_3 whereas psychocentrics would opt for D_1 as they mainly care about the existence of abundant facilities. Tourism preferences may also be affected by cultural distance as travellers from culturally close origin areas may search for different characteristics in a destination compared to tourists who have a different cultural background (Flogenfeldt, 1999). Finally, and unlike traditional microeconomic theory, the characteristics framework allows for the emergence of a new tourist destination (introduced with its own characteristic ray), as well as for the market exit of existing ones (as a result of poor product offering over time).

2.3 The social context of tourism decision-making

Both the expenditure budget and people's preferences are key variables underlying tourism demand, as shown above. The role of expenditure in determining demand has been examined in the single equation and system of equations models of tourism demand. However, traditional demand theory does not explain how preferences and tastes are formed and change or the process by which decisions are made in the context of the social environment. Research has, instead, concentrated on individual self-interest with regard to the allocation of income to consumption and saving, and the choice of products which consumers purchase.

The emphasis on the individual is reflected in the early tourism literature. For example, propositions by Maslow (1954, 1968), who posited a hierarchy of needs as creating the motivation for the individual to undertake certain activities, have formed the basis of studies in the 1970s and 1980s on the motivation to purchase a holiday, and non-economic investigations of motivation have been undertaken by, for example, Gray (1970), Plog (1973), Schmoll (1977), Crompton (1979) and Dann (1981). Motivational studies would benefit from being more strongly focused and, although some empirical testing of the hypotheses propounded has been pursued, attention has concentrated mostly on the reasons for and destinations of travel and the marketing implications (Pearce, 1987).

A common theme running through the theoretical studies of motivation is the need by individuals to escape from their daily work and home regime and to seek new experiences which can be met only by travelling: the so-called 'push factors' make people engage in tourism activities, while the 'pull factors' raise their interest in favour of a particular destination. The direction of research has led to a categorization of forms of tourism and types of tourist. Gray's (1970) 'wanderlust' category suggests a desire to get away, while his 'sunlust' category implies a need which cannot be experienced in the home environment. In his psychographic segmentation of tourists, Plog (1973) considered the behavioural dimension rather than simply the motivation to travel. He argued that tourists' patterns of behaviour are determined by psychological factors so that the 'allocentric' is more adventurous and self-confident while the 'psychocentric' prefers familiar and reassuring locations and social interactions. Iso-Ahola (1982) largely encompasses earlier theoretical models by Schmoll (1977), Crompton (1979) and Dann (1981) by identifying three principal motivations for travel: escape, seeking psychological benefits in a different physical environment and social interaction. Another branch of research took the form of models incorporating expectations (for example, Rosenberg, 1956; Fishbein, 1963), which investigated the degree to which consumption of particular goods and services resulted in increased levels of individual satisfaction.

These non-economic studies of motivation by social psychologists and, in some cases, geographers make a useful contribution to the development of economic models of tourism demand on two counts. First, they seek to explain the reasons for behaviour which economists observe only from preferences which are revealed in terms of expenditure on goods and services in the market. In this respect, the study of motivation assists in making more accurate explanations and forecasts of the level and pattern of tourism demand. Second, the approaches complement the relatively recent and empirically orientated branches of economics – experimental economics and economic psychology – which have shifted attention towards the social context of decision-making and shed light on the determination of preferences and tastes for tourism consumption. These approaches, while acknowledging the contribution made by other disciplines to explain tourism demand and behaviour, are set within an economic framework and can help to explicate and predict tourist consumption behaviour. The contribution of experimental economics and economic psychology lies in both extending the possible range of variables determining tourism demand and the analytical method, whereby the approach is generally inductive rather than deductive. Moreover, a number of the insights which they have provided have been incorporated into mainstream economic models. Experimental economics emulates the scientific laboratory method by conducting experiments to investigate consumer decision-making and is particularly appropriate for situations in which data relating to key interrelationships is not otherwise available. Economic psychology (also discussed in Chapter 4) provides a conceptual as well as methodological contribution by recognizing that perceptions, information processing, attitudes, expectations, motivation, preferences and tastes, which are largely unexplained or ill considered in utility-maximization theory, are important and susceptible to measurement.

Economic psychologists argue that explanation of the process by which decisions are made requires investigation of the social context of decision-making. It is argued that the social environment strongly influences the level of consumption and choice of products at the microeconomic level, as well as the ethos of consumption and saving at the macro level. In his relative income hypothesis, Duesenberry (1949), echoed by Liebenstein (1950), suggested that the level and pattern of consumption of a particular group is determined less by current or future income than by the level and pattern of consumption by other, usually higher income groups. The concept of a demonstration effect implies that consumers look to reference groups and imitate their consumption patterns. Here, economic psychology incorporates ideas from sociology (group theory) and social psychology (attitude postulates), shifting the analytical focus to the microeconomic level.

Support for these views is provided by reference to the historical context of tourism growth. Veblen's (1899) theory of conspicuous consumption and Liebenstein's (1950) snob and bandwagon effect explain how lower-income groups have followed the holiday patterns of the rich. The classification of tourist types, such as the allocentric and psychocentric, and the life cycle of resorts may reflect the influence of social reference groups. The allocentric, innovative tourist seeks new experiences and some less adventurous tourists, generally though not exclusively in lower-income groups, subsequently copy this behaviour. Such changing tastes have led to changes in the patterns of tourism behaviour. For example, demand for some mass-market destinations has decreased, as observed in traditional seaside resorts in the UK, illustrating the phase of decline in a resort cycle. Thus, social factors, including comparisons made by consumers with other consuming units, are seen as important in more psychologically and sociologically based approaches.

Economic psychology and sociology consider economic socialization to be an important determinant of behavioural patterns, preferences and tastes. Economic socialization (Stacey, 1982; Jundin, 1983) concerns the ways in which children are, in stages, introduced to and develop skills with respect to their consumption, from knowledge of money and possessions to social differentiation and socio-economic understanding. The concept of consumer socialization considers the specific roles of key individuals, for example parents, in forming children's attitudes to levels and patterns of consumption. The relationship of gender roles and the process of interaction within the family to consumption decisions are also important.

Research has shown that spouses' roles in consumer decision-making differ with respect to the goods or services being considered for possible purchase and the stage in the decision-making process at which each partner's views are particularly influential. For most goods and services, the more technically complex the item and the associated information required, is likely to be a male-dominated decision (Davis, 1970; Burns and Ortinau, 1979). However, there has been little investigation of non-durable and durable tourism consumption processes within or between social groups. Regarding holiday decisions, Filiautraut and Ritchie (1980) found that men's views tended to dominate, but Qualls (1982) found that decisions were made jointly by both partners.

In his review of holiday decision-making, Kirchler (1988) showed that other variables influencing consumption were age, stages in the life cycle, class and income and nationality. In traditional role and older households, consumption decisions were controlled by the husband. In middle-income households, more consumption decisions were made jointly. With respect to social class, decision-making was more male-dominated in the lower and upper classes. The unsurprising conclusion of the few studies undertaken to date is that the person with economic power dominates the initiation process and outcome of spending, but in changing social and economic circumstances different people hold this position. Oppermann (1995) discusses travel behaviour and subsequent choice of tourism destinations based on three time settings, namely the period after the Second World War, the various stages in a person's life and successive intergenerational preferences. His empirical findings are supportive of the argument that choice of travel depends on age and generation (i.e. to denote when people were actually born, e.g. before the Second World War, in the 1950s and 1960s baby boom years, or in the generation post 1980s) and has changed significantly over the last fifty years. Tourism areas are classified into four cohorts based on their popularity across the individual life cycle and different generations. Destinations in the most promising group (+,+) are in an advantageous situation as they are popular not only with older people (a + for age) but also with new consumers (another + for generation); conversely, areas in the worst cohort (−,−) seem to face a period of permanent concern; alternate-sign resorts should stress their strong points and face their disadvantages.

Experimental economics can shed light on the holiday expenditure decision as an interactive process within the consuming unit, as well as between that unit and other social groups, reflecting the social context of tourism demand and behaviour. The methodology can incorporate game theory (discussed in Chapter 4), in which possible outcomes are determined by the goals and strategies of the participants. For example, one or more members of a household who wish to purchase a beach holiday may collude, while others who seek a more active cultural experience may bargain with them in an attempt to achieve an alternative goal. Kent (1990) used a quasi-experimental approach to examine the process of decision-making, with the added dimension of certain members of the household occupying positions of dominance. Such positions may be achieved and reinforced through the possession of information not available to other members of the group.

Decision-making in situations of risk and uncertainty comes within the purview of both experimental and mainstream economics and could be applied to holiday decisions which involve substantial expenditure spread over long periods of time, such as the capital outlay on holiday homes. These decisions occur in a context of uncertainty concerning future income, relative prices, inflation and interest rates, and could be investigated using an experimental approach in which various scenarios are constructed and the choices made observed. Some social psychologists argue that even in circumstances in which the consumer could decrease uncertainty by obtaining more information, once an acceptable level of information is attained the search for and processing of additional information ceases. This may occur well before the consumer has attained the maximum amount of information which he or she can comprehend, and is akin to Simon's (1957) notion of the satisficing consumer. It also accords with the concept of bounded rationality as opposed to fully rational behaviour. The result may be that consumers are not consistent in their decisions, explaining the phenomenon of preference reversal, also indicated by the findings of system of equations models of demand discussed in Chapter 3 and which sometimes occurs in game theory simulations of consumer choice.

In the context of international tourism consumption choices, the employment of the experimental method has questioned the assumptions of models in which expenditure is deferred, for example the assumption that choices made in one time period are independent of those made in another (Loewenstein, 1987). It, thus, supports the case for analysis of the intertemporal nature of decision-making. Additional investigation of the ways in which individuals and groups obtain, select and use information in their consumption decisions has been undertaken within the field of marketing (for example, Middleton and Jackie, 2001; Kotler and Keller, 2008). Increased use of the insights provided by other disciplines can enrich economic theories of tourism consumption decisions. For example, at the micro level it has been hypothesized that tourism demand depends on the economic power of particular individuals and/or previous consumption by social reference groups. It may be possible to estimate quantitative measures of such variables and to include them in the tourism demand function. This would not only correct for omitted variables bias, but also permit the calculation of the magnitudes of the effects of these variables on demand. It can be concluded that the issue of preference formation for tourism consumption, which is particularly relevant to time series analysis of tourism demand, requires considerable further investigation.

2.4 Conclusions

This chapter has considered the conventional micro-foundations of tourism demand. One of the reasons why it was useful to do so is that appropriate modelling at the macroeconomic level requires some understanding of the micro-foundations and it is at the national, macroeconomic level that most tourism demand estimation has been undertaken to date. Macroeconomic modelling of tourism demand has incorporated, for example, assumptions about separability between consumption and labour supply decisions and the inter-temporal separability of consumption. The discussion in this chapter suggests that these assumptions may require modification, indicating that decisions to purchase tourism may be taken in conjunction with labour supply decisions (Figure 2.1) and may also be interrelated over time (Figures 2.8, 2.9, 2.10) depending on the available destination characteristics (Figure 2.11). It has also indicated that decisions to purchase tourism are related to decisions to consume other goods and services (Figures 2.2, 2.5, 2.6, 2.7) and that choices of particular types of tourism are made in the context of the possibility of purchasing other types (Figures 2.3, 2.4). If tourism demand is related to other types of consumption both at one point of time and over time, such interrelationships should be taken into account in tourism demand modelling, although most past studies have not done so.

Further issues arise in relation to the estimation of tourism demand models. A particularly important one is that of aggregation, which concerns the conditions under which relationships which hold at the individual level apply at more aggregate levels of groups or nations. For example, different individuals and groups have different consumption preferences, so that there are difficulties in identifying preferences at more aggregate levels. This raises problems concerning the choice of model to use for estimating demand at the macroeconomic level. The incorporation of inappropriate preferences in the model can produce inaccurate results. For example, some macroeconomic models involve estimating equations which permit aggregation over groups of commodities, but which involve questionable assumptions about preferences, such as equal proportionate responses of demand to proportionate changes in income or relative prices. Such models can provide misleading implications concerning individual or group behaviour. There may be additional problems concerning the interpretation of the estimated equations, for example when there have been changes in the composition of the population, level of unemployment, liquidity constraints and sporadic variations in the purchase of particular goods so that tourism might be consumed in one period but not the next, as in the case of destinations which experience periods of social unrest. It is clear, therefore, that the theoretical discussion of tourism demand has raised a number of issues which are relevant to empirical studies of tourism demand, which will be examined in Chapter 3.

3 EMPIRICAL STUDIES OF TOURISM DEMAND

3.1 Introduction

The aim of this chapter is to consider the models which have been used to estimate tourism demand in the context of both the theoretical discussion of the previous chapter and developments in consumer demand theory which will be introduced as the chapter proceeds. Empirical studies can help to explain the level and pattern of tourism demand and its sensitivity to changes in the variables upon which it depends, for example income in origin areas and relative rates of inflation and exchange rates between different origins and destinations. Such information is useful for public sector policy-making and the private sector. However, accurate estimates can usually be obtained only if the theoretical specification of the underlying model is appropriate. Hence, explicit consideration of the consumer decision-making underpinning empirical models is important in ensuring that the estimates which are provided are neither inaccurate nor misleading in their policy implications.

This chapter will initially explain and critically evaluate two approaches which have been used to model tourism demand, i.e. the single equation and the system of equations models. The single equation model which is considered in the second part of the chapter has been used in studies of tourism demand for numerous countries and time periods and posits that demand is a function of a number of determining variables. The estimated equations permit the calculation of the sensitivity of demand to changes in these variables. The discussion of the approach is centred on identifying its strengths and limitations and considers the associated implications for future research. In contrast to the first approach, the system of equations model requires the simultaneous estimation of a range of tourism demand equations for the countries or types of tourism expenditure considered. The system of equations methodology attempts to explain the sensitivity of the budget shares of tourism demand across a range of origins and destinations (or tourism types) to changes in the underlying determinants. In the third part of the chapter, the theory underlying the system of equations model is explained and the model evaluated. Directions for further investigation are, again, suggested.

The fourth part of the chapter considers the relevance of the two approaches for estimating tourism demand to methodologies for forecasting demand and evaluates alternative forecasting models. Then, the fifth part of the chapter builds on the characteristics framework analysed in Chapter 2 to discuss the importance of hedonic price modelling and discrete choice analysis in tourism demand from a disaggregated perspective. By the end of the chapter, it should be clear that a much tighter link between economic theory and empirical studies of tourism demand is necessary.

3.2 The single equation approach to estimating tourism demand

Tourism demand can be analysed for groups of countries, individual countries or states, regions or local areas. Demand can also be disaggregated by such categories as the type of visit (for example holiday and business tourism), and the type of tourist (covering nationality, age, gender and socio-economic group). On another dimension, its analysis can be related to particular types of tourism product, for instance sports tourism or ecotourism, or to specific components of the tourism product, such as accommodation and transportation, discussed in Chapter 5. Throughout this chapter the concept of tourism demand which is examined refers to the entire bundle of services which tourists purchase: transportation, accommodation, catering, entertainments and related services.

3.2.1 The methodology, its advantages and limitations

The single equation approach involves, first, theorizing the determinants of demand and, subsequently, using the technique of multiple regression analysis to estimate the relationship between demand and each of the determinants. A demand function might be written as

$$D = f(x_1, x_2, \ldots , x_n) \tag{3.1}$$

where D is tourism demand and $x_1 \ldots x_n$ are independent (explanatory) variables which determine demand. The theoretical issue is, therefore, to identify which independent variables should be included in the equation and the functional form (such as linear or log-linear) which is appropriate for the estimation of the equation. The application of the theory is relatively easy, subject to data availability, by means of computer packages which are designed to estimate the relationship between tourism demand as the dependent variable and the independent variables determining it. Such packages can test whether each of the hypothesized independent variables plays a significant role in determining demand and the extent to which each significant variable explains changes in demand.

The single equation approach has a number of advantages. Apart from being relatively easy to implement, it can provide useful information; once regression analysis has been used to examine the relationship between tourism demand and the determining variables, the extent to which a change in any of them alters tourism demand can be quantified by the calculation of the relevant elasticities. For example, the income elasticity of tourism demand for a destination is a measure of the extent to which demand changes as the result of changes in the income of tourist origin areas. The elasticity value can be calculated for different durations of time, thereby showing the difference between the short-run and long-run responsiveness of demand to changes in the variable under consideration. This may be useful for policy purposes, indicating, for example, the time within which any appropriate countervailing policy adjustment should take effect. Elasticity values can be estimated not only at the national level but also for different groups or nationalities of tourist or types of tourism products, which may differ considerably from the values estimated at more aggregate levels. For example, Gray (1970) pointed out that the price elasticity of demand for holiday tourism is higher than that for business tourism, while wanderlust tourists are likely to have a lower price elasticity of demand than sunlust tourists.

In part, the variations in the elasticity values estimated in the different studies of tourism demand are not surprising, since the values usually refer to different origins and destinations, time periods and measures of demand, such as tourism revenue per capita or visits per capita. However, a number of the estimated values may be inaccurate, owing to inappropriate specifications of the demand equations upon which they were based. In particular, the testing of demand theories has relied heavily on secondary data and econometric approaches in which the random error term contained the variables unexplained by the model. Thus, a central feature of much tourism demand modelling is that some of the variables influencing consumer behaviour have been ignored. Moreover, it is rare for studies to include the full range of test statistics, despite the provision of such statistics in computer regression packages. Thus, the degree of confidence which may be placed in the estimated results is uncertain. In some cases, results which the test statistics indicate as insignificant have been discussed as though they have been of equal importance to those which were shown to be significant, as pointed out by Johnson and Ashworth (1990). In other cases, such econometric problems as heteroskedasticity (where the errors in the regression equation do not have a constant variance) have been ignored.

Having the above in mind, it may be argued that the single equation model approach is subject to various limitations. If the tourism demand equation is mis-specified because of the inclusion of inappropriate variables or the omission of appropriate ones, the results obtained from estimation are likely to be spurious and can lead to inappropriate conclusions and policy recommendations. It may also be the case that, in some circumstances, the tourism demand equation should be estimated in conjunction with the tourism supply equation and/or labour supply equation. Failure to estimate the equations simultaneously may also produce biased results. These considerations further reinforce the case for a strong theoretical basis for the specification and estimation of tourism demand.

3.2.2 Model specification and the dependent variable

Most economic studies have used the single equation methodology to explain tourism demand, usually at the national level, as demonstrated by the reviews by Archer (1976), Johnson and Ashworth (1990) and Sheldon (1990). An example of a tourism demand function, where all variables occur at a given time period, t, is:

$$D_{ij} = f(Y_i, P_{ij/k}, E_{ij/k}, T_{ij/k}, DV) \qquad (3.2)$$

where D_{ij} is tourism demand by origin i for destination j, Y_i is income of origin i, $P_{ij/k}$ is prices in i relative to destination j and competitor destinations k, $E_{ij/k}$ is exchange rates between i and destination j and competitor destinations k, $T_{ij/k}$ is the cost of transport between i and destination j and competitor destinations k, DV is a dummy variable to take account of special events such as sporting events or political upheavals.

The majority of studies use either a log-linear (where both the dependent and the explanatory variables are expressed in logarithmic terms and the estimated coefficients are constant elasticities) or a linear (where all variables are expressed in levels and the estimated coefficients show constant marginal effects) specification (Franses and McAleer, 1998), mostly in the context of a time-series sample. More recently, however, a number of studies, such as Ledesma–Rodriguez et al. (2001), Naudé and Saayman (2005) and Garin-Muñoz (2006), have also introduced panel data analysis, which effectively combines time series and cross-sectional data to address multicollinearity problems and increase the available degrees of freedom.

With respect to the dependent variable (tourism demand) it is useful to refer to the UNWTO definition, according to which a tourist is a person travelling to and staying in places outside his/her usual environment for more than 24 hours but less than one consecutive year for leisure, business and other purposes (UNWTO, 2000). If a person stays in another place for less than 24 hours then he or she is classified as an excursionist. Tourists and excursionists together comprise the group of visitors. Tourism demand may be measured in a number of ways. In terms of volume statistics, it may be estimated by the number of tourist arrivals at specific entry/exit points (usually frontiers and public transport stations or terminals) and/or at registered tourism accommodation establishments (such as hotels, motels, tourist apartments, villas, etc.). Nevertheless, the collection of relevant statistics is not an easy task, especially for domestic tourists who do not cross frontiers, may use their private car (instead of public transport) and may stay with friends and relatives or at their second homes in rural or seaside locations. For this reason, domestic tourism demand statistics are often collected using visitor surveys, whose results, however, are potentially subject to inaccuracy.

On the other hand, international tourism demand may be easier to estimate as people cross frontiers and usually stay at tourism accommodation. Despite the difficulties imposed on international tourism by the existence of visa requirements in certain cases, these facilitate the collection of accurate statistics. When such restrictions do not exist and people can freely cross frontiers, as for example within the Schengen Area in Europe, then the collection of tourism arrival statistics at frontiers may at best rely on visitor surveys when the knowledge of tourism nationality is of importance. This is because statistics on international public transport services usually refer to passengers classified by the itinerary's country of origin and not by their nationality. For example, statistics of airport X in Greece may show that 3,000 people arrived from the UK in 2008; however, this does not necessarily mean that 3,000 British tourists flew to destination X as some people among the 3,000 may not have British citizenship. In this context, it may be better to estimate tourist arrivals at registered tourist accommodation as in many countries it is a requirement to fill a relevant registration form (which often includes citizenship details) when checking in. Moreover, analysts may refer to the number of nights spent in tourist accommodation as a more accurate measure of tourism demand since this accounts for the duration of the sojourn and excludes Visiting Friends and Relatives (VFR) tourist flows (Lim, 2006). Tourist accommodation statistics provide an accurate picture only to the extent that there are very few illegal (i.e. not registered) lodgings.

Tourism demand may also be measured using value statistics mostly in the form of tourism expenditure and/or receipts. These may be quoted in nominal or real terms both at aggregate and per capital levels. For international tourism, data may rely on foreign exchange statistics provided by the central banks of both the origin and the destination country. The validity of these data as an accurate reflection of tourism consumption patterns depends on whether foreign exchange transactions are properly (and legally) recorded or not (UNWTO, 2000). It should be also borne in mind that tourists may pre-pay for their holiday at the origin in the form of an all-inclusive package therefore spending only a limited amount of money at the destination, as discussed in Chapters 6 and 7. For domestic tourism, data on expenditure and receipts can be mostly collected using household or visitor surveys, which may prove inaccurate due to a number of reasons, including recall bias and memory effects (Frechtling, 2006). The same problem seems to emerge also in the case of countries forming part of a monetary union, such as the Eurozone. For example, measuring the expenditure of German tourists in Italy may prove a very difficult task as both countries use the Euro as their national currency. To address these problems, Frechtling (2006) proposes a number of alternative expenditure models such as the expenditure ratio, the cost-factor, the seasonal-difference and the supply-side judgemental model.

Having the above in mind, and in the light of the theoretical discussion in Chapter 2 and recent developments in the macroeconomic literature relating to consumption, explanatory variables which are likely to be significant determinants of tourism demand will now be considered. Income and inter-temporal consumption theory are examined initially. The discussion is then extended to take account of relative prices and exchange rates, which are important in determining the effective relative price of tourism at the international level. Additional variables which may be relevant in the tourism demand equation, notably transport costs, marketing expenditure and dummy variables for special events, will also be considered.

3.2.3 Income and the inter-temporal demand for tourism

The analysis in Chapter 2 showed that changes in income can have important effects on tourism consumption, so that it is clearly a key variable within a tourism demand equation. The discussion also examined the theory underlying the timing of tourism consumption. However, a key problem with most studies of tourism demand which have used the single equation model is the absence of an explicit theory of consumer decision-making. Consequently, an explanation of the process by which tourism demand occurs over time and the role of income within it has not been provided. Until the late 1990s, with few exceptions, notably Syriopoulos (1995), the majority of studies assumed that demand depends on current income but not on past or expected future income. By making this assumption, such research has ignored the debates within the economics literature concerning inter-temporal decision-making and the issue of whether consumers are backward-looking or forward-looking. Therefore, in those studies which have assumed demand to depend solely on current income, the relationship between tourism demand and its determinants may have been mis-specified and the calculated elasticities may be inaccurate. Nonetheless, advance in dynamic econometric techniques primarily based on the Autoregressive Distributed Lag Model (ADLM) and the Error Correction Model (ECM) – to be discussed later – have integrated inter-temporal choice issues into their analysis (Song and Li, 2008).

In fact, inter-temporal choice theory can help to explain tourism demand at both the macroeconomic and the microeconomic levels. The theory can take account of the fact that demand decisions are often made in the context of imperfect information, unforeseen events, expectations about the future and liquidity constraints which limit current consumption. The theory of inter-temporal choice allows consumption to depend on any combination of current, future and/or past income, so that the assumption that it depends solely on current income becomes a special case within a more general model. Inter-temporal choice theory argues that people decide how to allocate their consumption between present and future periods so that those with preferences for high current relative to future consumption are said to have a high rate of time preference, while the reverse is true for deferred consumption. The value of consumption in any given time period depends on the rate of time preference, the value of current income receipts and possibly on past and expected future receipts. The explanation of current consumption requires the formulation of a theory of expectations with respect to future income receipts because, if people anticipate a change in their income in the future, they are likely to alter their current consumption. In this case, consumers are 'forward-looking' and actual changes in consumption are a function of predictable changes in future income. The implication is that in order to understand changes in total consumption and its tourism component, it is necessary to explain (or model) future changes in income.

A range of theories of expectations of future income have been proposed. Much research on consumption has assumed that expected changes in future income are based on past changes in income. For example, it has been postulated that expectations of changes in income accord with an adaptive process, whereby expectations are adjusted in relation to the past values of consumers' income. Thus, consumption may be a function of the discounted present value of expected future income, predicted on the basis of past values of income, so that consumers are backward-looking. Alternatively, consumers may anticipate that their future incomes will change in a way which differs from past patterns because of what are called innovations in income, for instance changes in the benefit and tax system. Such innovations may affect expected future income and, hence, its discounted present value, leading to a change in consumption if they persist. Innovations in income vary within and between different countries and time periods and may be incorporated into anticipated future income differently because of varying expectations and contexts.

It is also necessary to take account of the differing amounts of information which are available to consumers in a particular context. The theory of rational expectations, from new classical macroeconomic theory (Sargent and Wallace, 1976; Maarten, 1993), argues that consumers obtain and use all available information when making decisions, so that actual outcomes are consistent with predicted outcomes. According to the theory, unpredictable changes in income lead consumers to revise their expectations about future income, causing changes in the discounted present value of their income and corresponding consumption changes. However, empirical evidence questions this theory of expectations formation (for example, Flavin, 1981; Blinder and Deaton, 1985; Jappelli and Pagano, 1989; Campbell and Mankiw, 1991). One possible reason is that consumers are imperfectly aware of innovations in incomes, so that their consumption response is small. New Keynesian macroeconomists posit that consumers' actions may be limited by market imperfections. For example, rigidities in wages may prevent individuals who wish to work from obtaining employment, thereby limiting their income. Some consumers may be unable to increase their current consumption owing to liquidity constraints associated with restrictions on their ability to borrow against their future income. Borrowing constraints may occur because of asymmetric information, whereby the consumer has knowledge of a future increase in income but the lender is unaware of the future increase or is doubtful that it will occur. Uncertainty about future incomes may also reduce consumption by inducing an increase in precautionary savings to cater for the possibility of future decreases in income.

The majority of studies of tourism demand, which have taken account of current but not past or expected future income in the demand function, are consistent with liquidity constraints on borrowing. Such studies are also consistent with consumer behaviour which is neither forward-looking nor backward-looking. Thus, the dependence of tourism expenditure solely on current income can be considered as a particular case within the range of consumer behaviour which can occur. If tourism consumers are backward-looking, the tourism consumption equation should include lagged values of income, weighted according to the degree to which current consumption is determined by past income, with proximate years usually having the highest weights. If consumers are forward-looking, in the absence of liquidity constraints, the tourism consumption function should include terms to take account of the process by which consumption changes in accordance with consumers' expectations of the discounted present value of their future income. This requires information concerning income changes over time and the ways in which people formulate their expectations and incorporate available information into their predictions of future income. Thus, the foregoing discussion indicates the ways in which the tourism demand function can be specified to reflect the relationship between demand and income which is particular to each origin country. The main issue is of course to test the theories empirically.

Most econometric studies use GDP or GNP in nominal, real or per capita terms as a proxy for income in the origin country (Lim, 2006). In terms of estimated elasticity values, an early study of US and Canadian outbound tourism (Gray, 1966) found that the per capita income elasticity values ranged between 4.99 and 7.01, implying that a rise in income of 1 per cent results in an increase in tourism expenditure of between 4.99 per cent and 7.01 per cent. Artus (1972) found that the income elasticities of demand by European tourists for international tourism varied between 1.36 (Switzerland) and 3.84 (Austria). Witt and Martin (1987) found that the per capita income elasticity values for UK travel to European countries ranged between 0.34 (Cyprus, Gibraltar, Malta) and 2.91 (Netherlands) for independent air travel, and between 0.86 (Spain) and 6.35 (Greece) for inclusive tour air travel. The income elasticities for tourism in eighteen European destination countries estimated by Tremblay (1989) ranged between 0.33 (UK) and 11.35 (Portugal). Little's (1980) study of US demand in ten destination countries found income to be an insignificant determinant of demand. Many more studies have provided additional estimates for a range of origins and destinations, as shown by Archer (1976), Johnson and Ashworth (1990) and Sheldon (1990). According to Lim (2006), income elasticities seem to range between 0.033 (Ffrench, 1972) and 14.32 (Lee, Var and Blaine, 1996).

3.2.4 Relative prices, exchange rates and tourism demand

The examination of tourism demand, so far, has concentrated on the ways in which income can affect demand. However, Chapter 2 demonstrated that tourism demand depends not only on its own price but also on the prices of other goods and services, while the choice of different types of tourism also takes account of their relative prices. Moreover, tourism can be either a substitute for, or a complement to other goods. Although tourism demand equations should incorporate these relative price variables, they have not been taken into account in empirical studies. First, price indices for tourism are often unavailable so that the retail price index has usually been used, although, with a notable exception of the study by Martin and Witt (1987), rather limited research has been undertaken to investigate its suitability as a proxy. Studies by Jorgensen and Solvoll (1996) and Kulendran and King (1997) are examples of research using the package tour cost as a measure of tourism prices. Second, most research has overlooked the possibility that decisions to purchase tourism may be taken in conjunction with decisions to purchase other goods and so the prices of the latter have not been taken into account. Third, many studies have either ignored the fact that consumers choose between a range of tourism products and destinations, or have included the prices of a range of alternatives without providing a well-argued rationale for the range selected.

With regard to international tourism, since the consumer's country of origin is a possible site of demand, the exchange rates between the origin and a range of other destinations are also likely to be relevant. Relative prices and exchange rates between tourists' origin country and their destination have been included in studies of tourism demand, often as separate explanatory variables (for example, Artus, 1972; Little, 1980; Loeb, 1982; Quayson and Var, 1982; Martin and Witt, 1988; Lee et al., 1996) but sometimes in the form of effective exchange rates (nominal exchange rates adjusted for differences in relative inflation rates). The latter may be more appropriate in the long-run (Syriopoulos, 1995). Occasionally, the prices and exchange rates of other competing destinations have also been incorporated.

As in the case of the income elasticities, studies have yielded a wide range of elasticity values for the relative prices and exchange rate variables included in the estimating equations. For example, the price elasticity values for tourism from the UK ranged between −0.23 (Austria) and −5.60 (Greece), and for tourism from West Germany ranged between −0.06 (Spain) and −1.98 (France) (Martin and Witt, 1988). Artus (1972) estimated price elasticities of European receipts from international travel of between −0.37 (Sweden) and −4.95 (Netherlands). In a study of international visitor flows to Australia, Divisekera (1995) estimated a price elasticity as little as −0.15, whereas on the other hand Lee, Var and Blaine (1996) have found an elasticity equal to −7.01 in the case of Philippine tourism demand for South Korea. Tremblay (1989) estimated exchange rate elasticity values of between 0.63 (West Germany) and 4.60 (Portugal) for tourism receipts by European countries. US demand for tourism was also associated with a range of exchange rate elasticity values, varying between −0.58 (Mexico) and −3.15 (Canada) as shown by Little (1980), while Loeb (1982) estimated values of between 0.8 (Italy) and 4.07 (UK). Moreover, Webber (2001) estimated the value of the exchange rate elasticity as high as −12.01 in the case of Australian tourism demand for Malaysia. However, relative exchange rates were insignificant determinants of tourism demand in Okanagan (Quayson and Var, 1982) and Singapore (Gunadhi and Boey, 1986). The range of elasticity values estimated may be due to the markedly differing circumstances of the origins and destinations under consideration but, as in the case of the income elasticities, some may result from different specifications of the equations which were estimated.

In general, there has not been much discussion in the literature of whether it is theoretically sound to include relative prices and exchange rates as separate determinants of the demand for tourism at the international level or whether effective exchange rates are appropriate. It can be argued that, over the short run, the rates of inflation and of changes in nominal exchange rates differ, so that tourists take account of relative prices and exchange rates separately in their decision-making. A counter-argument is that tourists are unaware of inflation rates in overseas destinations and take account only of nominal exchange rates. Alternatively, effective exchange rates may be the relevant explanatory variable. One justification for this view is that most tourists pay for their tourism consumption in their own currency and the prices which they are charged take account of both differences in relative prices and exchange rates. Thus, the prevailing methods of pricing and paying for tourism consumption are key considerations as to which variables to include in the estimating equation.

3.2.5 Lagged variables

Studies of tourism demand have usually included the current values of relative prices and/or (effective) exchange rates as explanatory variables in the estimating equation. However, given that tourism purchases are made in advance of their actual consumption, lagged rather than, or in addition to, the current values may be the appropriate independent variables, depending upon the pattern of tourism consumption in the country in question. Expectations of future changes in prices and exchange rates are less likely to be significant determinants of demand than past rates, given most consumers' lack of information and uncertainty about future movements in them. Consumers' awareness of prices and exchange rates for other destinations is likely to vary between different origins and destinations, and possibly over time, so further investigation of the relationship between tourism demand, relative prices and exchange rates is necessary. This could encompass such aspects as the possibility that consumers are subject to money illusion with respect to foreign currency prices, in that they may be unaware of their domestic currency values in the short-run. The possible effects of changes in interest rates in altering the timing of tourism purchases, outlined in Chapter 2, could also be subject to empirical study.

It is likely that the demand for tourism in a particular destination in a given period depends upon demand in the previous period. This is because the demand for tourism in an alternative location may be deterred by consumers' lack of experience and sometimes knowledge of it. Thus, it is commonly assumed that the more information consumers have about a destination, the greater is the demand for it. The effect of increased information can be taken into account in the estimating equation by including a lagged dependent variable whereby demand in the current time period is affected by the previous level of demand. This is also consistent with the hypothesis that some consumers develop the habit of making repeat visits to particular destinations (Witt, 1980; Witt and Martin, 1987; Martin and

Witt, 1988; Darnell et al., 1992; Syriopoulos, 1995) and is similar to the effect of habit persistence (time non-separability) in aggregate consumption expenditure (Braun et al., 1993). Habit may be explained by the discussion within the tourism literature of the psychocentric tourist, who has a preference for familiarity as opposed to new experiences and destinations; hence, there is a positive relationship (coefficient) between current and past demand. Conversely, a significant negative coefficient may indicate that tourists are allocentric, seeking new experiences in new destinations. Another explanation of a negative coefficient is that previous visits revealed some undesirable feature of the destination. With respect to the lags in the adjustment process, the psychocentric tourist may exemplify adjustment with a long lag or, indeed, no adjustment at all, whereas the allocentric may exhibit very rapid adjustment.

Clearly, the responsiveness of current demand to previous demand may change over time. For example, consumers may initially increase their demand for a destination, owing to greater information about it and/or to the acquisition of a habit of visiting it, resulting in a positive coefficient. However, consumers may subsequently decrease their level of expenditure at the destination owing to increased knowledge of the most cost-effective ways of obtaining products and services within it, for example, cheaper forms of local transport or cheaper restaurants and hotels, for a given level of quality (Godbey, 1988). This would result in a negative coefficient. The possibility of habit persistence and wider information availability have rarely been tested in empirical studies but were found to have a significant positive effect on tourism demand by former West German and UK tourists (Witt, 1980; Witt and Martin, 1987) and for south Mediterranean countries (Syriopoulos, 1995). Moreover, Alegre and Juaneda (2006) used three theoretical pillars in a study of destination loyalty, namely the importance of reputation, information asymmetry and behaviour of consumer. Using the Balearics as a case study, they confirmed their hypothesis that repeat visitors tend to spend less than first-timers. Nonetheless, emotional attachment to a particular tourist destination and active seeking of service quality may result in higher expenditure patterns.

From an econometric model specification perspective, lagged variables can be best considered using an ADLM of the following form (Song and Turner, 2006):

$$y_t = a + \sum_{i=1}^{n} \sum_{j=0}^{q} b_{ij} x_{it-j} + \sum_{j=1}^{p} c_j y_{t-j} + e_t \qquad (3.3)$$

where y is tourism demand (dependent variable), n is the number of explanatory variables x, p and q are the lag lengths of the dependent and explanatory variables respectively, e is the disturbance term which is assumed to have all classical properties, t is time period and a, b, and c are coefficients.

In the ADLM, the lag length is determined in the context of a general-to-specific methodology (discussed later) using suitable statistical criteria such as SBC (Schwarz–Bayesian Criterion) and AIC (Akaike Information Criterion). The ADLM can be algebraically manipulated to give the ECM. In particular, and by assuming the existence of only one lag ($p = q = 1$), the ECM is given by:

$$\Delta y_t = b_0 \Delta x_t - (1 - c_1)\left[y_{t-1} - m_0 - m_1 x_{t-1}\right] + e_t \tag{3.4}$$

where $m_0 = a/(1-c_1)$ and $m_1 = (b_0+b_1)/(1-c_1)$. The coefficient b_0 is known as the impact parameter, $(1-c_1)$ is called the feedback effect, m_0 and m_1 are the long-term response coefficients and the mathematical expression in the square brackets is termed error correction mechanism. The ECM is of particular econometric importance as it can simultaneously encapsulate both the short- and the long-term effects. It avoids problems of spurious correlation as the differenced variables are usually stationary (i.e. they have a trendless pattern) similarly to the $[y_{t-1} - m_0 - m_1 x_{t-1}]$ expression when the variables y and x are co-integrated (Hendry, 1995). Syriopoulos (1995) was the first to apply the ECM in tourism but over the last decade a growing number of researchers have explicitly used this dynamic specification (Song and Li, 2008).

3.2.6 Transportation

The price of transportation is another variable which some studies have included as an explanatory variable within the tourism demand equation. The case for and against doing so is complicated. On the one hand, the retail price indices, which have usually been included, in practice do not take explicit account of the price of transport between the origin and the destination, so that there is a case for the inclusion of a separate transport price variable. Moreover, its cost is such a significant proportion of the total price of a holiday that changes in it may induce a switch of mode or of destination choice altogether. On the other hand, the definition of tourism demand which is usually explained is the total bundle of tourism components which are purchased (accommodation, entertainments, other service provision and transportation), so that the own price variable which is included in the estimating equation should take account of the price of all these components. Therefore, the inclusion of a separate price of transport should not, in theory, be necessary. Moreover, even if a case can be made for the inclusion of a transport price variable for a particular set of origins and destinations, the form in which it might appropriately be included is not evident. Nor is it clear which other destinations might be substitutes for or complements to a given destination and, hence, which transport prices might be considered in addition to that for the particular origins and destinations under study. A further issue concerns the specific transport prices which are appropriate candidates for inclusion since, within as well as between most forms of transport, there are different fares, which vary according to such criteria as the pre-booking time, travel schedule and length of stay. The differences between traditional scheduled, low-cost, charter and 'bucket shop' fares for airlines are obvious examples, reliable time series data for the latter being unavailable (Divisekera, 1995).

Given these considerations it is, perhaps, not surprising that the transport price variables which have been included in studies of tourism demand have often been insignificant, as in the case of US travel to a range of countries (Little, 1980; Stronge and Redman, 1982) and tourism in Okanagan, Canada (Quayson and Var, 1982). Other studies have found a significant negative relationship, for example Kliman's (1981) study of tourism demand in Canada, where the elasticities ranged between −0.94 (Italy) and −3.09 (Portugal), and Tremblay's (1989) study, which found values varying between −0.48 (Belgium) and −4.17 (Sweden). Overall, it is apparent that consideration of the price of transport as a possible determinant of tourism demand should be treated with far more caution and be the subject of far more detailed theoretical and empirical investigation than has been the case to date.

3.2.7 Other variables

Other variables which have been hypothesized as determining tourism demand are expenditure on marketing and dummy variables reflecting atypical events, such as sporting attractions or major political changes. Studies of Barbados by Clarke (1981) and Turkey by Uysal and Crompton (1984) found marketing expenditure elasticity values to be significant but less than unity. On the other hand, Papadopoulos (1987) has estimated a marketing elasticity equal to 1.61. Divisekera and Kulendran (2006) studied the effects of advertising on Australian tourism demand to derive long-run elasticities ranging between 0.04 (in the case of American tourists) and 0.65 (for Japanese tourists). When relatively high advertising elasticities are combined with highly price-elastic demand, this provides support for the information theory according to which advertising raises awareness of substitutes and hence leads to a rise of demand price elasticity (Nelson, 1974). The inclusion of dummy variables for special events does not pose such problems and dummy variables for the Olympic Games and EXPO 67 in Canada were found to have relatively small values of 0.35 for the Olympics (Loeb, 1982) and 0.49 for EXPO (Little, 1980). Political events can have greater effects on demand, as in the case of tourism demand by Indonesians for Singapore when tensions between the two countries were associated with an elasticity of −1.5 (Gunadhi and Boey, 1986).

3.2.8 Implications for research on single equation models

The earlier discussion of single equation models of tourism demand has highlighted a number of issues for future research. These can be classified under the broad headings of theory, aggregation and estimation. The theoretical discussion in Chapter 2 indicated the importance of examining the relationship between total consumption, paid work and unpaid time, on the one hand, and between tourism consumption and the consumption of other goods and services, such as consumer durables, on the other. It also demonstrated the role of income and relative prices in tourism consumption. Empirical research to date has not incorporated a thorough examination of these issues. It has generally been assumed that tourism consumption is separable from decisions to engage in paid work or consume other goods and services. Moreover, the nature of the inter-temporal relationship between tourism consumption and changes in income and effective prices has received little attention. The role of expectations has also been neglected.

A number of studies of demand have included explanatory variables in the estimating equation on a fairly ad hoc basis. This procedure may have resulted in mis-specification of the equation and biased results. One solution which has been suggested as a way of overcoming bias stemming from the omission of relevant independent variables is the 'general to specific' methodology proposed by Davidson et al. (1978), Hendry and Mizon (1978) and Hendry (1983). The methodology involves the inclusion in the estimating equation of all possible relevant independent variables and the subsequent exclusion of those that are insignificant determinants of demand. Differences between the short- and long-run responsiveness of tourism demand to its determinants can be taken into account by the inclusion of lagged (previous period) values of a number of the independent variables and the use of an error correction mechanism, so that the short- and long-run elasticity values can be calculated.

The issue of aggregation is particularly important and has not been sufficiently tackled by empirical studies (Song and Li, 2008) which have estimated the demand for tourism as an aggregate commodity, for example UK demand for holidays in Spain, without considering the demand for the components of which it consists or the demand by specific groups of consumers such as families or elderly people. The general failure to consider the nature of the tourism demand equations for different types of tourism products or for different individuals or groups means that there is virtually no evidence about the micro-foundations of aggregate tourism demand equations. Therefore, it is not possible to ascertain whether the demand equations which have been estimated were appropriate for the case under consideration and, hence, whether the estimated results were accurate. The conclusion is that the issue of aggregation merits considerable further attention.

Given appropriate specifications for the tourism demand equations, the problem of possible econometric unreliability of results is relatively easily resolved by means of the inclusion and examination of test statistics. However, biased results could also result from the estimation of a single equation for tourism demand in contexts in which the simultaneous estimation of a tourism supply equation and/or a labour supply (paid work) equation is also relevant. If demand and supply are interrelated but this is not taken into account within the estimation process, the econometric problems of identification and simultaneity arise and result in incorrect estimates. An identification problem could occur because, for instance, of supply constraints, such as a shortage of accommodation or aircraft seats. In the context of supply limitations, it is necessary to consider whether demand and supply are interdependent. If so, the demand and supply equations should be estimated simultaneously. Overall, it is clear that there is scope for further investigation of tourism demand, based on rigorous theoretical examination of the appropriate forms of the equations to be estimated, including their dynamic structure which would show the way in which tourism demand changes over time. Such investigations would enable tourism demand equations to be tailored to suit the specific circumstances of the case under consideration.

A final issue which requires consideration is that of expenditure on tourism-related durable assets, for example a holiday home, timeshare, caravan or boat. Holiday homes or timeshare differ from most durable goods in that they are likely to appreciate or, at least, retain their real value over time and, thus, have a significant effect upon the consumer's wealth. Such purchases, therefore, might affect the future pattern of income and expenditure. For example, ownership of a second home will tend to commit a consuming unit to continuing tourism expenditure in the location of the residence (Williams, King and Warnes, 2004). Moreover, the property may be used as collateral for further borrowing and expenditure, not necessarily related to tourism. The role of tourism property as a particular form of durable has not been investigated, although, by implication, this type of asset has been embodied in wealth, thereby influencing consumption (Caballero, 1993). Empirical research incorporating an explicit theoretical framework could investigate, for example, the determinants of changes in the stock of tourism-related durables and the possible existence of cyclicality in expenditure on them. Estimation of the changes over time in the value of the stock of tourism durables, for different areas, would provide useful basic data series.

3.3 The system of equations models of tourism demand

System of equations models of tourism demand can be used to estimate the demand for tourism for a number of destination countries by consumers from a range of origins. The models have a strong theoretical foundation, having been formulated using microeconomic theories of demand. The objective has been to develop a model which permits generalization on the basis of the behaviour of an individual who is representative of normal behaviour, so that the equation which is used to estimate tourism demand by consumers in aggregate has an appropriate grounding in individual behaviour.

3.3.1 Assumptions, methodology and applications

In traditional economic theory, the individual is normally viewed as a 'rational economic person', who is assumed to want more material goods and to behave in an optimizing way to maximize his or her own utility. The individual makes decisions within the framework of a market, within which it is assumed that prices adjust to eliminate excess demand and supply. System of equations models of tourism demand generally assume that individuals make decisions according to the 'axioms of consumer choice'. These state that an increase in price results in lower demand (negativity), the sum of individual expenditures is equal to total expenditure (the adding-up condition), a proportional change in expenditure and all prices has no effect on quantities purchased or the budget allocation (homogeneity) and the consumer's choices are consistent (symmetry). If the axioms are valid reflections of behaviour at the individual level, generalizations to the aggregate level are more likely to be appropriate, permitting the specification of an estimating equation for aggregate demand which, rather than being ad hoc, can be justified in terms of economic behaviour.

According to system of equations models of tourism expenditure allocation, decisions are made by a 'stage budgeting process' based on the separability assumption discussed in Chapter 2. The consumer first allocates his/her budget among broad groups of goods and services such as tourism, housing and food, and then to sub-groups, for instance holidays in Europe, the USA and other regions of the world as sub-groups of tourism, followed by allocation to individual items, for example, different countries as holiday destinations. One model commonly used to estimate the allocation of consumer expenditure between a range of goods and services or between a number of countries is the Almost Ideal Demand System (AIDS) model developed by Deaton and Muellbauer (1980a, 1980b). The model incorporates the axioms of consumer choice and a stage budgeting process.

The allocation of expenditure between the items under consideration, such as different tourist destinations, may be calculated by means of multiple regression analysis, which provides estimates of the sensitivity of each item's share of total expenditure to a number of independent variables, particularly prices. A typical equation assuming a linear approximation (also known as LAIDS) is given by:

$$w_i = \alpha_i + \sum_{j=1}^{n} \gamma_{ij} \log p_j + \beta_i \log(x/P^*) + u_i \tag{3.5}$$

where w_i is the share of the budget of residents of origin j allocated to tourism in destination i, p_j is the price level in origin j, x is the budget for tourism expenditure by residents of origin j, P^* is the Stone price index calculated as $\log P^* = \sum_{i=1}^{n} w_i \log p_i$ which takes account of prices in the destination areas, Σ denotes the sum of, u is the disturbance term and α_i, β_i and γ_{ij} are coefficients. Hence, the model takes account of the role of the expenditure budget and of prices in determining tourism demand, in accordance with the theoretical discussion in Chapter 2. The AIDS system of equations model has been used to examine the demand for tourism from west European countries (such as the US, UK, former West Germany, France and Sweden) to south Mediterranean destinations (such as Italy, Greece, Portugal, Spain, Turkey and former Yugoslavia) (Syriopoulos and Sinclair, 1993; Papatheodorou, 1999). It was assumed that consumers first allocated their budget between a range of aggregated categories of goods and services, all types of tourism consumption being one category. Consumers subsequently allocated their tourism expenditure between major regions of the world and, having decided on their preferred region, allocated their expenditure on different countries within the region. The allocation of expenditure within the south Mediterranean region represents the final stage of this decision process and was estimated using an equation similar to Equation (3.5) above.

Nonetheless, Equation (3.5) is essentially static; to address dynamic issues an augmented version of the AIDS model has been recently introduced to explicitly consider the implications of the error correction mechanism. Such an extension (denoted as EC-LAIDS) may take the form (Li, 2004):

$$\Delta w_i = \alpha_i + \sum_{j=1}^{n} \gamma_{ij} \Delta \log p_j + \beta_i \Delta \log(x/P^*) + \lambda_i \mu_{it-1} + u_i \tag{3.6}$$

where μ_i is the error correction term, which measures feedback effects and is estimated from the corresponding co-integration equation of each budget share equation and λ_i is a negative scalar. So far, however, the use of the EC-LAIDS model specification remains relatively scant (Li, 2004; De Mello and Fortuna, 2005; Mangion et al., 2005).

System of equations models of consumer demand have been used to explain the allocation of a tourism expenditure budget between different destinations (White, 1982; O'Hagan and Harrison, 1984; Smeral, 1988; Syriopoulos and Sinclair, 1993; Papatheodorou, 1999; De Mello et al., 2002, Divisekera, 2003; Durbarry and Sinclair, 2003; Li, 2004) and different types of tourism expenditure (Fujii et al., 1987; Sakai, 1988; Pyo et al., 1991). The objective of these models is, thus, different from that of the models discussed in the previous section, which were concerned with explaining aggregate tourism expenditure but not its distribution by place or forms of expenditure. Based on Li et al. (2004), Tables 3.1–3.3 compare the results from different AIDS studies (Syriopoulos and Sinclair, 1993; Papatheodorou, 1999; De Mello et al., 2002; Li et al., 2004) on UK tourism demand.

In particular, the results indicate that tourism expenditure elasticities vary considerably between but also within destinations as shown in Table 3.1.

For example, tourism expenditure elasticity for Portugal was estimated at 0.04 by Papatheodorou (1999) compared to 1.580 by Syriopoulos and Sinclair (1993). Expenditure elasticities for Spain are almost always greater than one; on the other hand, half of the estimated elasticities for Italy are below unity. The elasticity values relate to the changes in the origin countries' shares of the tourism expenditure budget and indicate that Spain, for example, would gain a large increase in its share of the tourism expenditure budget as the result of an increase in the size of the budget, while Italy would receive a relatively smaller increase. Short- and long-run elasticities derived from the EC-LAIDS model do not differ substantially. Similarly, Table 3.2 provides the compensated price elasticity values.

The responsiveness of demand to price rises in destinations appeared particularly high in Portugal as the elasticity values range between −1.05 and −2.85; Li et al. (2004) report a very high elasticity value for Greece (−2.75), whereas Papatheodorou (1999) found a relatively low value (−0.93) for the same country. The lowest value is provided by Li et al. (2004) for

Table 3.1 *Expenditure elasticities estimated by different AIDS studies*

	M1	M2	M3.1	M3.2	M4 – SR	M4 – LR
France			0.63	0.81	1.12	1.09
Greece	1.05	0.80			1.20	1.20
Italy	0.88	1.05			1.00	0.90
Portugal	1.58	0.04	0.82	0.95	1.05	1.24
Spain	0.90	1.15	1.20	1.15	1.04	1.06

Notes

M1 refers to Syriopoulos and Sinclair (1993), M2 to Papatheodorou (1999), M3 to De Mello et al. (2002), where the whole sample is separated into two periods denoted as M3.1 and M3.2 respectively, and M4 to Li et al. (2004), where SR denotes short and LR the long-run. Source: Li et al. (2004).

Table 3.2 Compensated own-price elasticities estimated by different AIDS studies

	M1	M2	M3.1	M3.2	M4 – SR	M4 – LR
France			−1.76	−1.54	−0.53	−1.17
Greece	−2.54	−0.93			−1.91	−2.75
Italy	−1.24	−0.77			−0.65	−0.93
Portugal	−2.69	−2.85	−2.16	−1.71	−1.05	−1.16
Spain	−0.72	−0.65	−1.26	−1.40	−1.32	−1.52

Notes

M1 refers to Syriopoulos and Sinclair (1993), M2 to Papatheodorou (1999), M3 to De Mello et al. (2002), where the whole sample is separated into two periods denoted as M3.1 and M3.2 respectively, and M4 to Li et al. (2004), where SR denotes short and LR the long-run.
Source: Li et al. (2004)

the short-run elasticity of France (−0.53), implying that price competitiveness is a rather insignificant determinant of tourism demand in this case. Finally, Table 3.3 reports compensated cross-price elasticities; some destinations appear as substitutes (i.e. the elasticity has a positive sign), others as complements (i.e. the elasticity has a negative sign).

De Mello and Fortuna (2005) and Mangion et al. (2005) are other examples of studies which estimated an EC-LAIDS model. In addition, Li et al. (2006) introduced time varying parameter modelling (TVP) into the EC-LAIDS

Table 3.3 Compensated cross-price elasticities estimated by different AIDS studies

	France	Greece	Italy	Portugal	Spain	
France						M1
						M2
				0.09	1.45	M3.2
		0.19	−0.16	0.08	**1.15**	M4 – LR
		0.26	0.01	−0.08	**0.78**	M4 – SR
Greece			−0.23	0.56	1.22	M1
			−0.98	1.06	1.16	M2
						M3.2
	0.48		−0.03	**−0.43**	0.23	M4 – LR
	0.65		**−0.51**	0.29	−0.09	M4 – SR
Italy		−0.04		−0.83	0.41	M1
				0.25	1.09	M2
						M3.2
	−0.35	−0.03		**0.33**	−0.11	M4 – LR
	0.01	**−0.46**		**0.34**	0.18	M4 – SR

Table 3.3 (Continued)

	France	Greece	Italy	Portugal	Spain	
		0.53	−4.40		4.08	M1
						M2
Portugal	0.44				1.27	M3.2
	0.35	**−0.80**	**0.69**		−0.05	M4 − LR
	−0.35	0.53	**0.71**		−0.50	M4 − SR
		0.20	0.37	0.70		M1
						M2
Spain	1.73			0.23		M3.2
	0.77	0.06	−0.03	−0.01		M4 − LR
	0.52	−0.02	0.05	−0.07		M4 − SR

Notes

M1 refers to Syriopoulos and Sinclair (1993), M2 to Papatheodorou (1999), M3 to De Mello et al. (2002), where the whole sample is separated into two periods denoted as M3.1 and M3.2 respectively, and M4 to Li et al. (2004), where SR denotes short and LR the long-run. Source: Li et al. (2004) – elasticities in bold are significant at 5% level.

model to deliver a TVP-EC-LAIDS specification. TVP essentially assumes that the estimated coefficients of the explanatory variables in the demand model are also simultaneously determined over time by an autoregressive stochastic specification, known as transition equation (Song and Witt, 2000). Elasticity values were also estimated in Fujii et al.'s (1985) application of the AIDS model to different types of tourism expenditure in Hawaii, which provided expenditure elasticity values approximating to unity. The compensated own-price elasticity values, measuring the responsiveness of the budget share for a particular type of expenditure, such as food, to a change in the price of food assuming constant real expenditure, were generally less than unity. This indicated that the budget shares were insensitive to changes in prices. The uncompensated own-price elasticity values, which allowed for a change in real expenditure because of the price changes, were not significantly different from zero, showing that price changes had little effect on the budget shares.

In addition to the AIDS-related research, system of equations modelling is also associated with two specifications, namely Vector Autoregression (VAR) and Structural Equation Modelling (SEM). The VAR specification regards all variables as endogenously determined; it regresses present values of all the variables in question on previous (lagged) values of the same group of variables in the system (Song and Witt, 2000). On the other hand, SEM assumes that all variables can affect each other in a reciprocal manner. Turner and Witt (2001) introduced SEM to identify the causal link between three categories of tourism flows (visiting friends and relatives, leisure and business) with explanatory variables. Song and Li (2008) encourage further research using this econometric technique.

3.3.2 Advantages, limitations and research implications

The system of equations model has the advantages of incorporating an explicit theory of the consumer decision-making process and of being formulated in a way which is consistent with aggregation from the individual tourism consumer (representative agent) to the macroeconomic level. The approach avoids most of the charges of biases in the results, stemming from an inappropriate theoretical base. Improvements in the estimation of the dynamics of consumer demand are being incorporated in current research and should permit the methodology to incorporate inter-temporal decision-making. The model provides estimates of expenditure, own- and cross-price tourism demand elasticities. These elasticities are estimates of the sensitivity of tourism demand, as shares of total tourism expenditure, to changes in expenditure on tourism, the price of a particular type of tourism or tourist destination and the prices of other tourism types or destinations which may be substitutes for or complements to the tourism type or destination under consideration, as explained in Chapter 2.

Estimated elasticity values have implications for business strategies and policy-making. For example, low expenditure elasticity values for particular tourist destinations may be a cause for concern in that the benefits from increases in expenditure which occur over the long-run will go to alternative destinations. This indicates that it is necessary to examine the reasons for low elasticity values and what, if anything, can be done to render the destination more desirable. High own-price elasticities of demand, significantly greater than unity, may be a source of concern in countries with relatively high rates of inflation and/or depreciating exchange rates, but could be a means of increasing tourism receipts in contexts in which it is possible to improve price competitiveness. The cross-price elasticities of demand, which indicate complementarity or substitutability between tourism destinations or types of tourism, provide information which may be useful for tourism marketing campaigns by both businesses and public bodies. Such campaigns could be undertaken in conjunction with other countries or producers in the case of complementarity. In addition, holiday packages involving complementary destinations may appeal to potential customers.

In the system of equations approach, the AIDS model is generally considered to be the most flexible form for representing consumer preferences. However, it assumes that consumption and paid work decisions are made separately, in contrast to the discussion included in Chapter 2 which showed that such decisions may be taken simultaneously. In comparison with the single equation approach, it is less flexible in that all the estimating equations must incorporate the same independent variables and functional form for the equations to be estimated simultaneously. Thus, for example, dummy variables referring to political changes or sporting events, which may be relevant to the determination of tourism demand only in one country, should be added across the system of equations, even if their incorporation does not make sense for the other destinations. Moreover, the lag structure of the equations also has to be standardized. Although the model can be used to test the axioms of consumer choice which are supposed to characterize consumer behaviour, the results commonly indicate violation of the homogeneity and symmetry axioms of consumer choice, thereby casting some doubt on the assumption of rationality of the representative consumer upon which the model is based.

The empirical evidence that consumer decision-making does not always accord with the axioms of consumer choice indicates that some of the assumptions concerning consumer decision-making may require modification. For example, consumers may make decisions on the basis of bounded rationality, obtaining and using only limited amounts of information, or may behave in a way which provides them with a satisfactory rather than maximum level of utility, known as satisficing behaviour (Simon, 1957). However, proponents of the theory of utility maximization could posit an all-encompassing definition of utility, whereby some consumers' means of utility maximization involves spending limited time and effort on collecting and processing information, so that they have self-imposed constraints on their knowledge of all the possibilities available to them. Furthermore, it may be argued that the concept of a rational economic person, maximizing his or her own utility, represents a narrow view of consumer behaviour (Sen, 1979). The concept of rationality may be broadened to refer to a person who takes account not only of personal preferences but also of the preferences of others when making decisions. This broader concept of a 'social person' appears to accord with 'green' tourists, who are concerned about the environment, welfare and culture of the communities they visit. It also encompasses tourism consumption decisions in which individual tourists consider the consumption behaviour of other members of a consuming unit such as the family or an external social reference group, when making their own decisions (Kent, 1991). Decisions may be made within a context of market failure in the form, for example, of asymmetric information whereby some tourism consumers have more information (and sometimes power) than others, some are subject to constraints on borrowing and in which there are non-priced externalities and public goods. Some empirical studies have indicated the significance of differences in constraints, such as the ability to borrow on the actions of different households within the economy (Hayashi, 1985; Jappelli and Pagano, 1988; Zeldes, 1989).

Empirical research has also indicated that economic relationships which hold, over time, at the micro level can differ from those at the macro level. For example, changes in income are often negatively correlated with changes in their own past values at the micro level (MacCurdy, 1982; Abowd and Card, 1989; Pischke, 1991), but positively correlated at the macro level (Deaton, 1992). Moreover, some explanatory variables which are significant in time series consumption functions estimated at the micro level, for example household size and characteristics, are usually omitted from empirical studies at the macro level. The use of aggregate data alone may give rise to biased estimates of price and income elasticities, although micro-based models are not always superior (Blundell et al., 1993). In general, whereas diversity and difference are an issue at the micro level, they are eliminated by aggregation at the macro level.

Aggregate relationships between tourism consumption and income may differ from the relationship between tourism consumption and the income of individual or groups of tourism consumers not only because of income distribution and related effects (Drobny and Hall, 1989) but also owing to differences in the availability of information. For example, an individual may not know, or respond to, information about changes in macroeconomic policy which affects predictions of income and consumption at the aggregate level. Such differences are a further indication of the problems of generalizations based on the concept of an individual tourism consumer. System of equations models have the advantage of an explicit theoretical framework and can provide a large amount of useful information about tourism demand insofar as their assumptions of a representative individual who behaves according to the axioms of consumer choice are met. However, in some contexts their underlying assumptions may require modification, and improvements in the models' ability to take account of the short- and long-run dynamics of tourism consumption are also required. System of equations and single equation models of tourism demand can hide much of what is interesting about decision-making by individuals or groups of tourism consumers. As Sheldon (1990) has pointed out, there is a need for more investigation of tourism demand at the microeconomic level.

3.4 Forecasting tourism demand

Single equation and system of equations models of tourism demand can be used as one of three main methods of forecasting tourism demand and are termed econometric forecasting models, the other two being qualitative methods and univariate and multivariate prediction methods. Forecasts of tourism demand are, of course, of interest to members of the tourism industry, as well as to governments and tourism associations. The econometric approach involves estimating tourism demand equations using the relevant explanatory variables, and forecasting demand by including the likely future values of the variables in the equation. The accuracy of forecasts based on econometric models thus depends upon the underlying models which explain tourism demand so that more accurate forecasts can be obtained by improvements in them. The previous discussions of tourism demand modelling are relevant to forecasting in that they indicate the ways in which the models might be improved. For example, inter-temporal theory of tourism consumption indicated that tourism may be demanded by forward-looking consumers who discount the value of their expected future income, which involves modelling their expected income over time. If tourism demand depends on current and/or past income, alternative models of future income may be relevant. The econometric approach to forecasting tourism demand also involves modelling the future values of the other variables upon which demand depends, notably relative prices and exchange rates. The occurrence of some events, such as the Olympics or EXPO, is known in advance and can be included in forecasting equations.

Although it is important that developments in demand modelling are incorporated into forecasting models, the choice of model may be subject to considerable disagreement. This is particularly problematic when alternative models provide very different results and implications. Considering, for example, the role of income in determining tourism demand, it has been assumed that, over time, income is 'trend stationary', meaning that income grows according to a predetermined trend, with deviations from the trend having a constant mean and variance (Lucas, 1977). In this case, innovations in income affect the deviations from the trend and tourism demand does not respond highly to the innovations. An alternative theory is that income is 'difference stationary', so that it grows according to a stochastic trend (Nelson and Plosser, 1982). In this case, innovations affect the growth path of income, rather than deviations from it, and tourism consumption is highly responsive to the changes in measured income. The two theories have different implications for the nature of the business cycle and the effects of government intervention in the economy. It is not clear which model is more appropriate, owing to the common problem within economics of disproving a theory (Lakatos and Musgrave, 1970). One criterion for the choice of model is to use the data for the case under consideration to estimate the alternative models and to select that which provides results which are not only economically plausible, but also econometrically superior. The forecasts obtained from different models can, subsequently be compared with the actual data, providing a retrospective measure of the models' validity.

In contrast to econometric models of tourism demand, univariate methods involve predicting tourism demand solely on the basis of the past values of demand, without investigating the causes of the past values.

Univariate methods are, therefore, inappropriate for those who require explanations of the level of tourism demand and who wish to know the responsiveness of demand to the likely future values of particular variables or to possible alternative values of them. Non-causal, univariate forecasting methods include: calculation and prediction of the moving average of tourism demand; exponential smoothing; trend curve analysis, involving the projection of the trend of best fit; decomposition methods which take account not only of the trend in the data, but also of seasonal and irregular effects and the Box–Jenkins univariate method (Autoregressive Integrated Moving Average, also known as ARIMA). The last is usually written as ARIMA (p, d, q) where p, d and q are non-negative integers which show the order of the autoregressive, integrated and moving average parts of the model respectively. In its general specification, the ARIMA (p, d, q) model takes the following form:

$$\left(1-\sum_{i=1}^{p}\gamma_i L^i\right)(1-L)^d X_t = \left(1+\sum_{i=1}^{q}\delta_i L^i\right)u_t \qquad (3.7)$$

where L is the lag operator, γ_i are the coefficients of the autoregressive part of the model, δ_i are the coefficients of the moving average part and u_t is the disturbance term, which is assumed to be an independent, identically distributed variable, sampled from a normal distribution with zero mean. As an example, the ARIMA (0, 1, 1) model is specified as:

$$\Delta X_t = u_t + \delta u_{t-1} \qquad (3.8)$$

where Δ denotes difference. The augmented Dickey–Fuller test may be used to determine the order of integration of the time series (Song and Witt, 2000) and the exact structure of the ARIMA model can be determined by the autocorrelation function and partial autocorrelation function. The selection of the most appropriate models for forecasting is based on a number of diagnostic statistics, including the adjusted R^2, AIC and SBC (Harvey, 1993). The Box–Jenkins multivariate method and Structural model (Unobserved Components Autoregressive Integrated Moving Average, UCARIMA model), applied to tourism in Spanish regions by Clewer et al. (1990), are mainly extensions of the non-causal category of models. However, they involve elements of causality in that variables other than the past values of tourism demand are permitted to influence the forecast values of demand. Papatheodorou and Song (2005) used an ARIMA model to forecast international tourism flows at a world and regional level for the period 2001–10. Performance differs substantially among the UNWTO regions, fluctuations are sharp and negative tourism growth patterns are not unusual.

Qualitative forecasting methods such as the Delphi approach (Seely et al., 1980; Moeller and Shafer, 1987; Green et al., 1990a) or scenario writing (BarOn, 1979, 1983; Schwaninger, 1989) constitute a third approach to forecasting. Qualitative methods incorporate expert opinions of likely outcomes and possible alternative scenarios and are used in the absence of reliable time series data. One of the problems associated with qualitative forecasting is that the approach tends to be subjective and the assumptions upon which the forecasts are based are not always made explicit and justi-fied. Hence, the methodology is, in many ways, the antithesis of that which is used in econometric forecasting models, which pay attention to the economic reasoning which is involved in the assumptions which are made. On the other hand, qualitative approaches may be particularly advantageous when long-term forecasts are required, often in the context of a high level of uncertainty. Comprehensive discussions of univariate and qualitative forecasting methods are provided by Archer (1976), Witt and Martin (1989), Harvey (1993), Silverman (2004).

Forecasting models have been applied to tourism in a variety of loca-tions, including Barbados (Dharmaratne, 1995), Hawaii (Geurts and Ibrahim, 1975; Geurts, 1982), Florida (Fritz et al., 1984), Puerto Rico (Wandner and Van Erden, 1980), the Netherlands (Van Doorn, 1984), Spain (Clewer et al., 1990; Gonzalez and Moral, 1996) and a range of west European countries, North America, Japan and Australia (Means and Avila, 1986, 1987; Witt et al., 1994; Smeral and Witt, 1996). Studies which have com-pared the accuracy of different forecasting methods have indicated that, of the quantitative methods, no one technique is necessarily superior to another (Martin and Witt, 1989; Song and Li, 2008). Univariate methods have sometimes provided more accurate forecasts than econometric models and have the advantage of ease of estimation. Econometric meth-ods, on the other hand, provide information about the causes of the future demand, relating to the likely future behaviour of tourism consumers and the economies of tourism origin and destination countries (Makridakis, 1986). The forecasts obtained from econometric models can be adjusted in the light of changes in the economic circumstances of the countries con-cerned and may be particularly useful for tourism-related policy-making, especially if they can focus on the identification of turning points in the business cycle and the minimization of directional change errors (Song and Li, 2008).

3.5 The characteristics framework revisited

Forecasting demand is best understood in a dynamic context involving time series rather than cross-sectional data. Still, the latter can offer valuable insights when used in other types of demand analysis, including empirical applications of the characteristics framework. In fact, the characteristics approach has been predominantly associated with the so-called 'hedonic price analysis', which uses classical techniques to decompose the equilibrium (that is, demand and supply market clearing) prices of the goods into their constituent parts. In other words, it provides a method of measuring the explanatory importance of a set of characteristics for the explicit valuation of a product (Lancaster, 1971; Triplett, 1975). Interestingly, however, and in defiance of the pure hedonic price framework, several studies have found statistically significant good-specific effects (Dickie et al., 1997). On these grounds, therefore, and in line with data envelopment analysis (Tongzon, 2001), the hedonic price analysis may provide a very good benchmark for studying competitiveness and efficiency matters: negativity of a good-specific coefficient may imply a bargain ceteris paribus, whereas a positive sign can be related to low value for money (in terms of product characteristics). In the majority of hedonic studies in tourism, the price competitiveness of specific operators and destinations is assessed in the context of holiday packages (Sinclair et al., 1990; Clewer et al., 1992; Taylor, 1995). In addition, Papatheodorou (2002b) introduced a second-step correlation analysis further to explore the source of price differentials of Mediterranean holiday packages. He distinguished between core and peripheral resorts and found that the behavioural patterns of the two territorial groupings differ substantially in certain features. More recently, Falk (2008) estimated a hedonic price model for ski lift tickets in Austria by also considering neighbourhood spillover effects (Palmquist, 2005) and provided a ranking of the various ski resorts in the country according to their characteristics of quality.

From a modelling perspective, a typical hedonic price specification may assume a Cobb–Douglas functional form, namely (Papatheodorou, 2002b):

$$P = a \prod_{m=1}^{M} \prod_{i=1}^{I_m-1} C_{im}^{c_{im}} \prod_{j=1}^{J-1} D_j^{d_j} \prod_{k=1}^{K-1} F_k^{f_k} e^u \tag{3.9}$$

and its logarithmic version:

$$\ln P = \ln a + \sum_{m=1}^{M} \sum_{i=1}^{I_m-1} c_{im} \ln C_{im} + \sum_{j=1}^{J-1} d_j \ln D_j + \sum_{k=1}^{K-1} f_k \ln F_k + u \tag{3.10}$$

where P is the price of the tourist product or service in question (e.g. holiday package), C_{im} is a scaling factor for the ith characteristic (e.g. small, large) in the mth group of product or service features (e.g. accommodation size), D_j is a location scaling factor for destination j, F_k represents an operator scaling factor for travel firm k (if applicable), a is a constant, c_{im} is a dummy variable that takes the value of one if the product or service possesses the attribute i in group m or zero otherwise, d_j is a dummy variable that takes the value of one if the product or service refers to destination j or zero in all other cases, f_k is another dummy variable (1 if travel firm k; 0 otherwise),

M refers to the total number of groups of product or service features, I_m is the total number of characteristics in group *m*, *J* represents the total number of destinations, *K* is the total number of operators, *e* is exponentiation, *u* is an error term distributed according to the classical assumptions and $\Pi(\Sigma)$ is the mathematical symbol of multiplication (summation).

In addition to hedonic price modelling, the characteristics framework may also be validated in the context of discrete choice analysis. Papatheodorou (2003d) developed a model adopting a random utility approach that accommodates the observed behavioural inconsistencies within a rational choice framework (Ben-Akiva and Lerman, 1985). The tourist is assumed to select always the destination associated with the highest utility; however, the modeller does not know the enjoyment level with certainty and therefore considers utility as a variable that is random among individuals. As a result, the tourist's utility function contains an additive idiosyncratic constant that differs among destinations for each tourist and differs among tourists for each destination. In this random taste heterogeneity model the locational equilibrium is described in an essentially stochastic manner, even if the behaviour of the representative (randomly selected) tourist is fully deterministic (Anas, 1990). Having the above in mind, the probability P_i that a traveller chooses destination *i* among the whole group of *I* is given by:

$$P_i = prob\left[V_i + \varepsilon_i \geq V_n + \varepsilon_n\right], \ \forall i \in [1, I] \tag{3.11}$$

where $n = 1 \ldots\ldots I$ and $n \neq i$

In words, P_i is equal to the probability that the total utility (systematic and stochastic) related to destination *i* is higher than the utility related to any other destination *n*. In (3.11), V_i represents the indirect utility function, i.e. the solved-out form of the utility function. Moreover, the joint distribution of the random variables yields choice probabilities that are multinomial logit:

$$P_i = \frac{e^{\mu V_i}}{\sum_{n=1}^{I} e^{\mu V_n}} \tag{3.12}$$

For low values of μ, consumer heterogeneity is very significant and randomness dominates in the actual destination choice. In fact, for $\mu = 0$ each resort is chosen with an equal probability I/I, i.e. the importance of the characteristics in the deterministic part of the utility is minimal. On the other hand, as $\mu \to \infty$ differences in probabilities are fully explained by differences in the levels of indirect utility *V*.

Morley (1994) was one of the first researchers to apply discrete choice modelling in tourism from an empirical econometric perspective. More recently, Alegre and Pou (2006) used a binomial logit model to study the determinants of the length of stay in the demand for tourism. In particular, they regressed the length of stay (encapsulated by a dichotomous variable taking the value of zero if the stay was up to one week and one if it lasted for longer) on a number of traveller characteristics such as the age, the labour status, the nationality, the type of tourist accommodation and catering board, a number of travel motives, etc. In most cases, the explanatory power of these characteristics is strong enough to explain the observed changes in the duration of stay.

The hedonic price analysis approach and choice modelling models are considered further in the context of the valuation of non-traded resources in Chapter 10. There, the hedonic method is applied to the characteristics or attributes of environments that determine the revealed preferences of consumers as householders that influence property and land values. Choice modelling is used in conjunction with the contingent valuation method in expressed consumer preferences for the willingness to pay to retain environmental resources in their current state or to improve them, or accept compensation where the quality of such resources is likely to be impaired or put to alternative uses.

3.6 Conclusions

This chapter, along with Chapter 2, has attempted to show how economic analysis can contribute to explaining tourism demand and to evaluating empirical studies which have been undertaken. Economics has the advantage of a theoretical framework which goes beyond the descriptive categorization of tourist types, a shortcoming of some past work. The theory has the additional ability to identify variables which determine demand and to provide the basis for quantification of the short- and long-run sensitivity of demand to changes in these variables. The discussion has shown how some economic theory has been incorporated within models which have attempted to estimate tourism demand, particularly at the national level of aggregation. However, it has also indicated that much demand modelling to date has been ad hoc, with inadequate micro-foundations. In addition, the discussion has argued that empirical studies might benefit from theoretical contributions from branches of economics other than its mainstream one. The potential of such theoretical analysis and developments has not yet been fully realized as researchers have tended, so far, to restrict investigation to the effects of measurable variables for which data, though often of a secondary nature, are most readily available and which are amenable to the prevailing economic methodology of econometric modelling. In general, the effects of expectations, information availability and other variables which are not easily quantifiable have been neglected and the social determinants of decision-making have frequently been ignored.

Developments in the analysis of tourism demand have, nevertheless, been occurring via contributions arising from improvements in economics. The microeconomic theory of consumer demand is one example of an advance in economics, providing a framework for explaining tourism demand which can be applied at both the microeconomic and macroeconomic levels, although the analysis could be broadened to incorporate the social context of decision-making. Hence, there is considerable scope for extending research on demand beyond the current emphasis on modelling demand at the national level. The nature and diversity of tourism demand, within its temporally and spatially specific contexts, requires further investigation. For example, the effects on demand of the nature and amounts of information used by different categories of tourism consumers and the constraints to which they are subject merit further research. Insights from a range of disciplines can aid the formulation of the particular questions to be investigated and hypotheses to be tested which might, otherwise, be overlooked. For example, social psychologists have conducted research into the motivations underpinning tourism demand and sociologists have posited that social relationships, both within and outside the household, influence holiday choice.

It is clear that microeconomic studies of tourism demand could provide many interesting findings about tourism demand by different social groups, based on disaggregate models which are specific to the case under consideration. Economic models of the decision-making process and demand for tourism could be formulated for socio-economic classes, gender, race and age groups. The social context of decision-making could be taken into account, at least in part, by the inclusion in the estimating equation of additional explanatory variables which are significant determinants of demand. Some quantitative information about social preferences could be provided, for example, by investigation of the extent to which the demand for tourism in a given destination by a particular class is determined by the past demand for the destination by the same class or by the past demand for an alternative destination by a higher income class (the 'demonstration effect').

The choice of unit which is used to measure tourism demand is important. For example, demand by households does not reveal interactions which occur within the household and which are likely to be related to inequalities in economic power. For instance, the issue of whether tourism consumption decisions are made on an individual basis, imposed by the wage-earner on other members of a household or negotiated between members of a group, has been ignored in most past studies. Thus, consumers' decisions may be made within a context of social pressures, and alterations in the social context can result in changes in the pattern of tourism consumption.

Preferences are likely to vary by gender, class and race and information about the effects of gender differences may be obtained by comparison of results obtained from tourism demand equations for men and women.

Women's tourism consumption may be a function of their partner's income in the case of women without paid employment, but of their own or of the joint income in the case of women with paid work, and the presence of children is also likely to be relevant. Models of tourism demand could be extended to investigate differences in the ways in which tourists from different socio-economic classes or races and young and old people allocate their expenditure within the destinations of their choice. The investigation of variations in the demand for tourism by different socio-economic groups over the economic cycle could also be undertaken. In those cases for which data are unavailable, interviews can provide useful qualitative information and can indicate theories which merit further investigation.

Demand can, of course, become effective only when backed by income, and the consumer's budget and level and pattern of tourism demand are crucially dependent upon the underlying distribution of income and wealth. Changes in the distribution occurring via, for example, changes in government expenditure and taxation, cause changes in tourism demand. Microeconomic studies covering different socio-economic groups within the population could shed light on the likely effects of distributional changes and comparative country studies could provide some indication of the effects of alternative distributions.

The analysis of the demand for tourism by social groups could be extended by the estimation of disaggregate econometric forecasting models. As in the case of tourism demand analysis, aggregate models of the nation state need not be the main focus of attention, but research could encompass forecasts of demand for different components of tourism by different socio-economic categories, based on specific models of their behaviour. The demand for tourism by the different groups is related to the characteristics of the holidays or business trips which they require and the information and opportunities which are available to them. Analysis of the demand for different components of tourism, including transport, accommodation and other service provision, is therefore a key focus for future research. It would help to overcome a number of the theoretical and empirical problems which are implicit in tourism demand studies and which have resulted from aggregating a range of diverse services into the composite item 'tourism'. Chapters 4 and 5 will provide the basis for doing so by examining the different components of tourism, the markets in which they are supplied and the pricing and output strategies of the firms within them.

PART II
The economics of tourism supply

4 MICROECONOMIC FOUNDATIONS OF TOURISM SUPPLY

4.1 Introduction

Tourism supply is a complex phenomenon because of both the nature of the product and the process of its delivery. Principally, the product cannot be stored, cannot be examined prior to purchase, it is necessary to travel to consume it, heavy reliance is placed on both natural and human-made resources and a number of components are required, which may be separately or jointly purchased and which are consumed in sequence. Tourism is a composite product involving transport, accommodation, catering, natural resources, entertainment as well as other facilities and services, such as shops and banks, travel agents and tour operators. Many businesses also serve other industrial sectors and consumer needs, thus raising the question of the extent to which producers can be considered as primarily suppliers of tourism. The many components of the product, supplied by a variety of businesses operating in a number of markets, create problems in analyzing tourism supply. It is therefore convenient to consider it as a collection of industries and markets and to examine it using not only the neoclassical paradigm but also other schools of thought. This approach allows the analysis not only to cope with the complexities of the tourism product but also to take account of developments in economic concepts, theories and methods, with special focus on industrial economics and the issues with which it has been concerned.

The main objective in this and the following chapter is to provide a wider and more advanced exposition and application of economic principles to tourism supply, other than that offered in textbooks on the economics and management of tourism predominantly associated with neoclassical analysis. In fact, such an approach may prove useful for identifying different types of markets and giving a number of valuable insights into aspects of firms' conduct and performance. However, phenomena such as imperfect competition, particularly oligopoly and uncertainty, as well as the need to explain market dynamics, are inadequately explained by traditional analysis. The development of industrial economics and other schools of thought has attempted to fill the gaps left by traditional approaches and illustrate the theme of this book, not only by indicating recent theoretical developments and advanced economic analysis but also by helping to explain tourism supply. Within industrial economics, two main approaches have been used to examine the behaviour of firms in different types of markets. The first is the structure, conduct and performance (SCP) paradigm, which has played a major role within empirically oriented studies of firms in the manufacturing sector. Although it has been subject to some criticism, particularly in more theoretically based analysis, the paradigm remains a useful framework and appears to be relevant to a complex service industry such as tourism. The second more recent approach is game theory which is used to analyse the strategies that firms adopt in relation to the actions and probable reactions of their competitors. Game theory has been applied extensively in oligopolistic situations and has vastly improved the comprehension of firms' interactions and their outcomes in dynamic situations. It seems particularly important for understanding the behaviour and strategies of tourism suppliers.

Having the above in mind, this chapter first reviews briefly the approaches adopted in economics with regard to central issues concerning the market environment and the ways in which this can influence the structure and behaviour of tourism firms. The main schools of thought which are relevant to industrial economics and can assist in the explanation, as opposed to the description, of tourism supply and changes in it are discussed. Therefore, elements of the Austrian school and behavioural, evolutionary, institutional and psychological economics are outlined. The SCP paradigm is then explained in detail, encapsulating among others the basic tenets of the theory of the firm concerning output, costs, pricing decisions, revenues, profits and losses under the various competitive structures of the markets within which consumers and producers interact. Economic models of different types of market structure provide explanations of the operation of firms under well-defined conditions, each type of market structure being distinctly identifiable. Although in reality such conditions may be approximated to rather than attained, the models are, nevertheless, useful in going beyond mere description by providing explanations of firms' behaviour and predicting the short- and long-run outcomes of different market situations. This facilitates the identification of factors likely to be of importance in tourism supply, particularly with respect to the nature and extent of inter-firm and inter-sector competition and the consequent implications for consumer welfare. The SCP paradigm is structurally intertwined with policy-making, hence relevant issues such as competition and regulation policy, taxation and subsidization and crisis management are subsequently discussed in the context of tourism. The SCP discussion will conclude with an evaluation of the contribution and limitations of the approach.

The role which game theory can play in explaining the strategies of tourism firms in a dynamic framework and the associated changes in the structure of tourism markets will be examined subsequently. Different types of games are identified and alternative scenarios are evaluated to highlight complexity in corporate rivalry. Finally, the chapter concludes by providing an overall assessment of the various theoretical approaches on tourism supply. In this way it prepares the reader for Chapter 5, which presents the economic profile of all the major tourism industries, considering among others the most recent developments at a policy level.

4.2 The industrial economics background

Research on the economics of industry undertaken outside mainstream analysis has moved the debate on the competitive structure of the market into a different arena. There are two distinct but related strands to this work. One is concerned with the accommodation of the dynamic nature of markets, while the other has concentrated on the representation of the characteristics and circumstances of firms, industries and markets.

4.2.1 The Austrian school and evolutionary economics

The Austrian school, associated with scholars such as Hayek, Menger, Mises and, initially, Schumpeter (disavowed by neo-Austrians), is notable for its concentration on the process of competition as opposed to the static equilibrium analysis embodied in much conventional market structure analysis (Boettke and Leeson, 2003). These scholars acknowledge that change and uncertainty are endemic and that those involved in industry have to make decisions within this context. However, they argue that over time, with experience and the benefit of increased knowledge of processes and opportunities, better decisions are made in succeeding periods, thus tending to produce more competitive market conditions. In this sense there is a degree of what Hayek (1949) calls emerging 'order' but not necessarily equilibrium. Latter-day researchers who can be viewed as constituting the neo-Austrian school are Kirzner (1973), Reekie (1984) and Littlechild (1986). This view of industrial activity is paralleled by what Nelson and Winter (1982) refer to as 'Evolutionary Economics', in which procedures for carrying out required actions evolve with the accumulation of knowledge.

In particular, each organization operates within a selected environment of exogenously and endogenously determined industrial conditions, which affect functional sustainability and the potential for growth or contraction. Within this framework, the firm builds on several organizational routines, which facilitate the performance of common tasks. However, the subsequent achievement of dynamic economies of scale and scope is usually associated with lock-ins and inflexibility to address effectively the stochastic and structural changes of the industrial milieu (Clapp, 1986). To circumvent this problem, the firm should be in a constant search for subroutines related to the incremental, or even drastic organization and production modifications that are required for inter-temporal profitability. These effective innovations constitute the basic growth mechanism of Schumpeterian competition and are associated with the firm's history and search effort, in terms of both financial and human capital (Schumpeter, 1996). Clearly not all research and development yields marketable products or usable processes, such as 'just-in-time' and 'total quality control' systems, and the costs of their discovery and adoption are subject to much uncertainty. In fact, these advancements are stochastic by nature, as success can only be probabilistically assessed. In the positive cases, however, these improvements initiate a self-reinforcing growth mechanism, as the increased profitability and experience facilitate the undertaking of new plans, which enhance efficiency even further. In this sense, ill-considered innovations may cause a catastrophe in the existing structure and result in an abrupt concentration of the market configuration. Though this evolutionary framework was originally designed to explain patterns in the secondary (manufacturing) sector of the economy, it is applicable to services with only minor modifications. In particular, innovations in the latter are mostly associated with changes in the organizational structure and distribution networks, rather than with major technological advancements. In this sense, they are less costly; on the other hand, however, they are not patentable and can be easily replicated, at least in theory. Consequently, catastrophe in the existing structure requires the simultaneous presence of other factors, such as first mover advantages (due to consumer inertia, brand reputation or regulatory schemes), or idiosyncratic firm characteristics, such as good entrepreneurial skills. In fact, most of these features are apparent in the tourism industry. The emergence of a 'dual dualism' in tourism markets and spaces (Papatheodorou, 2004) as a result of an evolutionary process is discussed in Chapter 5.

The development of evolutionary economics has run in parallel with that of institutional economics and, in part, the thinking of the Austrian school. It has been concerned primarily with treating endogenously the way the beliefs, norms and customs of society might lead to the formation of institutions which facilitate corporate behaviour. Theories of the evolution of institutions are based on those employed in biology and mathematics involving natural selection. Hirshleifer (1982) reviewed these when modelling social and institutional change, arguing that game theory can indicate the outcome of cooperative and non-cooperative behaviour to show the desired human traits which would be required to devise effective institutions. The relevance of these notions to economic organizations such as the firm is to underpin the importance of the attitudes and objectives of entrepreneurs, managers and workers in determining the operation of industries and markets. An indication of this emerges in the examples of non-cooperative games presented towards the end of this chapter. Tourism supply is an interesting case because it displays, through the instability in some sectors, a measure of immaturity with respect to the evolution of its business organizations so that outcomes are often less predictable than in mature and stable markets, also reflecting a possible distinction between manufacturing and service sectors.

In emphasizing the dynamic nature of the market system and its institutions, the Austrian and Evolutionary Economics schools echo Marx (1967). In his analysis of competition, Marx was concerned not only to reveal its process but also to ascertain the outcome in terms of the impact on the distribution of income and wealth and the allocation of resources. Followers of Marx have tended to underplay his perception of competition and accentuate his view that industrial production would become more concentrated, thus focusing on the potential for exploitive behaviour, particularly concerning labour (Kalecki, 1939). This has had some influence in the industrial economics field, where monopoly has been studied.

4.2.2 Transaction cost and agency theories[1]

While mainstream economics has been a formidable tool with much empirical relevance, it has been subject to criticism (Eggertson, 1990) because it does not explain the rationale for different forms of economic organization and the effects of social rules on behaviour and outcomes. Institutional economics, although essentially adhering to the basic tenets of neoclassical theory, introduces information, time and transaction costs and property rights as constraints on the attainment of business objectives. It also emphasizes the need for empirical testing of hypotheses concerning such constraints. To this extent, the stance of institutional economics is fairly close to that adopted by much of industrial economics.

The institutional perspective on business activity is a relatively new one and for this reason there is as yet no clear agreement on its principles. Because it is still at the exploratory stage, there are differences of opinion as to its postulates. For example, some institutional economists reject the profit maximizing and rationality principles, replacing them with Simon's (1957) satisficing concept. The danger is that abandonment of too many principles leaves industrial economics with no theory, so that any investigation is merely descriptive. Notwithstanding its infancy and the problems of the choice of theoretical foundation to adopt, institutional economics offers insights into the impact of variables hitherto ignored and it also underlines

[1]The authors are grateful to Dr Irini Dimou for her contribution in the preparation of this section.

the need for economic analysis to consider market dynamics and the uncertainties encountered in the business environment. In this respect it is consistent with game theory analysis. It also links conventional analysis with behavioural approaches which attempt to relate human activity within business organizations to economic outcomes. Furthermore, it contributes to explanations of why market failure (see Chapter 9) occurs, a factor of importance when environmental issues in that chapter are examined, especially regarding property rights.

a. Transaction cost theory

Important contributions to the analysis of market imperfections in the form of information and transaction costs have been made by Coase (1960), Williamson (1985, 1986) and North (1990) and are reviewed by Stiglitz (1989) and Williamson (1989). These are the costs incurred in searching for and procuring information, which helps to reduce uncertainty and in executing transactions. They are likely to be substantial in some spheres, such as where a firm assembles a product from many sources, a feature of tour operators marketing package holidays. Such imperfections provide an incentive for economic integration between firms, as in the case of the World of TUI, as a means by which firms can internalize the costs which they would incur by operating as separate units.

Transaction cost theory adopts a contractual approach to the organization involving an evaluation of the comparative costs of planning, monitoring, adapting and enforcing a certain transaction. The basic principles highlighted by this framework are comparability (i.e. one form of organization is always compared to an alternative one) and feasibility (alternatives should be realistic not hypothetical). To retain only the feasible alternatives, transaction cost theory makes a number of behavioural hypotheses primarily related to the concepts of bounded rationality and opportunism (Williamson, 1996). The former is based on Simon's (1961) definition that people comport themselves with intentional rationality albeit to a limited degree; this is because economic actors have a limited ability to foresee and to acquire knowledge and skills. When bounded rationality is combined with environmental and/or behavioural uncertainty, organizations should abandon the market mechanisms and focus on internal governance; this gives the parties to a transaction the opportunity to react in a sequential, adaptive way without incurring the hazards of opportunism that spot contracting would pose.

On the other hand, opportunism refers to the inclination of economic agents to break their promises. According to the theory not every person acts opportunistically; nonetheless, it is difficult to tell ex ante who is trustworthy and who is not. Williamson (1985) defines opportunism as 'self-interest seeking with guile' and suggests that the definition includes actions such as lying, cheating or violating agreements. Opportunism poses a problem especially when a business relationship is characterized by a small numbers condition (Williamson, 1985) or supported by specific assets, whose value is significantly reduced outside the particular context (Rindfleisch and Heide, 1997). Internal organization is less vulnerable to opportunism when a small numbers condition exists ex ante or arises during contract execution, since hierarchy is less prone to disputes between parties and is able to resolve most of them by appeal to fiat.

Transaction cost theory analyses each transaction from three aspects, namely asset specificity, the degree and type of uncertainty and the frequency with which transactions occur. Asset specificity refers to the extent that physical or human assets are locked into a particular use and the degree to which they can be redeployed without sacrifice of substantial productive value (Williamson, 1996). It becomes a major issue of concern because it introduces bilateral dependency into the relationship, when combined with the above mentioned behavioural hypotheses and in the presence of uncertainty. Hence, the identity, reputation and willingness for continuous collaboration of the parties involved in a specific transaction are of crucial importance in the contracting process. Uncertainty refers to situations arising from random acts of nature or changed consumer preferences (state contingent kind of uncertainty), lack of communication (secondary uncertainty) and most importantly situations characterized by bilateral dependency. Finally, frequency is related to the degree and the recurrence with which a transaction occurs; this is also a relevant dimension, as the cost of a complicated governance structure could not be recovered in case of a single transaction, whereas it would be justified under situations of high transaction volumes.

As already mentioned, the transaction cost theory identifies the transaction instead of the individual or the industry as the basic unit of analysis and assesses a number of features under different governance structures, namely market, hybrids and hierarchies. The key differences between these structures are associated with the form of contract law that supports them, the adaptability of each mode and the use of incentive and control instruments that characterize them. The hybrid mode lies between the market and the hierarchy, being a more elastic and adaptive form of governance compared to the market one, but more legalistic and less adaptive than hierarchy. In the case of autonomous markets, where there is no dependency between buyers and sellers and the identity of the parties does not matter, classical contract law is sufficient, since the terms of the specific transactions are explicitly specified ex ante (Williamson, 1996). Nonetheless, when future contingencies cannot be forecast at the outset, contracting should be associated with an additional governance structure, namely neoclassical law and excuse doctrine. Hierarchy is a more adaptive mode and the type of implicit contract law suitable to this kind of organization is forbearance; this is a more flexible arbitration form, which allows the dispute outcome to be determined by mitigating factors not considered when disputes are resolved in courts. The rationale for forbearance rests on the fact that parties to a dispute have knowledge that is either impossible or too costly to reveal in a court; hence the parties either settle their differences on their own or let the hierarchy decide on unresolved situations.

In addition to contract law that applies to each form of organization, adaptability is another central problem of economic organization. Williamson (1996) suggests that adaptation can be classified into two types, namely autonomous, which has spontaneous origins and can be successfully applied in spot contracting, and coordinated, which is required in cases of contracts characterized by strong disturbances and therefore can be applied in cases of a more hierarchical governance. These adaptation advantages, however, are not obtainable without costs; in fact, internal organization weakens incentive intensity and causes bureaucratic costs to rise. Within hierarchical governance, incentive intensity is regarded not as an objective but as an instrument. As for the hybrid mode, this is characterized by a moderate degree of adaptability and incentive intensity features. Finally, in the market mode, high incentive intensity is prevalent.

b. Agency theory

Transactions theory also facilitates the study of the relationship between principals and agents, a relevant feature in this chapter. A principal–agent relationship exists where one party's welfare depends upon the actions of another. The principal is normally affected by the actions of the agent. Jensen and Meckling (1976) define the agency relationship as a contractual arrangement between two (or more) parties, where one, designated as the agent, acts on behalf of, or as a representative for, the other, designated as the principal, who delegates some decision-making authority to the agent to act in a certain domain of decision problems. Agency theory is primarily concerned with two basic problems that arise in the agency relationship. The first problem emerges when the principal's and the agent's goals are incompatible and it is difficult for the principal to evaluate the agent's effort, behaviour and actions. The second problem is related to risk sharing and arises when the principal and the agent have different attitudes toward risk; these may lead the agent to take different actions from those preferred by the principal (Eisenhardt, 1989).

The agency literature deals mainly with the first problem, which usually takes two different forms: adverse selection and moral hazard. The former arises in the case of pre-contractual information asymmetries. This term comes from the insurance industry, where people purchasing insurance are not randomly selected from the population but rather constitute a group of persons with private information about their personal situation; as a result, they may receive a higher-than-average level of benefit payments under the insurance policy. As discussed by Akerlof (1970) in his 'Market for Lemons', the asymmetric information condition may lead to the collapse of markets, when combined with unconstrained opportunism, since there would be no price at which the quantity of a good provided to the market by sellers would be equal to the quantity demanded by buyers. In general, markets may experience serious operational problems when there is private information which is difficult to verify. Interestingly, the person who has the private information may lose just as much or more than the person who does not (Noreen, 1988).

On the other hand, moral hazard usually refers to lack of effort from the agent's part, i.e. the agent is shirking. This term also originated in the insurance industry, referring to the tendency of people with insurance to change their behaviour in a way that leads to larger claims against the insurance company. Moral hazard problems may arise in any situation where a person is tempted to take an inefficient action or to provide distorted information because this person's interests are not aligned with the group welfare and because the action cannot be accurately monitored. Therefore, the principal's problem is to design a contract that rewards the agent according to the outcome, taking into account any tendency the agent may have to make decisions non-optimal for the principal.

Regarding the second problem in the agency relationship, i.e. risk sharing between principal and agent, this would still exist, even in the case of symmetric information between the two parties. Since both principal and agent are assumed to be risk averse there is a need to share the risk related to the outcome of the agent's actions. Indeed, if the agent were risk neutral he or she would bear all the risk; the principal would retain a fixed amount of the outcome and give the remaining to the agent, who would not suffer from any dilution of incentives. Nonetheless, since the agent, like all individuals, is averse to sufficiently high risks, 'the solution of preserving incentives by assigning all risks to the agent fails as soon as the risks are a large compared to the agent's wealth' (Arrow, 1985: 45). Having the above in mind, the agency theory aims to identify the most efficient contract for the agency relationship, given the various hypotheses made regarding human nature, i.e. whether a behaviour-oriented contract (e.g. salaries) is more efficient that an outcome-oriented one (e.g. commission, stock options).

To summarize, both the transaction cost and the agency theory differ from the neoclassical theory of organization which regards the firm as a production function, since the former considers the firm as a governance structure (Klein, Crawford and Alchian, 1978; Williamson, 1985, 1996), while the latter treats it as a nexus of contracts (Jensen and Meckling, 1976). The transaction cost theory traces its origins to explanations of vertical integration, whereas agency theory was originally concerned with corporate control. While some researchers believe that these two theories are based in different paradigms, others suggest that the differences are exaggerated. In addition, some scholars consider the theories to be complementary and use them jointly as a basis for their conceptual framework (Contractor and Kundu, 1998).

c. *Application of transaction cost and agency theories to tourism*

In spite of their significance within the industrial economics literature, the transaction cost and the agency theories have received only limited attention in the context of tourism. Tremblay (1998) studied the economic organization of tourism, emphasizing the development of network relationships as a means to manage change and acquire competencies in technology and marketing. More recently, Chen and Dimou (2005) used both theories to provide a suitable rationale behind the alternative expansion strategies of hotel companies (such as franchising, management contract and ownership), while Lamminmaki (2007) applied the transaction cost economics approach to study outsourcing in Australian hotels. The theories could be also applied to highlight the relationship between tour operators and travel agents: if the latter reach target numbers of holidays sold, override commissions are paid by the former. Incentives are also provided for devoting a certain proportion of racking space for the principal's brochures and direct reservation links which cut costs. In addition, issues of asymmetric information are important in the context of airline safety (Papatheodorou and Platis, 2007), as the carrier and/or the flight crew usually have a much better knowledge of the condition of the aircraft compared to the passengers, who may have to rely primarily on what they are told; similarly, passengers may not be aware of the real reasons behind a flight delay or the financial robustness of their preferred tourism supplier. Market regulation and safeguarding of competition also involve significant transaction costs mostly associated with monitoring and enforcement (Wolf, H. 2004).

4.2.3 Reflection and synthesis

An interesting feature of the evolutionary and institutional branches of industrial economics and the associated concepts of transaction costs, principal–agency analysis and the innovation process is that they identify market phenomena which are external to individual businesses and industries.

The issue emerging in these areas of investigation which is relevant to the development of industrial economics is the extent to which the problems can be internalized to minimize their adverse impact, in terms of costs, on a firm's or industry's operation. This raises a matter of fundamental importance in economic supply analysis, namely the organizational arrangements which are necessary. In short, it poses the question as to why firms exist. Are they merely entities for transforming inputs into outputs or are they production–distribution units which, in order to operate efficiently, require organizational structures by which their objectives are attained? Moreover, is it possible that managers and employees set their own objectives which the organizational structure also facilitates? The latter notion acknowledges the force of the longer-established behavioural analysis of firms and individuals within them. Such issues cannot be pursued here but a not implausible inference to draw is that although the theoretical development of evolutionary and institutional economics, transaction costs, principal–agency and innovation concepts is relatively new, a degree of consensus, if not convergence in economic thinking, is occurring (Dietrich, 1994). This might suggest that a more dynamic and unified theory of organizations will eventually emerge.

Research in industrial economics does indeed demonstrate a concern with establishing a firmly rooted theoretical base (Davies et al., 1989; Schmalensee and Willig, 1989; Basu, 1993; Martin, 1993; Armstrong and Porter, 2007). This is the justification for using such terms as New Industrial Economics and New Industrial Organization. It is not entirely coincidental that these movements were reinforced by the emergence of game theory as a means of analysing business strategies. Not only did the application of game theory allow for the incorporation of uncertainty and asymmetric information, but also it made it possible to construct more dynamic models, thus increasing their explanatory power. Game theory has reawakened interest in the long-established duopoly form of oligopoly models such as those constructed by Bertrand, Cournot and Stackelberg (Martin, 1993), included in mainstream microeconomic texts, so that the development of oligopoly theory is now a dominant element in industrial economics. This analysis has generated research into a number of issues, such as entry/exit conditions, price wars, predatory pricing, the role of advertising, cooperative contractual arrangements between the different elements of supply and demand and collusion. The re-emphasis on theory, however, has been criticized for failing to provide empirically testable models of industrial behaviour. Nevertheless, there has been renewed interest in specific case studies, relative to the earlier bias towards more generalized econometric analysis. Moreover, greater attention is being paid to the scope and impact of intervention, for example anti-monopoly legislation, market regulation, taxation and subsidies as argued in the last section of this chapter. The role of policy and its impact on welfare have been examined within public choice theory (for example Buchanan, 1968).

Despite the apparent conflict between the desire of theorists to develop general models of supply and the insistence of empiricists on establishing how firms actually behave, industrial economics, and with it the comprehension of what determines the pattern of supply, has made great advances. There is a better understanding of the interrelationships of suppliers within particular industries, such as between intermediate and consumer goods producers and between principals and agents. Models now provide better reflections and explanations of reality. However, most progress has been made in developing market behaviour models out of their duopoly origins, again incorporating the application of game theory and accommodating such notions as product differentiation, market segmentation, price discrimination and the reactions by firms to their rivals' price and non-price competitive strategies. Before proceeding to examine the ways in which the different approaches can shed light on behaviour in tourism markets, it is useful to consider briefly the behavioural theories of the firm which have been developed outside mainstream economic analysis.

4.2.4 Behavioural models of the firm

Concurrent with the development of new industrial economics but hardly acknowledged by it was the growth of behavioural models of the firm, studied within the sub-discipline of economic psychology. Behaviourally oriented theories do not rely as heavily on two major assumptions of conventional analysis: first, that business decisions are identical to those of individuals within firms, and second, that theoretically derived optimizing positions are representative of actual behaviour. Economic psychology perspectives on businesses question the axiom of rationality, perceiving it as a hypothesis which should be subject to empirical testing. The main thrust of such studies of the firm is that analysis of business behaviour produces observations which generate, through replication, broad generalizations which aid the construction of theories, through an inductive process.

Business behavioural researchers argue that decision-makers may not apply clear choice models when they lack knowledge of possible alternative courses of action and are not certain of their outcome. Preferences may be inconsistent and they may not possess firm guidelines or rules by which to make decisions. In this sense, the assumptions of conventional economic theory are challenged, as is apparent in the observations of pioneers in business behavioural theories such as Simon (1955), March and Simon (1958), March (1962), Cyert and March (1963), Simon (1979), and Cyert and Simon (1983). It is hypothesized that business people lack the information and time to optimize their activities. They are, therefore, aware of only a small number of options from which to choose. Thus, a predominant issue which behavioural researchers have investigated is the institutional structure and process of decision-making (see, for example, reviews by Slovic et al., 1977; Ungson et al., 1981; Kahneman et al., 1982). Of interest have been the ways in which problems are identified, the process of learning by experience, perceptions of probable outcomes, attitudes to risk, the allocation of attention to multiple problems and activities, and organizational adaptation.

Although economic psychology, as an empirically based approach to industrial organization, concentrates on behavioural issues, it is not in direct conflict with neoclassical optimizing theory but has modified the theory by relaxing the more abstract assumptions. Its principal contributions, however, have been to widen the scope of analysis by including the human behavioural element and to strengthen the links between theoretical and empirical studies. To an extent, therefore, industrial economics and the economic psychology of the firm have grown out of traditional theory and run in parallel with each other. Recent developments in these two areas of economic analysis, with the greater emphasis on the testing of theoretical models in an empirical context, have increased their potential for explaining the structure and operation of tourism supply. Many of the issues examined by the theoretical approaches are manifested within tourism markets. Furthermore, some aspects of the structure and operation of tourism markets, especially the persistence in some markets of overcapacity and disequilibrium, question the explanations, predictions and relevance of some industrial economic and economic psychology models. Thus, it would be informative to examine the rich empirical evidence which a study of tourism supply could yield.

4.3 The structure–conduct–performance (SCP) paradigm

The SCP paradigm (Chamberlin, 1933; Bain, 1956; Mason, 1957) dominated industrial economics until the 1980s. According to the paradigm, it is the type of market structure within which firms operate that is the ultimate determinant of their conduct and performance, measured by such criteria as profitability. Market structure variables such as the number of buyers and sellers and degree of market concentration are assumed to be relatively stable. As an empirically based analytical framework, the SCP approach accepts that market structures usually differ from the benchmark of perfect competition, so that if firms set prices in excess of marginal cost, there is a prima facie case for government intervention and the implementation of measures to bring about increased competition. The SCP approach is, thus, policy-oriented.

The SCP paradigm is a feasible means of analyzing complex markets comprising firms of differing sizes, in which varying degrees of concentration and/or integration occur and market power can be exercised (Brozen, 1971; Schmalensee, 1972; Demsetz, 1974; Cowling and Waterson, 1976; Peltzman, 1977; Spence, 1977; Clarke and Davies, 1982; Dixit, 1982; Lieberman and Montgomery, 1988; Tirole, 1988). The approach has the advantage of providing a clear framework which avoids unstructured descriptions but instead permits the examination of markets in terms of the analytical categories of market structure, firms' conduct and performance. The ability of the SCP approach to encompass many of the elements of tourism supply is indicated by Figure 4.1, based on Scherer (1970) and derived from Mason (1957). In addition to showing the main characteristics of structure, conduct and performance, Figure 4.1 makes reference to public policy and its impact on firms' behaviour. The principal modification to the conventional SCP

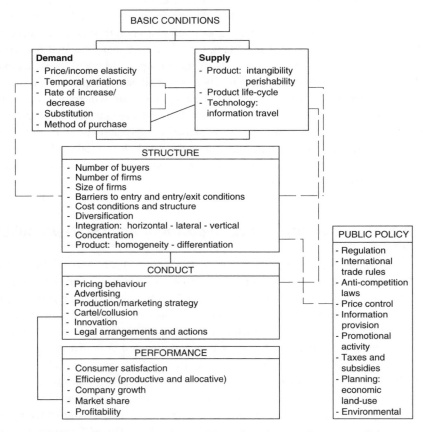

Figure 4.1 The structure–conduct–performance paradigm applied to a service industry

diagram, for the purpose of applying the analytical framework to tourism, is the omission of elements concerning the supply of tangible products, only peripherally relevant to a service sector, and decreased emphasis on the welfare considerations of performance, which are not the focus of this chapter.

Two key measures of competitive conditions are the number and size of firms and indices of concentration. Additional variables which can be considered, included in Figure 4.1, are the number of buyers, entry and exit conditions, cost conditions, product differentiation, diversification and inter-firm integration. Firms' conduct relates to pricing behaviour, advertising, marketing, research and development and innovation, sometimes in the context of tacit collusion or more formal cartels. Although innovation is partly seen as the result of exogenous technological change, it is also driven by competitive conditions in markets; for instance, in tourism the pressure on suppliers to install central reservation systems or electronic means of advertising products or payments transmission in order to reduce costs. To protect an innovation or to control the sale of products, companies resort to such strategies as licences and contracts organized through the legal system, and such arrangements often reinforce entry barriers and, thus, the structural characteristics of markets.

Performance can be considered in terms of consumer satisfaction, efficiency of operations, the growth rates of firms and industries, firms' market shares and profitability. Within tourism supply, short-term performance measures have sometimes been paramount, such as in the UK package holiday sector, where the concern has been for the growth of sales and market shares, often at the expense of efficiency and profitability. Performance is affected by public policy, especially changes in regulation, international trading arrangements and competition laws, which have played an important role in tourism. Price control exerted a strong influence in the transport sector in the past, such as in the setting of international air fares, which governments implicitly, if not overtly, supported. Furthermore, the promotion of tourism by public sector bodies and the provision of subsidies and/or tax incentives have had a marked impact on production, for example in relation to the supply of tourist accommodation.

The interrelationships between structure and the conduct and performance of firms within particular markets, as well as the basic conditions of demand and supply and public policy, are indicated by the connecting lines in Figure 4.1. The solid lines show the causal links as originally posited in the economics literature, running from structure to conduct and performance. However, the more recent formulation of the model does not deny the possible influence of conduct and performance on structure, thereby allowing its endogenous determination. In this respect conduct, where it embodies the actions of managers and employee in firms, reflects the importance which institutional economics attaches to the influence of human behaviour on outcomes. Therefore, later considerations of SCP analysis suggest, as indicated by the broken lines in Figure 4.1, that not only performance and conduct, but also public policy, affect market structure. For example, deregulation has influenced the conduct and performance of airlines, bus and coach companies and, with the passage of time, their market structure as trends towards concentration emerged.

Having the above in mind, the analysis proceeds with an examination of the main market structures and their implications for conduct and performance. The value of the theoretical constructs is that they impart rigour to the analysis of real life situations. In this context, this section of the chapter reflects subsequently on key features of tourism supply. An analysis of some major policy issues related to the SCP framework is then discussed, followed by an overall evaluation of the paradigm in the context of tourism markets.

4.3.1 Market structures and their implications for conduct and performance

The extreme case of a highly competitive structure, perfect competition, is first examined. Some modifications to the standard assumptions are then introduced using the concept of contestable markets. This is followed by an analysis of monopoly, the limiting case of an uncompetitive market. The discussion then focuses on the less restrictive cases of monopolistic competition and oligopoly, which more closely represent the conditions found in most tourism industries.

a. Perfect competition

The model of perfect competition provides a benchmark, illustrating the limiting case of a market involving a very high level of competition. A large number of firms and consumers are assumed to exist so that neither producers nor consumers can affect the price of the posited undifferentiated product. It is also assumed that there is free entry into and exit from the market, implying no entry or exit barriers. An example which might approach these conditions is the existence of the numerous producers of snacks, meals and drinks which are sold in streets and on the beaches of tourist destination areas in relatively poor countries, although there may be a spatial separation of such sellers which affects the degree of competitiveness.

The cost and revenue conditions for an individual supplier (firm) in a perfectly competitive market in the long-run are illustrated in Figure 4.2. As much as the seller wishes can be sold at the price P*. Each additional unit of output (such as a snack or drink) is sold for the same price. Therefore the extra revenue which the producer obtains from selling every additional

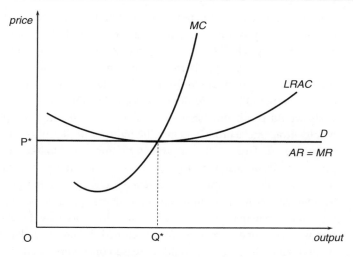

Note: P* denotes equilibrium price; Q* equilibrium quantity of output; *MC* marginal cost; *LRAC* long-run average cost; *MR* marginal revenue; *AR* average revenue, *D* demand.

Figure 4.2 Production by a firm in a perfectly competitive market in the long run

unit of output, i.e. the marginal revenue, is equal to the price and is also equal to the average revenue from selling each unit of output. A cost structure is assumed where both the marginal cost of producing each additional unit and average cost per unit first decrease as output rises and then increase owing, initially, to increasing returns to scale and subsequently to decreasing returns. For example, snack and meal producers purchase a certain amount of fuel for cooking and if they were to produce only one meal, the cost of the meal would be relatively high. Since a given amount of fuel can be used to produce more than one meal, the marginal cost of producing each additional meal declines and the average cost of producing the meals also declines. At some point, more fuel, or even another appliance or cook, is necessary to produce more meals, so that both the marginal and average cost rise. Producers are unwilling to produce for a price less than the prevailing market price, P*, as they would make a loss. Hence, the section of the marginal cost curve to the right of Q* is the producer's supply curve, since it shows the quantity of output which the producer would like to supply at each price in the short run. Effectively, this means that the quantity supplied will be increased only if prices rise.

The optimal point of production from the supplier's viewpoint occurs where the marginal cost is equal to the marginal revenue, corresponding to an output, Q*, in Figure 4.2. This is the point at which profits are maximized, as can easily be demonstrated numerically. Output below Q* is associated with a marginal revenue in excess of marginal cost, so that producers wish to increase production because they can increase profits. Conversely, at levels of output above Q*, marginal revenue is less than marginal cost so that producers experience reduced profit and would seek to decrease production to the profit maximizing level of output. Hence, the supply curve for the industry is horizontal at the price, P^*, over the long-run. If costs were to increase, for example if fuel became more expensive, producers with higher average costs would go out of business if they were just breaking even at output Q^*, i.e. earning only normal profits. An overall decrease in costs would result in a short-run outcome of supernormal profits which would attract more producers into the industry until profits returned to their normal level. Thus, within a perfectly competitive market, the prevailing price may result in supernormal profits or losses in the short run. However, there is a tendency towards a break-even price, equal to marginal cost where average cost is at a minimum, and consumers appear to benefit. This raises the issue of whether tourism markets which involve low levels of competition whereby supernormal profits are earned should be rendered more competitive and whether this would increase consumers' welfare. There is no unequivocal answer because, in the context of economies of scale where an increase in the firm's output is accompanied by a fall in the average cost of each unit of production, it is possible for imperfectly competitive markets to be more efficient than those under perfect competition. This issue is taken up below.

b. Contestable markets

It has been argued that although most real world markets are not perfectly competitive, a significant number give rise to similar economic outcomes. Baumol (1982) introduced the concept of contestability to take account of this outcome. A contestable market is characterized by insignificant entry and exit costs, so that there are negligible entry and exit barriers. Sunk costs, which a firm incurs in order to produce and which would not be recoupable if the firm left the industry, are not significant. Because of reasonably efficient information flows, the same supply conditions and technology are available to all producers. It is posited that producers are unable to change prices instantaneously but consumers react to them immediately.

The key insight of contestability is that new and existing firms find it possible to challenge the position of rivals through pricing strategies. Thus, firms in contestable markets operate similarly to those in perfectly competitive markets in that they charge approximately the same price for a given product. Although economies of scale and scope may arise, incumbent firms are unable to charge a price exceeding average cost because this would attract competitors into the market. Rivals would not be averse to entry because of the low sunk costs and low entry/exit barriers. Hence, contestable markets may be beneficial to consumers. For example, independent tour operators who are not vertically integrated with an airline, accommodation chain or other facilities are governed by many of the conditions prevailing in this type of market, especially ease of entry and exit and minimal economies of scale. Evans and Stabler (1995), in considering the UK air inclusive tour market, discuss such behaviour using the categories of 'second tier' and 'third tier' tour operations. Fitch's (1987) and Sheldon's (1986) earlier work but also Davies and Downward's (2001) later research also pointed to tiers of tour operations. The same kinds of conditions apply in travel agencies where, notwithstanding the presence of multiples in the sector, at the outlet level operational costs are not markedly below those of independent firms (Bennett, 1993).

The contestable market thesis is consistent with the Chicago School of thought, according to which resources are allocated optimally by competitive forces so that market power and pricing in excess of marginal cost are eliminated by free entry into the industry. The Chicago School views perfectly competitive, or at least contestable, markets as the dominant market structure which emerges over the long run so that consumers benefit and government intervention is regarded as unnecessary. The Chicago stance was adopted by some economists in Europe, as indicated in the reviews by de Jong and Shepherd (1986) and Hay and Morris (1991). There has been a long-running controversy between the Chicago economists and proponents of the SCP paradigm, who do not adhere to the view that markets are generally competitive over the long run (Martin, 1993).

c. Monopoly

Monopoly represents the opposite extreme of perfect competition. Unlike producers in a perfectly competitive market, the monopolist has considerable control over the product price and level of output. Given normal demand conditions, in order to sell additional output, the product price must be lowered so that average revenue falls and consequently the marginal revenue per additional unit of output sold decreases as the amount sold rises. The relationship between the two forms of revenue is such that marginal revenue decreases by twice as much as the average revenue (when the demand curve is linear) and hence the marginal revenue curve lies below the average revenue curve, as illustrated in Figure 4.3. Following economic suppositions regarding cost structures, returns increase and then decrease; therefore the marginal and average cost curves initially decline and subsequently rise.

Profits are maximized at a level of output where the marginal cost of production equals the marginal revenue and marginal cost is rising (or in cases where increasing returns prevail, marginal cost is falling less rapidly than marginal revenue). As shown in Figure 4.3, the average cost of production is lower than the price charged giving rise to supernormal profits

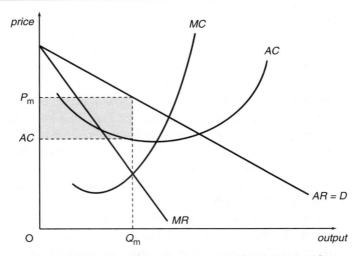

Note: P_m denotes monopoly price; Q_m monopoly quantity; MC marginal cost; AC average cost; MR marginal revenue; AR average revenue; D demand; the shaded area represents super-normal profits.

Figure 4.3 Production by a monopoly

(shown by the shaded area), above the minimum required to keep the firm in the industry. Thus, consumers are paying a price in excess of that which might emerge in a more competitive market. This raises the issue of whether the monopoly should be allowed to trade unhindered or be regulated or reorganized as a competitive industry.

Various components of tourism supply in different countries are deliberately organized as monopolies. For example, domestic air flights were monopolized by the state airline in the past and railway networks sometimes operate as a single industry. This may appear paradoxical as it would appear to run counter to the interests of consumers. However, two interesting outcomes may occur under monopoly as opposed to competition, which illustrate the debate concerning the relative advantages and disadvantages of each type of market structure. The first case involves a comparison of the equilibrium price and output combinations under competition and monopoly respectively, in which it is assumed that a competitive industry is monopolized without any change in production conditions. The outcome is illustrated in Figure 4.4, which assumes that each competitive firm has the same costs and therefore, for simplicity, the average cost curve can be excluded. It is clear that the price P_c is lower and the quantity produced Q_c is higher under competition than under monopoly. If the industry were monopolized, profit maximization would occur where the marginal cost MC is equal to the marginal revenue MR, resulting in a higher price P_m and a lower output Q_m. Thus, consumers would be worse off under monopoly than under competitive conditions. This could occur in the tourism accommodation

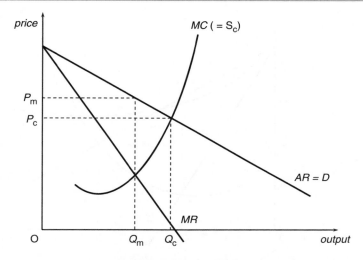

Note: P_m and P_c denote monopoly and competitive price respectively; Q_m and Q_c denote monopoly and competitive quantity of output respectively; MR is marginal revenue; MC is marginal cost; S_c is supply curve in perfect competition; AR is average revenue; D is demand curve.

Figure 4.4 Production by a perfectly competitive industry and a monopoly

or intermediaries sectors, but more particularly in the transport industry, where small independent airlines, bus, coach or ferry operators are amalgamated as, to an extent, has occurred with deregulation (discussed in Chapter 5); after an initial influx of new entrants, larger enterprises have taken over many of the smaller firms and exercised monopoly powers.

In the second case, production conditions differ according to whether the industry is operated as a monopoly or under competition as large economies of scale can apply when production is undertaken by a single firm. This applies in the case of a so-called natural monopoly in which both the marginal and average costs are lower over the range of output which could be purchased in comparison with competition. In these circumstances, consumers could benefit in terms of a lower price and higher quantity of production even if the monopolist made supernormal profits. This is illustrated in Figure 4.5, which shows the price and quantity combinations under an unregulated monopoly, P_m and Q_m. If the monopoly were regulated so that the price charged was equal to the marginal cost of production at P_c, consumers' welfare would increase owing to the lower price and higher output, Q_c. However, the government would need to subsidize production since the price would be lower than the average cost of production, so that the supplier makes a loss. For example, most rail systems operate at a loss in terms of not covering average total costs because they are viewed as providing a public service and may be required to charge prices at which they endeavour to cover operating (marginal) costs and contribute to meeting fixed costs. Even if the monopoly were not regulated consumers might benefit, in

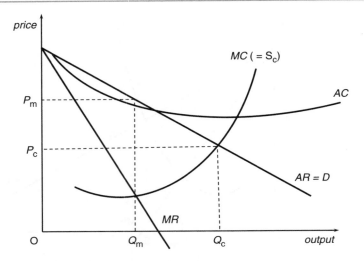

Note: P_m and P_c denote monopoly and competitive price respectively; Q_m and Q_c denote monopoly and competitive quantity of output respectively; MR is marginal revenue; MC is marginal cost; S_c is supply curve in perfect competition; AR is average revenue; D is demand curve.

Figure 4.5 Production by an unregulated and regulated natural monopoly

the long run, from product and process innovations resulting from the rein-vestment of some of the profits gained by the monopolist, including the nec-essary research and development. In other words, although the monopoly is characterized by allocative inefficiency (as price is always greater than mar-ginal cost), it can potentially be dynamically efficient. Under competitive conditions, the availability of funding for research and development is some-times problematic. Therefore the issue of whether monopolies in tourism supply, as well as in other sectors of the economy, should be allowed to exist or be regulated is complex and varies according to the differing circum-stances of particular industries. The absence of a general conclusion con-cerning the relative merits of the different market structures strengthens the case for specific empirical investigations of the tourism sectors in question.

The difficulty of predicting a determined outcome can be aptly illus-trated by reconsidering the examples of state-sanctioned monopoly rail and airline markets. Many governments have partially or totally privatized a number of services, which were hitherto perceived as natural monopolies. The rationale has often been that, notwithstanding economies of scale, they could be operated more efficiently as private sector industries. Privatization has often been accompanied by deregulation to encourage new entrants and increase competition. However, paradoxically, the eventual effect has often been to bring about greater concentration. One important reason is that smaller new entrants are unable to fully exploit economies of scale and therefore fail. It is still too early to establish empirically what the structure of the airline, bus, ferry or rail sectors will eventually be. Effective regulation may be needed to prevent abuses of monopoly power in the form of high prices and the extraction of supernormal profits or restraint of competition. In practice, governments usually define monopoly as occurring when a single producer accounts for a specific, relatively high percentage of production of a particular product and monitor the operations of the firm in question in an attempt to ensure that consumers are not disadvantaged.

d. *Monopolistic competition*

Monopolistic competition is a type of market structure, often associated with retailing, which is intermediate between perfect competition and monopoly. It is similar to that of perfect competition and contestable markets in that there is ease of entry and exit in the long run. However, it differs in that suppliers have some control over the price for which they sell their product and, thus, also over their price/output combination and associated market share. Nevertheless, the pricing and output decisions of an individual supplier do not have a significant impact on those of another, because it is normally assumed that there are many suppliers and no substantial degree of concentration. There are also usually only limited economies of scale, unlike the cases of monopoly and oligopoly. In tourism monopolistic competition applies in many respects to the hotel accommodation sector, which is characterized by many suppliers who provide products which are close but not exact substitutes, so that there is some degree of product differentiation reinforced, as in retailing, by the spatial separation and location of businesses.

In the short run, suppliers within a monopolistically competitive market can charge a price which provides them with supernormal profit. Firms produce where their short-run marginal revenue MR_{SR} is equal to their marginal cost and charge a price which exceeds their average cost, as illustrated by the price and output combination, P_{SR} and Q_{SR}, in Figure 4.6. However, in the long run, supernormal profits, combined with the virtual absence of entry and exit barriers for the industry, attract new competitors so that existing firms experience a decline in the demand for their products. This is illustrated by a shift to the left of the average revenue (demand) curve, AR_{SR}, and the short-run marginal revenue curve, MR_{SR}, until they attain their long-run positions, given by AR_{LR} and MR_{LR} with an equilibrium price P_{LR} and output Q_{LR} respectively.

Over the long run, demand decreases until the break-even point, where average revenue, shown as the price P_{LR}, is equal to the average cost of production AC_2 so that there is no further entry into (or exit from) the industry. Although output has contracted and supernormal profits have disappeared, the price charged is still greater than the marginal cost of production.

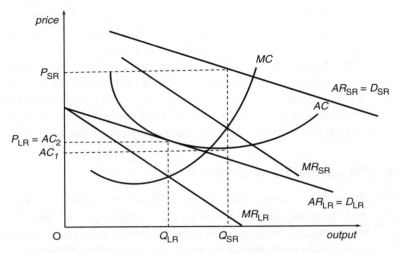

Note: P_{SR} and P_{LR} denote equilibrium price in the short- and the long-run respectively; Q_{SR} and Q_{LR} denote equilibrium quantity in the short- and the long-run respectively; MR_{SR} and MR_{LR} denote marginal revenue in the short- and the long-run respectively; AR_{SR} and AR_{LR} denote average revenue in the short- and the long-run respectively; D_{SR} and D_{LR} denote demand in the short- and the long-run respectively; MC is marginal cost; AC is average cost.

Figure 4.6 Short-run and long-run production by a firm in a monopolistically competitive market

Hence this form of competition appears less efficient than perfect competition although the wide variety of products provides consumers with greater variety of choice. There are a number of examples of tourism businesses in this position such as the smaller firms in the contestable segments of the accommodation and transport sectors, where the extent of their market limits the possibilities of operating at a level which reduces their costs (see Figure 4.6). This might explain why smaller airlines, bus and ferry operators eventually get taken over as they cannot compete at prevailing prices, especially during a price war.

e. Oligopoly

An oligopolistic market structure occurs when a small number of producers dominate the industry, the international airline industry being a case in point at a route level. Each firm has some control over its price and output decisions and there are some barriers to entry and exit. The key characteristic of oligopoly is the interdependence between producers so that each firm's price and output decisions depend, in part, on those of its competitors. One well-known example of such interdependence, which has become a standard case in economics, is that of the oligopolist's perceived kinked demand curve, which shows the likely outcome for a firm, should it contemplate a change in its price in the absence of changes in cost or demand

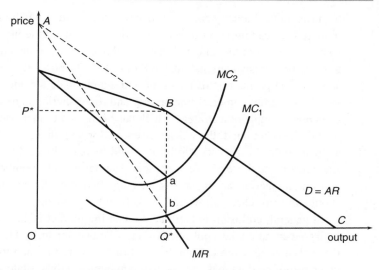

Note: *P** denotes equilibrium price; *Q** equilibrium quantity of output; *MC* is marginal cost; *MR is* marginal revenue; *AR* is average revenue, *D* is demand.

Figure 4.7 Production by an oligopolist when prices are sticky

conditions which affect the whole of the industry. It knows that if it decreases its price its competitors will follow suit, demand for its own product becoming more inelastic, so that it will not increase its market share by doing so. Conversely, if it increases its price competitors will maintain their own prices so its demand is more elastic and it will lose market share. Thus, the prevailing market price is the profit maximizing price for the firm. This is illustrated in Figure 4.7.

The equilibrium price and quantity of the oligopolist are P^* and Q^*. If the firm increases its price above P^*, it experiences a large decrease in demand as its competitors refuse to increase their prices in line, so that the average revenue (demand) curve for the oligopolist is relatively elastic (flatter) between A and B. If, in contrast, it decreases its price below P^*, it increases its sales only marginally because its competitors also tend to decrease their prices, so that its demand curve is relatively inelastic (steeper) between B and C. The firm's marginal revenue curve is, similarly, relatively elastic up to the quantity Q^* and relatively inelastic for quantities in excess of Q^*. Having the above in mind, the firm faces a kinked demand curve and a discontinuous MR curve. A shift in the MC curve from MC_1 to MC_2 in the discontinuous portion ab of the MR causes no change in the profit maximizing price or output. Thus, the prevailing price P^* and output Q^* tend to be stable in the absence of collusion among oligopolists.

Ideally, individual oligopolists prefer prices to be set at a level which maximizes the joint profits of all producers in the industry. In effect, if all the firms in the industry could act together, the result would be similar to a monopolist, leading to an increase in price and a decrease in the quantity sold, in contrast to the outcome that would occur under competitive conditions, as previously demonstrated in Figure 4.4. The possibility of obtaining supernormal profits by this means provides a rationale for inter-firm collusion. For example, inter-airline pricing and route-sharing agreements have been one example of a strategy which has been used to increase joint profits. On the other hand, there is an incentive for an individual firm to cheat as, if it could do so successfully, it would increase its profits and market share relative to those of its competitors. The stability of a number of pricing and output-sharing arrangements may be accounted for by the fact that should all firms cheat, all would be likely to be worse off.

In general, collusion is facilitated when the market is characterized by high concentration, i.e. controlled by a few large companies. This is because the existence of a limited number of firms renders the monitoring of the competitors' actions easier – hence, companies will hesitate to cheat (if they are members of a cartel) or actively compete (if they are part of a rival competitive fringe) fearing that they may be discovered and punished accordingly either directly (if the cartel can impose sanctions on its members) or indirectly (through a price or a capacity war). In fact, collusion is more likely to be sustained if the short- and long-term repercussions of this punishment are large compared to any potential immediate benefit from cheating or competing. Moreover, cartel stability is enhanced when 'noise' is small and the market fundamentals are simple, i.e. the product is relatively homogeneous; the corporate rivalry dimensions are limited; demand is essentially predictable and static; economic integration among companies is non-existent. On the other hand, when the market structure is fragmented, the penalties of cheating are small and the product and market features are complex, then cartel collapse and coordination failure among producers are highly probable.

Oligopolists can alter their output as well as the prices charged and take account of their competitors' possible reactions when they make their price and output decisions, as was shown in the case of the Bermuda resort hotels market (Mudambi, 1994; Baum and Mudambi, 1995). The possible strategies and responses which can occur have been examined by means of game theory. This is one of the most rapidly developing areas of analysis within industrial economics and is examined later in the chapter.

4.3.2 Key features of tourism supply

The theoretical models analysed in the previous section can be used to identify a number of criteria regarding the competitive structure of markets such as the number and size of firms; the degree of market concentration and the level of entry/exit barriers; the existence of economies and diseconomies of scale and scope; the problem of capital indivisibilities, fixed capacity and the associated fixed costs of operations; the use of price discrimination and product differentiation as a business strategy; the choice of alternative pricing policies and the pursuance of economic integration between firms. For example, the number and size of firms and level of entry/exit barriers indicate the degree to which oligopoly or monopoly powers may be exercised by tourism firms. Similarly, the extent of market concentration or of price leadership signifies potential limitations on the level of inter-firm competition. These criteria can be examined in the context of specific tourism sectors in different countries, providing insights into the types of market structure which prevail and the nature of inter-firm competition. This is useful because of the implications of different market structures for the extraction of profits above the break-even level and, hence, for consumer welfare. These criteria will now be analysed in further detail.

a. Number and size of firms

Where many small firms exist, the inference is that the market is competitive. A small number of firms, in contrast, indicates an oligopolistic structure, with monopoly being the limiting case of dominance by a single firm. Evidence on the numbers and sizes of firms in the transport and intermediary sectors is relatively easily ascertained where there are requirements regarding registration and regulation and trade associations exist. For example, the International Air Transport Association (IATA) and International Civil Aviation Organization (ICAO) publish statistics on airlines, while in the UK and many other countries, there are intermediary trade associations representing tour operators and travel agents which allow estimates of the numbers of intermediaries to be made. However, the size, diversity and fragmentation of the accommodation sector mean that, except for hotels at an international level, ascertaining the number of establishments can be very difficult.

The size of firms can be measured by a number of variables, ranging from number of employees, sales revenue and number of unit sales to capital employed. In the accommodation sector, the number of rooms offered by each establishment or the total number of bed-nights achieved are useful measures. The number of holidays sold is a feasible measure for intermediaries. Examples relating to transport are the number of passengers carried or passenger kilometres, which can be compared for different carriers, and the proportion of the total market or segment served. In the case of air transport, the Airline Business (2008) survey statistics reveal that in 2007 Delta Airlines carried by far the largest number of passengers (109.2 million) whereas American Airlines had the greatest passenger kilometres (222,719 million), followed by Air France-KLM (207,227 million) and Delta Airlines (196,403 million). In the UK intermediary sector, Civil Aviation Authority (CAA) data showed that in 2006 the largest tour operator in the air inclusive tour (AIT) market was the TUI Group; when combined with the First Choice Group (given that the two merged in 2007), it jointly accounted for nearly a quarter of the total number of holidays sold (CAA, 2007).

The composition of tourism sectors within and between countries differs in terms of the number and size of firms. In some countries, a domestic monopoly exists in the railway sector. In others, an oligopolistic structure prevails; for example, in Australasia, the air transport sector has been dominated by a limited number of airlines. Within the accommodation sector as a whole, the large number of establishments selling differentiated products is suggestive of monopolistic competition, while some bus and coach sectors have been highly competitive. However, it is also clear that although in some sectors a few large firms command a substantial market share, the remainder of the market consists of many highly competitive small firms. Hence, it is possible to conceive of a spectrum of competition with strongly competitive conditions coexisting with oligopoly in some tourism markets, with a degree of interaction between them. For example, innovations by small companies may be subsequently taken up by larger entities, as in the case of the timeshare concept or activity holidays (Stavrinoudis, 2006). Thus, tourism markets are complex and may consist of different sub-sectors characterized by alternative competitive structures.

b. Degree of market concentration and level of entry/exit barriers

The degree of concentration is a further indicator of the competitiveness of markets. High concentration is suggestive of oligopoly or monopoly, while a low incidence implies a high level of competition. Measures of concentration can be broadly divided into those which focus on the number and size of firms and those which consider the implications of variations in size. Size may be defined in several ways, as mentioned above, the indicator used depending partly on the availability of data and partly on the nature of the business. The Lorenz curve can be used to rank, in cumulative form, the smallest to the largest firms as a percentage of the total number of firms trading in the market. The Gini coefficient represents the magnitude of the divergence from exact equality in size of firms, so that a value of zero denotes that all firms are of equal size and, ostensibly, highly competitive conditions prevail, whereas a value of unity equates to pure monopoly (Papatheodorou and Arvanitis, 2009). The concentration ratio ranks firms from largest to smallest, so that a high percentage score indicates a market in which monopoly prevails, while a low score suggests that many firms serve the market. More sophisticated measures attempt to take account of the greater impact of larger firms on the market by weighting their market share more heavily. Well-known examples of such measures, widely applied in the industrial economics field, are the indices proposed by Herfindahl and Hirschman (Hirschman, 1964) and Hannah and Kay (1977), evaluated by Davies (1989a). The degree of market power can be measured by the Lerner index (Lerner, 1934), which considers the discrepancy between market price and marginal cost, where zero denotes equality and a highly competitive market and a value approaching unity indicate high profits and a highly concentrated market (Papatheodorou and Platis, 2007). The extent to which a higher degree of market concentration is accompanied by profits in excess of the break-even level is determined, in part, by ease of entry into and exit from the sector and the level of sunk costs involved in operating the firm. Traditionally, attention has concentrated on 'innocent' entry and exit barriers determined primarily by technological requirements, principally the scale at which firms can operate efficiently. However, firms may also engage in strategic entry deterrence in an attempt to maintain relatively high prices and supernormal profits over the long run.

Davies and Downward (1998) introduced the following panel data econometric model to assess, among others, the impact of market concentration on the return on sales of British tour operators (Davies and Downward, 2006):

$$ROS_{it} = \sum_{i=1}^{63} b_i D_i + b_{64} Conc_{it} + b_{65} MS_{it} + b_{66} Unem_{it} + u_{it}$$

where ROS is return on sales; D refers to firm-specific dummy variables; $Conc$ is the value of the Herfindahl–Hirschman index; MS is market share; $Unem$ is unemployment; u is the error term, $i = 1,...,63$ is the number of involved companies and $t = 1,...,5$ is the number of annual periods (1989–93). As is acknowledged by the authors, however, this model may suffer from an endogeneity problem as companies may actively seek to increase their market share to the detriment of profitability at least in the short run. Moreover, higher prices may lead to a rise in profitability, new market entry and thus lower concentration; alternatively, higher profitability may induce companies to invest in new capacity, raising barriers to entry and hence increasing concentration in the market. In addition, and according to the Chicago school proponents (Demsetz, 1974), a positive association of concentration with profitability may reveal efficiency rather than the exercise of market power. This is the case when production is characterized by economies of scale and the low-cost incumbent firms operate at full capacity so that any new production is accommodated by new, higher-cost entrants whose marginal pricing results in an efficiency rent for the low-cost firm (LECG, 1999). Partly to account for some of the above reasons, Davies and Downward (2001) used a reverse regression with MS as the dependent variable and all the others as explanatory ones.

The market shares equation has a larger explanatory power compared to the return on sales one. In neither case, however, were the market power or efficiency hypotheses validated at an aggregate level (Davies and Downward, 2006). When the total sample was segmented into two groups, one comprising the largest seven and the other the remaining small companies, the econometric results provided evidence of a stable oligopoly among the large tour operators; on the other hand, small firms managed to increase profits by eliminating competition and raising market concentration. In all cases, the firm-specific dummy variables proved statistically significant, highlighting the role of corporate strategies undertaken by individual tour operators. Nonetheless, had the individual advertising-to-sales ratio been introduced as a proxy for product differentiation in the equations, the significance of the firm-specific effects could have been lower.

c. Economies and diseconomies of scale and economies of scope

In order to relate economies and diseconomies of scale to models of market structure, it is necessary to look again at economic theory. The essence of economies of scale is that supply costs per unit of production decline as inputs increase and output expands. Clearly, as long as average unit costs are falling, there is an incentive for a firm to go on enlarging. The point where unit costs start to rise again, i.e. the lowest point on the average cost curve, is where diseconomies set in and the firm would no longer expand. In economic theory the issue is whether diseconomies are reached at a relatively small or large size of firm. The former implies many small firms operating in a highly competitive market, trading at prices which equate with lowest average cost, while the latter suggests the converse.

Thus, in conjunction with the number of firms, their size and the level of barriers to entry, evidence of the extent of economies or diseconomies of scale can indicate market structure (Lyons, 1989). In order to establish whether either exists in tourism, it is necessary to look at the cost structures of firms of differing sizes. In some transport markets, particularly air and sea travel, there are technical reasons which explain why larger aircraft and ships are more efficient, in terms of both operation and maintenance, therefore reducing per unit passenger costs. Furthermore, there are economies of scale in production through specialization. Increasing size in the accommodation sector also yields some economies of scale with respect to the management, staffing and servicing of a greater number of rooms and in the travel agency sector overhead costs can be spread over additional outlets.

Economies of scope occur where additional products share common inputs. For example, in advertising and marketing new products, suppliers can draw on existing resources. Thus, airlines, bus, coach and ferry companies may add new routes, more fully utilizing the means of conveyance. Travel agents and tour operators can serve new market segments using existing reservation systems and staff, while in the accommodation sector the market may be extended via the supply of self-catering in addition to serviced holidays or access to the establishment's facilities by non-residents. To an extent, therefore, opportunities for economies of scope are possible where there are capital indivisibilities or spare capacity.

Evidence of unexploited economies of scale or scope, or potential for them to occur in the future through technical change, would indicate the likelihood of increased firm size and greater market power, so rendering the market less competitive. Conversely, diseconomies of scale, or supply where widely differing input proportions are possible, would enhance the long-term viability of smaller firms and create a structure which is more competitive. Although large firms can gain economies of scale and dominate some markets, this does not preclude many small firms from being successful in the same market. The latter can often serve specialized segments which are not susceptible to large-scale operations and some activities remain small scale because the markets are local or national rather than international; enlargement of their activities would generate diseconomies of scale. In general, the principal transport sectors are characterized by the possibilities of economies of scope. It is in the accommodation and intermediary sectors where greater polarization has occurred, e.g. large hotel chains coexisting with small family accommodation establishments and vertically integrated mass tour operators sharing the market with niche independent ones (Papatheodorou, 2004).

d. Capital indivisibilities, fixed capacity and associated fixed costs of operations

Many tourism sectors utilize capital equipment which is indivisible, aircraft being obvious examples. The firm often possesses a set amount of equipment and thus fixed capacity over the short run and incurs fixed costs in maintaining it; it can be altered only in the long run when re-capitalization occurs. As capital is used, variable costs arise, for example, in the form of fuel and staff costs in the case of air transport. In the short run, economics argues that a firm, even if it does not cover its total costs, should continue to trade as long as it covers its variable costs. However, in the long run the firm must generate sufficient revenue to cover both its fixed and variable costs if it is to remain in business. There may also be sunk costs which are not recoverable if the firm leaves the sector. They normally relate to indivisible inputs and can affect the contestability of the market as they deter entry into a sector if wholly irrecoverable. For example, if an airline, bus or ferry operator finds a route uneconomic or a hotelier in a particular location finds demand to be lower than the break-even level, each would, ideally, wish to execute a costless withdrawal. This depends on whether assets can easily be redeployed or, through marketing and pricing strategies, market demand increased to fill space capacity.

It is useful to analyze the structure of costs to ascertain both the absolute level of costs and the proportion which is fixed, as opposed to the variable ones. In the airline sector, fixed costs are high in absolute value if skilled aircrew are taken as a fixed input and aircraft need to be stored and maintained if not operational. The variable costs are those which would be avoided if the aircraft were withdrawn from service, mainly fuel, landing and navigation charges and some on-board consumables and aircrew costs. For rail systems, the infrastructure and rolling stock create high fixed costs and relatively low variable costs. In absolute terms, bus and coach services are not characterized by high fixed costs, but in relative terms, particularly for the smaller operator, they are significant.

In the accommodation sector, the fixed costs as a proportion of total costs are much lower than in the transport field and the value of fixed costs associated with small accommodation establishments are often low. In the intermediary sector conditions are rather different, depending upon whether tour operators and travel agents operate their own aircraft and accommodation. In the case of small operators who are not vertically integrated, fixed costs are usually low in relative and absolute terms. Evidence of high fixed costs coupled with fixed capacity in some sectors is indicative of the emergence of large firms which can exploit economies of scale, increasing the degree of concentration.

Operations subject to high fixed capacity and fixed costs require relatively high occupancy rates or load factors in order to meet both fixed and variable costs; the distinction between profits and losses occurs at occupancy rates around such break-even points. Cyclical and seasonal variations in tourism demand exacerbate profit volatility, especially where aggregate fixed capacity is a feature of independent decisions made by many suppliers, such as in the accommodation sector. Strategies in the short run are aimed primarily at covering fixed costs, involving attempts to capture trade from rivals. In the long run, consideration is given to means of raising demand in order to increase load factors or occupancy rates, particularly in off-peak periods. The characteristic of high sensitivity of profits to occupancy rates and load factors engenders various responses by suppliers.

e. Price discrimination and product differentiation

The pricing and output policies practised by tourism firms provide some indication of market structure. A large number of firms producing relatively homogeneous products sold at the prevailing price in highly competitive markets approximate to the model of perfect competition. Taxi services in many tourist destinations are a case in point. In imperfectly competitive situations where firms have a degree of control over the market, pricing and output strategies can be implemented. There is considerable interdependence in the pricing and output policies of oligopolists as they have to take account of their rivals' actions when determining their own strategy. In order to extend their share of the market, firms operating under monopolistic competition endeavour to differentiate their products from those of other firms. However, given the appropriate conditions (discussed below), firms operating under all types of imperfect competition can benefit from price discrimination, which involves targeting specific groups of consumers. Thus, its existence indicates an imperfectly competitive market.

Price discrimination is based on the idea that many consumers, who purchase goods and services in markets where a single price prevails, enjoy a welfare gain known as consumers' surplus because they would have been willing to buy at higher prices, sometimes referred to as their reservation price. Suppliers who recognize this can discriminate between purchasers and charge higher prices to those with higher reservation prices, and vice versa. There are various degrees of discrimination depending on how many groups of customers can be identified. In the extreme case it is hypothetically possible to exercise perfect (or first degree) discrimination where every consumer pays a different price. Firms gain increased profits provided that the price at which additional units are sold is greater than the cost.

The most likely situation in which price discrimination can be employed is where different consumers have distinct price elasticities of demand, permitting the supplier to charge higher prices to those whose demand is inelastic, while others whose demand is more elastic are charged lower prices. The supplier wishes to divide the total number of units of the product available between the different groups of consumers so that marginal revenues for each group are equal. This determines the price each group should be charged, enabling the firm to maximize profits, and is illustrated in Figure 4.8.

Consumers with an inelastic demand (Market X) change their quantity demanded for the product by only a small proportion in response to a proportionate change in price and their demand is shown by the average revenue curve, D_X in the figure. The corresponding additional revenue which the firm receives for each extra unit of the product sold is shown by the marginal revenue curve, MR_X. The average and marginal revenue curves

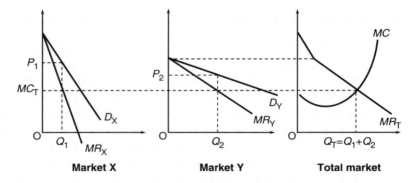

Note: *P* denotes price; *Q* is quantity of output; *MC* is marginal cost; *MR* is marginal revenue; *D* is demand.

Figure 4.8 Price discrimination between consumers with different price elasticities of demand

for a consumer with more elastic demand (Market Y) are flatter and are given by D_Y and MR_Y. The marginal revenue curve for all consumers is indicated by the line as $MR_T = MR_X + MR_Y$ (in Total market) and is obtained by adding horizontally the marginal revenue gained from each group of consumers. The marginal cost curve, MC, shows the additional cost incurred by the firm for producing each extra unit of production. The total profit maximizing output occurs where $MC = MR_T$, giving rise to output $Q_T = Q_1 + Q_2$ and a marginal cost MC_T. The firm divides its total output between the two groups of consumers, selling quantity Q_1, determined where $MR_X = MC_T$, at a relatively high price P_1 to consumers with an inelastic demand. A higher quantity Q_2 determined where $MR_Y = MC_T$, is sold at a lower price P_2 to consumers with a relatively elastic demand.

There are a number of examples of discriminatory pricing in tourism encapsulated in the context of yield or revenue management (Philips, 2005). It is possible to observe price discrimination in air, ferry and train travel and in the accommodation sector. For example, airlines perceive their passengers as falling into three broad groups – first class, business and economy – in which first and business class seats are filled mostly by business travellers. In ferry operations, commercial demand is generally more price inelastic than holiday demand. For rail travel, the division between commuting and pleasure demand is particularly significant, so that low prices are offered at times of low weekday and weekend demand in an attempt to attract travellers with a relatively high price elasticity of demand, while business people who often need to travel in peak periods are charged relatively high prices. In the accommodation sector, the larger hotels which have spare capacity offer weekend bargain breaks, charging guests with inelastic demand higher prices during the working week. Airlines, bus, coach and ferry companies also follow this practice in the off-season.

Market segmentation and product differentiation are also common practices under imperfect competition. Product differentiation can take the form of vertical differentiation between different quality products so that, for example, some tour operators attempt to specialize in providing luxury holidays in expensive locations. It may also involve horizontal differentiation via the supply of a range of product types, such as the provision of holidays for mass market demand as well as for the upper segment of the market, young people, elderly people and a wide range of special interest groups. The strategy of branding aims to raise consumer awareness of and demand for particular product types and advertising performs a similar function.

Few smaller firms have either the capacity or the market power to exercise significant control over prices so that serving a specific segment or differentiating their product is the only option open to them. Their concern is often to create rather than extend a market. Effectively they look for an unfilled niche and attempt to meet a specific demand in what may be a relatively small market segment. To this extent the product is automatically differentiated. It is large firms, enjoying economies of scale and scope and spare capacity, which attempt to differentiate their products through branding in addition to engaging in price discrimination. In the accommodation sector many international hotels serve the top end of the market, very often for business travellers, while budget hotels, for example Travelodge, cater for families. In some cases, such as the Accor group, both markets are targeted under different brand names.

f. Pricing policies, leadership and wars

Pricing strategies are central to oligopoly and may be used in attempts to maintain or increase market share and/or to eliminate excess capacity. The strategies of Laker Air in the 1970s, Virgin in the 1990s and Ryanair from then onwards are prime examples as the advent of deregulation has engendered further price competition among airlines on key routes. As the economics of oligopoly predict, price-cutting by one airline provokes a response from others. This occurred in the case of Trans World Airlines (TWA) when it cut fares within the United States in 1991. The matching of these fare cuts by rivals with a lower operating cost base partially explains the airline's bankruptcy. Price-cutting has also occurred within other transport sectors, such as bus, coach and ferry services. In the first instance it is very much the outcome of deregulation; in the last, the price 'war' on heavily used routes has been driven more by inter-sectoral competition despite the existence of intra-sectoral competition between companies. A feature of some interest is that many air, bus and coach sectors have experienced significant volatility because of structural changes which have destabilized trading conditions and prices. Ferry services are in a slightly different position because the entry of new companies is uncommon and there has often been tacit or overt collusion between the main ferry companies, giving reasonable stability in fares. This was also the case prior to deregulation in air travel where air fares were more formally controlled by IATA, resulting in a virtual cartel. In the accommodation sector, price wars have been less evident because there is greater differentiation of establishments in the market in terms of type, quality and location. Where establishments are clustered, such as the larger hotels in holiday resorts, near airports or in city centres, there may be implicit price-fixing because operators have easy access to information on competitors' charges and can set their own prices accordingly. Having said that, advancements in the Internet have redefined the geography of information, enabling price transparency and comparison at a global level.

The almost classic demonstration of the oligopoly case in tourism supply is package market tour operations. Here, price leadership, wars and attempts to increase market share by the larger players are strategies adopted since the advent of mass tourism in the 1960s. The evidence for the UK, including an historical review, can be found in Davies and Downward (2001) and Evans and Stabler (1995), following the studies of competitive structure and strategies by Sheldon (1986) and Fitch (1987). Evans and Stabler (1995) predicted some of the structural changes that subsequently occurred, such as the entrance of corporate (B2B, i.e. Business-to-Business) tour operators, often part of a conglomerate, which emphasize the gaining of profits rather than market share. There are also discernible trends in demand towards more specialized holidays. The mass package market is dominated by the top tier of approximately ten tour operators, who have increasingly diversified into specialized types of holidays to exploit economies of scope. At the other end of the firm size spectrum are small specialist operators which can survive through niche marketing, although there has been a high birth and death rate of companies. It is the second tier of thirty or so firms which has lost part of its share of the mass market to the top ten operators and the more specialized segments to the smaller firms.

Price leadership and pursuit of increased market share is mainly undertaken by large groups through longer-term strategies such as mergers or takeovers. Increased integration between firms appears to be a logical outcome of tourism markets where there is a long chain of supply. Large tour operators in Europe are vertically integrated, particularly with travel agencies and airlines, while in the USA there are strong linkages between travel agencies, the accommodation sector and transport modes, with the initiator tending to be the transport company. Reciprocal agreements, franchising and leasing constitute forms of integration other than ownership ties. The relevance of integration to competitive structure is that its existence suggests information advantages, cost savings and higher profits achieved by enlargement of the firm and the possibilities of increased market power, associated with oligopoly and even duopoly or monopoly. These issues are further explored in the following section.

g. Economic integration between firms

Tourism firms may respond to competition from domestic or foreign rivals by means of integration. There may be common ownership between firms or, alternatively, firms may have contractual links with each other with various intermediate arrangements between these positions such as management contracts or franchising. Three main types of integration occur between firms. Vertical integration involves the coordination of production by firms supplying different types of output within a production sequence, for example TUI's tour operations, TUIfly.com airline and Nordotel hotels. Horizontal integration consists of economic linkages between firms which supply the same type of output, as in the case of international hotels chains such as America's Sheraton or the Hong Kong-based group, Hong Kong and Shanghai Hotels. Conglomerate integration involves economic linkages between firms producing different types of product, as is the case with TUI's involvement in container shipping activities.

VERTICAL INTEGRATION

Vertical integration can take the form of common ownership of firms supplying different components of tourism supply but has followed different patterns in different countries. Within the UK and Spain, for example, joint ownership of travel agents, tour operators and charter airlines has been common and airlines such as British Airways (BA) and the German Lufthansa have had equity holdings in hotels in some destination countries. German tour operators generally rely more heavily on scheduled flights than UK tour operators but, unlike most UK and Spanish operators, are characterized by some ownership links with destination country hotels. Italy's national flag carrier, Alitalia, used to have a majority shareholding in the tour operator Italiatour and the state railway had almost total ownership of the largest travel agency in the country. There is also some common ownership of tour operators, airlines and accommodation by French firms.

The advantages of vertical integration include the reduction of transactions costs and improved synchronization of tourists' inter-country transportation, accommodation provision and entertainments (Bote Gómez et al., 1989). Vertical integration facilitates information acquisition (Arrow, 1975), the provision of inputs at known prices (Oi and Hurter, 1965) and can reduce uncertainty about future demand (Carlton, 1979). It may also be used as a strategy to increase market power (Hymer, 1976). For example, by vertically integrating with a travel agent, a tour operator may order its new affiliate not to stock brochures of rival firms. This constitutes an entry barrier as new operators can only sustainably expand their business if they decide to incur the extra cost of acquiring a high street presence. As argued in Chapter 5, the Internet has dramatically changed the business environment by enabling disintermediation and the direct contact between the tourist producer (or the tour operator) and the consumer. Nonetheless, there is always a reputation effect as many of the powerful online tourist companies today became known, thanks to their association with conventional, high street brands.

In any case, however, a vertically integrating company should seriously consider potential disadvantages, including an increase in fixed costs, foreign investment risk, reduced flexibility of operation and 'dulled incentives', whereby ownership ties with an inefficient producer dull the incentive to search for a more efficient supplier. Vertical integration, in particular, can constitute the raising of an exit barrier (Harrigan, 1985), so that a firm finds it difficult to cease to supply a product for which demand has declined. A topical example is that of ownership of tourist accommodation in a destination which is experiencing long-term problems of terrorism or other forms of violence or is in decline, being on the downward slope of a destination life cycle curve. Thus, many tour operators and airlines have been reluctant to purchase hotels, even in major tourist destinations. However, foreign direct investment may be a means of pre-empting entry into an industry by other firms (Smith, 1994) or of increasing market concentration, so that the benefits of cooperation are attained without collusion. Hence, common ownership may result from strategic decisions by the firm rather than location-specific advantages.

In its extreme form, vertical integration in tourism involves common ownership of the travel agencies in which tourists purchase their holidays, the tour operators which assemble them, the airline or other form of transport in which tourists travel and the hotels in which they stay. Tourists may also eat and drink imported food and beverages, so that the destination receives very little foreign currency from their stay. Conversely, in cases where the destination's airline and locally owned accommodation are used and local products consumed, both the value added in the destination and the foreign currency earnings accruing to it are higher. Thus, different types of integration affect the international distribution of returns from tourism. The case of Kenya exemplifies the possible distribution of expenditure on package holidays between the destination and origin countries. Taking a fourteen-day package holiday in a beach location in April and December of 1990, Kenya's share of UK tourists' expenditure was only around 38 per cent if the tourists travelled to and from the country in a UK airline (Sinclair, 1991b).

Losses were incurred in the form of payment for items imported for tourist consumption – approximately 30 per cent of expenditure on drinks and less than 20 per cent of expenditure on food. Kenya's share of spending on a fourteen-day beach and safari holiday was around 66 per cent. The higher figure is partly due to the fact that tourists on beach and safari holidays often travelled by air between Nairobi, the usual departure point for safaris, and coastal airports and were required by the Kenyan government to use the national airline. Kenya's share was considerably higher, approximating 80 per cent, if tourists travelled to and from the country using Kenya Airways. Hence, tour operators' use of origin- or destination-based airlines is a particularly important determinant of the destination country's share of tourism revenue. If tour operators and airlines are commonly owned, as is often the case, operators make preferential use of their own airlines and the destination's share of tourist expenditure declines.

HORIZONTAL INTEGRATION

Like vertical integration, horizontal integration has been widespread within the tourism industry and acquisitions, mergers and takeovers have increased the degree of industrial concentration so that a smaller number of firms control a larger percentage of total supply. Firms may integrate horizontally with the objectives of increasing efficiency by means of scale economies and of increasing market power, enabling firms, potentially, to raise prices and profitability. Integration may be a means of increasing the firm's growth and market share, increasing barriers to entry, increasing the market valuation of the firm and obtaining easier access to finance (Prais, 1976). It may also take place in order to decrease uncertainty, providing the firm with greater control over its environment (Newbould, 1970; Aaronovitch and Sawyer, 1975). Horizontal integration can take forms other than common ownership, including joint reservations arrangements between airlines, such as BA's comprehensive code-sharing agreement with American Airlines in the context of the oneworld strategic alliance, or management contracts or franchising agreements between hotels. For example, Marriott International topped the international list of managed hotels with 962 properties out of its total 2,999, while it used franchising for 1,922 hotels in 2007 (Gale, 2008).

CONGLOMERATE INTEGRATION

This type of integration can fulfil the function of decreasing the risk of volatility in earnings as the returns from all the different types of activities within the conglomerate are unlikely to be positively associated; thus, profitability and shareholders' returns are stabilized. Conglomerate integration can also decrease the cost of capital to the firm and a particularly profitable activity may, at times, subsidize one which requires further investment. The main argument against the need for conglomerate integration is that hedging activities and effective portfolio diversification should be better undertaken by the individual investor instead of being pursued at a company level. In fact, conglomerate integration involves the risk of making a company lose its primary emphasis on the core product, with possibly detrimental effects on service quality and brand value. Moreover, the investing community may become confused as to which particular product, or service market a company addresses. In this context, a company may decide to divest its interest in non-core business to concentrate subsequently on its main product. This may be one of the reasons why TUI decided to sell the majority share of its Hapag–Lloyd container shipping affiliated company in late 2008 (Retter, 2008).

4.3.3 Policy issues

As shown in Figure 4.1, the SCP paradigm has important implications for policy-making; conversely, public policy may also affect market structure, conduct and industrial performance. This section of the chapter studies three major areas of policy-making, which are important for tourism, i.e. competition and regulation; taxation and subsidization and crisis management. The first emerges as a natural offspring of the SCP analysis discussing the measures that should be taken to avoid the abuse of market power by producers. The second is also significant as it highlights the substitution effects that may arise as a result of changes in prices. Finally, crisis management should be understood in the context of public policy planning to pre-empt adverse situations and effectively address them when they occur. Environmental issues are another crucial area of tourism policy-making; these are extensively discussed, however, in Chapters 8 to 10.

a. Competition and regulation

Corporate rivalry and competition among existing or potential market participants is usually thought to improve allocative and productive efficiency, reduce X-inefficiency (i.e. company slackness) and possibly set the fundamentals to innovate and hence achieve dynamic efficiency (Papatheodorou, 2006b). On these grounds, competition is expected to raise consumer well-being often to the detriment of producers. Nonetheless, as the former are usually assumed to outnumber the latter, overall consumer welfare is likely to increase. If markets fail to deliver a socially optimum configuration and show a tendency to restrict competition, then the public (i.e. local, regional, national and/or supranational) authorities should intervene to reinstate order by implementing an appropriate competition policy. This should predominantly aim at monitoring and controlling the activities of dominant companies to reveal if these abuse their market power by exercising anti-competitive practices. Moreover, it should pay attention to the creation of new dominant firms as a result of mergers and acquisitions. To serve the public interest, competition policy should be implemented by independent authorities with no vested interests either among industrialists or state bureaucracy. Inquiries and appraisals should be transparent based on detailed relevant information and specific procedures (Morris, 2000). Most importantly, however, competition policy should focus on correcting potential deficiencies related to market conduct and performance: structural elements should only be of secondary importance. In other words, market power is not necessarily a bad thing; what really matters though is its abuse (NERA, 2001).

This view is not necessarily in contrast with the standard SCP approach, which uses structural characteristics as the benchmark for its analysis; after all, even the SCP framework acknowledges that oligopolistic markets are characterized by a variety of conducts and performance ranging from price war to perfect collusion depending on whether the market participants fail or succeed in coordinating their activities. On these grounds, attempts to define the market under consideration each time are valid and may provide useful insights in terms of whether it is worth it for competition authorities to spend valuable resources to further investigate the case of a potential abuse. If a market proves to be very fragmented no inquiries may be deemed nec-essary; on the other hand, if it is very concentrated, an investigation may be appropriate to reveal if there is price excessiveness in case of market domi-nance or to disapprove related merger activities. To define the market both demand- and supply-side substitution should be studied (NERA, 2001). The former is related to the estimation of own- and cross-price elasticities of demand and goes beyond the examination of similarities based solely on physical characteristics. The existence of vertical product disparities (i.e. different qualities) means that effective substitution may occur even in the case where goods are sold at different prices. Similarly, supply-side substi-tution refers to the existence of economies of scope in production, the importance of sunk costs as a potential barrier to entry and the time lag between the decision to enter the market and the actual production. For example, to examine demand substitution for the service of an air carrier between tourism origin A and destination B, competition investigators should calculate cross price elasticities with respect to all other air (and pos-sibly other transport) carriers, which serve this particular route. On the other hand, an analysis of supply-side substitution should consider how easily (in terms of monetary outlays and time) another airline (not currently offering this service) may enter the specific route to compete with the carrier in question.

From an empirical perspective, market definition is closely associated with the performance of the Small but Significant Non-transitory Increase in Price (SSNIP) or the 5–10 per cent test. This aims at examining the effects of a hypothetical permanent rise in the prices of the investigated company by 5–10 per cent (Niels and van Dijk, 2006). In particular, the SSNIP test identifies the size of all possible substitutions to see whether such a price rise is profitable or not: in this context, it is a test for non-exploited market power. High cross-price elasticities are consistent with a high degree of substitutability; hence, a 5–10 per cent price increase would prove non-profitable as consumers would divert to other products or services perceived as belonging to the same market. Conversely, if a 5–10 per cent price rise is profitable, then this may provide evidence that the markets are separate, i.e. there is room for a dominant company to abuse its market power.

Nonetheless, the SSNIP test may produce biased results by leading to a wide definition of the market. This can make dominant firms with abusive behaviour appear as innocent ones with a small market share. This caveat is also known as the 'cellophane fallacy', named after the 1956 Du Pont case in the United States (Niels and van Dijk, 2006), and stems from the basic microeconomics concept that a monopolist would never produce in the inelastic part of the demand curve: this is because a price rise in this case would always be profitable as total revenue would increase and total cost would fall. Having the above in mind, a company exercising its market power may be already charging high prices associated with low levels of production; performing an SSNIP test on that price level may then result in the estimation of high diversion ratios, masking in this way the firm's restrictive conduct. On these grounds, in the case of dominance investigations, competition authorities should not perform SSNIP tests based on the prevailing price levels but instead they should use as a benchmark the price that would prevail in the case of active competition in the market, i.e. a price directly associated with the level of costs. On the other hand, the performance of an SSNIP test to identify potential problems in a merger case is more valid as this is forward-looking (i.e. the merger has not happened yet), hence the current level of prices may be used as a point of reference to examine whether the merger would result in a price hike or not (NERA, 1999).

In any case, however, merger appraisal can be quite complicated as, in addition to market definition issues, a holistic cost–benefit analysis should be undertaken by competition authorities to provide a solid rationale for the approval or refusal of the merger. In particular, while conglomerate integration poses no real threat to competition, a vertical merger may create problems of market foreclosure and avoidance of regulatory constraints through the use of exclusionary practices. Nonetheless, what seems to be of major concern is the case of a horizontal merger as this can seriously impede competition because of unilateral and coordinated effects (NERA, 1999). Unilateral effects enable the merged entity to raise its prices without fearing diversion to rivals, as some of the alternatives are now eliminated since consumer choice is restricted. Coordinated effects arise because the removal of market competitors and the increase in concentration facilitates explicit or tacit collusion among the remaining players. On the other hand, and as argued in the previous section of this chapter, economic integration may generate efficiencies, which may be passed on as cost savings to consumers, therefore improving their welfare. Moreover, business activities which harm competitors are not necessarily harmful to competition per se. When the announcement of a merger raises reaction among rival firms, this may be an indication that the resulting efficiencies of the merger will put other companies at a competitive disadvantage. This can be good news for consumer welfare, especially if rivals decide to become proactive by innovating and increasing their productivity. Conversely, if a merger proposal is applauded by rivals, this may signal that the merger could render future tacit collusion easier to the detriment of consumer wellbeing. Many traditional carriers seem to have welcomed the merger between Air France and KLM in 2004, arguing that this paved the way for market consolidation and the subsequent realization of efficiency gains in the currently fragmented airline industry. Although this may be true, competition authorities should remain alert to ensure that these cost savings will actually result in a rise of consumer welfare because of lower prices.

In conclusion, competition policy is of primary importance in oligopolistic markets to ensure that a dominant position is not abused and to certify that the benefits of merger activities outweigh their costs. Still, there are occasions where competition policy may prove ineffective to address market failure, simply because there is no actual or potential rivalry among firms, as in the case of an institutional or natural monopoly. On these grounds, market regulation may act as a substitute policy. In principle, there are three types of regulatory regimes. The Rate of Return (ROR) regulation pegs pricing to costs by ensuring a 'fair' yield on the initial investment (Armstrong, Cowan and Vickers, 1994). It aims at avoiding the emergence of excessive pricing well above the marginal and average costs, which would be likely to occur in a monopolistic setting. Nonetheless, ROR regulation provides no incentives for cost reduction and rationalization, as the monopolist can always ensure that any rises in costs can be passed on to the final consumer in the form of approved higher prices. As a result, ROR regulation also suffers from a moral hazard problem, inducing monopolists to over-invest in unnecessary infrastructure and increase capacity to such an extent that any potential future attempt to open the market to other entrants proves futile.

To avoid the problems associated with ROR regulation, a price-cap regime may alternatively be introduced (Armstrong, Cowan and Vickers, 1994). A typical price cap may be defined as $CPI - X + Y$, where CPI is the consumer price index, X is a productivity factor and Y is related to cost increases as a result of investment or other issues. In other words, the regulated company is allowed to raise its price in accordance with the general price level and specific cost elements; nonetheless, it is also expected to realize efficiency gains and increase its productivity over time. Price-cap regulation may provide important incentives for cost reduction as the company is allowed to keep any efficiency gains beyond X as extra profit. For example, if CPI = 5 per cent, X = 3 per cent; and Y = 2 per cent, then the company is allowed to raise its prices by 4 per cent; but if the company manages to reduce its costs by 5 per cent (instead of the required 3 per cent), then it will be rewarded with an extra 2 per cent of return. In addition to a 'default' price cap, which is set as a blanket to all regulated companies, alternative contracts may be also signed. In this case, the original price-cap conditions are modified to take into account individual particularities of the parties involved. Price-cap regulation has been extensively used in Britain to regulate the performance of various utilities, including airports (Graham, 2008b). Its usefulness as a policy tool, however, depends on quality assurance and monitoring. In fact, a company can easily reduce its cost structure by compromising the quality of its service provision to the detriment of consumer welfare. In other words, effective price-cap regulation should be accompanied with the parallel setting of specific quality indicators to be strictly observed by the regulated entity.

In spite of its contribution to the resolution of market failure problems, the set-up of a price regulation mechanism may prove a costly exercise which consumes valuable resources in the process of negotiation, monitoring and policy enforcement (Wolf, 2004). On these grounds and using similar arguments to the case of contestable markets, self-regulation may be preferred as a third alternative option to ROR and price-cap regulation. In this context, the business entities voluntarily refrain from abusing their market power in case of drastic action, such as tight regulatory control and penalization of their activities by the state, if the opposite proves the case. In other words, if the public authorities can pose a credible threat, then this form of 'shadow' reserve regulation may prove quite effective (Toms, 2001). In reality, however, political lobbying, under-the-counter negotiations and lengthy litigation processes may provide substantial room for companies to avoid self-regulation. Thus, the existence of a clearly predetermined regulatory framework may be preferable.

b. Taxation and subsidization

An important policy issue to consider within tourism is the imposition of taxes and the granting of subsidies. Taxes are usually classified into direct and indirect ones; the former are imposed on income and wealth, the latter on consumption and expenditure. Direct taxation on wage income is equivalent to a reduction of wages. Hence, and building on the analysis of Chapter 2, both income and substitution effects are likely to emerge. If the income effect dominates, then people will be willing to work longer to attain their previous level of consumption and leisure time will decrease. This may have a negative impact on leisure tourism in terms of aggregate time spent on this activity; nonetheless, this should not be necessarily the case with tourism expenditure. In fact, people may now decide to acquire travel experiences within shorter periods of time, substituting longer holidays for mini-breaks. Moreover, they may decide to use faster transport modes to save time. On the other hand, if the substitution effect is stronger, then people will now spend less time working. Leisure time will increase, with possibly positive repercussions for the length of leisure tourism stays, but total disposable income will decrease; as a result, tourism expenditure may also go down. This may have an adverse effect on expensive tourism services (such as those provided by five-star hotels) but may be welcoming news for cheaper substitutes (such as self-catering tourism accommodation establishments).

It is also important to know whether the imposed income tax is progressive, proportional or regressive. In the first case, the marginal is higher than the average tax rate, implying that those earning higher income will bear a higher tax burden. If this population segment has a higher propensity to travel and engage in leisure tourism activities compared to lower income earners, then progressive taxation may have negative implications for tourism. Nonetheless, if progressive taxation induces high income earners to reduce only their savings instead of their overall consumption levels, then tourism expenditure may not be adversely affected. A fortiori, if the state authorities use income from progressive taxation to subsidize social tourism activities of the poorer, then overall tourism expenditure may actually increase. In the case of proportional taxation, the marginal is equal to the average tax rate, i.e. taxation has a neutral effect on the distribution of income. Regressive income taxation implies that the marginal is less than the average tax rate, resulting in higher income inequality; in this case, the opposite effects of progressive taxation are likely to occur. Based on the principle that 'taxes should be levied on an ability-to-pay basis, rather than on the benefit principle' (Hughes, 1981: 199), most countries implement a system of progressive income taxation.

Direct taxation affects tourism in an indirect way, i.e. through the trade-off between work and leisure. On the other hand, indirect taxes imposed on tourism goods and services may affect tourism consumption in a direct manner. The UNWTO (1999b: 7) notes that 'over recent years there has been growing concern from practitioners in the tourism industry worldwide that taxation in the sector has proliferated'. Tourism activities and tourists may be taxed for three main reasons (Jensen and Wanhill, 2002). First, tourism taxation provides an opportunity for state and local authorities to generate substantial revenue with limited political cost, especially in the case of inbound tourists who do not form part of the electorate body, i.e. 'taxation without representation'. Second, it partly rectifies the problems emerging from the inability to apply the 'user pays' principle in certain cases. For example, police and other public services are funded through direct taxation of citizens but may also be used by tourists at no extra cost; hence, revenue from tourism taxation may indirectly cover part of the expenditure incurred for the provision of these services. Third, tourism taxation may be imposed to account for the negative production or consumption externalities associated with the sector; this argument is extensively discussed in Chapters 9 and 10, in which the use of taxes, subsidies, regulation and relevant policy instruments are examined in relation to the environmental social costs and benefits.

Tourism taxation may take the form of a lump-sum tax, which is a fixed amount of money paid by the consumer or producer irrespective of the level of actual consumption or production, as in the case of an entry tax (usually associated with the issuance of a visa for foreign tourists) or an airport departure tax. Lump-sum taxes increase the fixed costs of a tourism firm but have no effect on its marginal cost. In the short run, there is no effect on price or quantity in any market. In the longer term, there is no effect on a monopolist unless its profits fall below normal levels and the company shuts down. If the firm stays open, then the monopolist pays the tax. On the other hand, if the industry is perfectly competitive, firms will be making less than normal profit after the imposition of the lump-sum tax. As a result, some firms will be induced to exit the market; the industry will contract and prices will rise until the entire tax is passed on to the consumers. Lump-sum taxes are undoubtedly easy to implement and have a low cost of administration. Nonetheless, they are essentially regressive (as they form a larger share of the poorer tourists' total expenditure) and may significantly affect relative prices, to the detriment of cheaper products. For example, low-cost airlines have argued against the lump-sum character of penalties imposed by the European Commission to protect air travellers in the case of denied boarding, substantial flight delays or cancellations (Regulation 261/2004), saying that these fines account for a much larger share of the low-cost airline's ticket compared to a traditional carrier's fare. On the other hand, however, the European Commission has defended its position, arguing that the inconvenience caused to passengers is the same no matter how much they paid for their ticket (EC, 2004).

Specific or flat rate taxes involve a fixed amount paid per unit sold. In the context of a hotel, this tax could be a set amount per tourist per night (e.g. €1). Such taxes have implications for marginal evaluations and result in an upward shift of the supply curve (or equivalently a leftward shift of the demand curve) in a parallel manner. Incidence is the term used to describe how the tax burden is shared between consumers and producers. If the demand (supply) curve is perfectly inelastic, consumers (producers) pay all the tax. If the price elasticity of demand is greater than the elasticity of supply, producers pay a larger portion of the tax, but if the price elasticity of demand is less than the price elasticity of supply, then consumers pay a larger portion. Similar to the case of lump-sum taxes, specific ones are regressive (albeit to a lower degree) and make cheaper goods and services become relatively more expensive. Moreover, they are associated with a deadweight (welfare) loss reflecting the parts of consumer and producer surpluses which are lost and not expropriated by the government as tax revenue. Specific taxes may create a negative sentiment among tourism businesses and tourists as in the case of the specific eco-tax introduced in the Balearic Islands in April 2001 but subsequently abolished in October 2003 (Gago et al., 2009).

Ad valorem tourism taxation, imposed as a percentage of expenditure or turnover, has also important implications for marginal evaluations and welfare loss. Nonetheless, it does not affect relative prices within tourism (although it obviously may shift consumption away from tourism altogether) and hence may be preferable at least from an economist's point of view. The Value Added Tax (VAT) is the most widespread example of an ad valorem tax and has a number of advantages such as the effective handling of regressiveness through zero rating; the avoidance of consumer choice distortion if implemented across all goods and services and the difficulty of evading it; and its greater flexibility and adaptability to fiscal conditions compared to income taxes (Jensen and Wanhill, 2002).

In essence, the imposition of both direct and indirect taxation seems unavoidable at least in mixed economies with public sectors of significant size. In any case, however, taxation should be based on the principles of effectiveness, equity and efficiency (Holecek et al., 1994) to minimize economic distortions. In the context of tourism, attention should be paid to the cost of administering tourism taxes and the incentives that these create for tax evasion, as in the case of tourists who reside in non-licensed establishments, or tax avoidance, e.g. when visitors stay with their friends and relatives. Moreover, a rise in tax rates may undermine the competitiveness of a tourist destination and/or country in the marketplace, although this problem may be effectively addressed if the incremental tax revenue is redistributed to tourism producers in the form of subsidies for product and service improvements and/or to deal with the negative environmental consequences of tourism. Conversely, a reduction in tax rates may improve competitiveness in the short run but could potentially create sustainability problems in the longer term. Moreover, if other tourism destinations and countries decide to match this reduction in tax rates to avoid losing market share (as discussed previously in the context of the kinked demand curve), then the incremental effect on tourism flows will possibly be minimal.

As for the granting of subsidies, this may be seen and analyzed as the opposite case of taxation. The government may decide to subsidize poorer segments of the population to participate in social tourism activities or investors to build new tourism infrastructure and superstructure or improve the existing ones. Subsidies may also be set within the context of the wider complementarities arising in tourism. For example, a number of regional airports in Europe (owned by the local authorities) have effectively subsidized Ryanair, the largest low-cost carrier in Europe, to start services to and from them, believing that these outlays would be more than compensated by the generated non-aeronautical revenue (i.e. passenger spending in airport shops and restaurants) and most importantly the revenue resulting from the consumption of hospitality and other services in the surrounding areas (Papatheodorou and Arvanitis, 2009). Nonetheless, states should be careful when subsidizing consumers and/or producers as moral hazard problems and retaliation from other countries may cancel any positive effects. This issue is also addressed in Chapter 7 in the context of strategic trade theory. Moreover, it should be always born in mind that subsidies cannot be given without taking the overall balance between taxes and government revenue into consideration: running budget deficits over a long period of time is clearly not sustainable. Finally, the aggregate effects of both taxation and subsidization cannot be fully assessed and understood unless a full impact analysis is undertaken with the aid of multiplier, input–output and CGE models as discussed in Chapter 6.

c. Crisis management

A crisis may be defined as 'an undesired, extraordinary, often unexpected and timely limited process with ambivalent development possibilities' (Glaesser, 2006: 14). It is related to the concept of a disaster, which refers to 'situations where an enterprise (or collection of enterprises in the case of a tourism destination) is confronted with sudden unpredictable catastrophic changes over which it has little control' (Faulkner, 2001: 136). Crises may emerge as a result of wars and civil unrest; diseases and epidemics (such as SARS); public transport accidents (mostly involving airplanes, ships and trains); natural disasters (such as earthquakes, hurricanes and wildfires); terrorist attacks and political and economic adversities (such as the crash of global stock exchange markets in 2008). Due to its risk-averse nature and sensitivity to environmental, political, economic and social conditions, tourism may be seriously affected by crises in terms of both flows and expenditure. Moreover, tourism may be the target of human-induced crises (as in the case of terrorism) with disastrous effects on a region's economy. Sönmez (1998) discusses the relationship between tourism, terrorism and political instability, highlighting the interdependence of these seemingly unrelated concepts and focusing on the potentially detrimental role of information media with a perverse perception of social responsibility. For all these reasons, effective crisis and disaster management is essential to rectify problems and restore normality.

Faulkner (2001) and Ritchie (2004) develop an integrated framework to deal with crises and disasters in tourism, examining the required measures and strategies that should be taken by the involved policy-makers at different stages of the problem. Pre-emptive and proactive actions, contingency plans and efficient mobilization of resources are of critical importance. Moreover, reassessment, review and constant monitoring should be used to avoid losing focus on the problem. Among others, particular attention should be paid to the collaboration of the internal (e.g. employees) and external (e.g. media) stakeholders in the context of integrated quality management, which 'can be seen as a systematic quest for internal quality and external quality, i.e. economic improvement in the short term and local development in the long term' (EC, 2000: 15). In fact, crisis management cannot succeed unless there is strong leadership by a partner who clearly designates complementary roles to the various stakeholders; nonetheless, this assignment of tasks should be the outcome of consensual discussion in advance of the crisis to avoid conflict and panic when the incident occurs. In other words, a strategic planning mentality should prevail, with detailed approaches at a tactical level. After all, public relations exercises may prove futile unless the fundamentals also change in a positive direction.

Blake and Sinclair (2003) examine tourism crisis management from a quantitative perspective by focusing on the US response to the 9/11 terrorist attacks. Based on CGE modelling, the paper assesses the effectiveness of different policy measures to address the emerging problems and concludes by arguing in favour of sector-specific subsidies and tax reductions. Pambudi et al. (2009) use a CGE approach to evaluate the impact of the Bali bombing on the economy of Indonesia. According to their findings, the terrorist attack had a very negative impact on tourism-related industries; on the other hand, export-oriented and import-competing sectors managed to expand as a result of labour and capital diversion. It seems, therefore, that crisis management is inherently associated with risk diversification and the development of a portfolio approach in terms of entrepreneurial activities practised in a tourism destination.

An interesting example of effective crisis management is the action plan undertaken by the Hong Kong International Airport (HKIA) to confront the implications of the SARS epidemic in 2003 (HKIA, 2003). HKIA established clear guidelines to handle SARS cases and asked all its business partners to report incidents on a daily basis. Cleaning teams were set up and the prevention of cross contamination became a task of major importance: access was restricted to critical operation centres; backup facilities were established; large teams were split into smaller divisions and physical interaction of people was minimized. To rebuild the confidence of passengers, the HKIA took a number of measures, including the introduction of the operation SkyFit in May 2003, where all 45,000 airport staff agreed to self-check for fever and a similar initiative regarding a temperature check for all staff entering airside. To rejuvenate traffic flows, which came to a standstill during the epidemic, the HKIA provided recovery incentives to partner airlines such as discounts on landing charges and cooperative advertising of Hong Kong as a safe travel destination. The management of the SARS crisis proved successful and by the end of 2003 airport operations had resumed at their normal level. In retrospect, however, recovery would have been faster if appropriate prevention measures and better contingency plans had been developed (HKIA, 2003).

4.3.4 An evaluation of the SCP paradigm in the context of tourism markets

The SCP framework is well suited to examine the characteristics of market structure and firms' conduct and performance, and allows for considerable flexibility in terms of the wide variety of variables which it can accommodate. It also provides a useful framework in which to describe industries and markets as a basis for analysis. Major criticisms of the paradigm are that market structure is assumed and that it is set too firmly within a neoclassical static equilibrium framework, resulting in limited ability to accommodate market processes. It is certainly true that having grown out of conventional market analysis, the approach has been used to ascertain the conditions for highly competitive, oligopolistic and monopoly markets to exist and its application has sometimes been restricted to focus on these objectives. The SCP paradigm in its original form does not claim to show the process of change in markets so that it is not well placed to explain, in particular, how certain market structures come about, the impact of entry barriers, the size and number of firms and the effects of firms' growth rates on market structures. To this extent it reflects the origins from which it has developed. The problem of identifying causal linkages largely flows from the model's static equilibrium analysis and assumption of an exogenously determined market structure which affects firms' conduct and performance. This has inhibited consideration of the impact of conduct and performance on structure, so determining it endogenously.

More recent investigations which attempt to take account of the dynamic nature and uncertainty of markets have emphasized the cost structures of individual firms, including those associated with acquiring information and conducting transactions. In particular, the impact on the firm of the nature of the fixed assets required, the timing of the purchase of variable inputs, the linkages between inputs and outputs, scale and scope and the range of outputs have been investigated. Such studies have departed from the traditional SCP framework, which tends to concentrate on an entire industry and to focus on product as opposed to factor markets.

The SCP paradigm should not be viewed as a fully adequate analytical framework but as a starting point for examining key economic issues. Its strength is that it is an appropriate method for providing a market perspective, in contrast to one based on industry alone. It has the merit of giving a holistic view and identifies the wide range of variables which require examination. It highlights the importance of certain characteristics, for example entry conditions, which have a bearing on the number and size of firms and thus the likely contestability of the market, firms' conduct in terms of pricing behaviour and strategies and profitability. Nonetheless if, as posited, tourism markets are not only complex but also in disequilibrium, then more recent developments in the economic analysis of industry which accommodate their dynamic nature should also be employed. Indeed, economic explanations of the tourism industry and the strategies of businesses within it need not be confined to a specific framework of analysis, such as the SCP paradigm. It is likely that the range of schools of thought, discussed at the beginning of the chapter, can offer valuable insights into the operation of tourism markets. The dynamic processes according to which firms behave, changes in market structures and a number of the insights provided by diverse schools of thought can be taken into account by game theory, whose relevance to tourism will now be discussed.

4.4 Game theory and tourism

Conventional analysis of supply and market structures has proved inadequate for explaining strategic interrelationships between firms, particularly within the context of oligopoly, which is an increasingly prevalent form of competition. In contrast, game theory lends itself to many circumstances which arise in oligopolistic markets, particularly where uncertainty prevails, and can be used to examine many of the strategies which are employed by tourism firms. There are many situations in which decisions are interdependent and firms may gain by engaging in cooperative strategies, such as those which previously characterized the airline sector, or they may decide that competitive strategies are required, as in tour operations where a firm may gain an advantage by continually changing strategies. What is not in dispute, however, is that many firms in tourism supply sectors take account of the behaviour of other firms in the market when deciding upon their own strategies. The crucial question is when does it pay to collude as opposed to competing.

Game theory can be used both to explain behaviour and to predict the outcome of strategies regarding price-setting, product choice and differentiation, advertising, capital investment, mergers and takeovers, and entry deterrence. In observing firms' behaviour and interaction, it is also possible to understand how markets operate and evolve; this is an important attribute in analysing some tourism sectors which are newly emergent and unstable. It is of value not only to distinguish cooperative (collusive) and non-cooperative (competitive) games but also to vary the initial assumptions concerning participants' behaviour, objectives and knowledge of their rivals' reactions. Normally, in simple explanatory simulations, rational behaviour by competitors, profit maximization as an objective and, initially, perfect information and symmetry are assumed, so that firms have full knowledge of their rivals' cost structures, production levels, prices and so on, as well as the market conditions under which they operate. Such restrictive assumptions can be relaxed in more complex situations and game analyses.

4.4.1 Types of games

In addition to cooperative and non-cooperative games, it is possible to consider either 'one-shot' or repeated games and, furthermore, distinguish between those played simultaneously and others which are sequential. In some circumstances firms, mindful of the illegality of collusion in some countries, may initially conduct non-cooperative strategies but in the light of the experience of rivals' reactions establish stable patterns of behaviour which may result in tacit collusion. For example, maintaining high prices in the knowledge that rivals will do the same would confer advantages on all firms. One-shot games reflect situations where once and for all decisions need to be made, for instance whether to invest capital to produce a new product or to deter entry. Conversely, repeated games are indicative of circumstances where continual jockeying to gain short-term advantages would be appropriate. For example, pricing and product differentiation strategies may be changed frequently so that rival firms periodically engage in non-cooperative games. There are a number of instances in tourism where this occurs, such as airlines, ferry and tour operators engaging in price wars to increase market share, while product differentiation is attempted in order to extend a firm's total market or serve a new segment.

A distinction is made between simultaneous and sequential games because they represent different market situations calling for changes in strategies. In some circumstances, for example the Cournot duopoly model, firms decide on their levels of output at the same time, whereas in Stackelberg analysis one firm makes a decision independently, to which the rival reacts. This is an important factor in situations where the first mover may gain an advantage. In the context of tourism, simultaneous decisions are often made in the light of short-run changes in demand levels and patterns in the airline, ferry and hospitality sectors, especially where there is excess capacity. Sequential or reactive strategies occur in these sectors and in the tour operator market, where lead times are rather longer. For example, in the summer sun package holiday sector, the offer of discounts for early booking by one operator is almost certain to lead to retaliation by rivals if it appears that their market share will be lost.

Some sub-sectors of tourism supply are oligopolistic and therefore strategies concerning price-setting, level of supply, product differentiation and branding, market segmentation, advertising, innovation and entry deterrence are common. In addition, it is possible to consider circumstances where a key firm can exercise a dominant strategy, perhaps because of a particular attribute the firm has, such as a brand name commanding a degree of loyalty, which might give it an optimal strategy independent of its rivals. In order to illustrate the application of game theory to tourism, the cases of advertising, the pricing decision and entry deterrence are considered, using examples of strategic decisions which those in the tour operator sector of tourism supply might face.

4.4.2 The case of a dominant strategy: one-off and repeated games

The simplest condition to examine is where firms have a dominant strategy, i.e. an optimal position can be achieved by an enterprise irrespective of what its rivals choose to do. This basic notion in game theory is illustrated by the case of a tour operator deciding whether to advertise. The example presupposes that advertising is both competitive by each tour operator to capture a larger market share as well as informative to extend the market and increase the pay-off for both firms. Figure 4.9 shows a typical 2×2 matrix in a two person (firm) game representing the duopoly form of oligopoly, in which the pay-offs to tour operator X with respect to advertising or not advertising (rows) are given in bold while those for tour operator Y (columns) are in italics in parentheses.

Tour operator X will advertise as this is the best strategy (best net pay-off) which can be adopted irrespective of what Y does. If Y does not advertise the pay-off for X is 25, whereas if Y does advertise X will achieve a pay-off of 20. Likewise for Y, the pay-off is respectively 10 if both advertise and 15 if X does not advertise. It can be seen, therefore, that both will advertise where the total market pay-off is 30 (20X, 10Y) in the top left cell. This position is a stable one.

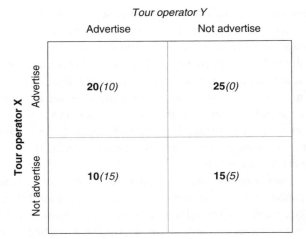

Figure 4.9 Advertising: dominant strategy case

If X does not have a dominant strategy then the optimal decision depends on what Y does. For example, in the advertising case given in Figure 4.9, if the pay-off to X is 40 if neither advertises (bottom right cell) then X's strategy is determined by what Y does. If Y advertises then so must X; nonetheless, if Y does not advertise, neither should X because the pay-off is much greater in adopting this strategy. Therefore, X must guess what Y will do. As Y has the same dominant strategy as before it will obviously advertise because whatever X does this is its best action. If X correctly guesses Y's action, which it should given the assumptions of the model, then a stable equilibrium can still be attained. When a dominant strategy does not exist, perhaps where advertising would extend the market rather than increase market share for individual tour operators, the question as to whether a stable equilibrium can be reached is raised. This is especially so if the assumptions of rational behaviour and correct interpretation of rivals' strategies and actions are relaxed. Game theory can show that there might be several equilibria or that there are none. Thus, it is possible to explain why instability may occur in certain industries in which particular circumstances prevail.

The example of the dilemma facing firms in deciding on price in each trading period can be employed to illustrate the point. At about the turn of the year tour operators launch brochures for the mass summer sun market for the coming season. Given the extent of the market and the fixed capacity of airline seat and accommodation capacity, which might become excess capacity in the shoulder months, individual tour operators recognize that consumers not only are price-sensitive but also may withhold purchase in the hope of late-availability bargains. Tour operators may consider, therefore, offering holidays at low prices to induce consumers to book early and so fill capacity. However, given the tight margins within which tour operators trade, such a strategy may result in low profits or even losses. The preferred strategy would be for all to charge a high price but there is no certainty that rivals would adhere to an implicit understanding either at the beginning of or throughout the season.

Repeated games can represent this real-life situation and feasibly predict what the outcome is likely to be, the main objective of simulating the market in this way being to identify the most robust strategy. Consider Figure 4.10, which includes negative as well as positive pay-offs and is akin to the 'prisoner's dilemma', a case where each participant has a dominant strategy which, if pursued, produces a joint profit less than that obtained if both collude.

As in Figure 4.9, X's pay-offs are again shown by the first figures in bold (rows) and Y's are indicated in italics in parentheses (columns). If both charge a high price each gains a pay-off of 10 (20 in total), whereas if both charge a low price each obtains 2, giving a total of 4 (the top left and bottom right cells respectively). The matrix also shows that if X charges a high price and Y a low one X suffers minus 10 while Y gains 20 (a net total of 10). Conversely if X charges a low price and Y a high one the positions are reversed. Obviously the preferred strategy for both is to charge a high price but the issue is whether one or the other will 'break ranks' and charge a low price in the hope of a greater gain. As long as implicit cooperation occurs, both will charge a high price. If one then charges a low price the other will follow until such time as one goes back to a high price. Provided each acts rationally, can ascertain the other's strategy and believes any threats are

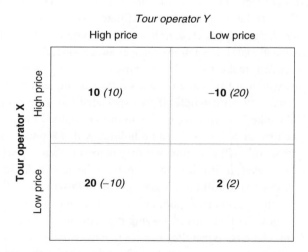

Figure 4.10 The pricing issue

real, a consistent pattern is likely to emerge so that tacit cooperation occurs. In other words, the rivals will realise that the short-term benefits from cheating are lower than the (discounted) longer-term benefits of collaboration. This is known as the 'Folk Theorem' of game theory (Rubinstein, 1979). However, this outcome seems to be at odds with what is being observed in the UK tour operator sector (Evans and Stabler, 1995; Taylor, 1998), where price wars continue to occur, i.e. non-cooperative behaviour persists. The likely reason is that other factors, such as variations in demand and costs, low entry barriers, excess capacity and the presence of many firms, preclude the establishment of cooperative strategies. Uncertainties introduced by such phenomena make it difficult to create the required stability. This kind of analysis of price takes the theory of oligopoly forward from the static hypothesis of the kinked demand curve depicted in Figure 4.7.

4.4.3 Sequential games

Game theory can also accommodate other market conditions and situations, such as first-mover advantages, threats, including those referred to above regarding entry and the consequent change in numbers of firms; these are factors which if stabilized might, in turn, stabilize markets and therefore prices. It is not possible to examine these cases in detail but some further exposition of the implications of relaxing the assumption of simultaneous action by firms is of value.

The Cournot model of duopoly considers the case where firms decide how much output to produce simultaneously, while Bertrand's approach examines a context in which firms engage in simultaneous price-setting. In contrast, the Stackelberg model posits that one firm sets output before the other. The Stackelberg approach facilitates the analysis of first-mover advantage, for example in research and development and investment, entry deterrence and marketing actions, such as advertising. Sequential games, as they are called, make game theory dynamic as it is possible to trace the processes which are initiated once a strategy is implemented. Examining product choice as an example, if two specialist tour operators, each unaware of the other's intentions, decide to introduce one of two new types of holiday, either an activity or cultural holiday with a limited market, then it is likely that both will lose money if they market the same product simultaneously. However, if firm X is able to market the activity holiday first, then Y will introduce the cultural holiday and so both will gain. This is essentially what the Stackelberg model posits, where the size of output can be determined by the first-mover, leaving the remainder of the market to the rival. Firms can ensure that they can obtain an advantage as first movers by pre-emptive strategic moves, i.e. by influencing rivals' choices. This can be done only if the firm has a reputation for doing what it states it will do; lack of commitment to a strategy suggests to rivals that subsequent announcements or threats as to a particular course of action are empty. Entry deterrence is another key strategy because, if successful, it confers greater monopoly powers and the potential to increase profits. Therefore, the incumbent firm has to convince potential entrants that it would be unprofitable to enter. This can be demonstrated by reconsidering the pricing decision as to whether to charge a high or low price. If the potential rival decides to enter the high-priced market, the incumbent firm will gain more by maintaining a high price. However, should the incumbent firm threaten a price war which would bankrupt the new entrant, and/or expand capacity to exploit economies of scale to sustain the possible price war, the deterrence commitment is substantiated. Figure 4.11 illustrates the sequence of moves and the impact on pay-offs.

In Figure 4.11, Scenario 1, the first row indicates the pay-offs at both high and low prices to the potential new entrant, X, shown in bold, and incumbent Y, shown in italics in parentheses, should X decide to enter the industry, incurring sunk costs of €20 million, it being assumed that the market is shared equally. The second row, in italics, indicates the pay-offs to incumbent firm Y if it takes action in response to X's intended entry and successfully deters it without having to incur increased costs.

Suppose firm X predicts that Y will maintain a high price in order to gain the most it can of the current €50 million total market pay-off; then X will get €5 million (€25 million half-market share minus €20 million sunk costs) and Y will retain €25 million (top left cell of the matrix). However, should Y decide to engage in a price war, i.e. charge a low price, then assuming a total market pay-off of €30 million, X will suffer a loss of €5 million (€15 million minus €20 million sunk costs) and Y will receive €15 million (top right cell of the matrix). If Y successfully deters entry by X, then, at a high price, it gains all the pay-off available of €50 million (bottom left cell). Should Y threaten a price war and lower price to signify its commitment to deterring entry, then it gains the total market pay-off of €30 million (bottom right cell).

In Figure 4.11, Scenario 2, it is assumed that Y, in order to substantiate its threat to deter entry, is obliged to undertake a €15 million investment programme to increase its capacity. This may allow it to exploit economies of scale, so reducing unit costs, thereby enabling it to sustain a price war should it be necessary to charge a low price. If X nevertheless decides to enter the industry and Y elects to charge a high price but is unable to extend its market or increase profits, the result is shown in the top left cell, where X is in the same position as before but Y obtains a €10 million pay-off (€25 million minus the €15 million investment cost). If Y predicts that by charging a low price it will sell more and at least recoup the investment cost, the result will be as shown in the top right cell, where X loses €5 million and

Scenario 1

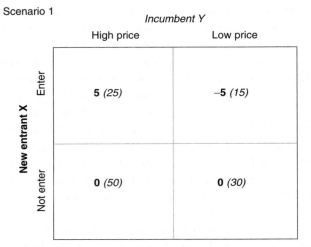

Assumptions:

- Entry costs of X equal €20 million; market shared equally
- High price: maintained by Y after entry of X; X also maintains high price
- Low price: price war by Y; X also charges low price

Scenario 2

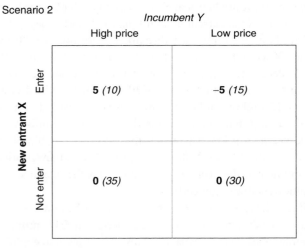

Assumptions:

- Incumbent Y invests €15 million to increase capacity; sales and profits unchanged at high price
- Incumbent Y increases sales and profits at low price to offset cost of increased capacity

Figure 4.11 Entry deterrence strategy

Y retains the same pay-off as before of €15 million. Of course, if Y has completely misjudged the market and fails to sell more, then a zero pay-off is suffered, i.e. the €15 million share is wiped out by the €15 million investment cost; this is not shown in Figure 4.11, Scenario 2. Clearly, if X decides not to enter the industry, Y suffers a reduced pay-off of €35 million (bottom left cell) if a high price is charged and the market is unchanged so that the €15 million investment cost reduces the total pay-off (€50 million minus €15 million). In the bottom right cell the pay-off for Y is shown as €30 million if the investment pays for itself, whereas if Y has misjudged the market the net pay-off will be €15 million.

The example in Figure 4.11 serves to show that many uncertainties arise when a would-be entrant attempts to break into a market dominated by an incumbent firm. The outcome for both entrant and incumbent largely depends on the extent to which deterrent threats are credible, the cost of gaining entry or deterring it and the changes in sales and profits with changes in prices. It is clear, however, that should the entrant and incumbent collude to charge a high price, the result is more beneficial to both than a price war.

Within tourism supply, as argued earlier, entry costs are often relatively low and the opportunities for pre-emptive investment are limited so that deterrence strategies are unlikely to be successful. Nevertheless, the example given here serves to illustrate one important aspect of tourism development in an international context. Investment supported or subsidized by the state can give an advantage to a country. For example, airline, ferry and train operations and, to an extent, the serviced accommodation sector are characterized by substantial economies of scale. By subsidizing these, more rapid expansion can be encouraged. This might prevent foreign firms from entering the market, thereby allowing the domestic sectors to charge higher prices and/or obtain larger sales. This notion is significant in the case of developing countries which choose tourism as the engine of economic development and is considered further in Chapter 7.

4.4.4 Reflection and synthesis

This relatively brief and simple exposition of game theory, using a few illustrative examples, indicates the direction in which economic analysis of industrial operations and market behaviour is moving. Attention has tended to concentrate on competitive interaction between firms because this reflects the current position in most tourism markets. Game theory can accommodate many other circumstances, for example collusion where firms produce a low output and sell for a high price, akin to the outcome under a monopolized market structure, or instances where greater uncertainty prevails and there is a considerable number of possible outcomes (multiple equilibria). In more complex cases, the analysis involves determining the probability of specific pay-offs, which accords much more closely with dynamic real-life situations.

Game theory can also accommodate different theories concerning the process by which firms reach a decision about the strategies which they adopt, put forward by the schools of thought discussed at the beginning of this chapter. A range of modes of behaviour is feasible and the Chicago School's neoclassical tenet of optimizing behaviour in competitive markets is a specific case within this range. The Behavioural School's hypothesis that decisions are made according to established norms and rules and the Evolutionary Economics School's view that behaviour evolves in accordance with past practice can also be incorporated within the range of possible modes of decision-making. The concept of updating can be used to describe consistent decision-making based on learning about other firms' behaviour (Martin, 1993) and economic psychology can help to explain the process of expectation formation. Principal–agent models shed light on behaviour by owners and managers respectively, permitting them to have different expectations and reactions. Hence, game theory is capable of analysing a variety of modes of decision-making and behaviour by firms in different sectors of tourism, in different countries and contexts.

4.5 Conclusions

This chapter has examined tourism supply using the SCP paradigm and game theory and has also endeavoured to examine in greater depth a number of economic issues in tourism. Recent developments in the industrial economics field were introduced in order to provide additional insights into the operation of tourism markets. Underpinning the discussion is acceptance of the need to perceive tourism markets as dynamic and frequently in disequilibrium.

The SCP paradigm can be used to describe tourism markets and to highlight key features as a necessary first step towards the analysis of market structures, behaviour and performance. The advantage of the SCP paradigm is that it gives an overview showing the interrelationship of many elements. These can be added to or adapted to reflect the circumstances in different markets. However, the approach is not well suited to examining the process of firms' decision-making and behaviour, for which recent developments in game theory are more appropriate.

Game theory has the ability to encompass insights from various schools of thought and to provide a dynamic analysis of tourism firms, industries and markets. Thus, it can incorporate the Austrian and Evolutionary schools' critiques of static equilibrium analysis of markets. It can also consider the role of technological change, which Schumpeter highlighted as of key importance in the growth process, by showing how firms use research and development and the associated changes in technology as a competitive strategy vis-à-vis other firms. Such strategies can alter the structure of the market over the long run. Uncertainty and asymmetries in information availability, for example between market incumbents and potential entrants, can also be accommodated. The theory acknowledges that, when making business decisions, most firms consider the likely responses of competitors. Inter-firm differences in the extent of knowledge, the degree of cooperation or non-cooperation can be taken into account and simple one-off strategies or sequential adjustment to different moves by competitors can be examined. The approach can indicate the credibility of signalled intentions and the probability that threat strategies will be successful. It has the merit of ensuring that the range of strategies available to firms and the associated range of possible outcomes are made explicit, showing how firms' conduct in terms of the strategies which they use and the structure of the market are determined simultaneously. The theory incorporates the most recent developments in industrial economics and, because of its widely encompassing framework and ability to analyse the processes of change, is clearly appropriate for examining tourism market structures, conduct and performance.

Equipped with all the above theoretical foundations, Chapter 5 now presents the profile and characteristics of the main tourism sectors from an industrial economics perspective.

5 THE ECONOMIC PROFILE AND CHARACTERISTICS OF THE TOURISM SECTORS

5.1 Introduction

Discussion of tourism supply has been conducted largely outside economics, with the result that coverage of economic issues is patchy and there is no coherent overview. Many texts directed at practitioners, particularly in the hospitality sector, touch on supply and associated issues in discussing financial structures, management, marketing, quality and training. At a very specific and applied level, studies of the planning, development, operation and performance of such enterprises as hotels, guesthouses, holiday villages, ski and timeshare resorts and theme parks have been undertaken. At a more general level, attempts have been made to model tourism locations, for example by geographers who aim to identify key factors determining growth or decline. With the exception of transport sectors, economists' interest in the industry has been peripheral, largely arising from research into industrial structure or organization, for instance multinational enterprises in the hotel and airline sectors (Dunning and McQueen, 1982a). Economists' tendency to neglect the service sector in general, and tourism in particular, creates difficulties because there are large gaps in the empirical evaluation of parts of the industry.

Having the above in mind and while some reference to specific tourism sectors has already been made in Chapter 4, it is necessary to identify and describe the nature of tourism supply in further detail. The approach adopted in this chapter is to outline the sectors that have been recognized in the tourism literature. Those which appear representative of the market structures in tourism are described concisely and the key factors indicating their competitive status are identified, such as the degree of concentration, entry and exit conditions, pricing strategies, profit levels, product differentiation, cost structures and capacity and interaction between firms.

The problem with any classifications of the supply components of tourism is how broad or narrow they should be. Such categories as transport and accommodation are very broad and benefit from disaggregation into submarkets with different structures and modes of operation. The classification of tourism markets adopted here largely follows the convention used in the tourism literature (for example Holloway and Taylor, 2006; Cooper et al., 2008), with some modifications to the nomenclature employed. In particular, section 5.2 deals with transport for the tourism sector, where emphasis is primarily given on air and cruising, although other sub-sectors such as rail and bus/coach are also discussed. The accommodation sector is treated as a tourism one par excellence in section 5.3, with a focus on its serviced part, although reference is also made to the concept of timesharing and to self-catering establishments with limited ancillary facilities. Section 5.4 discusses travel intermediaries, analysing predominantly the business of tour operations and travel agencies, whereas section 5.5 stresses the importance of natural and built attractions.

5.2 The transport sector

Given the wide range of transport modes, each possessing its own particular features and competitive characteristics and structure, transportation is best examined as a number of sub-markets. For example, the main commercial modes of air, cruise, bus/coach, ferry and rail suffer problems arising from indivisibility and associated fixed capacity and high fixed costs, periodicity and seasonality. Nonetheless, there are considerable differences in their respective market structures and conditions. Their relative importance also differs in terms of the numbers of passengers carried, revenues generated and the degree of substitution possible. For international travel, air far outstrips bus/coach, sea and rail. Air travel has shown extremely high rates of growth since the 1960s, fed by technological change, with potential for continued growth in the future. The appeal of cruising is partly related to the all-inclusive nature of the product, while car ferry operations have benefited from the introduction of the roll-on roll-off (ro-ro) vessels. Bus, coach and rail traffic have experienced relative decline in the face of the growth of private motoring.

The structures and competitive conditions of transport sub-markets are strongly influenced by the interrelationship between certain modes and the severity of the constraints imposed by regulation. There is also competition within modes, such as in the airline sector, where there is fierce competition on many short- and long-haul routes. Similar circumstances apply in the bus and coach market. In the ferry market, high volume routes tend to be subject to keen competition. There is a significant symbiotic relationship of modes. For instance, air travel depends on both bus/coach and rail for transfers to and from airports. Likewise sea crossings rely on road and rail transport for a large proportion of their traffic, while the cruise-liner market is now largely based on air travel to and from the point of departure of the vessel. The car hire sector is clearly dependent on airlines for much of its trade, often negotiating fly-drive packages as a means of increasing business. This complementarity, while it is an important feature of the transport market, does not suggest that there is not intense competition within and between modes. With respect to competition between modes, the opening of the Channel Tunnel linking the UK and France has demonstrated the fierce competition of ferry, rail and air, although the merging of some ferry operations raises the possibility of an oligopolistic market structure. In some European countries, such as France and Germany, air and rail travel are in competition on domestic routes, especially where rail systems have been updated for high-speed trains. Bus and coach travel and rail also compete, the former offering a low-cost alternative to the latter's shorter travel times.

Regulation of transport has occurred for two main reasons, safety and the preservation of a market for a national or state-owned or supported mode. Interest in the impact of economic regulation has centred more on deregulation and market liberalization with the aim of fostering greater competition. However, it has become apparent, as will be shown, that such an action often has unintended and opposite effects, as has been manifested in air and bus/coach transport. Moreover, the extent of regulation varies even within a specific mode. Again, in air transport, parts of the market are more strictly controlled than others, creating considerable variations in competitive situations.

5.2.1 Air transport

The operation and (de)regulation of air transport has been widely investigated in the context of both business and holiday travel (Levine, 1987; Doganis, 2005; Papatheodorou, 2008). Evidence of its conditions and structure illustrate a number of tourism and economic issues also arising in other transport markets, especially the economic environment created by regulation and deregulation and the resulting level of competition. An allied factor is the process of the privatization of a number of airlines and airports (Graham, 2008a). In the following sub-sections, the economics of airlines, airports and information technology are discussed, highlighting their implications for tourism.

a. Airlines

The cost structure of airlines is quite complex; operating costs and efficiency are related to technical factors with respect to the distance to be travelled, the size and type of aircraft and the payload, as well as the support services such as marketing and the reservations system. Moreover, capital costs are likely to be affected by whether aircraft are bought outright or leased, new or second-hand. Direct and indirect costs, such as airport charges and ground handling, are outside the control of the airline. Prima facie, airlines are high fixed cost enterprises with fixed capacity so that they need to attain high payloads to break even. Short-haul flights, involving fewer hours in the air and more take-offs and landings, are relatively more expensive from a unit cost perspective than long-haul carriage. Furthermore, aircraft are designed to fly specific distances as efficiently as possible so, for example, employing an Airbus A380 jet on European routes of up to 2,500 km is uneconomic. In general there are economies of scale stemming from technical efficiencies in operating large aircraft provided the routes operated on warrant it. In addition they depreciate less quickly (Doganis, 2002).

Uncertainty is a significant factor because of the rate of technical development of aircraft, for example with respect to aerodynamics and engine efficiency. The market served influences the economics of airline operations in a number of ways. For scheduled services, to maximize the payload and revenues, demand has to be ascertained in advance to discover its composition with regard to the numbers of business and economy customers, for example. The approach being adopted by airlines is termed yield or revenue management, which is essentially related to third degree price discrimination (analysed in Chapter 4), although in its most developed form, it is akin to perfectly discriminating monopoly pricing. Pricing policy also needs to take account of the route, destination and stage stops, the pressure of competition and longer-term level of demand. Chartered services reduce market uncertainty because it is possible to ascertain the payload well in advance. However, the distinction between these two main types of services has become more difficult to discern, as charter airlines have moved into the scheduled market and vice versa. This has occurred because travel on what were holiday charter routes has become so regular as to warrant scheduled services. Conversely, some under-used scheduled services, perhaps operated as a social service when airlines were state-owned, may be served more irregularly by special charters if not set within the framework of Public Service Obligations (PSO) schemes.

Building on the theory of contestable markets discussed in Chapter 4, the advent of deregulation and privatization has dramatically changed airline operations and has had some interesting economic consequences (Papatheodorou, 2002a). The US experience, where deregulation occurred in 1978, set the precedent for the stepwise air transport liberalization in Europe, where the process was completed in 1997 with the creation of the European Common Aviation Area (ECAA), which includes the European Union countries and a number of neighbouring states (Zenelis and Papatheodorou, 2008). In particular, the liberalization process in the continent officially started with the First Package of measures introduced in 1987.

This created a framework granting the right to airlines to develop their pricing policy freely up to a certain degree. In 1990 the Second Package emerged with only slight amendments to the previous one. The introduction of the Third Package in 1993 (and its full application by April 1997) resulted in three major changes. First, European carriers were granted full traffic rights within ECAA, including cabotage, i.e. the right to transport passengers and/or cargo within two points of another country. For example, British Airways now has the right to carry passengers between Paris and Marseille in France. Second, European airlines were given total freedom to set their own fare levels (unless these were proven to raise serious anti-competitive concerns) and, third, procedures regarding operational licences and flying certificates were harmonized across Europe. Capitalizing on the liberalization of its internal aviation market and the 2006 decision of the European Council, the European Union and its Member States are currently in the process of renegotiating existing bilateral agreements with third countries to align them with the Community Law (Papatheodorou, 2008). Most importantly, in 2008 the EU and the USA established an Open Aviation Area, which is forecast to generate additional traffic of 25 million passengers in the first five years of its implementation, leading to €12 billion in benefits for the consumers and the creation of 80,000 new jobs (EC, 2007a). Other countries in the world have also taken significant steps to liberalize their air transport market domestically, at a bilateral but also at a multilateral level. An example of the latter is the effort undertaken by the African countries to create open skies among them based on the principles highlighted in the 1999 Yamoussoukro Decision (Arvanitis and Zenelis, 2008).

As a consequence of the underlying competitive dynamics, rivalry in the air transport sector has certainly intensified since new carriers moved into the market, entry having been made easier. This tends to suggest that entry costs, particularly capital ones, are lower than casual observation indicates. Chartering, leasing or buying aircraft second-hand facilitates ease of entry, as does negotiating reservation, ground handling services and maintenance agreements with specialist contract firms or established and larger airlines. In the short run the increased competition has brought fares down and extended the market, also forcing airlines to be more sensitive to their customers' needs. The reduction of fares as a competitive strategy has been particularly encapsulated by the low-cost carriers (LCC), a new breed of scheduled carriers which appeared in the 1990s (building on Southwest Airlines' original success in the 1970s in the USA). They widely consolidated their position at a global level since the beginning of the new millennium: Ryanair, easyJet and Air Berlin are the best known LCC in Europe; Southwest and Jet Blue in the USA; Air Asia in South-east Asia; Jet Star and Virgin Blue in Australia.

Although it is difficult to provide a specific definition for the LCC, their business model has a number of characteristics that clearly distinguish them from the traditional scheduled carriers (such as American Airlines and Air France–KLM). Lawton (2002) tried to identify these operational differences, arguing that a typical LCC offers a single class product (based on a dense seat configuration) with no complementary in-flight catering and/or entertainment services and relies almost exclusively on direct marketing (through the Internet and call-centres) for its distribution to avoid the cost of travel agent intermediation. The LCC fly point-to-point services bypassing busy and expensive hubs. Moreover, they usually operate from/to regional, less congested airports, which offer faster turnaround times (landing and take-off) and charge lower fees. Furthermore, the development and utilization of e-ticketing procedures was another time and monetary cost-saving achievement that made LCC popular among the public. With respect to market positioning, the LCC mainly introduced new destinations, but also engaged in serving city pairs already included in the flight programmes of existing traditional scheduled and charter airlines. Nonetheless, by exercising aggressive marketing policies they managed to enlarge their market share (Zenelis and Papatheodorou, 2008).

To address the competitive threat of LCC, traditional scheduled carriers tried to incorporate several operational 'no-frills' tactics without (substantially at least) degrading the service quality of their product. In some cases, traditional airlines introduced LCC affiliates to avoid brand dilution: typical examples include Go (introduced by British Airways and later sold to 3i and then to easyJet), Buzz (introduced by KLM and later sold to Ryanair) and Jet Star (an affiliate of Qantas). Likewise, charter carriers (which had already a low-cost structure) added elements of flexibility in their model (such as the sale of seat-only packages) to compete more effectively with the LCC. The continuous effort of all types of airlines to minimize operational costs transforms their operational policies and characteristics; as a result, the distinction between the various airline business models has become increasingly blurred at least in the short- and medium-haul market. The only exception to this trend has been the emergence of all-business class airlines such as Eos and Silverjet in the second half of the 2000s; however, most of these carriers survived for a limited period of time only and eventually went bankrupt as a consequence of the economic recession in 2008.

The emergence of strategic airline alliances is another important development in the post-deregulation period. In particular, an alliance involves the cooperation of two or more companies aiming to achieve jointly a competitive position and enhance their profitability by sharing various resources such as brand assets, market accessibility, the use of advance technology and the dissemination of know-how in important operational and other areas (UNWTO, 2002). Alliances are likely to be the preferred model for collaboration when organic growth is difficult (e.g. due to the limited size of home markets, the existence of infrastructure constraints and the level of marketing and other costs to develop operations abroad) and growth through mergers and acquisitions faces regulatory and other impediments. Strategic allies remain independent from each other but collaborate to extend their destination coverage and gain marketing advantages through various agreements such as code-sharing, which allows a carrier to introduce its code in a flight operated by a partner, irrespectively of whether the former serves the city pair in question or not: as a result, airlines can extend their commercial presence in markets that they do not actually service. In this way, strategic alliances enable the creation of a 'seamless network' experience for participating passengers at a global level as flight schedules may become more frequent and coordinated to allow smooth interline connections in long-haul travel. Consumer preference of allied over non-allied carriers may then lead to an improvement of capacity utilization of the former and the reduction of their unit costs. Further savings may be achieved by means of joint marketing campaigns and the common use of airport facilities (e.g. business lounges). Moreover, the allied carriers may exercise their negotiation (i.e. oligopsonistic) power to reduce the cost of supplies such as jet fuel and spare aircraft parts (Iatrou and Oretti, 2007).

At present, the world airline market is dominated by three strategic alliances, namely oneworld (led by British Airways and American Airlines), Star Alliance (led by Lufthansa and United Airlines) and SkyTeam (led by Air France–KLM and Delta Airlines). These alliances jointly account for 60 per cent of the world traffic (Iatrou and Tsitsiragou, 2008). To capitalize further on their network advantages, the participating carriers also collaborate in their frequent flyer programmes (FFP). These are loyalty schemes, which reward frequently flying passengers with miles that can be subsequently redeemed mainly in the form of free flights, class upgrades and services offered by other business partners such as hotels and car rental companies. In exchange, the carriers obtain access to a wide passenger database with unique marketing advantages; additionally, they effectively manage to raise the switching cost to other carriers (as in that case the passenger would lose the opportunity to accrue miles) and are able to command a price premium for their services. As expected, the intrinsic value of an FFP depends on network breadth; hence, reciprocal FFP agreements in the context of strategic alliances may offer a major advantage to the customer who seeks global coverage (Papatheodorou and Iatrou, 2008): for example, a passenger who is a member of Lufthansa's Miles & More programme may accrue miles by flying with other Star Alliance members such as United Airlines (with extensive presence in the USA), Thai and All Nippon Airways (with large networks in Asia). For this very reason, the strategic use of FFP by strong carriers may raise serious competition challenges as it essentially forecloses the market to new potential entrants with limited networks and/or induces existing smaller airlines to exit or merge, raising overall concentration in the market (Papatheodorou, 2006c).

In fact, market instability and concerns over the possibility of dominant position abuse are some of the negative effects of deregulation and liberalization in the airline industry. Legacy carriers of the past such as Pan Am in the USA, Swissair and Sabena in Europe and Ansett in Australia were forced out of business due to severe financial problems – a number of charter and low-cost carriers also suffered a similar fate. The benefits to consumers may be short-lived as the number of airlines will almost certainly decrease in the long run through mergers and reciprocal agreements and the failure of smaller, newer entrants. This concentration and consolidation could possibly lead to higher fares as the larger companies take control of key routes and gain the market power to dictate price levels and limit choice; for example, the Office of Fair Trading in the UK imposed a fine of £121.5 million on British Airways for fixing prices in long-haul passenger fuel surcharges (Office of Fair Trading, 2007). The development of the hub and spoke pattern of air travel where long-haul flights are concentrated on major hub airports gives opportunities for smaller feeder airlines to create a spoke market niche, although the larger airlines may determine the terms on which such routes will be served. There are additional concerns where airlines cut costs in order to remain competitive as the quality of the service may be lowered, less attention being paid to maintenance and safety margins (Papatheodorou and Platis, 2007).

The evidence from this brief description of what is the best researched tourism sector is that although a domestic monopoly or oligopoly structure has been common in the past, with a single state-supported airline or a small number of competing airlines, deregulation and liberalization have made some markets competitive in the short run. In the international market some routes are competitive, being served by many carriers. Most of the others are served by at least two carriers, indicating an oligopolistic market, although a few routes are served by a single carrier, which may be tempted to exercise monopoly powers. This does not undermine the ability of the theory of the firm to explain the market structure; it signifies a need to divide the sector into sub-markets and to consider each separately. The sector, in common with the others reviewed, is in a fluid state, reflecting deregulation and the changing demand for foreign air travel.

b. Airports

Airlines and airports are structurally interdependent: the former cannot operate outside an airport, while the latter exist because the carriers wish to fly there. In fact, airlines and concessionaires (primarily related to rented airport property and shops/restaurants within a terminal) are two important airport customer groups; however, what really matters is the passenger. Airlines choose airports based on various criteria such as their catchment area and potential demand; capacity constraints, environmental restrictions and competition from other airlines; network compatibility and ease of transfer connections; level of airport and other fees and availability of discounts; range and quality of passenger and maintenance facilities and marketing support (Graham, 2008b). Similarly, passengers' choice of airports is affected by both airline and airport related factors. In terms of the former, what seems to matter most is the number and features of airlines serving a specific airport, i.e. network of destinations covered; frequency of services; availability and timings of schedules; image and reliability of airline; the participation of a carrier in a strategic alliance and its policy regarding frequent flyer programmes, etc. With respect to the latter, the cost and ease of surface access to the airport is important in conjunction with the car park pricing, the range and quality of commercial facilities and the overall image of the airport (Graham, 2008b).

The demand for air travel and tourism has a great impact on the nature of the airline services and the airport aeronautical revenues (e.g. landing fees, aircraft parking). It also largely determines the prosperity of airport shops and the associated non-aeronautical airport income. The International Air Transport Association (IATA) often blames the airport operators for profiteering by charging fees not related to infrastructural costs (Field and Pilling, 2003). This may be true; however, the airlines were the first to dissociate cost and prices by applying sophisticated revenue management techniques: not all passengers pay the same price for their air ticket even for the same level of service. In this context, sunk costs play a predominant role. Empty seats are lost forever; therefore, effective revenue management is essential to the airline profitability. For example, late applicants for a particular flight may be able to secure a discounted fare if there is spare capacity. Similarly, the airport should be compensated for exercising its risky and irreversible option of investment and the airline should pay for retaining its flexibility to move elsewhere (Papatheodorou, 2003b).

The issue is, of course, whether airlines can actually relocate to another airport. This might be difficult in congested areas like Greater London. Ironically, these supply constraints might benefit the incumbent airlines that possess valuable slots in busy airports. The allocation of slots at all of Europe's most congested airports is made by virtue of Regulation (EC) 793/2004, which amended Council Regulation (EEC) 95/93. The current Regulation introduced new and significant provisions, particularly on the definition of a slot. More specifically, Regulation 95/93 defined a slot as 'the scheduled time of arrival or departure', thus identifying it with a runway movement; in this way, it failed to acknowledge that at some airports, the runway is not the only binding restriction. In addition to runways, capacity limiting factors that need to be taken into consideration at an airport, when coordinating slots, include airport: capacity (ramps, taxiways, weather, gates, landside limits, such as terminals and road access); airspace capacity (airspace design, controller workload); demand (peak pricing, hub and spoke networks); and environmental (community noise, emissions) issues. New Regulation 793/2004 changes the slot definition, referring to a permission to use the full range of airport infrastructure necessary for a flight (Kostis, Papatheodorou and Parthenis, 2008). Slot allocation has traditionally relied on 'grandfather rights' on a 'use-it-or-lose-it' basis: in other words, the airlines that used an airport's slots in the past are allowed to do so both at present and indefinitely in the future provided that they actively operate at least a certain percentage of these slots every year (80 per cent in the European Union). Such a rule may potentially constitute a significant barrier to entry for new airlines and raises anti-competitive concerns. For this reason, the European Commission and air transport experts from Europe, and especially the UK, have suggested that slots should be better allocated through transparent commercial mechanisms, including primary trading (e.g. auctions) and secondary trading mechanisms instead of through exclusively administrative rules (Boyfield et al., 2003; Civil Aviation Authority, 2006; Mott MacDonald, 2006; Airports Council International, 2007).

Due to the above mentioned supply constraints and the existence of natural monopoly elements in their operations, a large number of airports around the world face economic regulation based on principles highlighted in Chapter 4 (e.g. the introduction of price caps). This is especially the case for private airports (such as London Heathrow, owned by BAA) and those operating under a private–public partnership regime (such as Athens International Airport in Greece). But, would an unregulated airport necessarily abuse its power against the airlines? The answer essentially depends on the opportunity cost of the airport operators. Higher airport fees might increase aeronautical revenue; nonetheless, in their effort to pass along these charges to the passengers, the airlines may harm the airport revenues from non-aeronautical services: higher air fares might induce people to travel less and/or reduce their spending in airport shops. In the end, the final outcome depends on market sensitivity and relative elasticities. The current trend towards airport commercialization (Graham, 2008b) will probably reduce the relative importance of aeronautical revenues and consequently the incentive for an airport operator to overcharge the airlines. A switch from a single till (where all airport activities are economically regulated) to a dual till (where only the pricing structure of aeronautical activities is regulated) in congested airports such as London Heathrow could lead to an effective redistribution of scarcity rents (arising from the inadequacy of slots) from the airlines to the airports (CAA, 2000). This could be harmful for incumbent airlines, but the overall credibility of an overcharge threat depends on all parts of the airport business. In this context, a dual till regulatory system may not necessarily result in a substantial loss for the airlines if it provides an incentive for increased airport commercialization and investment in service quality. A fortiori, a better understanding of the symbiotic relationship between airlines and airports could reduce further the regulatory constraints (Papatheodorou, 2003b).

To highlight this point, the corporate strategy of low-cost airlines, and particularly of Ryanair, may be considered. This Irish carrier flies to small, secondary unregulated airports in Europe that in some cases offer seemingly neocolonial contractual terms to attract the carrier. The rationale behind such preferential treatment is clear. Low sustainable fares encourage leisure tourists to visit these areas; moreover, the enhanced accessibility by air creates new business opportunities (Papatheodorou and Lei, 2006). Both factors may contribute substantially to regional economic development, employment and income creation due to the existence of multiplier effects, as discussed in Chapter 6. Consequently, the market potential of the airport increases due to a dynamic and prosperous catchment area. Subsequent contractual renegotiation with the airline can be made on better terms. Such agreements have been recently blamed, however, for distorting competition. The Strasbourg Administrative Court ordered the end of the commercial partnership between the city airport and Ryanair in 2003 following the lawsuit of Britair, the subsidiary of Air France (E-Tid, 2003). Not surprisingly, Ryanair had strong support from the Strasbourg authorities in this legal battle given the phenomenal recent increase of tourist flows: 'from October 2002 to July 2003, Ryanair carried more than 125,000 passengers between London and Strasbourg, i.e. five times more than Air France over a comparable period in previous years' (Strasbourg and Bas-Rhin Chamber of Commerce and Industry, 2003). A similar case in Charleroi Airport near Brussels generated a lengthy legal process, which started in 2002 and eventually ended in December 2008 with the verdict of the Court of First Instance annulling the 2004 decision of the EC that Ryanair illegally received state aid through its contractual arrangement with Charleroi Airport (Oxera, 2009).

Two issues emerge in this context. The first is whether such partnerships constitute state subsidies (French, Belgian) to a carrier, which, ironically perhaps in this case, is registered in another country (Ireland). The second issue is with respect to the introduction of 'most favoured customer' contractual clauses aiming at foreclosing market entry or inducing exit. By succeeding in the Charleroi case, Ryanair established a clear legal precedent for future agreements between airports and airlines. This creates a benchmark primarily for state-owned airports as they are able 'to compete on a level playing field with privately-owned airports all over Europe' (Travel Mole, 2003a). It seems, therefore, that the virtues of the symbiotic relationship between airports and the airlines are maximized when the former internalize the benefits of enhanced air transport accessibility, i.e. the wider impact on regional economic development. It may be argued, however, that state-aid rules regarding airport charges may prohibit taking into account such benefits for reasons of transparency and the need to replicate the private investment decision-making process. Such a logic, nonetheless, is defeated when considering the European Common Aviation Area Public Service Obligations scheme: airlines may be granted state subsidies (following a tender process) to initiate or retain services in peripheral areas. In this context, the EC developed specific guidelines on the financing of airports and start-up aid to clarify the whole issue (EC, 2005).

Still, and from a tourism destination perspective, policy-makers in cities or regions, which compete against each other to ensure support by major or low-cost carriers, should stay alert. In fact, this intensification of regional competition may result in a zero sum game, where the sole beneficiaries are the airline companies. The problem is even more serious when the likelihood of flow diversion or even complete airport abandonment is considered. Given the sunk nature of airport facilities, the abandoned areas are in a worse position than those suffering from 'agglomeration shadows', i.e. areas denied substantial autonomous development as a result of lying proximately (but not in the heart) of large urban centres (Papatheodorou and Arvanitis, 2009); whereas the shadowed places (also known as 'inner peripheries') remain greenfield free of developments, the abandoned areas lock in people and infrastructure with potentially detrimental financial and social results.

c. Information communication technologies

Information communication technologies (ICT) have revolutionized tourism since the mid-1990s (Sheldon, 1997, 2006). Their impact at a firm level has been both on operations and practices of strategic design, while their contribution into corporate rivalry has also proved significant. In the following, the three major innovations generated by ICT are presented, namely developments in Computerized Reservation Systems (CRSs), the Internet and dynamic packaging, along with their effects on tourism.

COMPUTER RESERVATION SYSTEMS

In particular, the CRSs were one of the first applications of ICT in tourism designed and developed to provide real time information on the availability and the cost of air travel services, giving also the advantage of instantaneous access to inventory and real time booking. The databases available to CRSs enable the companies to determine fares and availability in their services and products, to issue tickets, boarding passes and vouchers, and more generally to retrieve or provide valid travel information regardless of its complexity (Sigala, 2004). Accordingly, CRSs generate revenue by receiving a booking fee from the airlines and a subscription fee from the travel agents. On the other hand, they may provide a signing bonus and/or an incentive payment to the travel agents (EC, 2007b). The foundations of the CRSs date back to 1940 when American Airlines were using manual methods for their reservations such as 'Lady Susan' and 'Tiffany Card' and the subsequent introduction of the first computerized system in 1953 by IBM named SABRE (Semi-Automated Business Research Environment), which constituted the first CRS company ever. Following the example and success of American Airlines, many carriers invested in devising their own CRS to counter the increased competition from the companies using specific CRSs, as well as to further explore the advances of technology and the benefits of entering a new dynamically growing market.

The carriers' competition was quickly transferred to the level of service promotion and distribution. The primary benefits for airlines promoting their own CRS were twofold. The first, and most obvious, was the profit from the CRS service itself as sold to travel agencies and other carriers; the second was the promotion of their flight services and the exploitation of an oligopsonistic advantage gained as suppliers with increased bargaining power; this could further enhance their revenue from sales. Following the advancements in North America, the first CRSs in the EU were developed in the early 1980s, with Galileo and Amadeus being the best known. However, the increased competition with the American systems led all CRS vendors to horizontal integration strategies through mergers and acquisitions. That was partly due to the pressure for cost reduction by travel agents who could not work efficiently with a large number of systems. Eventually and to facilitate the job of travel agents, CRS entered operations also the hospitality industry with the distinctive term GDSs (Global Distribution Systems). Nonetheless, with the majority of CRSs being in a strong position against carriers, regulators and authorities in the 1980s were worried about the potential of market abuse by their vendors.

In fact, since their early applications, CRSs have suffered from bias. System vendors were designing them in such a way as to favour and promote their own flights. In addition, escalating fees or accessibility restrictions were the most common methods for carriers to exclude their competitors. More specifically, according to the US Department of Transportation (2004), this market power was arising as a result of the very high share of travel agents' revenue generated from air ticket sales in conjunction with the fact that almost all agents were using at least one CRS. In addition, the vendors of the latter forced air carriers to collaborate with every CRS, thus increasing in this way their distribution costs. In many cases, travel agents were choosing the CRS of the carrier dominating the region of their main commercial interest because of the level of service quality received. This was related to the practice of the CRS vendors of building in display bias in favour of their air services and/or those of other carriers. In addition, the structure of CRS favoured the regular update of the information received from its vending carrier as all changes were taking place on the same software platform (architectural bias). Finally, parent carriers increased booking fees to exclude small and new carriers from the market while they motivated agents to maintain exclusively their own CRS through special offers, agreements and premiums.

The increased power of the CRS vendors led the USA in 1984 to take measures to restore the smooth operation of the market (Buhalis, 2006). In particular, on a short-run basis, the US authorities forbade the immediate abuse of the parent carriers' market power. Other measures made participation of carriers in competing CRSs mandatory based on terms of equal fees and bonuses. The parity clause (which essentially obliged participating carriers to buy the same level of CRS service from all systems) was forbidden, while the authorities banned the limits that carriers set on contracts with travel agents, such as time limits and exclusiveness. In 1989, market pressures led the EU to issue a Code of CRS Conduct by introducing Council Regulation 2299/89 based on Article 81 of the Treaty of Rome (regarding competition). In line with the regulatory framework imposed in the USA, the Code of Conduct focused on the obligatory unbiased function of the CRSs, with emphasis on the fair provision and presentation of the information to all systems by the parent carriers. Moreover, the Code of Conduct forced CRS providers to inform and allow access to all carriers regarding system improvements and innovations. The measures proved effective while adjustments were discussed during the 1990s for a more efficient application of the rules set.

Interestingly, however, and similarly to the case of airlines, the gales of deregulation started blowing in the CRS sector too. In July 2004, the US Department of Transport deregulated the CRS market and Canada followed suit shortly afterwards. The rationale behind this decision was essentially based on two pillars. First, CRS providers were no longer in control of air carriers in the US market. Furthermore, the emergence of alternative, relatively costless, distribution channels like the Internet reduced significantly the exclusive use of the CRSs. The existing market regulation altered the terms of competition between the CRSs and the Internet, therefore, and to comply with fair trade and market rules policy, authorities decided to proceed with deregulation. The effects of the new liberalized regime became evident in a short period of time. Bargaining between carriers and CRS vendors resulted in an important decrease in booking fees (between 20 per cent and 30 per cent). In exchange, carriers guaranteed full access to the CRS in their database. Travel agents, on the other hand, agreed with CRS vendors to access their systems in exchange for a booking fee. At the same time, the reduction of financial premiums and subsidies reduced the profit margin of the travel agents. This is expected to lead the small and medium travel agents at least to raise their service fee. On the other hand, the development and exploitation of new alternative technologies by the large travel agents may lead to a significant reduction in the profitability of the CRSs.

Nonetheless, the CRS deregulation in the North American aviation market favoured the domestic carriers against the competing European ones. For this reason, the EU also started discussing measures to deregulate the market in the mid-2000s, given that approximately 75 per cent of the CRSs were not owned by air carriers while for the rest of them only a minority share was in the possession of airlines. Moreover, the increasing use of the Internet services limited bookings through CRSs and agents to 60 per cent (Buhalis, 2006). Eventually in 2009, the EU introduced Regulation 80/2009 revoking the previous one, acknowledging that 'technological and market developments allow for a substantial simplification of the legislative framework by giving more flexibility to system vendors and air carriers to negotiate booking fees and fare content' (Official Journal of the European Union, 2009: 1), although 'in the present market context it remains necessary ... to maintain certain provisions on CRSs, insofar as they contain transport products, in order to prevent abuse of competition and to ensure the supply of neutral information to consumers' (Official Journal of the European Union, 2009: 1).

INTERNET AND DYNAMIC PACKAGING

The Internet was originally developed to meet the needs of the US Armed Forces for processing information in a more efficient way. More precisely, it concerned a network of computers able to exchange information among its members. It was not until the 1990s, with the invention of the World Wide Web, that the Internet grew popular in the rest of the world. As a result, the spread of information to consumers brought a revolution to the tourism industry. For the first time, operators, carriers and other service providers found themselves in the awkward position of trying to justify and defend their prices to clients, while the latter were in a better position to evaluate their purchase since tourism could now be virtually experienced. In addition, empowered consumers were able to interact with each other, sharing travel experiences.

At a business level, the Internet substantially enhanced communications and operations with commercial partners and other organizations.

The emergence of new integrated management information systems has increased efficiency, reduced costs, improved flexibility and interactivity and set the grounds for the beginning of new and modern e-business type organizations. Advanced online yield management systems that adjust pricing to demand changes have significantly improved profitability in all tourism sectors. The most popular application of the Internet is Web 2.0, i.e. the new generation of web development designed to facilitate businesses worldwide adopting and expanding their interaction and activities. A multitude of services and applications to web-based communities are now at firms' disposal for commercial exploration. New developments introduced in Web 2.0 services include blogs (online journals distributed to other sites or users through RSS – see below); collective intelligence (systems including databases of groups especially useful for decision-making); mash-ups (programmes drawing data from various sites for the formation of a new service); peer-to-peer networking (networking through internet or closed networking for sharing files); podcasts (multimedia distribution through aggregators such as iTunes); RSS (referring to the possibility of subscribing to various services online); social networking (systems such as MySpace and Facebook that allow users to share personal data); web services (which ease the automatic communication and update of various systems on information and transactions); and wikis (which allow their users to contribute to and share knowledge with a published document or discussion).

Having the above in mind, a 2007 survey by McKinsey aimed to highlight and analyse experiences accumulated by business executives from their applications of Internet services. Interestingly, more than 75 per cent of respondents intend to increase or at least maintain their Internet interaction, encouraging further exploitation of social networking and Web services. While the new trends are in favour of a major growth of blogs and online social networking, companies seem to be more interested in automation and networking. Moreover, 42 per cent of the respondents aim to expand their companies' capabilities towards better understanding and use of Internet services. New firms originating from developing countries like India seem to be occasionally at the forefront of new technologies and trends, while executives from China and Latin America state that, in contrast with their cautious past spending on these services, they plan to further explore and invest in Internet services (McKinsey, 2007).

In the context of tourism, the Internet has introduced significant changes, including the broad use of websites and advanced comprehensive systems by the majority of service providers and tourism destination authorities; the increased investment in e-commerce facilities from tourism related firms, with online booking being of primary importance; and the extended use of the Internet by consumers to retrieve information and engage in online commercial transactions. On these grounds, dynamic packaging seems to play a very important role. The term dynamic, also known as Do-It-Yourself (DIY), packaging refers to the holiday package the components of which are assembled online by the consumer without the interference of a tour operator or a travel agent, based on real time inventory and pricing structures. Its popularity stems mainly from two factors: first, from its inherent flexibility, since the consumer is able to choose among various combinations of services and companies to create their package; and second, from the confidence and reliability received by the customer from their personal interaction with the prospective service providers, possibly by achieving better prices while saving the costs of intermediation. Illustratively, towards the end of 2006, the turnover of the independent holidaymaking sector was over 35 per cent larger compared to the traditional package holiday (EC, 2007c).

Commercial practices based on dynamic packaging appealed to airlines and especially low-cost carriers, which were actively seeking complementary sources of ancillary revenue and further ways to exploit the Internet's potential. Illustratively, a 2003 IT survey revealed that 32 per cent of the carriers under review offered excursions, while 33 per cent and 24 per cent offered hotel booking and car hire respectively (Airline Business, 2003) – percentages are now expected to be much higher as new services are constantly added. For example, Ryanair has capitalized on the popularity of its website to offer gift vouchers and Bingo online! Still, dynamic packaging raises important issues that need to be seriously considered by both consumers and airlines. The latter should not invest too many of their marketing resources in promoting solely their ancillary services since the core product (i.e. flying) is their effective distinguishing factor from the other travel service providers. In addition, selling cheaper to profit from complementary products is risky, as there is always the possibility that consumers could take advantage of the reduced flight price and purchase the remaining elements of their package from other companies. A final but not unimportant argument against the increased attention paid by carriers to dynamic packaging is the potential cannibalization of partnerships and established relationships with the travel agents.

From the passenger's point of view, dynamic packaging entails risks because of inadequate protection from hackers and malware capable of stealing personal data. Furthermore, there is always the danger of dealing with seemingly reliable companies which may eventually go bankrupt. Certain countries implement detailed schemes of consumer protection to avoid having travellers left stranded in remote destinations should their travel company collapse. The UK CAA operates the Air Travel Organizers' Licensing (ATOL) scheme, the regulations of which 'require companies selling air travel to hold ATOL licenses unless they operate the flights themselves, or provide an airline ticket or ATOL holder's confirmation straight away' (CAA, 2005: A1). Because of ATOL protection, more than 200,000 travellers were repatriated by the end of their vacation and more than one million were refunded between 1986 and 2005 (CAA, 2005), whereas in 2008, over 20,000 tourists received back their advance payments and more than 1,600 were given the opportunity to complete their vacation in spite of the financial bankruptcy of their operator (CAA, 2008). The ATOL scheme was previously based on bonding provisions but since 2008 these have been largely replaced by the ATOL Protection Contribution (APC) plan, according to which the Air Travel Trust of the CAA can 'collect contributions from ATOL holders, based on £1 for every passenger booked under an ATOL' (CAA, 2008: 6). The plan is financially supported by a credit facility with Barclays Bank, a guarantee of the UK government and an insurance policy with AIG UK (CAA, 2008). Nonetheless, and in contrast to traditional holiday packages sold, for example, by tour operators, dynamic packaging is not covered by ATOL financial protection primarily because scheduled carriers are not part of the plan. Thus, DIY travellers could in theory rectify this legal loophole by buying appropriate travel insurance online in conjunction with the other elements of their package. However, the majority of passengers seem to be unaware of the risks involved in dynamic packaging, raising concerns over information efficiency in the market (CAA, 2005).

5.2.2 Ferries and cruising

The ferry market clearly serves the private road, coach and rail transport modes and has grown strongly with the increase in surface holiday travel, particularly in Europe. New forms of conveyance, such as hovercraft, hydrofoil and catamaran craft, have reduced travel times and extended the market, increasing both intra- and inter-mode competition. Short routes involve high entry and operating costs, somewhat resembling short-haul air travel. There are economies of scale in increased vessel size and speed but these can be offset by the danger of increasing loading and unloading turn-round times and excess capacity in off-peak and off-season periods, akin to those in the air, bus, coach and rail market, giving rise to a complex fare structure and travel conditions. Companies rely heavily on on-board spending to bolster revenue. On longer-haul routes the problems of periodicity and seasonality are more acute and where remote island communities are served, state subsidies are required.

As far as the cruising sector is concerned, with an annual average growth of 7.4 per cent since 1990, this is one of the most interesting tourism activities, currently enjoyed by more than 12.5 million passengers worldwide, 80% of which is located in North America (Cruise Lines International Association, 2008). Still, from an academic point of view the cruise sector remains relatively unexplored (Dowling, 2006a). This is mainly due to its special characteristic, i.e. the explicit combination of transportation and accommodation services, which are usually studied separately in the context of transportation economics and hospitality management respectively. Nevertheless, the cruise sector possesses very important features from an industrial organization perspective, mainly because of its high level of market concentration: four large groups dominate the market worldwide, i.e. Carnival, POPC, RCCL and the Star Group, while in 2002 Carnival and POPC were allowed to merge, creating a dual listed company (TravelMole, 2003b). Jointly the four groups controlled about 85 per cent of North American capacity in terms of lower berths and about 60 per cent of ships in 2007 (Cruise Lines International Association, 2008). This may raise competitive concerns and is in sharp contrast with the market fragmentation still prevailing in the air transport industry.

As might be expected, the cruise industry sector has certain similarities with the aviation industry and the hotel accommodation sector (UNWTO, 2003). More specifically, cruising is characterized by very high fixed costs and the presence of significant economies of scale. Therefore, it is reasonable for companies to build large cruise ships, so as to reduce unit costs and be able to price their services competitively. Illustratively, 263,936 lower berths were located in 185 ships in North America, equivalent to an average of 1,426 berths per ship (Cruise Lines International Association, 2008). This, in turn, brings to the surface the problem of managing large numbers of passengers. On the one hand, an overcrowded ship might not be appealing, at least to certain types of consumers, whereas, on the other, the opportunity cost of an unsold berth is larger compared to an unsold seat in an aircraft or an unsold room in a hotel due to the usually long duration of a cruise (about 70 per cent of which last for at least one week) and the loss of the possibility of generating revenues by selling ancillary services on board. Thus, cruise liners need to develop special techniques for pricing and revenue management 'to attain maximum revenue while ensuring that the ship sails with 100 percent capacity' (Competition Commission, 2002: 147).

Concerning scale economies, these can be achieved through the increase in the size of the purchased vessels supplied by the shipping company or by increasing the number of ships in ownership. The large size of a cruise liner allows it to exploit a degree of oligopsonistic power over its suppliers regarding the purchase of fixed and variable cost elements, such as food and fuel. Moreover, cruise packages often include, other than transportation, accommodation and catering services, excursions and activities at their ports-of-call, where a cruise ship may stop en route to another port, often giving passengers the opportunity to disembark for some time. The bargaining power of the cruise liners along with the opportunities for synergies with other service providers in tourism are of major interest since they create profitable opportunities. It seems plausible to assume that global cooperation among all tourism companies would contribute significantly to efficiency improvements, with the desired reduction in average costs, while policy-makers should protect the market from power abuse.

The heterogeneity characterizing the cruise industry motivates differentiation both horizontally and vertically. In particular, corporate rivalry can take place at four major levels, i.e. the booking, the time of the cruise, the itinerary and of course the ship itself (Papatheodorou, 2006d). With respect to the ship, the emphasis is put on its size and age, its registration nationality and its personnel; concerning facilities, the number and quality of berths along with entertainment and catering services are significant. As far as the itinerary is concerned, the wider geographical area where the cruise takes place in conjunction with the homeport (i.e. the port where a cruise starts or ends) characteristics and the number of ports-of-call can vary greatly. The time of the booking is an extremely important factor also: low–high seasons along with scheduling of specific holidays and events give the potential to achieve an important degree of differentiation in cruising. In addition, bookings can be made through various channels, including the use of intermediaries (travel agents and tour operators) and direct distribution (Internet and call centres).

In spite of the gradual emergence of new players in the market, the cruise sector is characterized by the existence of significant barriers to entry. First, it is important to note the role of brand name and company reputation, which may reduce consumer uncertainty when experiencing cruising especially in the beginning; after all, many of the new cruise market entrants, such as Disney and TUI, have taken advantage of their existing reputation in other areas of tourism (Papatheodorou, 2006d). Another issue to consider is the significance of the travel distribution system. The fact that the cruise product is relatively complex and expensive, in conjunction with the demographic profile of the average passenger, who is over 50 years of age (Cruise Lines International Association, 2008), implies that, in contrast to the air transport and/or hotel accommodation sectors, distribution through Internet and dynamic packaging is still somewhat limited. In fact, direct cruise sales were estimated to account for 12 per cent of total bookings in 2006 (Mintel, 2007). In addition to their relatively good market positioning, travel agents play a major role in the promotion and sale of cruise packages. The large cruise companies cooperate closely with travel agents to achieve a satisfactory level of distribution of their services. Exclusiveness, along with the development of solid trustworthy relationships based on discounts and override commissions, gives the incumbents an advantage against newcomers to the industry and secures their position in the market. Still, training programmes, such as those provided by the Association of Cruise Experts, may play a creative role in educating agencies on the idiosyncrasies of the cruise product and possibly set the fundamentals for the successful entry of new businesses into the sector, which can manage to persuade the knowledgeable customers on their added value.

The shipbuilding process may also be identified as a possible barrier to entry. Large contracts between incumbent cruise firms and shipbuilders deny spare capacity among the latter, which could allow prospective entrants to place an order for a ship which can be fulfilled within a short period of time. A new liner may take at least three years to build (Klein, 2005) and as a result new companies may have to revert to the market of second-hand vessels, which can be very complex due to information asymmetries, as analysed in Chapter 4. But even if new companies manage to enter the market, they may have to face a price war as a consequence of heavy capacity building by the incumbents; inductively, companies may then decide not to enter the market at all, as shown in the game theoretic framework of the previous chapter. Moreover, and similarly to the case of airport slots, difficulty in accessing ports and related facilities may constitute a market impediment for newcomers. In particular, popular ports-of-call may operate at full capacity, unable to host new ships, and solutions such as port expansion may take a long time and generate severe environmental concerns. A fortiori, ports and local authorities may engage in zero-sum, beggar-thy-neighbour policies to attract cruise ships and reap the benefits from incoming tourism: large cruise liners may then take advantage of their oligopsonistic power to achieve substantial tax exemption and heavy discounts for the use of port services as shown in the case of British Columbia. Among others, as a result of the Alaska Cruise Ship Initiative, ships that are not allowed to dump their waste from Seattle to Alaska on US waters can do so in Canada due to the more relaxed policies of British Columbia. This is a prime example of the need for collaboration among all authorities to avoid a policy war to attract cruise ships against global social benefit (Klein, 2005).

The above discussion raises a very important issue, namely the economic significance of cruising for the local economies. On the one hand, and as an essentially all-inclusive product combining transport, accommodation and catering, cruising may have a limited impact on the various ports-of-call, as passengers may end up spending money only on souvenirs, especially if local excursions have been pre-arranged as part of their overall cruise package. In this case, the negative environmental impacts are likely to offset, any potential benefits. On the other hand, cruising may work in a complementary manner, i.e. people who first visit a destination as part of a cruise and like it may then decide to revisit it in the future as 'proper' tourists spending money on hotel accommodation and catering. Moreover, the gradual development of low-cost cruises pioneered by easyCruise may prove beneficial for tourism consumption at ports-of-call, as passengers may prefer to buy only the core product (i.e. accommodation in the cruise ship) instead of the all-inclusive package. Finally, cruise passengers are likely to spend more money in homeports, especially if they arrive at the particular port a few days before the cruise as part of a wider fly-cruise activity. The major assumption here is that a passenger embarking on a five-star cruise ship is likely to choose to stay in a five-star hotel (or) patronize a high quality restaurant to dine while at the homeport. For this very reason, the cruise industry lobbies in various tourist countries including Greece to press policy-makers to grant cabotage rights to the large transnational cruise liners. In fact, many of these liners register their ships with 'flags of convenience' such as Panama or Bahamas to avoid heavy taxation and the need to abide by strict legal frameworks regarding such regulations as labour conditions, life saving equipment, safety and security and other issues. As a result, they cannot design itineraries using Piraeus (the port of Athens) as a homeport since cabotage restrictions allow this commercial opportunity only to EU-registered vessels. Consequently, Piraeus can at best serve as a port-of-call, achieving only limited economic benefits (Hellenic Association of Tourist and Travel Agencies, 2006). The counterargument in this case is that this form of economic protectionism enables Greek-registered vessels and cruise liners to survive in the market: creating more choice for the consumer (who escapes pressure from the transnational cruise oligopolists), employment for the local people and tax revenues for the government. Moreover, a homeport war among local authorities may prove detrimental, as argued previously. It seems, therefore, that, as in many other cases in tourism economics, matters are quite complex and solutions are difficult to achieve unless based on detailed quantitative methodologies such as CGE, as discussed in Chapter 6.

5.2.3 Rail, bus and coaches

The structures of the rail, bus and coach sectors are similar to that of air travel and cruising in that they, too, experience problems of high capital costs, fixed capacity, peaked demand and the need for feeder routes to sustain profitable ones. Some state support and regulation characterize these modes. On the other hand, there are opportunities to exploit economies of scale and exercise price discrimination to maximize revenues and fill any excess capacity.

a. Rail

As a result of the latest advancements in transport technology, rail offers a good alternative to air travel in certain parts of Europe, especially for distances up to 500 kilometres. A typical example here is Eurostar, the high-speed train that connects London to Paris. However, setting aside the technological aspect, there are other reasons lurking behind the competitiveness of the rail sector in Europe today. In particular, the European Commission revealed its plans for the rail industry back in 1996, with a White Paper entitled 'A strategy for revitalising the Community's railways'. The Paper argued that European railways should operate on a commercial basis and that financial burdens of the past in the sector should be dealt with by Member States. Moreover, the Commission's view of a 'new kind of railway' envisages three broad axes of development, namely the introduction of market forces; the improvement of the mode as a public service with additional objectives related to economic development of the EU; and the inter-operability of railways among Member States.

What is evident from the above is that the European rail system needs to operate more competitively, more commercially and be able to operate on a self-financing basis. A vital clue for 'boosting competition' is the creation of separate organizational entities that would take over rail transport operations on the one hand, and the running and maintenance of rail infrastructure on the other. In that manner, the train operator would approach the rail infrastructure 'provider' to obtain network path allocation, licences and potential fees to operate in that particular network. Last but not least, Member States need to form regulatory bodies managing what is essentially competition.

As a result of these policy measures, the whole of the rail industry has gradually been affected at a pan-European level, with some Member States going further than others as far as rail liberalization is concerned. In Sweden, Bahnverket, the rail infrastructure owner, was established back in 1988. On the other hand, there is only one state-owned operator that runs most of the network, although tenders are being considered on some peripheral parts of it. All train operators pay annual charges based on marginal costs. As Hensher and Brewer (2001) point out, there has been an increase in rail infrastructure investment since the above took place. Greece has fairly recently separated track ownership and operator services, leading to the creation of the National Manager of Railway Infrastructure (EDISY SA). This infrastructure owner was created as per EU Directives 2001/12, 2001/13 and 2001/14, after their transposition into Greek legislation. The EDISY SA handles, among other things, conditions of access, capacity allocation (i.e. train path through the network) and access charges. It should be noted that for the time being there is only one state-owned operator, OSE. A typical example of a Member State that went all the way, applying both privatization and deregulation, is the UK. As White (2008) points out, following the Railways Act of 1993, most of the rail industry had been privatized by 1997, allowing for evaluation to take place. In line with EU directives, separation between infrastructure and operators took place, leading to the creation of Railtrack Plc (now renamed Network Rail) and about twenty-five different train operators. Therefore, the number of players in the market has critically increased, leading to a traffic rise, particularly on regional and inter-city services. As expected, however, this rise in traffic exercises substantial pressure on the existing railway network infrastructure; in the case of Britain, this is quite extensive but rather old compared to France or Japan. As far as tourism is concerned, Holloway and Taylor (2006) highlight that in spite of the general positive traffic trends there are a number of drawbacks. For instance, a potential passenger may hesitate to choose a particular train company due to the lack of an 'integrated marketing' approach among operators, which may confuse the travelling public. Another problem is the lack of investment originally promised by train operators, attributed in part to the implementation of cost-cutting policies.

In fact, cost-cutting issues emerge clearly from the economics involved in this new type of railway that the EC envisages. In the real world, the level of costs and the ability to keep them low represents a major challenge for most companies in the transport sector. The successful management of costs may distinguish a company most likely to win a franchise for a route from one that fails to match the potential financial performance of bids by competitors. In particular, franchising introduces competition to the market while at the same time awarding a monopolistic position to the successful bidder. This 'mix' of competition and regulation is mostly evident in the rail and bus industries as opposed to the air industry. There are two bidding processes, i.e. a bid for a contract on the basis of lowest consumer price and a bid for a contract on the basis of payment for the contract. In addition, franchising contracts can be divided into those involving ownership as opposed to ones concerned only with operating the assets. Each contract type has its own merits. Owning the rail stock, for example, may prompt companies to upgrade it at a later stage in view of a possible renewal of the franchising contract. Moreover, one must never forget the 'winner's curse' and how this may affect the bidding process and the subsequent risks related to the operation of a franchise, after agreements have been finalized. In fact, Alexandersson et al. (2008) assess that sometimes operators may bid for lower subsidy levels, aiming at a renegotiation of the contract half way through the franchise period.

The UK experience has demonstrated that, for the rail market, several franchise contracts had to be renegotiated and that almost half of them had failed in their original form (Alexandersson et al., 2008). More specifically, franchising in Britain led to major cost and traffic increases, thus implying the need for higher levels of subsidy. From a more theoretical perspective, Domberger (1985) argues that franchising leads to increased contestability in the market. By competing for the market, franchising promotes productive efficiency that can be seen as a reward for a franchise renewal. As Roden (2004) discusses, First Group was about to bid for a number of GNER routes in which both Virgin Trains and National Express Group were interested. Eventually, in December 2007, it was National Express that won the prestigious East Coast route, connecting London to major cities in the North. In fact, the National Express Group will run the franchise until March 2015; at present, the service covers 920 miles and carries over 17 million passengers a year.

The above discussion of the franchising process in the rail market clearly demonstrates the importance of costs to rail companies. In essence, Hensher and Brewer (2001) distinguish between economic and accounting costs. The former relate to the opportunity cost of inputs employed as a means of producing transport services. However, it should be noted that it is the opportunity cost of employing this input more profitably elsewhere that may give a cost advantage to the firm. On the other hand, the accounting cost for an input is not related to its opportunity cost, which could be revealed in a competitive market. Thus, in this new economic environment, the rail service is more difficult to cost than road services because of specific characteristics such as joint use of assets by different traffic flows, e.g. freight and passenger traffic with specific subgroups. The principal problem of the rail sector is the high fixed costs associated with the track, signalling system and stations. The crucial difficulty in deregulating and privatizing rail markets and operating them commercially is that uneconomic lines may be closed, although they have the potential to support the principal inter-city services on the same basis as the evolving hub and spoke structure in air travel.

b. Bus and coaches

In a similar manner, the bus and coach industry was privatized and deregulated in a stepwise manner by a number of EU and EEA Member States. The bus industry appears to play a vital role in transport and developments point towards more upgraded (in terms of quality) and cost-efficient services in the UK since 1986. As far as deregulation is concerned, White and Farrington (1998) point out that the government's intentions were the introduction of competition, efficiency gains, innovation potential, more user choice and fewer subsidy requirements. Local bus services outside London and Northern Ireland were deregulated in October 1986, launching a series of privatizations in the late 1980s and the beginning of the 1990s. Most controls related to price, quantity, timetabling, market entry and exit, etc. disappeared. This 'freedom' to operate allowed companies to expand into other territories if they wished to, subject only to safety and road traffic provisions. Operators had to register specific details such as route stops, timings and price range to the traffic commissioner and provide a 42-day notification prior to commencement of services (Mackie and Preston, 1996). With regard to the bus operations of Passenger Transport Executives (PTEs) covering public transport in England's largest conurbations and local councils, reform and empowerment took place, allowing them to support services deemed necessary on social grounds. Such support services would not be provided otherwise by the commercial end of the market.

In essence, the effects of deregulation on competition can be divided into those affecting tendered and those affecting the newly commercially run routes. According to Perrett et al. (1989) and Vickers and Yarrow (1988), competition did not present itself on a widespread scale, although increased levels of competition were observed in some PTE areas, notably Strathclyde, Manchester and Tyne and Wear. These urban areas demonstrated 'on-the-road' competition for some periods after 1986. Every operator in the commercial routes became able to register and run a service subject to some typical requirements outlined above. There were occasions where some operators registered as many routes as possible commercially, in order to deter entry, while others attempted to enter a geographical area by entering the most profitable routes. As for the tendered market, a number of academics believe that is close to and definitely more contestable than the commercial market as entry costs are substantially lower (Preston, 1991; Mackie and Preston, 1996; Toner, 2001). In fact, as Preston (1991) highlights, the tendering authority sets outs most of the specifications and undertakes any transport planning requirements as well. The outcome is that the entrant saves on publication costs and is provided with route information that would otherwise be acquired by consuming valuable company time and monetary resources. Mackie and Preston (1996) even mention advertising as an obligation of the tendering authority on some occasions. In a similar manner, Savage (1984) acknowledges that 'in this industry, the optimal solution is competition for the market rather than competition in it', and points to the direction of competitive contracting or franchising of services. The UK has a record of high use of competitive tendering and (with the exception of London) all results of the bidding processes must be published along with the exact number of bids, preferred bidder and even explanations about the decisions (White and Tough, 1995; Dodgson and Topham, 1998). As far as Norway is concerned, Mathisen and Solvoll (2007) admit that competitive tendering, after a long 'transitional period', has led to increased competition among the companies operating in the Norwegian bus market. In fact, by blending competition with tendering, Norway's regulatory regime led transport companies to reassess their plans and engage in mergers or 'cross-ownership' to limit the number of bids made and eventually set the bids higher.

In conclusion, it is evident that bus deregulation in a number of European countries, including some Nordic countries and the UK, has produced a completely new business environment under which transport operators may now operate commercially and subsequently make a profit or make a loss. Twenty years after the deregulation of the UK market, there are signs of competition in terms of quality and marketing related to bus services in the tendered market. In addition, there are advertising initiatives on some occasions, although this is usually the responsibility of the tendering authority, along with other route specifications (Mackie and Preston, 1996). Generally, changes in the quality of the vehicles operating the routes, branded ticketing and bus liveries and passengers subject to the, sometimes, intense marketing and advertising campaigns of various operators may be signs of product differentiation as well as potential sources of a long-term competitive advantage. Still, some sort of pressure is required so that more efficient operational practices are introduced in addition to innovations in marketing and branding. Although deregulation initially lowered prices on key routes, the entry of many small companies has been accompanied by greater concentration at the national and regional level of operations. Large companies not only offer a more comprehensive coverage but also are able to exploit economies of scale and scope in such factors as reservations, repairs and maintenance, administration and advertising. For example, companies like Arriva and Stagecoach have been successful as operators in both rail and bus/coach businesses. Furthermore, large firms aim at fully utilizing vehicles with respect to payloads and operational hours and serving only the most lucrative routes at higher fares, endangering in this way the long-run existence of the smaller companies. Consequently, market power may be vested in few hands. Having the above in mind, it is clear that a perfectly competitive environment is hard to find in the European bus and coach industry today; in any case, however, the tendering process meets many of the theoretical contestability conditions. Moreover, and as far as planning of bus networks is concerned, Guiver et al. (2007) highlight ecological and sustainability issues which must be considered in addition to revenue and traffic flow forecasts on particular routes.

5.3 The accommodation sector

Cursory inspection of the accommodation sector might suggest that a few large chains dominate the market, giving the impression of an oligopolistic structure. However, the service hospitality sector within holiday tourism is mostly fragmented in many small units where location and the spatial distribution of accommodation are important factors determining the degree of competition. Furthermore, the wide range and quality of accommodation, its multi-product nature (for example, camping, caravanning, holiday centres, timeshare, as well as serviced accommodation) and seasonal variations in demand introduce an additional dimension into the operation of the market. Accordingly, different forms of structure – perfect competition, monopolistic competition, oligopoly and even monopoly – might reflect the conditions of different elements of the sector, ranging from the serviced to the un-serviced self-catering segments. It is therefore interesting to go through these components that shape the market's mixed structure.

5.3.1 General characteristics

In general, the accommodation market is characterized by substantial fixed costs such as the cost of land acquisition and real estate development, which is relatively high in areas proximate to key attractions and accessible from urban centres of major market potential. Hence to reduce unit costs and achieve economies of scale, large accommodation complexes are built at a global scale. Evidence from the North American market justifies this view: whereas in 1948 only 44 per cent of US hotel rooms were found in establishments of one hundred or more rooms, this figure had risen to 58 per cent by 1977 and was close to 62 per cent by the mid-1980s (Go and Pine, 1995). Lodging facilities with less than fifty rooms gradually exit the market, especially in seasonal markets where additional investment is not sustainable. Moreover, there is evidence that some forms of accommodation can exploit economies of scale by the management of a large number of hotels. This partially explains the existence of chains that control many hotels, such as Wyndham, Choice, InterContinental and Accor. To an extent it also accounts for the concentration in the sector, the nature of ownership and location being other explanatory variables as discussed below. For example, in the USA, with a much higher proportion of corporate ownership, the concentration ratio of the four (eight) largest hotel companies was 22.4 per cent (28.8 per cent) in 2002 (US Census Bureau, 2005), while in eastern Mediterranean holiday resort destinations, comprised of family-owned businesses, this is much lower.

From a revenue perspective, location, size and quality disparities are directly reflected in the revenue composition. For example, city centre hotels record a greater proportion of revenue from the tenancy of rooms than countryside and resort-based properties. Similarly, letting in high-standard hotels represents only 50–60 per cent of the total revenue, i.e. a much lower percentage than in budget hotels with limited food and beverage sales. Not surprisingly, the accommodation services are the major contributor to gross profitability, as the respective departmental costs account only for the 25–30 per cent of tenancy income; on the other hand, these outlays represent more than 80 per cent of the earnings in the food and beverage sectors (Lawson, 1995). Nevertheless, it should be always borne in mind that the quality of the catering services determines the hotel grade and rates to a major extent; consequently, any careless cost reduction policies may have detrimental effects on reputation. Moreover, particularly in larger units offering a wide range of services, high fixed and sunk costs drive operators to attain high occupancy rates through strategies such as product differentiation and market segmentation.

These characteristics tend to involve elements of both natural monopoly and oligopoly. Most importantly, the hotel accommodation industry is characterized by quality differentiation; for example, some hotels concentrate on the luxury segment while others serve a budget clientele. Many seek flexibility by targeting the business market during the working week and the leisure sector at weekends. Disparities are also enhanced by functional specialization; for example, a city centre hotel is likely to include extensive facilities for meetings and conventions, whereas a resort spa establishment may provide leisure attractions combined with medical, fitness and convalescence services. In holiday resorts, the needs of different groups can be met over the year, for instance catering for skiers in winter and walkers in the summer, as occurs in Austria, France and Switzerland.

The self-catering sector is an interesting area of the accommodation market that has shown both high growth rates and great diversity. It has complemented the serviced sector by providing additional rooms whose clients are allowed the use of the facilities provided, and it has also been incorporated into purpose-built or redeveloped tourism centres offering many attractions and facilities. Holiday centres and villages and timeshare come into this category. There are a number of large players in the field, for example in the UK Butlins, Haven, Pontins and Warners, formerly low-cost concerns, have upgraded their product, while the Center Parcs organization is an innovative market leader with a number of properties also across Continental Europe (Center Parcs, 2009). Disney Vacation, Hapimag, Vacation Resorts International, villa owners clubs and timeshare are similar in terms of the holiday centre concept but differ in that they involve capital investment by consumers. In particular, timeshare, recently known as vacation ownership, offers its purchasers the possibility to enjoy annually and for a set period of time, the facilities and services offered by one or more resorts. The product often consists of the co-ownership of one resort and/or the right to use another. The product's main characteristics comprise the resort that is the tangible form of the right and the possibility to exchange the stay with an equivalent set in various places and time frames (Stavrinoudis, 2008). There are specialized companies, for example RCI (Resorts Condominiums International), that facilitate exchanges domestically and internationally between resorts. Timeshare is often referred to as one of the alternative forms of tourism which can be easily combined with other forms of tourism (e.g. cruises, meetings) allowing great advantages such as the stimulation of mutual demand and the enrichment of the final product offered (Stavrinoudis, 2006a). Still, for the timeshare market to function properly, tourism destinations must pay special attention to three issues, namely the institutional framework concerning the national legislation that regulates the product development; the operational features of the affiliated enterprises; and the overall level of tourism development of a destination and its characteristics (Stavrinoudis, 2006b).

5.3.2 Demand issues

In spite of the overall positive trend and the income-elastic nature of the product, the demand for hotel accommodation exhibits strong fluctuations. First, it is susceptible to the business cycles of the aggregate economy; while a stable and expanding economy has a positive impact on hotel performance, the reduction of business travel and entertainment during the periods of recession has typically adverse effects on the consumption and profitability of hotel services. Furthermore, inflationary pressures on labour, energy and construction erode profits, especially when the prevailing market conditions impede hotels from raising room rates proportionately. The stronger players may survive the economic crises and subsequently set the terms of corporate rivalry in the context of an oligopolized market.

Second, demand exhibits strong periodicity, as the number of rented rooms varies considerably between weekdays and weekends as well as among different seasons. In general, the counteracting flow patterns of the various markets, segments reduce the overall seasonality in city centre hotels: recreation (business) tourism is high (low) during summer and low (high) in the other seasons. Seaside resorts, however, exhibit high occupancy rates during the summer period only, whereas hotels in mountainous areas may have at most two high tourist periods. A third reason is that hotel demand is subject to external, exogenous shocks, such as natural disasters (e.g. hurricanes in the Caribbean) or political instability. Nonetheless, many of the large hotel organizations have access to advanced know-how and managerial skills that are essential for dealing with the above demand characteristics.

Finally, it should be borne in mind that shifts in modality patterns have serious repercussions for hotel accommodation: faster and more frequent transport services enable same-day return trips, whereas enhanced luxury standards in high-speed railways may result in a loss of hotel business tourism. For instance, the Deutsche Bundesbahn Inter-City Express offers business facilities on board on many of its routes (Deutsche Bahn, 2009). Furthermore, a number of pleasure market accommodation substitutes, such as resort timesharing, condominiums, second homes, recreational vehicles, camping grounds and cruise lines, may pose a significant problem for leisure hotels. On the same grounds, the advancement of teleconferencing and virtual reality (in a more futuristic context) may also cause some long-run rivalry in the hotel accommodation business (Williams and Hobson, 1995). Strategies of vertical integration, including agreements and alliances with large aviation companies and travel agents, reinforce the hotel groups' position against these alternatives to the accommodation sector.

5.3.3 Market dualism and its spatial implications

It is important to highlight that the hotel accommodation sector is characterized by a notable dualism observed in size (large–small), ownership (chain–independent), structure (oligopoly–monopolistic competition), quality (high–low) and location (centre–periphery). More specifically, the non-standard accommodation sector is characterized by a notable variation in architectural and operational styles. Establishments of this type are scattered in numerous locations, mainly serving leisure tourists who visit local attractions. In general, the average size of these hotels does not exceed 75 rooms and though repeat visits are usual, occupancy rates are rather low (Papatheodorou, 2000). The management has some traditional style household elements, as the hotels are usually family-run. Services are of limited range, particularly after meals and during the night, in order to economize on labour costs. As expected, the marketing strategy is ad hoc and there is no persistent advertising. Finally, the pricing policy is flexible, given the strong seasonal character of demand, the dearth of bargaining power against the tour operators and the intense intra-type competition; consequently, profitability is also rather low (TEPRO, 1985).

On the other hand, the international standards hotel sector is homogeneous, following mainly the American way of life as its service model. Establishments of this type are usually clustered near urban centres and resort attractions, catering for business travellers and wealthy vacationers. Their average size varies between 250 and 500 rooms for the US/Asian hotels and slightly fewer for the European; nevertheless, there is a substantial group of hotels exceeding 1,000 rooms (Papatheodorou, 2000). Repeat visits are unusual, but relatively high occupancy rates are achieved due to the use of sophisticated marketing techniques, heavy advertising and company loyalty schemes. The management is uniform, separated from the ownership of the hotel and in many cases dictated at a transnational level. There is a wide range of services usually on a 24 hour basis and labour costs tend, therefore, to be relatively high. Finally, in spite of several concessions, the pricing policy is rather inflexible, due to the low price elasticity of the associated business tourism, the bargaining power of the establishments' brand name and the various strategies of horizontal differentiation; as a result, profitability is kept at satisfactory levels.

Some of these large accommodation groups also attempt to enlarge their market share and control by takeovers and mergers along with franchising, leasing, management contracts and collaborative agreements strategies (Dimou, 2004). Economic integration between firms results in reduced fixed costs in terms of advertising, product development and training; while entry barriers may make it possible to increase occupancy rates by tapping new segments of the market. In addition to scale and scope economies achieved, hotel chains may benefit from technological advancements (Sigala, 2003) and experience along with know-how practised by high-quality personnel. All the above may result in the creation of a strong brand name, which generates customer satisfaction and yields marketing advantages. To summarize, the expansion of hotel chains increases significantly market concentration in the accommodation sector, with branding playing an important role in future competition. According to Deloitte (2006), while location is the most important factor influencing consumer choice in 97 per cent of occasions for leisure tourists and 93 per cent for business travellers, the existence of a strong brand is also important, accounting for 57 per cent and 54 per cent respectively.

The above discussed market dualism has also important spatial connotations. More specifically, large corporate-run hotels, mainly serving business travellers, tend to cluster in or around large urban areas, airports and on land transport routes. On the other hand, holiday hotels are more likely to be independent and more widely dispersed, although clustering still occurs, such as in resorts or locations which are principal tourist attractions. In this sense the accommodation sector is akin to monopolistic competition in the retail market, in which accessibility and complementarity, central tenets of urban economics, are the main characteristics (Balchin et al., 1988). As also discussed in Chapter 6, urban economic theory (Evans, 1985; Hoover and Giarratani, 1999) shows that key locations provide benefits that give a commercial advantage and therefore a higher turnover and profits, explaining why large firms can outbid smaller ones in high-cost areas. In effect, a market monopoly can arise because of a locational monopoly on specific sites; for example, the ownership of land in competitive locations gives the potential to the existing firm to prevent the entrance of competitors. It also helps to account for the pattern of accommodation in holiday areas like seaside resorts where smaller, lower quality hotels are pushed into secondary locations away from the seafront. This is explained in terms of spatial rent/property price gradients whereby, for individual businesses, location is determined by the property/land costs each faces, which are dictated, in turn, by the demand for specific sites by competing activities. Around airports, in urban centres and along seafronts, land costs are high, whereas farther away they are lower as demand for sites is less.

Thus, businesses with lower revenue and profits are forced into areas with lower rents/prices. Hotel chains with financial power, derived from higher levels of demand and a willingness to pay higher prices for accommodation by their guests, have the advantage of being able to access expensive land that gives them the possibility of monopolizing the market and so increasing their profits. However, although the intensity of competition is lower in particular destinations outside locations subject to such clustering, it is still significant in terms of the total market, given the number of holiday choices open to consumers and the size and quality range of accommodation available. For example, in the summer sun package market there is a large number and wide distribution of destinations and types of accommodation so that, in the face of static demand, the market structure is highly competitive.

5.3.4 Barriers to market entry and the revenge of the humble

Along with marketing strategies and their industrial and spatial implications, the accommodation sector is characterized by a number of barriers to entry sustained and enforced by the major corporate players. In particular, entry into the international standards hotel sector is associated with a considerably larger investment expenditure, particularly due to the escalating real estate and construction costs for the reasons given in the previous section. In addition to the need for heavy advertising campaigns, even if the proposed plan proves to be profitable, financial market imperfections and asymmetric information regarding solvency may oblige the new hoteliers to borrow money on rather unsatisfactory terms. But even if the high fixed costs problems are resolved, unless the appropriate location and space are found, the various investment plans are likely to be unsuccessful. Locational requirements increase the risk and unpredictable urbanization dynamics can seriously affect future demand and real estate values.

The hotel Global Distribution Systems (GDS) provide valuable information on consumer preferences and enhance yield management in a similar way to the airlines' CRS. In many cases, however, the setup of these systems can be difficult to implement. First, they involve considerable sunk outlays and, second, their benefits are accrued only by large network owners due to the presence of significant scale and scope economies. Moreover, the incumbent hotel chains may increase their benefits from network economies by negotiating strategic alliances with other hotel corporations. For example, until recently the American Radisson Hotels collaborated with the Norwegian SAS International Hotels and the Swiss Mövenpick Hotels International. According to the terms of the agreement, Radisson distributed the Mövenpick product in North America, while Mövenpick promoted Radisson in Europe (Go and Pine, 1995). Nevertheless, the subsequent termination of Radisson's collaboration with the SAS group led to the establishment of the Radisson Blu brand, which will gradually replace the existing Radisson SAS Hotels & Resorts name (Radisson Blu, 2009).

Furthermore, and similar to the case of airlines discussed earlier, many hotel corporations introduce loyalty schemes to reward their regular customers. The associated benefit for the consumer is substantial in the case of a large hotel network. For example, the Marriott Rewards is a Marriott Hotels, promotion programme, which gives loyal customers free accommodation at luxury hotels and resorts worldwide; each stay generates points, later redeemed in more than 250 ways including travel awards such as cruises (Marriott, 2009). By introducing these interchangeable point schemes the major companies in the tourist circuit tie their customers into using specific airline and accommodation networks and enhance inter-sectoral collaboration within a framework that is more flexible than vertical integration practices. Not surprisingly, a new entrant is hardly capable of offering the same choices.

Finally, many hotel corporations have followed a quality improvement strategy, mainly in order to enhance their global business clientele. Despite, however, the high price premium of these services and the prospects for substantial profitability, the growth limits of this market are certainly finite and some first saturation signs have already occurred in developed regions. On the other hand, by taking advantage of its established brand name, a hotel corporation may launch new chains and product lines that fill all potential niches of the quality spectrum and generate a new vehicle for growth. For example, if the mid-price market is saturated in a given location, Choice Hotels International can franchise a budget-price Comfort Inn, whereas the Accor Group may introduce a Formule1 mass-market product or establish a deluxe Sofitel Hotel. In this way, entry deterrence may be successful, since potential hoteliers may be unable to launch a perceptively different product.

In certain cases, however, independent and smaller hotel operators can successfully manage to survive in the marketplace based on their personalised services, their flexibility and their decision to form voluntary associations such as cooperative consortia to reduce overheads. In particular, although chain corporations may be more efficient than small individual hoteliers, a number of tourists and business travellers are attracted by the informal and traditional ambience of the latter. Cultural and geographical differences between origins and tourist destinations favour this aspect, especially in small cities and sea resorts. Furthermore, the recent trend towards product customization and the subsequent rise of special interests tourism (such as arts, heritage, education, ethnic, health and sports tourism) have a major impact on the sector, as it is no longer possible for any hotel company to understand and capture the entire travel market. Independent hoteliers are, therefore, in an excellent position to respond to special travel preferences by promoting their uniqueness and originality.

In addition, hotel chains often become bureaucratic, inflexible and wasteful as they grow larger. Their excessive reliance on standardized procedures may result in decision rigidity and inability of strategic reorientation. For example, although Holiday Inns' franchise strategy was a major business success until the early 1970s, their ignorance of subsequent changes in the industry caused the corporation trouble. On the other hand, because of their size, small independent hotels are potentially flexible. In fact, by preserving elements of pre-Fordist, artisanal behaviour, they are able to adapt themselves more easily to the contemporaneous post-Fordist production system (Ioannides and Debbage, 1998). In other words, the inherent complexity and diversity of flexible specialization gives rise to an organizational structure that may fruitfully encapsulate the numerous diverse practices of the traditional hotel sector (Poon, 1990).

It should be noted, however, that small size does not necessarily imply flexibility; unless receptive to network externalities and dynamic economies of scale, it may remain a source of inefficiency and structural weakness. On these grounds, the hotel consortia, which are oriented towards the small producers, may provide a substantial service. Branding and enhanced competition have induced independent hoteliers and smaller second-tier hotel chains to seek affiliation in centralized organizations, which provide them with a number of services without undermining their independent status. In addition to hybrids, Jones and Pizam (1993) distinguish four basic types of consortia, namely full ones like Best Western that provide marketing expertise and human resources assistance to a degree that resembles a proper hotel chain (Best Western, 2009); marketing consortia, such as the Small Luxury Hotels of the World (Small Luxury Hotels of the World, 2009); central reservation systems like Utell, which offer network facilities both online and through a single, toll-free telephone number (Utell, 2009); and finally referral consortia that seek hotel affiliation with the reservation system of a particular airline, such as the JAL World Hotels Programme (JAL, 2009). Generally speaking, the hotel consortia enable the associated producers to market themselves in the international clientele context, using methods and tools which are financially unsustainable from an individual perspective. In particular, the consortia create a critical mass to support marketing consulting, international sales offices and global distribution systems. Furthermore, by establishing a brand image, they are able to replicate large chain practices, enhancing business and customer loyalty. As expected, to build a reputation each consortium has to set specific, regularly inspected standards for its members; however, the successful candidates are, usually, numerous enough to keep the annual fees at acceptable levels.

In conclusion and as a result of the absence of regulation in the hotel accommodation sector, with the possible exception of certain architectural and environmental standards policies (Lawson 1995), market forces worldwide have been left to shape an evolutionary pattern based on multidimensional duality. Taking account of the significance of spatial factors in shaping market conditions, on a continuum of perfect competition to monopoly, the market structure of tourism accommodation tends to accord with the contestable – monopolistic competition positions. In business and resort centres, large hotels experience oligopolistic conditions but outside these areas the structure is closer to monopolistic competition. The picture is clouded somewhat by the intra-sectoral choice of type and quality of accommodation, with some segments, for example self-catering, being more contestable than others.

Transnationalism within the international standards hotel sector will probably expand, given the world policy towards liberalization and welcoming of foreign capital inflows. Moreover, the gradual saturation of the existing up-market niches in the developed countries will probably intensify the pursuance of brand proliferation strategies. In any case and as noted by Deloitte (2005: 8) quoting Andy Cosslett, CEO of Intercontinental Hotels Group, 'the traditional core markets of Europe and America still have plenty of growth potential and will remain a focus for us in terms of our existing resource, however we are also very active in other markets, particularly Asia, where we are investing in new resources to build our position'. Still, as land scarcity gradually becomes a serious issue, even in developing countries, regulators should be alert to the preservation of the common interest and benefit.

5.4 The intermediary sector

Tour operators and travel agents respectively assemble and retail holidays largely for the mass market. The role of tour operators is to supply holiday packages and to facilitate the link between the suppliers of travel, accommodation, facilities and services, both in origins and destinations, and the tourist. They procure the components of the product, usually by negotiating discounted prices, and retailing it through travel agents or directly to the customer. Tour operations are conducted in a number of ways: by an independent firm specializing solely in holiday assembly and marketing; as a subsidiary of a conglomerate business with diverse interests; as a division of an airline; or linked with a travel agency.

5.4.1 Tour operators

Apart from their principal – agent relationship, an issue of some significance in industrial economics, the structure of the intermediary market, for example in Europe and the USA, raises other interesting questions. There are some large players in the market in terms of both ownership and control and market share. In the USA, there were approximately 600 companies operating in the late 1970s but this number increased to over 1,000 by 1985 (Sheldon, 1986) and 1,500 in total in the early 1990s, of which 40 tour operators (3 per cent) controlled almost a third of the market. Similarly, in 2009, more than 2,350 were recorded (Manta, 2009a); moreover, the 46 active corporate members of the United States Tour Operators Association (USTOA), represent 150 brand names and generated a total sales volume of over $9 billion carrying over 11 million passengers in 2008 (USTOA, 2009). In the UK, the top ten tour operator groups jointly accounted for about 60 per cent of the 29.6 million seats of ATOL (Air Travel Organizers' Licence) licences granted in 2006/07. The four largest groups, i.e. TUI, Thomas Cook, First Choice Holidays and MyTravel, took 46 per cent of the air holidays licences in 2007 (CAA, 2007), while in 1994 the market share of the Big Four was much larger and almost 60 per cent (Evans and Stabler, 1995); in reality, however, market concentration in the UK tour operations market has risen as in mid-2007 TUI merged with First Choice and Thomas Cook with My Travel. Both TUI (which acquired Thomson, the largest UK tour operator, in the mid-2000s) and Thomas Cook are German-based companies, reflecting the transnational character of consolidation in the contemporary European tour operators sector. In addition to this marked concentration, there is also extensive vertical integration with respect to air travel, accommodation, travel agency and other leisure services.

Other features of tour operation of note are, despite the degree of concentration, the rate of increase in and the total number of firms. In the UK there were around 500 firms in 1985, which, by 1993/94, had nearly doubled when 1,000 companies were licensed (Evans and Stabler, 1995). In 2008, the Association of British Travel Agents (ABTA – representing both travel agents and tour operators) reported 1,421 active members compared to 1,515 in the previous year (ABTA, 2008). Of more significance, however, is the birth and death of tour operators, mostly of smaller firms. It has been estimated that only a third of those in business in the late 1970s were still extant in the middle to late 1980s in both the USA and Europe (Sheldon, 1994). In the UK, twenty-four firms ceased trading in the year 1992/93 and another twenty in 1993/94, notwithstanding which the total number of firms increased. Tour operator failures among ABTA members were twenty in 2008 (ABTA, 2008). Nevertheless, today, the number of tour operators remains high both in Europe and in the US.

The performance of tour operators is very sensitive to market conditions, particularly variations in demand arising from such factors as changes in exchange rates, economic recession in origin countries and inflation and perceived political instability in destinations. In the face of tourism development and the continuing extension of possible destinations, capacity has sometimes outrun demand. Furthermore, as in the hospitality market, seasonality is significant. Large firms rely on a high volume of sales with low margins, a significant proportion of profit being generated by investing receipts from advanced bookings. A high rate of sales makes it possible to achieve substantial economies of scale and scope through operating efficiencies, a wide knowledge of an extensive market and market power to gain large discounts from carriers and hoteliers, where past performance is the key. Provided fixed investment is low, as in the case of smaller and more specialized tour operators, the return on fixed investment can be relatively high. Nevertheless, many firms suffer losses over a series of years and when net profits are made they are modest, the return often being less than 4 per cent in the UK, although there is considerable variability. Today, possibly with the exception of Italy and Spain where the nature of the market is rather different because of late bookings, the average profit margin across the European continent is about 2 per cent of total turnover (FTO, 2009a). Corporate rivalry, the slowdown of the economy and the low margins are some of the factors that contribute to suppress the return even further.

In the inclusive package market there is intense competition to secure sales volume in order to generate cash flows. Consequently, discounting can be widespread at the launch of the subsequent season's packages to encourage early booking. Discounting can also occur at the end of the season to fill excess capacity. The expectation of discounting by potential tourists, who delay making a firm booking earlier in the season, exacerbates tour operators' problems because not only do they have to dispose of holidays at little above cost but also their cash flow is adversely affected. Discounting strategies are both a manifestation of the drive to maintain or increase market share and a reflection of the long lead times, often up to three years, required to launch new holiday types, increasing the likelihood of overestimating demand leading to over-supply. The average profit made on package holidays is extremely slender. Many firms in the retail consumer sector are able to record a net margin on sales turnover in the area of 10 to 12 per cent; nevertheless, the CAA has estimated that the representative tour operator's return on turnover in recent years has been barely over 3 per cent, generating about £8 profit on a £400 holiday package FTO (2009a). Even large companies indulge in price wars, the leaders in the market often initiating them to safeguard their market position. The emergence of low-cost carriers in addition to the new trends of dynamic packaging has introduced competition even from consumers who decide to create the package themselves.

A number of factors such as ease of entry and exit, the number of tour operators, fierce price competition, low margins and often significant losses all point to contestable if not highly competitive market conditions in the UK and many other countries. However, the degree of concentration of market share in the package holiday segment suggests an oligopolistic structure. Still, there is little empirical evidence concerning the extent of economies and diseconomies of scale and economies of scope, capital indivisibilities, fixed capacity and fixed costs. The larger operators in the top tier, which have invested in charter airlines and some travel agents, appear to enjoy economies of scale and scope, although such investment has entailed the problems of indivisibility and high fixed costs with the concomitant fixed capacity which needs to be filled. Average load factors on aircraft and occupancy rates in hotels are measures which indicate the extent to which such problems are overcome. However, the majority of tour operators are relatively small and these structural characteristics tend to be less important. Operators which organize the more specialized forms of tourism where the market is not extensive do not require high investment in fixed capacity and, indeed, would be likely to suffer from diseconomies of scale if the more personal bespoke service were undermined.

Tour operators' conduct involves recurrent price wars and high volume sales by the largest operators in the absence of collusion between them. Firms have differentiated their products in the context of segmented markets. Performance outcomes include moderate consumer satisfaction and moderate efficiency of firms' operations, with low profit margins and some bankruptcies. The sector has been subject to regulation although anti-competitive conduct has not been deemed a significant problem to date. As with the accommodation sector, the reality of the market is more complex than can be encompassed in any single theoretical model of market structure. It appears necessary, therefore, to consider different segments as being characterized by different competitive conditions. This, together with the apparent inherent instability, suggests a dynamic market which tends to show up the limitations of the neoclassical paradigm. In addition to providing a framework for the examination of market and firm characteristics, the SCP model predicts that structural features affect firms' conduct and performance. Owing to the important role which the structural base is hypothesized to play, two structural features which are of particular relevance to tour operations will now be examined. First, the role of entry and exit conditions for the sector will be discussed. Implications of market concentration will be considered subsequently.

a. Entry and exit conditions for tour operations

The high growth rate in the number of tour operators in the UK is indicative of low entry barriers. Capital costs are low and the identification of a niche and expenditure on promotion to bring the product to the attention of the potential tourist do not constitute major obstacles to entry into the market. It is necessary to convince purchasers of the viability of the company in terms of guaranteeing the security of payments made in advance. A significant cost of entry is therefore the bonding requirements, e.g. to participate in professional associations such as the USTOA or ABTA. Additional entry barriers may also occur. Salop (1979a, 1979b) distinguishes between innocent barriers and strategic barriers which incumbent firms erect. Even if perfect information is assumed, economic theory can show that there are advantages to incumbent firms. For example, incumbent firms may have erected pre-entry barriers by having established licensing and franchise schemes, created a differentiated branded product, secured consumer and supplier loyalty and contracted with hoteliers and facilities in the most desirable locations. These kinds of asymmetries between the positions of entrants and incumbents were originally seen by SCP advocates, such as Bain (1956), as entry barriers but more recently they have been interpreted as first mover advantages (Lieberman and Montgomery, 1988). Post-entry barriers such as economies of scale and scope and lower input prices can also be identified and may confer absolute cost advantages.

Strategic entry barriers are a conscious effort by established firms to deter new firms. Typical ploys are to adopt limit pricing, perhaps of a predatory nature, to differentiate products, increase advertising expenditure in a counteractive manner and seek improvements in efficiency or capacity to lower unit costs of production. Firms have also reinforced the strategic core of their business. The strategic core is the principal activity of a business enabling it to achieve its aims and objectives, taking account of the prevailing opportunities and threats and its strengths and weaknesses. In economic terms the core activity facilitates the attainment of increased economies of scale, or extended product differentiation and multi-market targeting to exploit economies of scope. In some cases, incumbents maintain spare capacity so that they can increase output and decrease their price rapidly if a new firm enters the market, making its position unprofitable and forcing it out of the market. The maintenance of spare capacity is, therefore, a strategy which involves a credible threat to potential entrants, backed by the 'punishment strategy' of a price war. Vertical integration involving investment in sunk costs, i.e. those which are large and not recoverable in the short run without involving the business in considerable expense, such as the purchase of aircraft or a chain of hotels, can also constitute a 'pre-commitment' which acts as a deterrent to entry and exit, especially where substantial cost savings are achieved. Entry deterrence strategies and their likelihood of being successful were examined within a game theory context in Chapter 4.

Tour operators have not vigorously pursued such strategies to deter entry per se. They have tended to be more concerned with increasing market share, accepting that they have insufficient power to exclude new entrants. In this respect, economic theory would suggest that sunk costs are low and similar for newcomers and incumbents alike. Baumol (1982) argued that low or recoverable sunk costs make markets contestable, concluding that if prices are competitive and not far above marginal cost, then monopoly and oligopoly structures are largely benign. Such a view, prima facie, appears to apply to the UK tour operations sector up to the late 2000s. It is clearly dominated, in terms of market share, by a few firms which do not always earn supernormal profits.

However, an assumption of low sunk costs in all sectors of a market is an oversimplification. It is more likely that the level of sunk costs varies, depending on the market segment into which entry is sought and also on the strategic group into which a particular firm falls. Entry/exit conditions do not necessarily relate to an entire industry or market but may be specific to sectors or segments within them. An example serves to illustrate not only entry/exit conditions but also their interrelationship with sunk costs and economies of scale and scope. Suppose a tour operator in a specialist segment with a limited market, say activity holidays, contemplates expanding into the mass package market. Exploitation of economies of scale and scope and so the reduction of unit costs may act as an entry condition which must be met. To achieve such economies the operator may be compelled to operate its own airline and hotels, thus incurring substantial sunk costs which act as a deterrent to entry.

The strategic group argument is that in a comparatively diversified market in terms of firm size and importance and products, there are marked similarities in the characteristics of a number of companies and their products, which suggest common interests and thus that they can be classified into identifiable and distinct groups. In tour operations it has been posited (Evans and Stabler, 1995) that there are three tiers of companies distinguished by size, capitalization, market share and range of products offered. It is useful to re-examine this feature in order to illustrate the concept of the strategic group. The first tier, consisting of around 10 operators in the United Kingdom, is more integrated, heavily capitalized and serves several market segments. It enjoys economies of scale and scope and wields considerable market power. The second tier, of around 20 to 30 firms, is more specialized but still has a significant share of the market. The remainder, in the third tier, are numerous, small and are often unincorporated, with low sunk costs. The apparent absence of barriers and deterrent action to entry bear out the contestability thesis even in the top tier of the market.

However, it does not account for the presence, in the first tier, of a high degree of concentration, another aspect of market structure which has been cited as indicative of conduct and performance and a significant characteristic of tour operations.

b. Market concentration

In recent years the tour operations sector has exhibited a rising market concentration, particularly in Europe and most notably in the UK (Papatheodorou, 2003a). In many European countries, the joint market share of the five largest tour operators is over 60 per cent (Toulantas 2001). As discussed earlier, in 2007 TUI and First Choice decided to merge, followed by Thomas Cook and MyTravel. This consolidation raised a lot of scepticism and was carefully examined by the EC before it approved the two cases. In particular, the Commission concluded that the remedies submitted by the parties were sufficient to safeguard competition and remove the serious doubts raised as a result of the increased concentration; on these grounds, it decided not to block the proposed mergers (EC, 2007c; 2007d). This is in sharp contrast to the Commission's verdict in a 1999 case regarding the proposed merger between Airtours (later renamed MyTravel) and First Choice, which was rejected on grounds of collective dominance, although this decision was annulled by the Court of First Instance in 2002; that merger never actually materialized (Papatheodorou, 2006c). This change in the Commission's stance is undoubtedly related to recent market developments and most importantly to the rise of informed consumers who purchase travel services online. At present, travellers can and do shop around to find the best deal for a foreign package holiday especially in the short-haul market. This is facilitated by the presence of a large number of web-based search engines which allow consumers to compare prices and availability within seconds. In addition, consumers show little brand loyalty and engage in high levels of switching in a marketplace where the competitive offering changes almost daily. Therefore, if the Big Two decide to raise their prices in the context of a wider coordination, they will most probably suffer from a significant sales diversion to the benefit of smaller tour operators and independent travel service providers.

Still, the accelerated trans-border growth of major European tour operators may result in a conflict of interests (Bastakis et al., 2004), which goes beyond national jurisdiction and entails issues of political economy in addition to industrial economics. In particular, it is very likely that the British or German competition authorities (and in general those of tourism origin countries) will implicitly encourage the emergence of a competitive oligopsony among their tour operators (Papatheodorou, 2006a). In this way, such tour operators can exercise their negotiationing power on the various foreign destination service providers (such as local hoteliers in the Mediterranean) to push their prices down so that any generated cost savings are subsequently passed to the origin consumers in the form of lower prices for their holiday package (Papatheodorou, 2001a). It is interesting to mention that all the reports of the UK Competition Commission on the conduct of British tour operators have systematically abstained from highlighting the possible detrimental consequences for destinations, which will then benefit only to a very limited degree as a result of tourism development. Destination policymakers are then likely to encourage local service providers to find ways to reduce the bargaining power of tour operators by differentiating and improving the quality of their product; by seeking collaboration with smaller tour operators; by developing enhanced Destination Management Information Systems (DMIS), which can enable disintermediation and direct contact between the origin consumer and the destination service provider; and by creating marketing and other consortia, which could possibly negotiate terms with tour operators at an aggregate level, replicating the conditions of a bilateral oligopoly. In any case, the success of these strategies will depend on the ability of the oligopsonist tour operators to play the various destinations against each other as well as on their willingness to develop longer-term relationships with destination service providers based more on service quality rather than reduced prices. At a policy-making level, the authorities of tourism origin countries should realize that the welfare of their consumers also depends on the satisfaction of destination service providers from the financial terms they receive; likewise, destination competition authorities should understand that the imposition of strict regulation against international tour operators may eventually harm local development. In other words, 'Corporate rivalry should be encouraged; however, coordination success is also needed: co-opetition, therefore, may be the way forward' (Papatheodorou, 2006a: 205).

5.4.2 Travel agents

Travel agents function as brokers in arranging all aspects of business travel and holidays but act as agents in that they represent principals, whether they be tour operators or the ultimate suppliers such as carriers, hoteliers, car hirers and insurance companies. The travel agency component of tourism supply is concentrated in that a limited number of firms with multiple outlets, some integrated with tour operators or carriers, dominate the market, particularly in the USA and lately in Europe. There is a need to generate a high turnover in the face of low margins, especially since commissions have significantly fallen over time as a result of disintermediation. This sector of the intermediary market has also experienced a rapid growth in numbers. In the USA there has been a dramatic increase in agency outlets since the early 1970s. With around 40,000 in existence in the late 2000s (Manta, 2009b), the great majority are single office agents, whereas only a limited number are branches of multiple firms. In fact, the size of the representative US travel agency remains small: 53.6 per cent of them employ between one and two full time employees, whereas another 28.8 per cent between three and five; 34.7 per cent of all US travel agencies produce less than $1 million turnover, followed by another 24.7 per cent which generate $1–$1.9 million; only 8.2 per cent of the companies achieve a turnover exceeding $10 million (American Society of Travel Agents, 2008). This situation is in sharp contrast with Europe, where the major multiples (100 or more branches) are of much greater importance despite the existence of about 80,000 tour operators and travel agents in 2008 (De Blust, 2008), reflecting the existence of low entry and exit barriers to the sector. In comparison with tour operators, the death rate record of travel agents is somewhat lower. Half the agents currently in business have existed for ten years or more. In terms of sales, however, the multiples predominate.

As in tour operations, economies of scale and scope characterize the travel agency sector but to exploit these fully in the UK, a company needs to own between 150 and 200 branches. The principal economies are gained in the provision of central, mainly back-office, services such as accounting and computing. However, the experiences of multiples in the UK in the late 1980s did not verify that scale economies were actually achieved where other smaller multiples were taken over. Because these struggling chains were not always profitable, the large multiples did not always increase their market share. What has become apparent is that the selective opening of new branches, sometimes in new forms such as in large stores or through franchising, has often been a better strategy than absorbing smaller enterprises (Liston, 1986). In any case, the majority of the large vertically integrated tour operators have significantly reduced their number of high street outlets since 2001 to economize on resources and centralize their activities. For example, Thomas Cook reduced them by 19 per cent, MyTravel by 30 per cent and TUI by 6 per cent – the trend was only bucked by First Choice, which increased the number of its outlets by 51 per cent (EC, 2007c). In the late 2000s, there were about 7,000 travel agency shops in the UK (FTO, 2009b).

Evidence provided by Liston (1986) indicates that travel agency, like tour operations in the UK, appears to be polarizing into the extremely large and the very small, although independent travel agents may cooperate to counter the dominance of the larger chains (Daneshkhu, 1997). The low entry costs contribute to explaining the large number of independently owned travel agencies but another important reason is that travel agency, as a retail activity, normally involves face-to-face contact with customers so that a more personal service is possible. Well-trained and skilled staff and close management control can be an asset to a business and engender a loyal clientele. The single independent travel agent or small local chain with a sound knowledge of the local market can perform better than the branch of a multiple, especially where specialization in certain services or types of tourism is required. In this sense, the large multiples with a uniform approach to selling holidays suffer from diseconomies of scale.

With respect to spatial factors, an advantageous location is an important determinant of success, but this is offset by the costs of prime sites because of the competition to occupy these from not only other travel agents but also other forms of retailing. Nevertheless, spatial separation of travel agencies lessens this competition and makes it possible for a firm to earn an adequate return on its investment. Though not conclusive, it partially explains the high proportion of independent agents in the USA, 60 per cent of which are located in suburban or small urban areas. Opinion is divided, both among academics (see, for example, Goeldner et al., 2005; Holloway and Taylor, 2006) and practitioners, as to the competitive structure of travel agents. Innovation, especially the introduction of information technology in the area of reservation systems and most importantly the Internet, may prove to have a mixed effect. Up to the late 1990s, principals could increase travel agency dependency and reduce competitors' impact by affecting the businesses that served them, and the terms on which they did so, by utilizing a system specific to their own operations as the only means of communicating with the company, for example Thomson's TOPS (Thomson Online Program System). Such practices not only constituted a restraint on trade, so acting against consumers' interests, but also increased agency costs because of the need to set up the means of accessing a number of systems. They also inhibited the long-run achievement of economies of scale by the agencies.

From the late 1990s onwards, advancements in Internet technology and online bookings empowered the principals, which subsequently decided to dramatically reduce their commissions to the travel agents and save money on distribution expenditure; illustratively, the majority of US air carriers have adopted a zero-commission policy since 2002 (Ioannides and Daughtrey, 2006). As a result, and according to ASTA (2008), the percentage share of airline sales by travel agents in the US went down by 53.7 per cent over the period 2000–2007; the same results hold for hotel bookings (−32.7 per cent) and car rental (−45.7 per cent). To face this development, many travel agents now charge service fees to their customers to survive financially. However, in the competitive marketplace, travel agents should convince their patrons that they can really provide a service of added value compared, for example, to what a web search engine may offer. The Internet has also led to the establishment of online travel agents such as

Expedia and Lastminute.com, which have become very popular among travellers in the USA and Europe. Based on the merchant model (according to which tourism providers sell their services to travel agents at a net price, enabling the latter to determine the retail price at their discretion), many of these companies managed to achieve significant profit margins and grow over time. Finally, the Internet has given the opportunity to traditional individual travel agents to work from home, reducing in this way the need to rent or own property for commercial purposes. Illustratively, 27.7 per cent of the members of the American Society of Travel Agents (ASTA) were home-based in 2007 compared to only 7.3 per cent in 2002; likewise, the percentage of those selling on the high street has fallen from 39.4 per cent to 28.9 per cent over the same period (ASTA, 2008).

The issue of whether trade associations perpetuate restrictive practices or promote greater efficiency in the intermediary sector is problematic. Those associated with travel agents have been able to secure as members a high proportion of those trading, largely because of the need to subscribe to schemes such as bonds as a guard against the failure of tour operators, as a prerequisite for the procurement of a licence or to reassure customers. ASTA in the USA and ABTA in the UK respectively are cases in point. Smaller businesses have also considered that these associations have given them a measure of protection, often by representing their interests to larger and powerful principals and public bodies. Of late, the function of these associations has been called in question as the cost of membership and protective schemes has risen sharply. Thus, many facets of the intermediary market are of current interest in economics, in particular the study of company strategies and operations in what would appear to be both oligopolistic and competitive situations. However, the interdependent relationship of travel agents and tour operators and the realization of the need for more effective lobbying at a national and European level is reflected in the merger between the FTO and ABTA which took place in mid-2008.

With respect to the regulatory framework at a European level, there is a rising trend towards the introduction of licence schemes for all travel agencies and tour operators regarding access to the profession. Nevertheless there is a significant number of EU Member States (including the UK) that have not yet introduced such a licensing system (De Blust, 2008). The Services Directive 2006/123 which will be implemented by the end of 2009 is expected to solve some of the issues puzzling the sector, such as the grounds for the establishment of the licence, the free provision of services and the imposition of restrictions for reasons of public security, health and environmental protection. Moreover, a formal consultation procedure to revise the existing Package Travel Directive in view of the recent developments in the industry commenced in 2007 (De Blust, 2008).

In conclusion, an oligopolistic structure prevails in the intermediary sector of the UK in that fewer than ten tour operators dominate the inclusive tour holiday market and determine the terms on which travel agents operate. This is particularly the case where tour operators have the potential to sell directly to the customer. Travel agents, therefore, are in the peculiar position of obtaining their inputs from an imperfectly competitive market but selling in one which is highly contestable. This peculiar structure again puts the theory of the firm to a severe test and the issue is one of the need for empirical investigation to establish the nature of the competitive structure of the travel agency sector, perhaps more in the context of an industrial economic analytical framework. Consumer protection seems to be a major issue as far as travel agencies' operation is concerned. There is a lot of scepticism regarding the direct purchase of travel products or services via different websites as this does not offer customer protection. In the same manner, travel agencies' licensing is an issue that needs to be resolved. It seems that the intention of both the industry and the regulatory bodies, within the European Union at least, is to set the common ground for a unilateral licensing system amongst travel agencies.

5.5 Tourism attractions

Tourism attractions can be conceptualized as multifaceted entities operating as business units in a competitive environment. Their market structure is close to monopolistic competition characterized by the existence of many buyers and sellers with good knowledge of the prevailing contestable market conditions and the provision of a differentiated product or service. Attractions are considered to be an integral part of the tourism product and their type, nature and purpose can make the difference in typical travel itineraries. A possible dilemma of tourist choice between two relatively similar destinations can be dissolved by differing appreciation degrees of the attractions found in the two areas. An attraction can satisfy various needs in the context of wanderlust tourism but also provide a place of entertainment for the local community. The complexity of the purposes served by an attraction is both an advantage and a drawback as a result of the challenges posed for the business entity owning or commercially exploiting the place to develop and apply appropriate marketing and pricing strategies.

5.5.1 Types of attractions

There are many ways to classify attractions depending on their ownership status, capacity, catchment area, permanence or type. Two main categories may be identified, i.e. natural attractions and built ones. The human-made attractions can be further classified into cultural, traditional and events-related, although boundaries are occasionally blurred. Clawson and Knetsch (1966) developed a sophisticated sorting of attractions based upon spatial criteria such as proximity to markets, level of rareness and intensity of use to make their categorization approach more flexible and precise.

In particular, the strong asset of natural attractions is the resource itself, while location and ease of access are considered to be of relatively minor importance. Coastlines and lakesides have proven to be the most popular natural attractions since water-related holiday packages and services are widely used for marketing reasons by the tourism industry. Europe's north–south tourism flows make it rather clear that in addition to the uniqueness and beauty of a natural attraction, there are other factors that influence the popularity of a destination such as the climate, the level of available infrastructure, etc. Many of the countries close to the equator are considered to be popular tourism destinations since their climate is mild throughout the year.

In many cases, areas surrounding natural attractions offer only a limited amount of tourism services; therefore, if there is an intense need to increase supply, this can only be realized at the expense of something else, usually the alternative use of the existing infrastructure or land. Prior to an investment in tourism, such alternatives have to be carefully evaluated in many ways to assess, among others, their implications for social welfare and the environment. In many cases the final decision concerning the land use belongs to the state, which can develop strict policies for the alternative uses of land. On occasions where the land is privately owned and there is conflict between the owner and the state, the public body in charge may decide whether to proceed or not with compulsory purchases and expropriations to preserve social welfare and protect the territory from over-exploitation.

The main problem that natural attractions have to deal with is their commercial viability since their maintenance costs cannot be always financially sustained based on their operational profitability. Seasonal use of natural attractions as opposed to the regular administrative costs that have to be incurred throughout the year should be taken into consideration by the authorities to preserve the social benefits of a place. Since market forces cannot ensure the viability of an attraction, the state should support its existence through appropriate measures. Besides, it is a common policy for governments around the world to own and financially support natural attractions since their existence is crucial for the welfare of the local population, the development of inbound tourism and the viability of several other sectors of the local economy directly or indirectly related to tourism. Parks in large metropolitan cities such as Central Park in New York City and London's Hyde Park are good examples of such natural attractions. In this context, state-owned natural attractions are usually regarded as public goods.

Sustainable tourism development is the term broadly used to describe the framework within which a natural attraction should be preserved and commercially exploited. Apart from intentional actions that could downgrade or destroy an attraction, excessive use is the most common reason for conservation measures to be taken. According to sustainable tourism principles, profit is desirable under the condition that the asset, of any kind, is kept integral and undamaged for the generations to come. Natural surroundings are of high priority since nothing non-natural could replace them, while any damage that might occur is considered to be almost irremediable. In the context of a beach, for example, the application of certain artificial modifications is often required to accommodate higher volumes of visitors. All interventions though have to be realized according to certain planning guidelines which take into consideration that over-exploitation and distortion are phenomena to be avoided. In several coastal destinations, local governments impose restrictions upon the level of accommodation and supply of services to keep visitor flows under control and preserve the environment.

Human-made attractions could face similar threats, though in these cases evaluation is based upon subjective criteria mainly related to issues of restoration and authenticity. More specifically, a human-made point of interest may be defined as a creation whose formation or design purposes are to attract visitors, such as museums and art galleries, theme parks and entertainment areas. Nevertheless, in many cases human-made attractions were originally built to satisfy specific community or social needs, but in the long run turned out to be of major historical or cultural interest: examples include religious sites, parks, gardens and monuments. In addition to such buildings as castles, churches, palaces there are many cases of old buildings that are no longer operationally viable commercially and have been transformed into tourism zones hosting leisure activities, exhibition centres and shopping malls. By the revival of such edifices a combination of economic development and upgrading of ex-industrial areas can be achieved, as in the case of the Meatpacking District of New York City (Greenwich Village Society for Historic Preservation, 2009).

5.5.2 Management issues

The management of attractions can be a rather complex issue since the attractions are numerous, operating under different ownership conditions, situated in diverse geographic environments and in need of sophisticated marketing approaches. The operational and administrative issues become more demanding and difficult to handle where the public sector is involved, even if the attraction is privately owned, bearing in mind that the state has simultaneously to deal with more than one point of public interest.

Where it is of national importance, the controlling administrative body has to develop a management plan that successfully incorporates each attraction's uniqueness and operational dynamics. According to Middleton (2002), a national strategy for managing attractions should incorporate several components. Among others the government body in charge should identify and study the demand- and supply-side trends related to the attraction's nature and type; study and check whether best practices developed in other similar sites could be incorporated in their administration plan; decide upon the management's type and funding issues of the chosen structure; contact national authorities to negotiate any tax concessions in addition to any available subsidies; and work in collaboration with the local authorities to include the region where the attraction is located in an active development project.

When planning the management structures and procedures, one of the most important factors to consider is the ownership status of an attraction since most of them are owned by the government, a public body or a public welfare institution. In this context, the willingness and flexibility of the owner to comply with the suggested actions for a viable and beneficial management of the attraction should be considered. Publicly owned attractions are privileged in a way since they are partly or entirely financed by government budgets that mainly spring from taxation. The problem of financially supporting the operation of an attraction in this way may be the different values assigned to the attraction by the taxpayers. The level of appreciation of a beach or a medieval castle varies among the local community and thus a golden rule of funding through taxation is not easy to find. In several developed countries where free market terms and conditions apply, museum visitors are charged for their admission. This approach is opposed by student communities, certain museums management teams and other social clusters who consider access to museums as a public service, admission therefore being free.

But what happens when an attraction cannot be self-financed? In this case, local or public authorities may pursue a cost–benefit analysis (CBA). Should the results of an impact study prove that the benefits arising from the attraction's operation via the multiplier approach are significant in terms of employment and income generation, then subsidies or public sponsorship from local authorities or other organizations can be of high significance. Apart from attractions that are publicly owned, there are several museums or galleries that belong to private organizations and thus cannot raise money from taxation to satisfy their operational and other costs. Most of these organizations try not to depend solely upon admission charges since they accept donations or receive membership fees. If their revenue sources cover maintenance costs then they might prefer not to charge entry fees since they normally have non-profit aims. Moreover, these associations might not qualify for public subsidies but they usually enjoy tax relief privileges and exemptions. Charitable funds are the most common example of this type of organization.

On the other hand, there are several attractions that are purely commercial and operate according to the free market philosophy, even if their activities are sometimes subsidized as their existence is beneficial for the prosperity of the local community. Most commercial attractions face seasonality problems and discontinuity in their revenue generation throughout the year. Since management and maintenance costs cannot be equivalently curtailed, sophisticated business plans have to be developed to ensure viability and operational completeness. In addition, amusement parks have to face changes in trends and social tendencies since what is in fashion today may be considered old fashioned and out of date tomorrow. Disney's amusement and theme parks, for example, try to attract multiple visits simply by periodically introducing new facilities and services to encourage customers to return some time in the future so that each visit is regarded as a new and original experience.

5.5.3 Cost structure and pricing policies

The cost structure of an attraction is similar irrespective of its type and purpose. Fixed costs (C_F) are considerably higher than variable ones, forcing attractions to seek high tourism volumes to reduce unit cost. In cases of high fixed costs, low numbers of visitors cannot even cover the repayment of the capital invested for the attraction to be fully and efficiently operational. In Figure 5.1 the R-line shows revenue from visitor admission fees, while the two parallel lines, C_1 and C_2, represent variable costs. In cases of zero visitors turn-out the cost that still has to be incurred is the fixed one, i.e. C_{F1}, C_{F2}. When the fixed cost is relatively low (C_{F1}), then to reach the break-even point (A) a number of V_1 visitors are required, while when fixed cost rises to (C_{F2}), then the break-even point (B) can only be attained if the turnout is respectively higher (V_2). As mentioned before, the cost requirements to keep an attraction operational are relatively high but sometimes the social cost of shutting it down might be significantly higher. As a result, public funding, tax exemptions, subsidies, bequests and donations can be of major importance for the attraction to remain functional and open to visitors.

As for the development of a pricing policy for an attraction, this can be quite complicated due to the many factors to be considered such as its type and ownership status, whether it is profit-oriented or not, etc. The management team has to study all these issues together with seasonality phenomena, long-term trends and national tourism policies to increase a pricing model that can ensure the attraction's viability and prosperity. To set the optimum admission fee, the management may wish to estimate first the

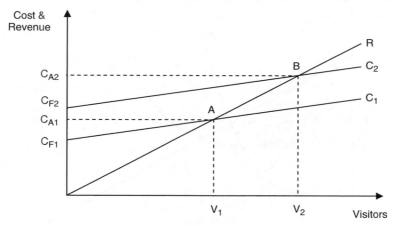

Note: C_F is fixed cost; C_A is aggregate cost

Figure 5.1 Cost structure of a tourism attraction

marginal cost that each ticket is expected to cover given the aggregate cost and the anticipated volume of inbound visitors. After the break-even level of charge is determined, the management team of the attraction may decide whether an extra amount should be charged for profit purposes or to generate a surplus for such requirements as maintenance of and reinvestment in the attraction. This surcharge is what usually varies among attractions according to their ownership status and objectives.

Based on Wanhill (2006), Figure 5.2 shows alternative pricing policies that can be implemented in the case of an attraction. More specifically, if the attraction enjoys monopoly power it may decide to charge an admission rate of P_1 where the expected size of visitors is given by V_1 on the demand curve DD. The aggregate profit is the difference between set price (P_1) and average cost (AC_1) multiplied by the number of visitors (shaded rectangle area AP_1AC_1C). Such a pricing policy ensures profit maximization while it limits the number of visitors to prevent damaging the attraction, for example by overuse where it is has fragile artefacts or environments. Nonetheless, admission rates at this level usually face severe public opposition and criticism, particularly where overcrowding and congestion might occur. The extreme opposite admission approach would be free entrance to all visitors, resulting in a maximum access of V_2. Such a socially oriented (non-pricing) strategy, however, may have significant drawbacks, such as the physical damage of the attraction already referred to above and the inability to generate revenue, thus making self-financing unfeasible.

In addition to the policies described above, an approach based upon productive efficiency principles implies a price of P_3. This corresponds to point E, where the marginal cost curve (MC) cuts the average cost curve

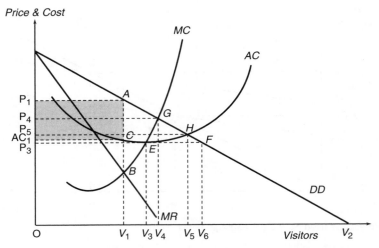

Note: P is price; MC is marginal cost; AC is average cost; DD is demand; MR is marginal revenue

Figure 5.2 Alternative pricing policies of a tourism attraction

(AC) at its minimum, resulting in V_3 visitors. The problem of this approach is that point E is located to the left of the demand curve, implying an amount of unsatisfied demand equal to EF: visitors may then need to queue to visit the attraction, being admitted on a first-come first-served basis. Point G (P_4, V_4), where demand (DD) and marginal cost (MC) curves intersect, represents a number of visitors V_4, which is allocatively (as P = MC) but not productively efficient. Similarly, a pricing policy based upon the break-even principle (intersection of DD and AC curves) corresponds to point H, where price P_5 generates V_5 visitors.

All pricing policies described above make the implicit assumption that visitors are charged the same price. Third degree price discrimination based upon yield management principles is a complex pricing strategy where visitors are classified into groups (children, students, elderly people, local residents, etc.) according to the admission fee they are willing and can afford to pay. The pricing structure development is brought to completion when these groups are determined and the respective charges are set. Moreover, low admission fees during the off-peak season may attract visitors who would never pay a visit under higher fixed pricing strategies. Sometimes, price discrimination techniques are even more sophisticated, such as in high or low season or throughout the week, by applying differential pricing by, for example, raising prices in the high season, or lowering them on weekdays or less busy hours during the day (for instance in the early morning or late afternoon or evening); a factor that would possibly determine daily pricing strategies, especially if the attraction consists of mainly outdoor facilities.

In conclusion, as for most services, a visit to an attraction involves satisfaction that is intangible in nature and cannot be stored or inventoried. Such an activity is information and experience intensive, while the visitor's physical involvement is an essential condition for the realization of the activity. This is a major issue that the management team has to incorporate into its plan in order to ensure the viability of the attraction. Since the popularity and quality of attractions are very important elements of the overall tourism product, both local and national authorities seek to implement appropriate management policies and techniques that would help them optimally handle visitor volumes throughout the year and adopt pricing policies that can ensure viability and continuation of an attraction's operation. A dilemma is the ownership status (private or public) since it is not easy to conclude under which an attraction would be most efficiently managed. It is common knowledge, though, that the competitiveness of a destination/attraction is largely dependent on how well the various stakeholders of the local economy cooperate jointly to provide the tourism experience based on the principles of comparative and competitive advantage. As is argued in Chapter 7, Integrated Quality Management may be the way forward in this case.

5.6 Conclusions

The outlines of the structure of the transport, accommodation, intermediary and other tourism sectors and the discussions of their key features reveal that market structures can be quite heterogeneous within each sector. They possess such characteristics as a wide range of competitive forms, market segmentation, product differentiation, high rates of entry and exit, some scale economies and significant variations in the degree of regulation. Moreover, there appear to be considerable variations in the structure of tourism supply in terms of the number of firms, their size and market share, particularly in the intermediary sector. These characteristics raise two questions. First, how far do the neoclassical theory of the firm and the competitive structures it has identified explain what is actually observed in tourism supply? Second, does heterogeneity within a specific sector suggest that the theory of the firm is inadequate or does it merely signify that the categorization of tourism sectors and markets adopted currently is too broad?

With regard to the first question, the test of the appropriateness of the theoretical concepts is whether they can explain actual market structures and predict the outcome of changes occurring in the industry or its markets. It appears that in each of the three main sectors considered, there are elements of contestability (whereby new or established firms can engage in price competition with existing firms) alongside the dominant market forms of monopolistic competition and oligopoly. This contestability aspect is not easily encompassed within the conventional theory of the firm because its static equilibrium analytical framework is not well suited to accommodating what is effectively a dynamic situation. Thus, while it is possible to explain the conditions of tourism supply sectors as being indicative of particular theoretical competitive structures, it is more difficult to predict the outcome of changes in the conditions. For example, in the transport sector the possibilities of exploiting economies of scale should give rise to increasing concentration and few firms, therefore denoting an oligopolistic structure, as appears to be the case in the air, bus, coach and ferry sectors. Furthermore, the theory would indicate reasonably stable prices, given the constraints which individual firms face in raising or lowering them because of possible reactions by rivals. However, deregulation has effectively reduced entry barriers, which has an impact running counter to that of economies of scale. It has created greater volatility in prices and led to an influx of smaller companies and greater competition in the short run. Although, after initially successful trading, there are signs of a decline in the number of firms because of failures by new entrants, particularly in the bus and coach sector, it is not certain that this will occur in the airline sector, at least in the very short run. Thus, the predictive power of conventional theory is called into question.

In the intermediary sector, particularly tour operations, increasing concentration is occurring in what conventional theory would posit to be a highly competitive market because of low entry/exit barriers and limited economies of scale. Paradoxically, the largest tour operators in what might be termed the oligopolistic segment of the UK holiday market have been engaged in fierce price competition and, over many years, margins and profits have been low and losses have even been sustained. Conversely, the many smaller specialized tour operators have avoided such competitive conditions. A similar scenario applies in the UK travel agency sector. The long-running pursuit of increased market share in the intermediary sector appears to be at odds with theory which assumes profit maximizing behaviour.

The second question as to the observed competitive heterogeneity of most tourism supply sectors and the relation to the theory of the firm is problematic. Analysis within the conventional framework tends to compartmentalize markets into clearly defined competitive structures, seldom acknowledging that sub-markets may be subject to different conditions. Does this matter? Is the theory undermined by the recognition that in particular markets different competitive forms coexist? The response by mainstream economists might be that the various segments should be treated as separate markets. However, this may not be appropriate for it has already been demonstrated that there is an interrelationship between the differing competitive segments within a single sector. For instance the largest firms within the transport sector do react to the behaviour of smaller and newer rivals. Accordingly, the further sub-division of tourism markets to meet traditional analytical requirements may not be appropriate.

While acknowledging that a full exploration of the applicability of the theory of the firm has not been undertaken, the illustrative examples indicate the limitations of its explanatory and predictive power in the context of tourism supply. The theory aids identification of the key variables determining individual firms' and industries' behaviour within each of the competitive forms and it acts as a rigorous foundation on which to develop more complex and differentiated models. Increasingly, the economic analysis of industry is recognizing the role of institutional structures and their possible evolution, together with the need to examine the dynamic nature of markets where information availability, uncertainty and transaction costs are significant. Attention, therefore, was turned to a more industrial economics-oriented perspective where a number of features of tourism supply, perceived as important to its structure and operation, are identified and examined.

The discussions of supply in this and the previous chapter have not taken explicit account of the international context of tourism markets, despite the importance of cross-country links between firms at different stages of the supply chain. The globalization of tourism supply, involving a variety of forms of economic integration, has implications for specialization in production and trade between different countries. Such issues as the contribution of tourism to income, employment and foreign currency earnings and the relationship between tourism and economic growth are relevant. These issues are explored in the following two chapters, prior to investigation of the wider environmental effects of both domestic and international tourism.

PART III
The economics of tourism at a national, regional and international level

PART III

6 TOURISM IN A NATIONAL AND REGIONAL CONTEXT

6.1 Introduction

The previous chapters analysed tourism from an essentially microeconomics perspective to provide a solid understanding of both demand- and supply-side issues. However, it is also important to address issues at an aggregate level and encapsulate the 'big picture' regarding the economic role of the sector. In this context, this and the following chapter discuss tourism in a macroeconomic context, first in this one at a national/regional and then in the next at an international level. In particular, the present chapter first focuses on measuring the size of the tourism industry in relation to the total economy. This is an important topic not only from a scientific and analytical point of view but also for policy-makers who need to have a relatively accurate estimate of the overall size of the sector before being exposed to the lobbying power of various pressure groups. Tourism satellite accounts (TSA) are extensively analysed in the chapter as they are broadly agreed to be the best methodological tool for this purpose, at least from an accounting perspective.

However, it is important to distinguish between the size of an industry and its aggregate contribution to the economy. In this context, the second section of the chapter discusses the economic impact of tourism, which is largely affected by structural economic characteristics such as the presence of linkages, leakages and displacement effects. From a quantitative approach, the calculation of the long established tourism multiplier is perceived as the main way to measure the impact of the sector; but developments in more general and sophisticated methods are beginning to supersede it. The chapter presents three alternative models, i.e. the multiplier, the input–output (I–O) and the CGE methods, with emphasis on the last. Instability of earnings is another feature of tourism with potentially important implications for its economic impact. In a dynamic framework, however, growth is of further significance: this topic is extensively discussed in the chapter, which highlights the contribution of new growth, regional and urban economics and economic geography theories alongside their policy connotations. In the same longer-term vein, the chapter then focuses on issues of economic transformation with a presentation of the structural and migratory repercussions of tourism as well as of the role of gender in tourism employment. Summary and conclusions then lead the way to Chapter 7.

6.2 The size of the tourism sector

The tourism sector is arguably of major importance in terms of both income and employment generation at a global level. Nevertheless, when it comes to the actual measurement of its size, a number of issues emerge associated with definitional caveats and classification problems. This is because tourism is only a partial industry consisting of sectors grouped together through demand complementarities rather than through supply of homogeneous goods and services. For example, the air transport and hotel accommodation industries are both regarded as part of the tourism sector in spite of the fact that the former is associated with servicing travellers while on the move, whereas the latter caters for the needs of people who are stationary (Smith, 1998). In fact, the two markets are functionally linked in terms of providing services to tourists in different phases of their journey.

To resolve this supply-side conundrum, the TSA concept has been developed. In particular, it is aimed at:

- presenting the constituent elements of a country's tourism sector and evaluating its size and impact; and
- collecting and analysing statistics related to all quantitative aspects of a country's tourism industry (Smith and Wilton, 1997).

The TSA can be compiled by applying specific tourism-related functional criteria on the System of National Accounts (SNA) of each country and its related Standard Industrial Classification (SIC). The SNA 'consists of a coherent, consistent and integrated set of macroeconomic accounts, balance sheets and tables based on a set of internationally agreed concepts, definitions, classifications and accounting rules' (United Nations Statistics Division, 2008a), whereas the SIC provides a detailed structure of sectors and their components in an economy. In its effort to harmonize data collection and enhance comparison of statistics across countries, the UN has proposed common national accounting methods known as the SNA93 documentation, where in Chapter 21 there is detailed discussion of satellite analysis and accounts (United Nations Statistics Division, 2008a). In fact, the methodology suggested by Chapter 21 has been subsequently incorporated in the TSA framework jointly proposed by the UNWTO, the Organization for Economic Cooperation and Development (OECD) and Eurostat (Spurr, 2006).

Moreover, the UN has developed a system of International Standard Industrial Classification known as ISIC. For example, within Section I (entitled 'Accommodation and Food Service Activities'), Division 55 refers to 'Accommodation', which comprises three groups, namely 'Short-Term Accommodation Activities' (group 551), 'Camping Grounds, Recreational Vehicle Parks and Trailer Parks' (group 552) and 'Other Accommodation' (group 559). Within group 551 there is Class 5510, which bears the same title as the group and includes 'the provision of furnished accommodation in guest rooms and suites or complete self-contained units with kitchens, with or without daily or other regular housekeeping services, and may often include a range of additional services such as food and beverage services, parking, laundry services, swimming pools and exercise rooms, recreational facilities and conference and convention facilities'. Class 5510 services are provided by 'hotels, resort hotels, suite/apartment hotels, motels, motor hotels, guesthouses, pensions, bed and breakfast units, visitor flats and bungalows, time-share units, holiday homes, chalets, housekeeping cottages and cabins, youth hostels and mountain refuges'. However, class 5510 excludes the 'provision of homes and furnished or unfurnished flats or apartments for more permanent use, typically on a monthly or annual basis' (United Nations Statistics Division, 2008b).

When constructing a TSA, one would expect Class 5510 to be clearly identified as part of the tourism sector as a whole. But why is it so; what makes Class 5510 a tourist one? The answer lies on the so-called 'tourism ratio', defined as the percentage of a sector's turnover which may be attributed to tourism demand. Each sector has a 'tourism ratio' and as a rule of thumb when this exceeds 15 per cent, then the sector is regarded as part of the tourism industry – in this context, the tourism ratio of Class 5510 is expected to be over 90 per cent (Smith, 1998). The tourism ratio may be calculated based on customer profile and expenditure surveys (i.e. to measure, for example, what percentage of a restaurant's turnover should be attributed to visitors instead of residents), information from service suppliers or expert opinions (UNWTO, 1999a). As expected, the tourism ratio is spatially and seasonally dependent. For example, the tourism ratio of the taxi sector is likely to be less than 5 per cent in London in the UK during the winter but over 15 per cent in Ibiza during the summer. As a result and in order to derive reliable tourism ratios and decide on whether a sector is part of the tourism product or not, it is important to rely on observations taken throughout the year (to address seasonality issues) but also to consider the sectoral orientation of the economy under consideration each time.

Based on the concept of the tourism ratio, the UNWTO classifies goods and services into tourism characteristic, tourism connected, tourism specific and tourism non-specific products. Tourism characteristic products are those which 'would cease to exist in meaningful quantity or that consumption would be significantly reduced in the absence of tourism, and for which statistical information seems possible to obtain' (UNWTO, 1999a: 36). These products are associated with hotel and other lodging services; second homes services on own account or for free; food and beverage serving services; passenger transport services; travel agency, tour operator and tourist guide services; transport equipment rental; cultural services; and recreation and other entertainment services (UNWTO, 1999a). Similarly, tourism connected products are those which 'are consumed by visitors in volumes which are significant for the visitor and/or the provider but are not included in the list of tourism characteristic products' (UNWTO, 1999a: 36). These two categories jointly comprise the tourism specific products as opposed to the tourism non-specific ones. Having the above in mind, the UNWTO then defines tourism characteristic and tourism connected activities. The former are undertaken by tourism characteristic producers in the context of tourism industries, defined 'as all establishments whose principal productive activity is a tourism characteristic activity' (UNWTO, 1999a: 42).

Stepping forward and moving closer to the main aim of the TSA, which is to measure the size of the tourism sector, it is important to address the issue of value added. This may be defined as the difference between the value of output and the value of intermediate consumption related to the factors of production (land, labour, capital and entrepreneurship) used as inputs in the production process. In other words, and to derive the value added, it is important to rely on input–output models usually associated with a number of spreadsheets where goods and services are listed horizontally in columns and industries are plotted vertically in rows (Smeral, 2006). In the 'make matrix', each cell shows the value of a given good or service produced by a specific industry in a specific period of time, whereas in the 'absorption matrix', each cell shows the value of every good or service consumed by a specific industry in a given period of time. Knowledge of the absorption matrix and total production in the economy enables the calculation of internal (or intermediate) consumption in the economy. Final demand can then be given in another spreadsheet (Smeral, 2006).

In the context of the TSA, it is important to distinguish between the Value Added (at basic prices) of Individual Tourism Industries (VATI), the Tourism Value Added (TVA) and the tourism GDP (TGDP). In particular, the value added generated by the provision to visitors of goods and services associated with the tourism sectors is considered by all three measures. Conversely, the value added generated by the provision to non-visitors of goods and services not associated with the tourism industries is not regarded by any of the three statistics. However, the value added generated by the provision to non-visitors of goods and services associated with the tourism industries is part of VATI but neither of TVA nor TGDP. Similarly, the value added generated by the provision to visitors of goods and services which are not associated with the tourism sectors is calculated in the TVA and TGDP but not in VATI. Finally, net taxes on products and imports incorporated in the value of tourism consumption (at purchasers' prices) are included in the TGDP but neither in VATI or TVA (UNWTO, 1999a).

Having the above in mind and according to the joint methodology proposed by UN, UNWTO, OECD and Eurostat, the Tourism Satellite Accounting system (UNWTO, 1999a; Spurr, 2006) consists of ten tables (spreadsheets):

1　**Inbound tourism consumption/expenditure** (related to the exports of tourism services), which identifies consumption of specific (i.e. characteristic and connected) and non-specific tourism products by same day visitors (i.e. excursionists), tourists and total visitors in terms of both gross and net valuation of package tours – gross valuation considers all components of the value of package tours, whereas net valuation covers only the components provided by resident suppliers;

2　**Domestic tourism consumption/expenditure** (part of total domestic consumption), which identifies consumption of specific and non-specific tourism products by same day visitors, tourists and total visitors in terms of both gross and net valuation of package tours;

3　**Outbound tourism consumption/expenditure** (related to the imports of tourism services), which identifies consumption of specific and non-specific tourism products by same day visitors, tourists and total visitors in terms of both gross and net valuation of package tours;

4　**Total tourism internal** (which synthesizes Tables 1+2) and **total tourism consumption/expenditure** (which synthesizes Tables 1+2+3);

5　**Production accounts**, which identify production of specific and non-specific tourism products by the domestic tourism industries, connected and other non-specific activities;

6　**Supply and tourism consumption of services** (arguably the most important table of the TSA), which identifies:

　a　production of specific and non-specific tourism products by the domestic tourism industries, connected and other non-specific activities;

　b　the tourism ratio of domestic tourism industries, connected and other non-specific activities with respect to the consumption of each specific and non-specific tourism product;

　c　the TVA calculated as the sum of the products of tourism ratios and the activity value added (by considering intermediate consumption); and

　d　the TGDP calculated as the addition of TVA and the sum of the products of tourism ratios and net taxes on products;

7 **Employment in the tourism industries**, which identifies the number of establishments, the number of jobs (male, female and total) and the status in employment (employees and other for male, female and total) in the tourism industries. The exact structure of this table has not yet been agreed;

8 **Tourism gross fixed capital formation**, which identifies net acquisition and/or improvements to tourism specific and other produced fixed assets by tourism industries as well as by public and private sector producers outside them. The exact structure of this table has not yet been agreed.

9 **Tourism collective consumption**, related to expenditure on collective non-market services such as tourism promotion, general planning and coordination, generation of statistics, administration of information bureaus, etc. at national, regional, local and total level;

10 **Physical indicators**, which identify the number of arrivals and over-nights by:
   a   inbound, domestic and outbound excursionists, tourists and visitors;
   b   means of transport (for inbound tourism); and
   c   forms of accommodation.

The table also classifies the establishments of characteristic and connected tourism activities according to the number of their employees (i.e. 1–4, 5–9, . . . , 250–499, etc.)

The completion of the above tables with reliable data enables TSA to measure scientifically a set of major economic variables related to tourism, such as the contribution and share of the sector in a country's GDP, fiscal and trade balance; the size of tourism employment and gross value added, etc. TSA also enhance statistical comparison of the tourism industry with other sectors of the economy based on approved accounting methods derived from the SNA. TSA identify industries and products which belong either to the 'core' or to the 'periphery' of the tourism sector and set the fundamentals for the construction of performance indicators, which are useful for both policy-makers and investors (Spurr, 2006). Still, TSA are expensive to derive and require a major commitment in monetary, time and human capital resources. Dearth of available data constitutes another important operational difficulty. Moreover, it should be noted that TSA measure only direct effects of tourism demand and do not deal with indirect and induced effects: in other words, they are an appropriate measure of the size of the tourism industry but not of the sector's overall contribution in the economy. Smeral (2006) considers this as a potential problem and proposes the extension of the TSA to incorporate further elements of input–output modelling. On the other hand, Smith and Wilton (1997) seem to perceive this as a factual virtue, arguing that the function of TSA should not be the measurement of impacts (discussed in the following section).

In any case, TSA seem to be a very promising tool for tourism analysis and countries which have implemented them (with Canada and France playing a leading role) have benefited from a major advance in tourism statistical information. Moreover, the gradual extension of the TSA in various directions, including the OECD 'employment module' (Dupeyras, 2002), capital formation (Miller, 2002) and implementation at a regional and/or sub-national level (Spurr, 2006), may lead to their further improvement and usefulness. For example, Diakomihalis and Lagos (2008) used a TSA methodology to evaluate the importance of yachting in Greece.

6.3 The economic impact of tourism

This part of the chapter will consider some of the effects which tourism can bring about, such as changes in income and employment, which are of concern to private individuals and firms as well as the public sector. Tourism creates income and employment directly in the sectors in which expenditure or tourism-related investment takes place and also induces further increases throughout the economy as the recipients of rises in income spend a proportion of them. Income and employment generation result not only from expenditure by foreign tourists, along with the associated increases in private investment and public expenditure, but also from domestic tourist expenditure, which often exceeds that by foreign tourists. It is useful to consider the effects of both types of expenditure within a national and/or regional context in order to take explicit account of the leakages in other regions or abroad in such forms as payments for tourism-associated imports and remittances of income, profits and dividends, which tend to be important in small, open economies.

The discussion first focuses on the structural elements which magnify or reduce the economic impacts of tourism, namely inter-sectoral linkages, leakages and displacement effects. Subsequently, alternative models to measure the importance of tourism economic effects are presented based on the concept of the multiplier. In particular, the Keynesian multiplier model, the I–O approach and the CGE framework are examined. The part concludes with an analysis of the repercussions of earnings instability for the economic implications of the tourism sector.

6.3.1 Linkages

The impact of tourism at a national and regional level depends undoubtedly on the existing and potential linkages between tourism and the remaining sectors of the economy. These linkages may be classified into forward, which examine the relative significance of tourism as a provider/supplier of downstream activities, and backward, which measure the relative significance of tourism as a consumer/demander of nationally and/or regionally upstream produced goods and services (Hirschman, 1958). A typical example of backward linkages would be the purchase of local food and fruit by a hotel restaurant or of local furniture to equip hotel rooms. On the other hand, forward linkages would involve, for example, the sale of airline seats to a tour operator who prepares holiday packages for the final consumer. It is possible even for front-line tourism sectors (i.e. those which are in direct contact with the visitor) to generate forward linkages if their services constitute an input for other business uses. Illustratively, the car-rental sector may not create forward linkages in the case where a car is rented by a leisure tourist. Nonetheless, if the sector rents a car to a business tourist who subsequently uses this transportation service for further downstream commercial transactions, then forward linkages do exist (Cai, Leung and Mak, 2006).

In their study of inter-industry linkages for Hawaii's top 20 tourism-related sectors in 1997, Cai, Leung and Mak (2006) found that the majority of industries had relatively significant backward, but only modest forward linkages: in particular, this group comprises sectors such as hotels, automobile rental, amusement services, golf courses, other general merchandise stores, apparel and accessory stores, recreation services, food services, museums and historical sites and bakeries and grain product manufacturing. Other state and local government enterprises had important forward and backward linkages, while investigation and security services, advertising and support activities for transportation had limited backward linkages but significant forward ones. Finally, sightseeing transportation, air transportation (although marginally), ground passenger transportation, miscellaneous store retailers, travel arrangement and reservation services and department stores had only limited linkages with the rest of the economy.

6.3.2 Leakages

When inter-sectoral linkages are strong, then the tourism impact is usually larger. This is because more money is expected to stay within the economy instead of leaking out, primarily in the form of imports (Sinclair and Sutcliffe, 1978). In some Caribbean countries, for example, the import content of beverages and cigarettes has been as high as 69 per cent and that of food 62 per cent (Cazes, 1972). Considering countries in other parts of the world, the import content of beverages and food in Fiji has been estimated as 45 per cent and 56 per cent respectively (Varley, 1978), while the overall import content of tourism expenditure in the Gambia was 55 per cent (Farver, 1984). In contrast, the import content of beverages in Kenya was approximately 35 per cent and that of food 10 per cent (Sinclair, 1991b) and there is scope for decreasing the import content in other countries, as illustrated by the case of food production in Lombok, Indonesia (Telfer and Wall, 1996). The corresponding values for most developed countries, for example Spain (Sinclair and Sutcliffe, 1988b), are also relatively low. In general, the import content is expected to be lower in larger countries with a diversified economic structure. In any case, leakages may be reduced through administrative measures, i.e. by imposing protectionism in favour of goods and services produced within the national borders. As argued in Chapter 7, however, such a strategy may trigger retaliation from other countries to the detriment of tourism. From an alternative and perhaps more viable perspective, destinations are advised to increase their competitiveness (by raising their value-for-money through cost reduction and/or service quality improvements) and/or to switch tourist preferences in favour of local produce through advertising and other practices. This may also have positive repercussions for local exports of specific goods abroad. For example, until the early 1990s Greek wine and dairy products (such as feta cheese and yoghurt) were difficult to find in British major supermarkets, such as Tesco and Sainsburys. Nonetheless, the dramatic increase of British tourists visiting Greece, together with the new social trend to bring 'a flavour of holidays back home', encouraged the gradual entry of Mediterranean cuisine into British homes and restaurants, increasing demand for Greek agricultural products. Moreover, and in the context of corporate social responsibility and fair trade (discussed in Chapter 7), large transnational hotel companies may make serious efforts to fortify their linkages with the local economy when not bound by contracts with international suppliers.

6.3.3 Displacement effects

The extent to which backward and forward inter-sectoral linkages are likely to affect income and employment depends, in part, upon the existence of the appropriate surplus resources within the economy. For example, airport capacity must be sufficient to meet additional demand by incoming and departing tourists, hotels and guesthouses must have less than full occupancy rates so that more tourists can be accommodated and labour with the appropriate skills must be available to provide the additional services required. If surplus capacity is not available, as in periods of high seasonal demand, additional demand results in rising prices and wages rather than increases in real income. Monetarists such as Friedman (1968) argue that if the economy has stabilized at the 'natural rate of unemployment', subsequently re-conceptualized as the non-accelerating inflation rate of unemployment (NAIRU), increases in demand give rise to an increasing rate of inflation rather than a rise in real output, even in the context of spare capacity. New classical economists (for example, Lucas, 1972; Sargent and

Wallace, 1976) add the assumption that people use all available information when making forecasts (rational expectations) so that short-run changes in real output occur only when they are mistaken. Hence, it is necessary to examine whether there is surplus capacity in the economy in order to assess the effectiveness of linkages and the potential for increases in demand to raise output. Moreover, it is important to study the way in which consumers and business people formulate their expectations about the future in order to determine whether increases in demand result in higher output or inflation. The assumption that spare capacity is available and is utilized to produce more output in response to an increase in demand is typically Keynesian and, as discussed in the next section, it provides the basis for the use of the multiplier methodology to estimate the value of income and employment which is generated by tourism. If surplus capacity is not available, the use of the methodology is likely to provide spurious results. Wanhill (1988) attempted to take account of this problem by introducing capacity constraints into his multiplier model.

It should be noted, however, that such displacement effects may not occur in case of economic growth. More specifically, when the business outlook is positive, investors may be willing to relax capacity constraints by investing in new tourism infrastructure and production practices which allow a better utilization of existing resources. For example, if the existing airport capacity is not sufficient, a new runway or terminal may be built: although in most European countries, this may be difficult to implement due to spatial constraints and environmental concerns; the situation in the Middle East or in countries such as India and China is very different. Moreover, in a European or North American context better air navigation techniques may relax the problem of constrained capacity: in late 2008, for example, Eurocontrol (the European Organization for the Safety of Air Navigation) and Airports Council International (ACI) Europe signed an agreement to jointly address airport congestion and reduce fuel emissions (Eurocontrol, 2008). As discussed later in this chapter, new growth theorists argue that economic growth is endogenously rather than exogenously determined: as a result, active supply-side policies but also demand-side public sector interventions can have a real effect in the economy through inter-sectoral backward and forward linkages. This effect may also have spatial connotations in the context of tourism urbanization and the development of resorts (discussed later).

Nonetheless, the importance of linkages in an economy may be reduced when some other effects are taken into account. For example, the theory seems to neglect the personal income distribution effects which result from the change in tourism demand. Each type of tourist expenditure is associated with different distributional repercussions and although the aggregate welfare gains may be significant, there may be adverse effects on particular individuals and groups, notably those who do not own the land used in the service sector (Copeland, 1991). Thus, a promotional strategy, which favours, for example, luxury coastal tourism, is likely to provide additional income for property developers but often results in negative externalities such as marine pollution which may adversely affect the livelihoods of poorer fishermen. Moreover, local people who are displaced from the areas of tourism growth may be inadequately compensated for their loss of land and therefore experience unfavourable wealth effects, as in the case of Santa Cruz in Mexico (Long, 1991). Such effects should be considered alongside the more readily quantifiable income and employment effects but are all too often ignored. If the probable costs and benefits, both economic and non-economic, of different types of tourism were made explicit, a choice between them could be made in the context not only of their estimated income and employment generation but also of their wider social and distributional consequences (Tsartas, 1992; Tsartas and Lagos, 2004), as well as their environmental impacts, discussed in Chapters 8–10.

6.3.4 Multiplier: concept and application

The evaluation of the economic impact of tourism from a quantitative perspective is essentially based on the concept of the multiplier, defined as the ratio between the value of sales, output, income, employment or government revenue generated and the initial change in tourist spending or tourism-related investment. In brief, the underlying logic of the tourism income multiplier is that the initial expenditure undertaken by the tourist will have not only direct implications for the economy but also a number of indirect and induced effects (Cooper et al., 2008). Consider, for example, the case of a tourist who spends 100 euros in a hotel. The direct (or first-round) effects of this tourism expenditure are related to the income staying within the destination after subtracting any savings as well as the cost of imported goods and services required by the hotelier to offer their product to the customer. Let it be assumed that this income is equal to €80. This is then partly spent by the hotelier on goods and services provided by local suppliers (e.g. furniture manufacturers, lawyers, builders), who use this money to fund subsequent rounds of consumption of locally produced goods and services, leading to a further boost of the local economy. Assume that as a result of these indirect effects an extra income of €50 is generated in the economy. Finally, the initial tourism expenditure creates induced effects associated with the remuneration of local factors of production (i.e. capital returns, entrepreneurial profit, labour wages and land rents), whose owners in turn spend part of their money to consume locally produced goods and services. If these induced effects are assumed to create €30 of local business turnover, then the total impact in the economy is €80 + €50 + €30 = €160, leading to a tourism income multiplier equal to $160/100 = 1.6$. Similar calculations would be made to estimate the employment multiplier. In general, the larger the linkages within the local economy and the smaller the leakages and the displacement (and/or negative externality) effects are, the larger the size of the income/employment multiplier is likely to be.

It might, therefore, be supposed that high multiplier values are more beneficial to the local economy than low values. However, it is necessary to consider both the multiplier value and the aggregate value of the income or employment generation, as it is possible for a high multiplier value to accompany a low value of total income generation, owing, in part, to a low initial level and change in demand. For example, the multiplier values associated with small-scale establishments may be high, owing to the high local content of tourists' expenditure on them, but the value of expenditure on them, both in per capita and in aggregate terms, tends to be low, so that the magnitude of income and employment generated is low. In contrast, the multiplier value associated with expenditure on higher category hotels may be low because of the higher import content of tourists' expenditure, but total income and employment generation is usually relatively high, owing to the high value of tourist expenditure on them.

The multiplier concept is also structurally intertwined with the theory of the accelerator, which relates net (that is, gross minus replacement) investment to the rate of change of national (or regional) income (Harrod, 1939). Rising national income requires net investment whereas falling national income may signify that even replacement investment might not be necessary. The accelerator model assumes that companies wish to keep a fixed ratio between the output currently produced and their existing stock of fixed capital assets. This is called the capital–output ratio or the accelerator. For example, a capital output ratio of 4:1 means that, for example, €4 of capital is required to produce €1 of output. Having the above in mind, an increase in national (or regional) income resulting from an initial tourism expenditure through the multiplier process may subsequently create a positive business climate and trigger new investment in tourism and/or other projects through the accelerator principle, leading to a self-reinforcing growth pattern. Conversely, low tourism multiplier values (i.e. below unity) could result in negative expectations about the future of tourism business in an area, inducing market exit due to corporate failure (Vlamis, 2007) or at best disinvestment, either actual (e.g. through demolition of existing hotels) or through depreciation of existing infrastructure.

There are three major models used to calculate the value of the tourism multiplier, i.e. the Keynesian multiplier model (and its ad hoc version), the I–O model and the CGE approach. The models will not be discussed in detail here as they have already been the subject of much attention (for example, Archer, 1973, 1977a, 1989; Sinclair and Sutcliffe, 1988a, 1988b, 1989;

Johnson and Thomas, 1990; Fletcher and Archer, 1991; Blake et al., 2003, 2006; Dwyer et al., 2003, 2006, Sugiyarto et al., 2003; Cooper et al., 2008). In any case, however, their major principles are analysed in what follows.

a. The Keynesian model

The Keynesian income multiplier model involves the use of a range of equations to estimate the aggregate multiplier values associated with different types of tourist expenditure or investment and the alternative definitions of income which can be measured: gross domestic product (GDP) or gross national product (GNP) at market prices or factor cost or disposable income (Sinclair and Sutcliffe, 1988a). A good example of a short-term Keynesian multiplier is given by Equation (6.1) (Archer, 1976):

$$k = \frac{1-L}{1-(c-c_j-t_ic)(1-t_d-b)+m} \tag{6.1}$$

where k is the value of the multiplier, L is the value of first-round leakages out of the economy; c is the marginal propensity to consume goods or services produced within a region (or country); c_j is the marginal propensity to consume goods or services produced outside a region (or country); t_i is the marginal rate of indirect taxation; t_d is the marginal rate of direct taxation and national insurance contributions; b is the marginal rate of government benefits (i.e. transfer payments), which are expected to vary inversely with income; and m is the marginal propensity to import goods and services used by the industrial, commercial and service sectors of the regional (or national) economy. Linkages (essentially associated with c) increase the value of the multiplier, whereas leakages (related to the remaining elements of Equation 6.1) reduce it. Interestingly, an increase in one leakage category may reduce the value of another: for example, a rise in t_d will reduce disposable income and consumption of domestic goods as services, leading to a decrease of government revenues from indirect taxation. Based on the multiplier effect an expansionary fiscal policy and/or a rise in the demand for private investment could have a stimulating effect on the tourism economy.

In addition to the above derived form of the Keynesian multiplier, ad hoc models may also be constructed to address the particular needs of specific studies. Such models are particularly useful in the context of regional analysis as they allow disaggregation at a high level of detail. Archer (1976) proposes the following ad hoc regional multiplier:

$$k = \sum_{j=1}^{N}\sum_{i=1}^{n} Q_j K_{ji} V_i \frac{1}{1-c\sum_{i=1}^{n} X_i Z_i V_i} \tag{6.2}$$

where k is the value of the multiplier; j is each type of tourist ranging from 1 to N; i is each type of business ranging from 1 to n; Q_j is the proportion of total tourist expenditure undertaken by the jth type of tourist; K_{ji} is the expenditure share of jth's type of tourist in the ith type of business; V_i is the direct and indirect income generated by one unit of expenditure by the ith type of business; c is the marginal propensity to consume; X_i is the share of total consumer spending by the residents of the region under consideration in the ith type of business; Z_i is the proportion of X_i which takes place within the region. Archer (1976) develops alternative formulas for the estimation of V for each ith type of business to subsequently derive specific multiplier values.

In terms of actual statistics, the multiplier value in the Spanish province of Malaga for GNP at factor cost was estimated at 0.72 and that for disposable income at 0.54. This implies that for every €1,000 of tourist expenditure, provincial GNP increased by approximately €720 and the disposable income of the local population rose by around €540. Thus, the calculated multiplier and income generation values also vary according to the definition of income which is being measured, although few authors have provided an explicit definition of the type of income which they are measuring. Multiplier values for the change in GNP or GDP are likely to be significantly higher than those for disposable income. The calculation of GNP or GDP multiplier values alone may be misleading because it is disposable income which remains in the hands of the local population after remittances, income tax and national insurance payments have occurred. Tourist expenditure and investment also vary over the tourism season and time, causing different related short- and long-run multiplier and income generation values (Sinclair and Sutcliffe, 1989).

In conclusion, Keynesian-type multipliers can offer interesting insights on measuring the economic impacts of tourism. Compared to I–O and CGE multipliers, they are usually easier and more straightforward to calculate; hence they may be more appealing to practitioners and analysts in policy-making departments. Still, Keynesian-type multipliers face a number of caveats. First, their micro-foundations are rather problematic due to their ad hoc nature and assumptions, i.e. they are not the product of a rational optimization process. As discussed in Chapter 4, individual agents are likely to be non-optimizers for a number of reasons; however, if such an approach prevails in quantitative analysis, then the robustness of its results may be seriously questioned as multiple answers can emerge by assuming different degrees of 'irrationality'. Second, and building on the Keynesian tradition, multipliers of this type seem to ignore the importance of displacement effects, assuming implicitly the existence of a perfectly elastic aggregate supply curve. While this may be possibly true at a micro-regional level and/or during periods of major unemployment, such an assumption becomes clearly restrictive at an aggregate level and/or when an economy experiences a boom. In any case and to pre-empt the discussion of Chapter 9 on cost–benefit analysis (CBA), Keynesian-type multipliers seem to only deal with the financial and employment benefits of tourism and not the costs falling on destinations and their inhabitants and natural environments. Finally, Keynesian-type multipliers are rather static in nature and unable to incorporate changes in the internal structure of an economy. On the other hand, this is an area explicitly considered by I–O models, as argued in the following section.

b. The input–output model

Like the Keynesian income multiplier model, the I–O model permits the calculation of the aggregate multiplier values for different definitions of income. However, it also provides estimates of the multiplier values for different sectors of the economy, such as food and drink provision, electrical equipment, textiles and cleaning materials. It therefore demonstrates the relative importance of the interrelationships between the different sectors at a given point of time (Zacharatos, 1986). It assumes the existence of a transactions table where rows show sales and columns purchases. The productive sectors of the economy (i.e. industries 1, . . . N) purchase inputs from other domestic industries (the inter-industry matrix) and primary factors of production (labour, capital, land) but also from abroad (in the form of imports), all subject to taxation. In addition to this intermediate demand, the final sectors of the economy (related to household consumption, private investment, government and exports) also purchase inputs subject to taxation (in the form of goods and services) produced by the domestic industries (productive sectors), the primary factors of production or imported from abroad. Intermediate and final demand jointly generate the total demand for inputs and simultaneously determine the total amount of output to be produced. As a result, an increase in final demand sales (due to a rise in tourist expenditure) will subsequently generate additional turnover for all the sectors of the economy in the form of both inputs and outputs. Having the above in mind and using simple matrix algebra, the following formulation can be obtained (Cooper et al., 2004):

$$\Delta \mathbf{X} = (\mathbf{I} - \mathbf{K}^*\mathbf{A})^{-1}\Delta \mathbf{Y} \qquad (6.3)$$

where $\mathbf{X}$ is the vector of total sales of each sector in the economy; $\mathbf{I}$ is the identity matrix; $\mathbf{A}$ is the matrix of inter-industry economic transactions; $\mathbf{K}^*$ is the matrix of competitive imports, which reduce the consumption/production of domestic goods and services; $\mathbf{Y}$ is the vector of final demand sales, and Δ shows change. In this context, a change in the tourist expenditure may result in the following change in primary inputs (Cooper et al., 2004):

$$\Delta \mathbf{P} = \mathbf{B}(\mathbf{I} - \mathbf{K}^*\mathbf{A})^{-1}\Delta \mathbf{T} \qquad (6.4)$$

where **P** is the vector of total inputs in the economy; **B** is the matrix of inter-industry primary inputs; and **T** is the vector of tourist expenditure spent in each sector of the economy.

The I–O model provides detailed information on the actual structure of the economy and hence generates multipliers, which take into account all the underlying effects at a direct, indirect and induced level. Still, the model is based on a number of restrictive (and similar to the case of Keynesian-type multipliers) assumptions, which may lead to positively-biased estimates (Dwyer et al., 2006). First, the components of the final demand are assumed to be exogenously determined, unless their change constitutes the assumed shock to the system. In other words, their values are provided in an ad hoc manner instead of being determined by specific behavioural equations as would be required from a microeconomic perspective. Second, the factors of production are assumed to be endogenously determined, implying a perfectly elastic aggregate supply curve; as discussed in the previous section, any increase in output in this case can be accommodated without additional opportunity cost or displacement effects. As a result, there are no upward pressures on prices which could discourage increases in final quantity demanded or trigger substitution effects among goods and services or production factors. The marginal rate of substitution (at a consumption level) and the marginal rate of technical substitution (at a pro-duction level) are assumed to be constant; real wages are fixed as well as nominal and real exchange rates. The additional assumption of a small country with no buyer power enables the use of fixed world price of imports (Dwyer et al., 2006).

Despite these weaknesses, the I–O model is still extensively used in tourism economic impact analysis. For example, the model has been applied to a wide range of countries, including Antigua (Pollard, 1976), the Bahamas and Bermuda (Archer, 1977b, 1995), Hong Kong (Lin and Sung, 1983), Korea (Song and Ahn, 1983), the Philippines (Delos Santos et al., 1983), Greece (Zacharatos, 1986); Palau, Western Samoa and the Solomon Islands (Fletcher, 1986a, 1986b, 1987), Mauritius (Wanhill, 1988), Singapore (Heng and Low, 1990; Khan et al., 1990) and the Seychelles (Archer and Fletcher, 1996). In any case, however, and to account for these caveats, a number of researchers have stepped forward to focus on CGE models.

c. The computable general equilibrium (CGE) model

General equilibrium theory aims to explain the behaviour of demand, supply and prices in the context of the aggregate economy where a multi-tude of markets exist for both goods and services and factors of production. General equilibrium theory is regarded as part of microeconomics based on constrained optimization principles, which embed the interactions among the various sectors of the economy; this is in contrast with partial equilibrium theory that lies extensively on the 'ceteris paribus' concept. For example, partial equilibrium theory would assume that a rise in the demand of a good would only lead to a price increase along an upward sloping supply curve; at best, partial equilibrium theory would also study the impact of this price increase on the demand of related substitute and complementary products. General equilibrium theory, however, would study not only changes in demand conditions but also how this price rise would affect the structure of production in terms of factors used and their remuneration. In a general theory equilibrium context, therefore, a rise in the demand of one good could have important implications not only for the price of that particular product but also for the shape and location of demand and supply curves across all markets. As expected, general equilibrium theory may involve a very complicated framework based on complex mathematical equations, whose analytical and/or numerical solution requires substantial computational power.

Until the early 1970s general equilibrium theory was essentially a theoretical subject. From then onwards, however, various efforts were undertaken to operationalize the theory and make it useful for applied economists and policy-makers. Scarf introduced applied general equilibrium (AGE) models in 1967 (Scarf, 2008); these have been subsequently replaced by CGE models since the mid- 1980s. In particular, a CGE model comprises a set of equations that describe the introduced variables as well as a supportive dataset. The equations abide by the general microeconomic optimization principles and foundations; in fact, CGE models may be built in such a way as to incorporate non-market clearing, imperfectly competitive markets, externalities, government expenditure, taxation, etc. The supportive dataset contains transaction tables (similar to those discussed earlier in the input–output model) at a detailed level as well as elasticities related to consumption (e.g. own and cross-price elasticity of demand, income elasticity of demand) and production (e.g. price elasticity of supply, elasticities of factor substitution).

All CGE models are characterized by the existence of a large number of potential variables compared to the available equations, implying that some variables should be treated as exogenous, i.e. not determined by the model. The actual decision on the variables to remain exogenous is called model closure and may raise robustness concerns, i.e. numerical results may become seriously dependent on the actual choice of endogenous variables. In many cases, variables such as technology and consumer tastes are treated as exogenous (Ginsburgh and Keyzer, 2002). Accordingly, CGE models may also be classified into comparative static and dynamic. The former model economic changes using a snapshot approach and do not provide a detailed description of the adjustment process to the new equilibrium. On the other hand, the latter are built in such a way as to embed explicitly the transitional process in the numerical solution. Nonetheless, such dynamic CGE models may pose substantial modelling and computational challenges. The General Algebraic Modelling System (GAMS) and the General Equilibrium Modelling Package (GEMPACK) software are extensively used by CGE modellers to derive numerical solutions (Brooke et al., 1988; Dixon et al., 1992; Ginsburgh and Keyzer, 2002; GAMS, 2008, GEMPACK, 2009).

In the context of tourism, CGE modelling has been used since the mid-1990s to measure its current impact on the economy (Copeland, 1991; Dwyer and Forsyth, 1994; Zhou et al., 1997 Adams and Parmenter, 1999; Alavapati and Adamowicz, 2000; Blake, 2000; Blake et al., 2000, 2003, 2006; Dwyer et al., 2003, 2004, 2006; Sugiyarto et al., 2002; Blake et al.; 2008; Wattanakuljarus and Coxhead, 2008; Pambudi et al., 2009) or even to forecast its potential contribution (Blake, 2005). In line with applications of CGE modelling in other sectors and in contrast with the input–output specification, the main components of final demand are assumed to be endogenously determined, whereas factors of production are usually regarded as exogenous, operating close or at full capacity according to the neoclassical tradition. Expenditure incurred by the government is usually variable, whereas tax rates and nominal exchange rates are fixed. Prices of tourism goods and remuneration of associated production factors are flexible; they play a major role both as inducers of substitution and quantifiers of the emerging displacement effects (Dwyer et al., 2006).

Having the above in mind, it is not surprising that multipliers produced by CGE models are usually much more modest in value compared to those generated by I–O models. In their study of the economic impact of a large event held in New South Wales, Australia, Dwyer et al. (2006) calculated a number of multipliers using both I–O and CGE techniques. For example, the income (value-added) multiplier for New South Wales (NSW), the rest of Australia (RoA) and Australia in total (Aus) was estimated at 0.8, 0.09 and 0.8 respectively using I–O modelling; on the other hand, the CGE multiplier values for the same regions were 0.4, –0.2 and 0.3 respectively. Likewise, the employment multiplier (per million of Australian dollars) was estimated at 10.2, 1.4 and 11.6 for NSW, RoA and Aus respectively using I–O, but only at 6.2, –3.7 and 2.5 for the same regions based on CGE modelling. In other words, the CGE multiplier values were not only shown to be significantly smaller compared to the I–O ones but even had a negative sign in the case of RoA. This raises concerns about the possibility of a zero-sum game involved in the organization of large scale events at a regional level, i.e. NSW seems to benefit to the detriment of the RoA. This is the reason why the CGE multiplier values are very small at the aggregate national scale.

In conclusion, CGE modelling offers an interesting alternative to both the Keynesian and the I–O model specifications with respect to the measurement of the economic impact of tourism at a national and regional level. Its reliance on a set of possibly restrictive assumptions may occasionally raise concerns over the validity of its results; in any case, however, CGE modelling seems to better encapsulate (short-run) displacement effects compared to the other proposed methodologies. Moreover, CGE modelling can explicitly consider environmental variables and hence provide an excellent tool when conducting CBA. As a result, and despite its costly implication, CGE modelling may provide useful advice to policy-makers, especially on decisions concerning the funding of major events and/or the construction of infrastructure associated with large sunk costs.

6.3.5 Tourism and instability of earnings

Tourism has been one of the main sources of foreign currency in many countries but is also perceived as a sector which is particularly prone to instability. In Kenya, for example, earnings from tourism were more important in absolute terms than those from any other export in the late 1980s but real dollar earnings, both total and per tourist, were lower during the 1980s than during most of the 1970s (Sinclair, 1990, 1992). Similarly, Papatheodorou and Song (2005) examined the evolution of various tourism time series over the period 1960–2000 in the six major UNWTO regions and the world collectively. Performance differs dramatically among the regions, fluctuations are sharp and negative tourism growth is not unusual in the case of international tourism receipts in both real and per capita terms.

Tourist expenditure and investment in tourism services vary between different locations and time periods owing to such causes as seasonality, one-off sporting or cultural events, which raise demand for a short period, or political instability, which depresses it. The associated multiplier effects on income and employment are negative where tourism demand decreases or, assuming surplus capacity, positive if it rises. The new levels of income and employment persist only if the change in tourism expenditure persists; if not, they occur on a one-off basis, after which income and employment return to their initial levels. Thus, tourism-related income, employment, foreign currency generation and the welfare of local inhabitants can vary considerably over time.

Instability in tourism receipts can have a range of adverse repercussions on destination countries (Rao, 1986). In addition to the simple multiplier effects on income and employment, a decrease in the rate of growth of income from tourism and other sectors of the economy can induce a fall in investment as argued earlier in the context of the accelerator theory.

Investment also depends on potential investors' expectations, which may be affected adversely by instability in foreign currency earnings and uncertainty about its cost and availability. A further repercussion is that purchases of capital goods imports may be deterred, which, in turn, will cause instability in the fiscal yield from import and export taxes as well as from income and indirect taxes, thereby possibly constraining government spending on infrastructure and services. If such effects occur, they are likely to lower the rate of growth, particularly in developing countries which have low levels of foreign currency reserves.

A somewhat different view of the effects of export earnings instability, put forward by Knudsen and Parnes (1975), is based on Friedman's (1957) permanent income hypothesis of consumption. According to their theory, increases in foreign currency receipts only have an expansionary effect on consumption if they are viewed as a permanent increase in income, as Friedman argued that consumption depends on permanent rather than transitory income. Knudsen and Parnes hypothesized that if earnings from exports are unstable, then increases in them are not likely to be perceived as permanent but as transitory income and are saved rather than consumed. The lower propensity to consume and higher propensity to save do, however, provide funds which can be channelled into higher investment, resulting in higher growth. In this way, export earnings instability is said to have an advantageous, rather than an adverse or neutral effect on the economy, the implication being that instability in tourism demand should not be discouraged. In contrast, new classical macroeconomists posit that changes in demand have no real effect unless they are unanticipated and have not been discounted in advance.

There is little empirical research on the measurement or effects of tourism earnings instability, partly owing to the problem of selecting an appropriate definition of instability but also to the fact that this topic is related to the wider issue of the appropriate specification of consumption, savings and investment functions (Deaton, 1992). Using Singapore as a case study, Wilson (1994) finds that export receipts exhibit a larger instability compared to tourism receipts. Sinclair and Tsegaye (1990) studied a range of developed, intermediate income and developing countries, using different measures of instability (including the mean or standard deviation of absolute deviations from the trend in export earnings, estimated using multiple regression analysis or a moving average) and found tourism earnings to be relatively unstable. The results indicated that, for most of the countries considered, variations in tourism receipts were not accompanied by offsetting variations in receipts from exports of goods, so that the net effect appeared to be an increase in instability. Those countries in which instability has adverse repercussions could implement policies to diversify the mix of tourist nationalities or types of tourism so as to stabilize net receipts or to reduce dependency on tourism by developing other forms of economic activity. The desired mix varies according to the preferences of policy-makers concerning the trade-off between higher values and higher variability in receipts and their perception of future trading conditions in all sectors of the economy. Within tourism, alternative combinations of nationalities or tourism types are associated with differences in the values of receipts (the 'return' from tourism) and in their variability over time (the associated 'risk'). Portfolio analysis, a technique used in the financial analysis of the risks and returns associated with investors' portfolios of stocks and shares, can be used to estimate the combination of nationalities or tourism types which would constitute policy-makers' preferred mix or tourism portfolio (Board et al., 1987); appropriate incentives can then be provided.

6.4 Growth

Debates about the effects of export earnings instability have been particularly concerned with the positive or negative consequences of instability for growth. Little attention has so far been paid to the topic of growth, as the discussion in this chapter has been conducted within a static context or within a comparative static framework in which an initial position is compared with a subsequent outcome, for example the levels of income and employment before and after a change in tourist spending. To address this issue, the present section of the chapter analyses the implications of new growth theories and new economic geography for tourism and discusses a number of relevant policy connotations. The emphasis here is on the concept of 'economic growth', which differs from 'economic development'. In particular, actual (or real) growth refers to an increase in goods and services produced by an economy within a specific time period; if the available factors of production are not fully and efficiently employed, actual growth may be achieved even if the total productive capacity of an economy does not increase. Potential growth refers to a rightward shift of the production possibility curve (or frontier) primarily as a result of improvements in technology, human capital and infrastructure. Economic development refers predominantly to changes in the composition of the productive sectors in an economy from the primary (extractive industries) to the secondary (manufacturing) and tertiary (services). It also refers to a consequent rise in the living standards across the population: economic development usually necessitates economic growth; nonetheless, economic growth will not necessarily result in economic development, when income inequality and other barriers prevent the diffusion of economic benefits to wider parts of the population. Distributional issues are beyond the scope of this section of the chapter – they are addressed, however, elsewhere in the book, particularly in Chapter 9, mainly in the context of pro-poor tourism and fair trade initiatives.

6.4.1 New growth theories

Economic growth has been a key area of attention within economics in recent years and new growth theories have modified much traditional thinking. The neoclassical view which dominated much of the literature until the early 1980s (Solow, 1956) argued that growth depended upon the supply of labour and capital, with any residual growth being determined by exogenous technological change (independent of the economic system). The economic context was assumed to be one of perfect competition, with labour and capital being subject to decreasing returns. Thus, each additional unit of labour or capital would make a smaller contribution to output than the previous unit, in the context of given quantities of other units of production. The main competing theory was put forward by Harrod (1939) and Domar (1946, 1947), who argued that growth was given by the ratio of the savings rate to the capital–output ratio; that is the input of capital required to achieve a given level of output, the required capital input being a constant proportion of output with a fixed capital–output ratio. If the actual growth rate exceeds business people's expected growth rate, they increase their investment and the actual growth rate increases further. The opposite would occur if the expected growth rate exceeded the actual outcome. Thus, within this approach, expectations play a key role in determining growth.

Economic growth is determined endogenously (within the economic system) according to new theories of growth (for example, Grossman and Helpman, 1994; Romer, 1994; Van der Ploeg and Tang, 1994). Capital is defined not only as physical capital in the form of equipment and machines (Km) but also as public infrastructure and as human capital (Kh), for example in the form of skilled labour. Economic growth can arise from investment in the broad definition of capital, including investment in knowledge. It is assumed that there can be substitutability between capital and labour and between different forms of capital, and that there are constant returns to the broad definition of capital so that there is no incentive to decrease investment in capital. Firms can operate under imperfect competition, thereby reaping supernormal profits in the long run from their investments. However, any given firm is unlikely to be able to appropriate all of the benefits from its own investment expenditure so that investment by one firm has positive external effects on other firms at the national and/or international levels as information is diffused about new products or methods.

New growth theories consider that growth can result from the accumulation of Km, involving education and training of present and future workers (Lucas, 1988, Azariadis and Drazen, 1990; Ventura, 1997; Benos, 2008).

Although the case for education and training to raise the rate of growth in tourism and hence the economy as a whole may appear obvious, it is only relatively recently, in the UK for example, that the need for training in tourism has been widely recognized and increased provision made (Christou and Eaton, 2000; Papatheodorou, 2003c; Sigala and Christou, 2003; Stergiou et al., 2008). Moreover, there is still some debate about the form of training which is most appropriate; consequently its quality varies considerably both within and between countries. 'Learning by doing' is a second form of knowledge accumulation (Arrow, 1962; Romer, 1986; Young, 1991; Cannon, 2000) and within the management literature there has been considerable debate about 'empowering' workers to make an increased contribution to improving firms' performance. In the tourism industry, some hoteliers make arrangements for members of staff, including managers, to work in other hotels which are considered to be of a comparable or superior standard. In some cases this involves sending staff to other countries, so that the ultimate objective is a cross-border transfer of knowledge and skills. At an international level, the European Union also encourages such practices through appropriate sponsoring, for example the HERMES (Harnessing Employment, Regional Mobility and Entrepreneurship in South-eastern Europe) programme funded by DG Enlargement and aimed at raising competitiveness, territorial socio-economic cohesion, geographical and professional mobility of workers and sustainable development in four frontier regions of Greece to address effectively the implications for labour markets (including tourism related ones) arising from the 2004–7 EU enlargement process (Research Institute of UEHR, 2008).

Research and development has been proposed as a third determinant of growth (Grossman and Helpman, 1990a, 1990b, 1991; Romer, 1990; Aghion and Howitt, 1992, 1998; Jones, 1999; Aghion et al., 2001) and can increase the variety and quality of products supplied. Like learning by doing, research and development can increase the general availability of knowledge beyond that of the individual firm undertaking the development. This may occur relatively quickly, as new tourism products are marketed, or more slowly if innovating firms attempt to appropriate the returns from the innovation, as in the case of computer reservation systems for holiday bookings.

Growth can also be facilitated by the provision of public infrastructure either in material form such as roads for local residents and tourists or in non-material form, for example health care (Barro, 1990; Barro and Sala-i-Martin, 1992, 2004; Baier and Glomm, 2001; Chen, 2006) and social security, which essentially provides a safety net and reduces uncertainty about the future in the population. The view that public provision of infrastructure can further the growth of the private sector contrasts with much of the thinking that was dominant during the monetarist era, when government spending was thought to crowd out private consumption and investment via increases in prices and interest rates. New growth theorists acknowledge that public sector provision may be financed by distortionary taxes which can be inequitable but show that the net effect on growth can be positive rather than negative. Within tourism, infrastructure is seen as facilitating the sector's growth and is referred to as the secondary tourism resource base.

The role of initial factor endowments in the new growth theory depends upon the degree of inter-regional spillovers of knowledge (de la Fuente, 2002). For example, if firms invest in a new product or process and knowledge about their investment is available to domestic but not foreign firms, the positive externalities of the investment are retained within national or regional boundaries. In this respect, the country or region which has the initial advantage in knowledge and investment retains a comparative advantage in products which incorporate this type of knowledge, irrespective of its factor endowments. If, on the other hand, there are no restrictions on access to knowledge, no country or region can gain a comparative advantage in supplying the product so that comparative advantage is, instead, determined by countries' factor endowments (the concept of comparative advantage is examined in the context of international aspects of tourism in the next chapter). In the case of the accommodation component of tourism, for example, hoteliers in some developing countries may be unaware of innovations in accommodation and service provision in wealthy countries, such as improved computing equipment and software. Even if they are aware of such innovations, many firms in developing countries are unable to take advantage of them owing to a shortage of investment funds and experience. Thus, what is important is the extent and speed of intra- and inter-country knowledge generation and transmission and firms' ability to take advantage of the knowledge which they obtain. As argued by Capó-Parrilla et al. (2007) and elsewhere in the literature (Brau et al., 2003; Neves-Sequeira and Campos, 2005), by specializing in tourism, destinations can move towards a trajectory of long-term growth; to sustain such a leap forward, however, it is important to focus on human capital in order to increase productivity and provide innovatory tourism services of superior quality and value-added. In this context, at a European level, most projects funded by the EU focus on the cross-regional and cross-country dissemination of knowledge to reduce inequality and promote growth.

6.4.2 New Economic Geography (NEG)

New growth theories bear a number of similarities to a seemingly unrelated sector, namely New Economic Geography (NEG), pioneered by the

Nobel Laureate Paul Krugman in the early 1990s (Krugman, 1991a; 1991b). This approach aims to formalize some parts of the extensive economic geography literature (Dicken and Lloyd, 1991) by focusing on the interaction of centripetal agglomeration forces with centrifugal diseconomies of scale; love-for-variety consumer preferences and increasing returns in production (Dixit and Stiglitz, 1977) play a central role in this context. A trade-off between intermediate input/final product differentiation and transport costs is shown to exist; this generates cumulative causation processes and results in the formation of regions, industrial districts and cities. As argued by Fujita and Mori (2005), NEG is based on four major pillars, namely spatial general equilibrium modelling; increasing returns and/or indivisibilities in production; transport costs; and location mobility of consumers and production factors.

a. Conceptual framework

To understand the concepts introduced by NEG it is important to distinguish first between internal and external economies of scale (Fujita and Thisse, 1997). The former are related to savings made at the level of an individual firm and emerge when the production process is characterized by the existence of large fixed (and occasionally sunk) costs compared to the variable ones: as a result, unit costs decrease as the level of production increases. Internal economies of scale are important to understand the efficiencies of large enterprises as opposed to the 'backyard capitalism' paradigm of neoclassical economics (Krugman, 1991a). On the other hand, external economies of scale step beyond the level of an individual company and emerge as a consequence of the connections that link various firms of the same or different sectors. These economies are explicitly associated with spatial clustering and agglomeration and comprise both technological and pecuniary (i.e. monetary) externalities (Scitovsky, 1954). The former (also known as the spillover effects discussed earlier) focus on the implications of non-market interactions on the consumer's utility and/or a company's production function. For example, the presence of a tourism department in a university located on a tourist island is beneficial for both university graduates (who can easily find a job) and potential employers (who get access to an educated labour force at low search costs). Monetary externalities stress the benefits of interactions realized through the mechanisms of the market and the clearing (i.e. exclusionary) role of prices and are explicitly related to the existing backward and forward linkages in an economy. For example, the presence of a large number of hotels in an area may render financially sustainable the establishment of furniture manufacturers in that area, as the relatively large demand for furniture allows them to reap internal economies of scale. Nonetheless, the establishment of furniture manufacturers reduces the input costs of hotels, enabling them to lower their prices, achieve higher occupancy rates, reduce unit costs and possibly increase capacity to the benefit not only of the hoteliers but also of the furniture manufacturers. This is closely related to the market potential concept (Harris, 1954) and the cumulative causation argument (Pred, 1966): as stated by Krugman (1995: 46), 'Firms want to locate where market potential is high, that is, near large markets. But markets will tend to be large where lots of firms locate. So, one is led naturally to a consideration of the possibility of self-reinforcing regional growth or decline.'

External economies of scale may also be classified into localization and urbanization economies. The localization ones are generated at an industry level as a result of business clustering and the creation of a particular entrepreneurial milieu (Porter, 1990): typical examples include the Silicon Valley which is characterized by the presence of a multitude of IT companies and Las Vegas, which is the world capital of gambling. Despite the intensification of competition arising from the establishment of one company close to another, the benefits of interaction in terms of knowledge, entrepreneurship and labour mobility of highly skilled workers can be large enough to justify this clustering and the subsequent formation of backward and forward linkages in the economy. Similarly, urbanization economies emerge at a wider level as a consequence of the aggregate business activity in a specific location and are related to the availability of specialized inputs (goods or services) and the existence of modern infrastructure. Tourism may play a particularly important role in this context. For example, the small population of a remote island may deny the existence of a hospital, a bank, a sewerage system or an airport; nevertheless, if this island subsequently becomes popular with tourists, then the new levels of demand may render all these services and infrastructure financially sustainable, improving in this way the living standards of the local community.

It seems, therefore, that the interaction of internal and external economies of scale may trigger a self-reinforcing agglomeration mechanism at both urban and regional levels. Still, limits to growth may exist as at some stage the various centrifugal forces may initiate a process of deglomeration. This is essentially related to the leakages and displacement effects discussed earlier. In particular, spatial clustering may lead to high levels of congestion, pollution and environmental degradation; as argued in Chapters 8–10, these may pose serious concerns on the sustainability of tourism development, especially when the carrying capacity of a specific destination is exceeded. Moreover, prices of immobile production factors (such as land, sunk infrastructure and possibly labour) and non-traded goods and services may soar, rendering their consumption unattractive. For example, high land rents in popular tourist resorts have a snowball effect, leading to substantial inflationary pressures throughout the local economy causing symptoms of a 'Dutch disease' (Nowak and Sahli, 2007). In addition, large agglomerations may pose administrative problems and phenomena of social pathology (such as an increase in crime rate, begging and illegal prostitution) whose solution requires the introduction of costly practices. For all the above reasons, residents may decide to quit an urban or regional cluster and tourists may choose to abstain from visiting popular resorts in favour of smaller spatial entities. Having the above in mind, and similarly to other theoretical constructions in economic geography, there should be an optimal size of agglomeration at which the various goods and services are produced at the highest efficiency level, maximizing the internal and external economies of scale (Weber, 2002). Nonetheless, the measurement of this optimal size may prove an extremely complex and difficult task.

Three broad categories of thought fit in the NEG framework (Fujita and Mori, 2005). The first one considers a two-region/two-sector economy, where circular causation results in the concentration of firms and workers both through forward (i.e. the supply of more varieties in the intermediate goods increases the workers' real income) and backward (i.e. a greater number of consumers attracts more firms) linkages. In this way, the increasing returns at the level of the individual firm surpass the urban territorial scale and hold in a regional context, giving rise to a core–periphery structure. Good examples of this line of thought are provided by Krugman (1991a, 1991b, 1995) and Puga and Venables (1996; Venables, 1996). The second strand pays attention to city formation. Differentiation in consumption and increasing returns in the production of intermediate goods result in the interrelation of labour division (i.e. farmers/industrial workers) and market size within cities; subsequently, the overall structure of the urban system becomes endogenously determined. Fujita and Krugman (1995) and Fujita and Mori (1996) suggest that as the population of a national economy increases continuously, new urban areas emerge periodically if a certain threshold is exceeded; in other words, when transport and congestion costs become unsustainable, the urban system is likely to experience radical changes. The original models produced an asymptotic structure characterized by equivalent, equidistant cities (Fujita and Thisse, 1997); more recent work, however, set the necessary economic fundamentals for the creation of a new urban theory which can explain issues of spatial hierarchy (Fujita et al., 1999), building on the central place theory advanced by Christaller (1933) and Lösch (1940). Finally, the third category of NEG relates agglomeration with international trade (Krugman and Venables, 1995) and may be essentially considered as a sub-category of the first group of models. International trade and its relation to tourism are extensively discussed in Chapter 7.

b. NEG and tourism

Papatheodorou (2003d, 2004) was perhaps the first to apply the NEG theory in the context of tourism. Interestingly, the focus on the tourism sector enables the expansion of the NEG research beyond its current frontiers. More specifically, tourism is a spatial activity par excellence. First, it involves a complicated territorial framework, as it requires a discrete choice and the physical movement of the customer from a place of origin to a tourist destination. Second, tourism urbanization depends on the availability of local resources, albeit in a dynamic way. In other words, while growth may put pressures on the physical environment, it may potentially have positive repercussions on built attractions and infrastructure; the synthesis of resources changes over time and an optimal level of expansion emerges. The latter is associated with investment in expensive sunk structure and the developers should be cautious not to undertake projects that endanger the long-term sustainability of tourism development. In addition to its spatial features, tourism plays an important role in the service sector economy and encompasses many different industries, where consolidation and strategic interaction among producers are apparent in multiple dimensions.

Papatheodorou (2003d) builds on the second strand of the NEG literature, i.e. city formation. He proposes an interesting mathematical model of tourism development by extending NEG with discrete choice (based on multinomial logit) and oligopolistic competition. The choice of tourists is examined simultaneously from a dual approach (discretely at the inter-destination level and continuously in the setting of intra-resort consumption) to jointly consider systemic and probabilistic factors. Alongside the discussion of dynamic issues, Papatheodorou (2003d) undertakes a welfare analysis by numerically comparing and contrasting the optimal spatio-temporal configuration under a social planner with the one produced by a decentralized regime; illustratively, while in the former case, the fifth resort is feasibly attained with 53,560 visitors in period 115, its sustainable emergence under a decentralized regime requires 569,588 travellers (i.e. 10.6 times the optimal number) and occurs in period 140. Unless protected, therefore, from reckless growth, tourism destinations may possibly suffer from irreversible degradation, becoming victims of their own success.

Similarly, Papatheodorou (2004) studies evolutionary patterns in tourism using NEG elements mostly associated with the first class of models, i.e. those which propose a core–periphery configuration, albeit from a dynamic perspective. Building on the concept of the tourism area life cycle (TALC) model (Butler, 1980; Cooper, 1992; Getz, 1992; Gordon and Goodall, 1992; di Benedetto and Bojanic, 1993;. Gordon, 1994; Douglas, 1997; Tooman, 1997), Papatheodorou (2004) suggests a new theoretical model where endogenous changes to the tourism business result in a dualism in market and spatial structures, where powerful industrial conglomerates (such as those examined in Chapter 5) share the markets with a competitive fringe and core tourism destinations co-exist in the tourism space with peripheral resorts. The model illustrates graphically the interdependence of market and spatial forces and examines connotations for destination development. The short-run analysis focuses on the relationship among origin regions, core and peripheral destinations, whereas the longer-term framework studies smooth and abrupt patterns in the evolution of tourism flows. In this way, Papatheodorou also takes into account the work of Debbage (1990), who examined the role of oligopolistic tourism firms and their strategies with respect to market share, innovation and diversification in his resort cycle model of the Bahamas, and Gordon (1994), who considered the effects of monopoly in accommodation and land ownership.

c. Evaluating NEG

NEG has undoubtedly helped substantially in bringing spatial issues into the foreground by raising the interest of mainstream economics. Its application in tourism remains rather scant until today, but this may change in the future as the theory becomes more widely known among researchers. Still and with regard to the spatial dimension of tourism, NEG has re-emphasized what the discipline of geography engendered in land economics, especially in the regional and urban field, in the mid-twentieth century. In this sense, the economic interest in spatial elements of commercial/industrial activity and residential location predated the focus of NEG by several decades; this is also the case for the study of the impact of transport systems and modes, land use and values, as well as of the effects of town and country planning with respect to regional and urban development.

Of particular importance to economists of the past was the potential to apply quantitative modelling to the interaction between the economic activities and systems in regions, cities and towns. In fact, regional economics is of relevance to tourism because it offers a theory that not only explains the sector's potential for development and growth, but also indicates the factors that govern its viability, such as the accessibility of the resort; the quality of and location of natural and human-made attractions; the required infrastructure; and the availability of facilities, services and labour. McCann (2001) offers an introduction to the theory of regional economics relating to the determinants of rates of growth and decline and supporting empirical evidence, including the nature of government regional policies and their effectiveness.

Given the focus of the NEG on the urban areas with regard to the spatial aspects of tourism, attention with respect to identifying the economic analysis in this field should be concentrated on its relevance to the issues raised above. Evans (1985) argues that the foundation of urban location theory in economics was originally in the pattern of residential location because of the large proportion of space it occupies; nonetheless, this analytical framework is also important to studying the impact on the location of all the other principal urban economic systems and activities: transport, labour markets, industrial, office and retail. The early von Thünen model (Dicken and Lloyd, 1991) considered the relationship between the land-value or rent gradient and transport costs: the former declines from the central business district to the outskirts of cities and the latter rises with increasing distance from the centre. From this basic model all the other aspects of urban theory and economic effects emerge, such as spatial clustering, the size, structure and hierarchy of cities and towns and changes in them that occur over time.

Of special interest in urban land economic analysis are the implications for tourism businesses, ostensibly in competition with each other, of the complementarity thesis that posits that they gain by clustering in the same location. Another relevant feature of urban economics analysis is the explanation it offers, for example, of the relationship between the accessibility, location, quality, size and services of tourism hotels and guesthouses that determine the prices charged, also discussed in Chapter 5. For example, the hotels in the best locations, say, facing the sea along the promenade and closest to facilities and services, where the demand is greater for good quality accommodation, can charge higher prices than those located away from the sea front; consequently, land values will be higher in such areas. New hoteliers trying to break into the hospitality sector may find these values too high for them to secure a prime location. Furthermore, land economics also can contribute to explaining the resort life cycle as it identifies factors, indicated above, that influence the growth and decline of urban areas.

In conclusion, the main elements of NEG such as central place theory; agglomeration economies; external and internal economies and diseconomies of scale; accessibility; spill-over effects; traffic congestion and spatial competition, particularly monopolistic competition and oligopoly, that are relevant to tourism have all been extensively examined also in the past. This is not to say that the value added of NEG is minimal; on the contrary, its contribution, at least at an analytical/mathematical level, should be explicitly recognized. Nonetheless, it does prove that, as elsewhere in economics, many of the good ideas of the present lie on innovative thoughts of the past.

6.4.3 Implications of new growth theory and New Economic Geography for tourism policy

It is clear that, in contrast to traditional neoclassical theory, new growth theory and New Economic Geography provides a possible role for government (van der Ploeg and Tang, 1994), although government intervention does not always appear beneficial (Barro, 1991). The measures which the government can undertake include direct expenditure on, or measures to encourage investment in, a broad range of human and physical capital. In the case of tourism, the natural (Kn) and built environment (Km) should also be taken into account. Such investment, along with the provision of a complementary institutional framework, may also succeed in shifting the country's comparative advantage towards targeted sectors, as discussed in Chapter 7. Improved provision of roads, airports, health care and disease eradication campaigns in developing country destinations increase both tourism demand and supply and may increase growth and spatial clustering. Tourists gain increased and/or improved accessibility to different areas within destinations and are confident that their health needs will be met, while private sector suppliers are encouraged to provide more accommodation and supporting facilities within reasonable proximity of the new infrastructure. Investment in Kn as well as Km and Kh should also be undertaken if growth is to be sustainable, as pointed out in the environmental economics literature considered in Chapters 8–10. Legislative measures can attempt to provide an institutional framework which facilitates sustainable growth and commercial, fiscal and industrial policy measures can be used to stimulate investment, such as in the provision or modernization of tourist accommodation.

Where domestically generated savings are insufficient to cover investment requirements, governments have sometimes encouraged additional investment by deregulating foreign direct investment in tourism enterprises, as in the case of Indonesia's Fourth Five Year Plan, Repelita IV, 1983/4–1988/9. The potential for international collaboration to shift the recipient country on to a higher growth path is, however, dependent upon an optimal degree of knowledge transfer to the host economy, as a low level has a small effect on growth while an excessive level crowds out local firms (de Mello and Sinclair, 1995). Thus, governments of some host economies have imposed conditions concerning the degree and nature of technology transfers by foreign investors, which are required in return for permission to invest in the country (Chakwin and Hamid, 1996), but have paid little attention to this issue in the context of tourism investment.

Higher standards of education and training, promoted and/or financed by regional/national governments or international organizations, are a further means of supporting ongoing growth. The case for investment in both human and physical capital has been acknowledged by governments and international organizations. For instance, centres for training in tourism in Bandung and Bali in Indonesia received aid, respectively, from Switzerland, the International Labour Organization (ILO) and the United Nations Development Programme (UNDP), and the State University of Udayama provided diploma and master level courses in tourism from the mid-1980s

(Pack and Sinclair, 1995a). Government institutions have supplied tourism training in India, aided by the ILO (Pack and Sinclair, 1995b), as in many other countries. Investment in physical capital for tourism has been undertaken or supported financially by national governments, and international organizations including the Asian Development Bank (ADB), World Bank, the UNWTO and UN have supplied financial and/or technical assistance for infrastructure provision, as well as for more general tourism development programmes (for example, ADB, 1995). As discussed earlier, the EU has also considered a variety of measures to support tourism and increase welfare (Commission of the European Communities, 1994, 1995a, 1995b, 1996).

If appropriate measures are implemented in both tourism and other sectors of the economy, they may assist not only in promoting endogenous growth and agglomeration but also in moving the economy on to a higher growth path. Examples include the Singapore government's encouragement of increases in investment, education, training and wages, which stimulated production of more technology-intensive, higher-priced goods (Chadha, 1991; Tan, 1992). Hence, identification of the conditions under which tourism could constitute a locus of specialization and source of higher growth and welfare in particular countries is clearly important, as also argued in Chapter 7.

6.5 Tourism and economic transformation

Studies using the multiplier methodology take account of the short-run effects of tourism on income and employment but the models are not appropriate for examining its longer-term repercussions. These affect both the structure and the spatial distribution of economic activity. Of major importance are issues related to development, migration and the role of gender in tourism employment. These are analysed in what follows.

6.5.1 Structural and migratory repercussions

One of the most obvious effects of tourism expansion is the rise in the importance of the service sector within the economy. In developing countries this is usually accompanied by a fall in the agricultural sector's share of gross national product, while many developed countries experience a process of deindustrialization in addition to declining primary sector production, to which the growth of tourism may contribute (Copeland, 1991; Adams and Parmenter, 1995). Surprisingly, little attention has been paid to the growth of the service sector in either context. The development literature concentrated, in the 1950s and 1960s, on the growth of manufacturing via the transfer of supposedly surplus, low-wage labour from

agriculture to industry. From the late 1960s onwards, relatively more attention was paid to agriculture as the problems of poverty and unemployment
in the cities demonstrated the inability of the manufacturing sector to assimilate the migrants from the land. It was not until the late 1980s that the need
to examine the service sector was generally recognized (UNCTAD, 1988),
although tourism was not often acknowledged in either developing or
developed countries. Interestingly, the World Bank decided to disengage
itself from tourism in the mid-1970s, blaming the sector for economic instability and volatility, thus not being suitable to be encouraged as tool for
poverty alleviation; nonetheless, this stance may change in the context of
the United Nations Millennium Development Goals (Hawkins and Mann,
2007). Other organizations such as the Overseas Development Institute
(ODI), a major NGO in Britain, actively promote tourism as a means for
poverty reduction in the context of the Pro-Poor Tourism partnership
(Pro-Poor Tourism, 2008). In particular, based on a 'sustainable livelihoods
approach' pro-poor tourism 'is not a specific product or sector of tourism,
but an approach to the industry' which aims to 'unlock opportunities for the
poor, rather than to expand the overall size of the sector' (ODI, 2001: 2).
Blake et al. (2008) study the relationship between tourism and poverty
relief in Brazil to show the importance of such distributional effects.

Empirical evidence about the role of tourism within the structural transformation of economies over the long run is limited, but indicates that the
common conception of the development process as a transition from an
agriculture-based economy to manufacturing and thence to services may be
misguided. In Spain, for example, it was the large-scale receipts from foreign tourism, as well as remittances from Spanish nationals working abroad,
often in tourism-related activities, which underpinned the country's indus-
trialization process (Bote Gómez, 1990, 1993). Much of the foreign capital
which flowed into the country was channelled into the tourism sector,
particularly purchases of property, which constituted 20 per cent of total
foreign investment in 1964 and varied between 59 per cent in 1974 and
10 per cent in 1991 (Sinclair and Bote Gómez, 1996). The changes in
land values which resulted from planning permission for construction,
particularly for tourist accommodation in coastal areas, along with the
profits made from the construction and sale of hotels, flats and villas,
resulted in large increases in the wealth of national and foreign property
developers. Interestingly, Spain has now become a significant investor in
tourism expansion abroad, as is illustrated by its growing participation
in the hotel sector in Cuba; illustratively, Sol Meliá, the leading Spanish
hotel group, controls more than twenty properties on the island (Sol Meliá
Cuba, 2008).

The case of tourism in Spain clearly demonstrates the rapid growth and
spatial concentration which can characterize tourism demand and supply.

Whereas in 1951, 1.3 million people came to Spain, by 2007 over 59 million arrived from abroad (UNWTO, 2008b). The vast majority of tourism demand occurs in the Mediterranean coastline, Balearic and Canary Islands, which account for three-quarters of demand for and supply of officially registered bed-places. Foreign tourists, in particular, have strong preferences for beach rather than rural and city tourism and the expansion of tourism in coastal areas created an important migratory movement from villages in the interior to meet the rising demand for cheap labour to work in tourism activities. It is only in isolated cases that rural tourism development has succeeded in decreasing the population outflow. The examples of tourism in La Vera in the west of Spain and Taramundi in the north illustrate the small-scale but significant effects that local tourism development programmes can bring about (Bote Gómez, 1988, 1990; Bote Gómez and Sinclair, 1996).

In some contexts, the expansion of leisure and tourism in rural areas has created considerable income and employment and increased the spatial dispersion of tourism, such as in Britain during the 1990s (Diamond and Richardson, 1996). In Egypt, in contrast, the growth of tourism in the interior has been constrained by well-publicized violence against a small number of tourists. Large-scale construction, with probable adverse environmental consequences, is occurring in the coastal areas of the north and the Red Sea, resulting in migration from the cities and towns to these relatively unpopulated areas (*Economist*, 1996). Although many of the newly created jobs have been taken by men, some Egyptian women have been able to obtain supervisory roles in large modern hotels. Thus, although the major gains from tourism have clearly gone to property speculators and construction companies, tourism has also altered the distribution of employment and income at the most basic level of individuals and households. Some of these effects will be discussed in the following section.

6.5.2 Gender structuring of tourism employment

There has been quite an extensive literature on gender studies related to tourism that has mostly been concerned with the role of women. Fairly early work in the 1990s tended to concentrate on the respective gender roles, particularly the impact of cultural and social structures in developing countries that governed the representation, status, employment opportunities, the nature of work and the remuneration. Later considerations have been the effect on women in destinations of sex tourism (Scheyvens, 2002), and with the growth of international tourism, the position of women as tourists from tourism generating countries (Bowes, 2004).

Tourism brings about changes in the pattern of employment, as well as in the distribution of income. Relatively little attention has been paid to labour market analysis of tourism and the following discussion focuses on research which has been undertaken on the gender structuring of tourism employment, while acknowledging that considerable further examination of labour market theories and tourism employment is required within the literature. Studies of employment in tourism have indicated the ways in which it, like other sectors of the economy, is structured by gender (Bagguley, 1990; Kinnaird et al., 1994; Adkins, 1995; Swain, 1995; Sinclair, 1991c, 1997a). In the UK in 1995, for example, 76 per cent of jobs in the transport sector were filled by men, while 62 per cent of those in accommodation and catering were undertaken by women (Purcell, 1997), indicating horizontal segmentation of work. Vertical segmentation also occurs as most top jobs are carried out by men (Guerrier, 1986; Hicks, 1990; Purcell, 1997). Since women's labour supply is concentrated in a limited number of occupations which are not characterized by a tradition of trade union activity, the wages which they are paid tend to be relatively low. The majority of seasonal, part-time and low-paid tourism work in many areas is undertaken by women (Breathnach et al., 1994; Hennessy, 1994; Sinclair, 1997b; Scheyvens, 2002), in combination with their childcare and domestic duties.

In developing countries and in rural areas in some intermediate income countries, many of the tasks which involve direct contact with tourists are carried out by men. This is exemplified in the case of Turkish Cyprus areas. Local women have traditionally undertaken behind-the-scenes work such as cleaning and bed-making, which are akin to their domestic roles, although women from Eastern Europe are employed as croupiers in casinos (Scott, 1997). The allocation of domestically related duties to women has also occurred in a range of other areas, including Greece (Castelberg-Koulma, 1991; Leontidou, 1994) and the Caribbean (Momsen, 1994). In some countries, gender norms are less restrictive so that women also work in tourism businesses, for example in Western Samoa (Fairbairn-Dunlop, 1994), Mediterranean countries (Scott, 2001) and Greece (Svoronou and Holden, 2004), but traditional expectations concerning the roles which it is appropriate for married women to undertake often constrain their activities, as in tourism-related work in Bali (Long and Kindon, 1997).

Some insights into the gender structuring of tourism employment may be obtained by examining the demand for and supply of labour in relation to prevailing societal norms (Sinclair, 1997a). In countries that are characterized by strong traditions concerning the roles deemed suitable for men and women, employers hire (demand) male workers to fill some posts and women to fill others, while men and women tend only to apply for (supply their labour in) positions which are viewed as appropriate for their gender. In other countries in which capitalist development has altered prevailing norms, employers are more indifferent concerning the gender of those who undertake different jobs, although applicants remain subject to modified expectations concerning gender roles. Hence labour demand and supply not only vary in relation to quantifiable variables such as wage rates, taken into account by mainstream labour economists, but also depend on norms and expectations which are not easily measured.

It is interesting to consider how international tourism can bring about changes in the structure of employment. In the Turkish Cyprus areas, for example, accommodation has traditionally consisted of small, family-based guesthouses used by domestic tourists (Scott, 1997). Social and financial transactions with tourists are usually undertaken by male household members, while women are responsible for cleaning and other household work. However, an influx of foreign tourists has been accompanied by the construction of large hotels in which many younger and more educated Turkish-Cypriot women have obtained employment. Gender norms in both the demand for and supply of labour in the new hotels differ from those prevailing in the guesthouse sector, providing more opportunities for local women. Therefore, it may be necessary to examine labour demand and supply in different sub-sectors of tourism supply.

Although tourism can provide local women with higher levels of income and greater independence within the household, it can also result in considerable problems (Chant, 1997). Prostitution tourism is an obvious example. This has provided work and income for many women, as well as men and children, often from poor rural areas, and has sometimes received implicit support from the state (Lee, 1991). However, it has involved harsh conditions for many of those involved. Campaigns to discourage sex tours have had some success in deterring tourists from travelling to countries known for their high levels of prostitution. However, they have also had the effect of encouraging 'trade' in women, as gangster syndicates have recruited poor women from developing countries to work as prostitutes in such wealthy countries as Japan (Muroi and Sasaki, 1997). The intermediaries extract high profits, whereas the women themselves are usually unaware, on leaving their homes, that they have been recruited to work as prostitutes rather than entertainers, receiving a very small share of the returns and often being unable to escape from their new destination. To address such issues a number of efforts have recently been undertaken. For example, the Task Force for the Protection of Children in Tourism is a UNWTO initiative aimed at protecting children from exploitation in tourism, including trafficking (UNWTO, 2008a).

The examples of tourism in the Turkish Cyprus areas and of prostitution tourism illustrate the role which both cultural norms and economic variables play in determining the nature of employment. International tourism results in the interaction of different, historically based and gendered cultural systems which are modified but not standardized by tourist inflows.

Hence, countries' and regions' specific cultural endowments, as well as their particular combination of land, labour and capital, determine the patterns of specialization, trade and employment in tourism and other activities.

6.6 Conclusions

This chapter has discussed a number of topics related to tourism in a national and regional framework. Starting with an analysis of tourism satellite accounts to measure the size of the sector, the chapter then studied the economic impact of tourism, primarily focusing on the concept of the multiplier and alternative methods to calculate it. Growth issues were then analysed with emphasis on endogenous factors (new growth theories) and spatial connotations (New Economic Geography) aiming to provide useful advice for policy-makers. Finally, the chapter dealt with longer-term effects of tourism in the context of economic transformation.

The discussion in the chapter has in some cases broadly distinguished between developed and developing economies and/or regions, in preference to the distinction between large and small countries and islands which has appeared in some tourism literature (for example, Conlin and Baum, 1995; Hampton and Christensen, 2007). What seems to matter primarily in this chapter's analysis of economic effects is the level of national income and wealth, the associated trading relationships and the proportion of GNP arising from tourism. For example, the extent of intra-industry linkages tends to be higher within developed economies and leakages from tourists' and local residents' expenditure tend to be lower, resulting in higher income and employment generation. In contrast, lower-income, developing countries or regions are subject to relatively high leakages from tourist expenditure and the associated income and employment generation are usually lower. Moreover, import leakages may rise over time as local residents experience the demonstration effect of tourist spending.

The short- and long-run effects of tourism may be examined using different theoretical approaches. Studies of the effects of tourism in changing the structure of employment illustrate the insights that feminist (Pritchard and Morgan, 2000), as well as a mainstream labour economics perspectives, can provide. Analysis of the relationship between tourism and economic growth involves conflicting theoretical viewpoints. According to Keynesian theories of growth, tourism results in increased demand, which induces higher investment and income. In contrast, neoclassical theories of growth imply that growth rates are not affected directly as tourism is unlikely to play a significant role in increasing labour, capital or technological progress, although it does provide additional foreign currency which may be used to increase the stock of capital. Endogenous growth theories suggest that increases in the levels of education, training and infrastructure for tourism help to prevent the marginal product of capital from falling, thereby contributing to ongoing growth. Higher demand for labour to work in tourism may also result in rising real wages, thereby stimulating investment in more capital-intensive production and maintaining growth. Moreover, and based on the theory of New Economic Geography, this process may also facilitate the achievement of significant external economies of scale and enhance tourism agglomeration and spatial clustering if properly managed. Therefore, applied research to investigate the relevance of the different theoretical perspectives is necessary. Studies could also be undertaken on the possible roles of governmental and international organizations in improving education, training and infrastructure, in conjunction with tourism firms in specific tourist destinations. Governments can play a key role in assisting natural as well as human and capital resources to play an optimal role in tourism development. This is particularly important since natural resources are tourism's primary input base and are often open access goods which are either un-priced or are priced at levels which lead to over-use and degradation, thereby leading to deglomeration and threatening the future of the tourism sectors they underpin.

7 TOURISM IN AN INTERNATIONAL CONTEXT

7.1 Introduction

Tourism has been one of the highest growth activities in the world since the Second World War, in terms of both expenditure and foreign currency generation. International tourism arrivals grew from 69 million in 1960 to 903 million in 2007, i.e. at an average annual rate of 5.4 per cent. Over the same period, international tourism receipts grew from 6.8 billion to 856 billion US dollars, i.e. at a rate of 10.8 per cent. Adding the expenditure related to international passenger transport (165 billion US dollars), total international tourism receipts amounted to over 1 trillion US dollars in 2007 (UNWTO, 2008b). The high levels of tourism expenditure have significant implications for tourist origin and destination countries, contributing to a worsening of the balance of payments of net tourism generator countries (such as Germany) and an improvement of net recipients (like Spain). Hence, tourism can raise or lower a country's dependence upon other countries and can be of particular importance to developing countries whose economies, apart from tourism, are based on primary products. While over the long-term both domestic and international tourism can make a significant contribution to a country's economic growth, its potential for generating income and employment within a destination may be constrained by the country's ability to supply the goods and services which tourists wish to consume (Sinclair, 1991a). Tourists' consumption of food and beverages which are imported from their country of origin, in hotels owned and managed by fellow nationals, is a prime example of the leakages of receipts from a destination, as discussed in Chapter 6. Therefore, since the pattern of tourism supply and associated distribution of receipts have considerable effects on countries' economies and welfare, it is useful to provide some explanations of the patterns which occur on a global scale. International economics can help to provide such explanations, which are the prerequisite for the formulation of policies designed to alter the cross-country structure of production. The discussion builds on the industrial economics analysis in Chapters 4 and 5 and the regional economics framework discussed in Chapter 6 by examining tourism supply in an international context.

This chapter examines possible economic explanations for international tourism, based primarily on trade theory. The chapter starts by discussing the traditional theories of comparative advantage of Ricardo and Heckscher–Ohlin in a context of competitive market structures. These models are essentially supply-driven and focus on resource endowments. From an integrated perspective, however, the role of demand (captured by similarity in tastes as well as by the love for variety) should also be explicitly recognized; more importantly, supply and demand should be jointly examined when product differentiation is important and co-exists with economies of scale and scope in production. The chapter deals with such issues to subsequently emphasize resource deployment, competitive advantage and destination competitiveness at an international level. In this context, Michael Porter's strategic framework becomes of particular relevance, especially when imperfectly competitive markets are prevailing. The importance of economic integration within a globalized tourism environment is then discussed; among others, the evolution of transnational corporations is assessed using the OLI (Ownership–Location–Internalization) framework. The chapter concludes by discussing international policy issues in tourism. Over the last few years, strategic trade theory has provided a good rationale for active industrial policies to be pursued in the international arena. While valid, this may potentially support the case for protectionism, which may result in lose–lose outcomes when the possibility of retaliation by other states is considered. Therefore, and to avoid beggar-thy-neighbour policies, states should learn how to collaborate and exchange ideas in international fora: on these grounds, the implications of the General Agreement on Trade in Services (GATS) for tourism are analyzed, with the aim of promoting globalization not only for the sake of the rich but primarily for the benefit of the poor.

Throughout the chapter the term 'developing countries' is used to denote the relatively poor nations which are elsewhere known as less developed or underdeveloped countries or the Third World, while the wealthy countries, which are sometimes termed high income or industrialized countries or the First World, are here referred to as developed countries. As Harrison (1992) points out, the choice of terminology is problematic. The terms used in this book are not intended to imply that one category of countries is superior to another or that all aspects of a move from developing to developed status are desirable. These issues will not be considered in the chapter but remain an important topic for debate.

7.2 Explanations of trade and tourism

The vast majority of international tourist flows are relatively short distance, involving travel between geographically proximate countries for business and holiday purposes. Nevertheless, a considerable number of tourists travel longer distances to both developed and developing countries and 'long-haul' tourism has experienced significant growth as more efficient, lower cost, air transport has been provided. Developed countries' foreign currency earnings from their exports of manufactured products are supported by service industry earnings, including inward tourism, whereas expenditure on overseas trips by their residents results in currency leakages abroad, often resulting in a net loss on the tourism account. Many developing countries, which have traditionally relied on earnings from exports of primary products, receive net currency inflows as the result of diversifying into tourism and others are attempting to gain additional receipts by increasing tourist flows from abroad. Tourism's image as a pot, if not of gold, at least of foreign currency, raises the question of why some countries have specialized in tourism and whether gains have resulted from the new pattern of production and trade.

7.2.1 Comparative advantage: Ricardian theory

One of the most well-known explanations of international trade is the Ricardian theory of comparative advantage (Krugman and Obstfeld, 2009). The theory's distinctive contribution lies in its main tenet that even if one country is more efficient in absolute terms in producing goods than another, short-run gains from trade can be obtained if it specializes in the production and export of the goods which it produces relatively efficiently, i.e. in which it holds a comparative advantage. If each country were to specialize, total output would be greater as a larger quantity of the goods could be produced for given inputs. The theory, thus, predicts that the pattern of trade is determined by differences in the relative efficiencies of production in different countries and that gains can result from specialization in production.

According to Ricardian theory, given competitive conditions, each country's domestic (non-trade) price ratio is determined entirely by supply-side conditions, the relative efficiency of production stemming from technology. The post-trade price ratio is determined by both supply-side conditions and by demand, based on consumers' preferences for the traded products. Departing from the static context of the theory, differences in the rates of growth of demand for the two products can result in a movement in the commodity terms of trade (defined as the price of exports divided by the price of imports); for example against the country producing and exporting the product for which there is a low growth and a low income elasticity of demand. The country is able to purchase fewer imports per unit of its exports and therefore income and welfare fall. Thus, specialization can possibly have disadvantageous effects over the long run as discussed later in this section.

7.2.2 Comparative advantage: the Heckscher–Ohlin theorem

The Heckscher–Ohlin (H–O) theorem posits that a country's endowments of factors of production (labour, capital and land/natural resources), rather than relative efficiencies of production, determine its comparative advantage (Agiomirgianakis et al., 2006). Thus, countries such as Tanzania, which have a large supply of labour and land as well as plentiful natural resources of wildlife, mountains and beaches, would appear to have a comparative advantage in tourism. The H–O theorem has been applied to the agricultural and manufacturing sectors and attention has generally focused on the endowments of labour and capital. Thus, a country which is relatively well endowed with labour is said to have a comparative advantage in producing and exporting goods which are produced labour-intensively, while a country which is capital-abundant has an advantage in producing and exporting capital-intensive goods. Samuelson (1948, 1949) took the theory a stage further by arguing, in the factor price equalization theory, that trade would have the effect of equalizing the returns to capital and labour across countries. This would occur as consumers demand products with relatively low prices stemming from low labour costs, thereby increasing the demand for labour and the wage rate. A similar process would occur for products incorporating relatively low capital costs. This leads to the Stopler–Samuelson theorem, according to which a rise in the price of a good will result in an increase in the return of the factor used intensively in that sector and a reduction in the return of the other factor as long as both production factors are fully and efficiently employed (Sahli and Nowak, 2007).

The H–O theory is useful insofar as it points to the role which the supply side can play in determining the pattern of international production and trade, thereby building on the closed economy analysis of tourism supply discussed in Chapters 4 and 5. It might, at first sight, be assumed that tourism provision is labour-intensive, so that countries which are well endowed with labour have a comparative advantage in tourism. However, Diamond (1974) pointed out that tourism can involve large inputs of capital as well as skilled labour, as in the case of Turkey, which is supposedly a labour-abundant country. Studies of the values of the incremental capital–output ratio (ICOR), which is the number of units of capital required to produce one more unit of output or income, in the tourism sectors of Kenya, Mauritius and Turkey provided estimates of 2.4–3.0 for Kenya (Mitchell, 1970), 2.5 for Mauritius (Wanhill, 1982) and 4.0 for Turkey (Diamond, 1974, 1977). Comparable estimates for the agricultural and manufacturing sectors of each country were 2.7 and 4.4 for Kenya, 3.3 and 3.9 for Mauritius and 2.3 and 2.1 for Turkey. Thus, the capital intensity of tourism varies between countries, and can also vary over time, at different stages of tourism growth. Since tourism is not homogeneous it is likely that it is relatively labour-intensive in countries with a large supply of labour and capital-intensive in countries which are capital-abundant. Estimates of the factor intensities of tourism production in different countries and time periods would be useful and could indicate tourism's potential for reducing unemployment in destination countries.

The relationship between tourism production and factor endowments is further complicated by the problem of measuring factor abundance and quality. Abundance can be measured in terms of either quantity, and hence supply alone, or value, in which case demand also enters into play as higher demand for the product results in a higher price and higher value. Its quality is more difficult to ascertain as, for instance, labour can be classified as skilled or unskilled and capital can be of differing vintages and efficiency. When the third factor, land, is considered, the situation becomes even more complicated. Land in the form of either natural resources or the built environment is usually a key component of the tourism product but it is not clear whether land is complementary to capital or labour or both. Land may be complementary to capital in some types of tourism supply. For example, some tourism products within an economy which is well endowed with labour may be capital-intensive. Game tourism for high-income groups visiting Kenya, where light aircraft are used to gain access to more remote areas, is an example. Alternatively, land may be complementary to labour, as in the case of trekking holidays in Nepal.

The issue of whether a capital-intensive form of tourism production can become more labour-intensive in an economy characterized by under- or unemployed labour or, conversely, whether tourism can become more capital-intensive in a context of low birth rates, is clearly important. In Spain, for example, traditional labour-intensive hotels have introduced more self-service in their restaurants as wages have risen. Changes in the factor intensity of production and comparative advantage can be advantageous or disadvantageous to a country and require consideration in a dynamic rather than static context. On these grounds, it is useful to consider the possible effects which international tourism can have on economic growth. One reason for doing so is that countries can become entrenched in specific patterns of production and trade, which may affect their long-run growth adversely or beneficially. For example, developed countries tend to specialize in producing and exporting manufactured goods with high income elasticities of demand, while developing countries produce and export primary commodities with low income elasticities. International tourism may reinforce this type of specialization as developed countries supply the high expenditure, high income generation, high growth components of tourism, while developing countries tend to supply environmental resources and accommodation and local transport, which may be subject to foreign ownership or control. Ironically perhaps, developing countries may also suffer from 'immiserizing growth' (Bhagwati, 1968), when growth leads to a deterioration of their terms of trade. According to the Rybczynski theorem, when the quantity of a production factor rises, there is a relative increase in the quantity produced of the product using this factor intensively (in a H–O context). This may subsequently lead to a reduction in that product's relative price (Sahli and Nowak, 2007). In this way, developing countries may become effectively locked in the so-called 'development of underdevelopment' (Frank, 1966).

The growth differences between developed and developing countries may become self-perpetuating as the former take advantage of economies of scale which are supported by trade (Rivera-Batiz and Romer, 1991). Developing countries are usually unable to benefit from ongoing innovation and, consequently, convergence of growth rates between countries fails to occur. The obvious implication for countries on low growth paths is to alter their comparative advantage by specializing in high-growth exports, including tourism products with a high income elasticity of demand. Thus, tourism provides a possible means for lower-income countries to escape from the low product quality, low expenditure and low-income pattern which generally constrains their development. It is, therefore, useful to consider comparative advantage in the context of economic growth theory in order to examine whether and how it can change over time, i.e. dynamic comparative advantage. New growth theories analysed in Chapter 6 and strategic tourism policy discussed later in this chapter may further highlight this point.

7.2.3 Comparative advantage: evaluation

To summarize, Ricardian theory is useful in indicating the gains which countries can make from international tourism if they are relatively efficient in tourism production and, hence, points to the importance of increasing production efficiency. The H–O theorem's emphasis on the role of countries' different resource endowments also helps to explain international trade and tourism. An obvious implication of the H–O theorem is that further research should be undertaken to investigate inter-country differences in the factors of production which are used in tourism and the ways in which countries might use their resources more effectively. However, the theorem's assumption that land, labour and capital are homogeneous can be called into question and relative factor endowments can change over time. For example, consideration of the differences between the natural and built environment resources used in tourism in different locations demonstrates the significant inter-country diversity of the supply of land and natural resources.

Neither Ricardian theory nor the H–O theorem pays sufficient attention to the role of demand; for example, diversity of demand may occur as preferences relating to the consumption of tourism and other goods, initially discussed in Chapter 2, differ between countries. Moreover, the assumption of competitive market conditions, while offering interesting insights into the determination of comparative advantage in tourism, precludes examination of features which characterize imperfectly competitive markets. Thus, the theories of comparative advantage concentrate on supply-side differences in relative factor efficiency and factor endowments but neglect such characteristics as increasing returns to scale, market power and control over pricing or output, product differentiation, transport costs, inter-country differences in demand, market segmentation, differences in factor 'quality' and differences in countries' access to information and technology. A number of these considerations are relevant to an explanation of the high level of international tourism flows occurring between countries with similar factor endowments and similar levels of income and wealth, as will be seen in the following section.

7.2.4 International tourism and imperfectly competitive markets: overlapping tastes

Many tourism markets depart from traditional assumptions of competitive markets and constant returns to scale, as was shown in Chapters 4 and 5. On the supply side, tourism markets are characterized by a multiplicity of structures and tourism products and, on the demand side, consumers demand a wide range of types of holidays. The fact that tourism supply involves products of particular types and consumers demand tourism products of particular qualities implies that insights from theories of international trade in imperfectly competitive markets can shed further light on international tourism production and trade. A number of features of such markets were previously considered in Chapters 4 and 5, in the context of a closed economy. They will now be discussed within the international economic arena, starting with Linder's (1961) explanation of intra-industry trade, i.e. two-way trade in products supplied by the same industry, such as exports of Volkswagen by Germany to France and German imports of Renaults.

Linder highlights consumers' similarity in tastes as a cause of trade. His explanation of intra-industry trade is complementary to theories which focus on monopolistically competitive market structures and also provides a rationale for the high proportion of total tourism flows between geographically proximate countries with similar levels of income and wealth. Linder argues that suppliers from a given country initially offer a variety of products to meet demand from the domestic market. In the case of tourism, these might be, for example, holidays in different natural or built environments, in a range of accommodation types and with a choice of different sporting, entertainment and other leisure activities. Suppliers from the country subsequently provide a range of products for export, for example holidays specifically designed to meet the preferences of foreign tourists. According to Linder, the more similar the demand for the products supplied by different countries is, the greater the likelihood of trade between them becomes.

The explanation is based on the fact that the nature of demand by consumers is determined by the level of per capita income within a country. Residents of developed countries with relatively high per capita income demand a range of higher quality products. On the other hand, people living in developing countries are more likely to purchase a range of lower quality goods. Consequently, there is more overlap in the range of product qualities demanded by people in countries with similar income levels. Conversely, overlap in the demanded and supplied product ranges between rich and poor countries is less. Hence, there is greater potential for trade between countries with similar levels of income than between those with dissimilar ones. This is because the former have more market segments in common than the latter. On the demand side, trade permits consumers in countries with overlapping market segments to take advantage of a greater variety of products; on the supply side, producers benefit from economies of scale and scope in production, which facilitate sales abroad.

Linder's theory helps to explain the high level not only of trade but also of tourism movements between countries which, in contrast to the postulates of the H–O theorem, have relatively similar factor endowments. The theory predicts the quality range but not the specific tourism products a country will supply. In fact, most intra-industry trade, including that in tourism, is in products which are differentiated by such means as branding, publicity and promotion, by a unique technology or by the use of a specific local environmental resource. Product differentiation provides producers with the ability to influence the prices of their products and to obtain supernormal profits in the short run, as indicated in Chapters 4 and 5 and in the discussion below. Such advantages are sometimes reinforced by restrictions on competition from foreign firms, for example restrictions on sales of package holidays by tour operators based in other countries.

7.2.5 Economies of scale and scope: product differentiation

Economies of scale and scope provide gains from international trade, even in the absence of inter-country differences in technology or resource endowments (Helpman and Krugman, 1993). Many firms within the tourism sector experience economies of scale, as indicated in Chapters 4 and 5. Suppliers wish to achieve decreasing unit costs not only in order to increase profitability but also as a means of deterring competitors from entering the industry. However, they may be unable to attain significant economies by tapping the demand for tourism from domestic residents alone because of the more limited extent of market demand. International demand for their products can enable them to increase their output sufficiently to attain the desired increasing returns to scale.

On the other hand, relatively small tourism firms, for example in low-income countries with a low level of demand, may be unable to compete on an international scale owing to their high initial costs and prices and may be eliminated from the market before they are able to gain sufficient economies of scale to compete effectively. In such situations there may be a case for short-run protection of the 'infant industry' until it has attained economies of scale on a par with those of larger tourism suppliers. The infant industry argument has provided a partial justification for government protection of some international airlines, although an element of national pride is also a factor (Raguraman, 1997). However, while some have increased their efficiency, others have taken advantage of their featherbedded status or been unable to achieve the economies of scope necessary for effective competition, as in the case of Cook Islands International Airline, which eventually went bankrupt (Burns and Cleverdon, 1995; Taumoepeau, 2008).

As argued in Chapters 4 and 5, there is an interesting trade-off between economies of scale and product differentiation. In the context of monopolistic competition, consumers' love for variety results in a long-run equilibrium price above the minimum average total cost; the duplication of fixed costs (when introducing a new product or service) and the limited market size for each individual product which preclude the achievement of scale economies. Trade, therefore, is of particular importance as the increase in total demand can either render financially viable the production of a larger variety of goods and/or enable the production of existing goods at lower costs. Thus, trade has the advantages of increasing variety and lowering prices (Krugman, 1980). The NEG literature (Krugman, 1991; Krugman and Venables, 1995) has extensively dealt with such issues to consider how internal economies of scale intermingle with external localization and urbanization economies to result in a self-reinforcing agglomeration process (Fujita and Thisse, 1997; Zhang and Jensen, 2007), possibly within a core–periphery framework as discussed in Chapter 6.

Consumers' demand for diversity of tourism products is met by product differentiation by the large number of firms in monopolistically competitive domestic markets, as in the cases of accommodation and entertainment. As the demand for tourism extends from domestic to foreign markets, some firms which supply products with characteristics similar to those sold by foreign competitors cease operations, as their rivals are better placed to take advantage of economies of scale. Others, however, are able to differentiate their products from those of their rivals, for instance by means of branding in the international hotel sector, and so remain in the market, increasing the range of products sold. Hence, the analysis of international competition under monopolistically competitive markets points to the role of product differentiation and targeting niche segments of the market, along the lines of producers of manufactured goods in Taiwan (Rodrik, 1995). Therefore, product differentiation, like the attainment of economies of scale and scope, can be used as a tourism business strategy. It also suggests that international competition can give rise to a more imperfectly competitive domestic structure as some firms are unable to compete successfully and so fewer, and by implication, larger firms remain within a country. Whether this change in market structure will necessarily result in more restrictive competitive conduct is, however, an open question.

a. Market power and competitive conduct revisited

In particular, and as discussed in previous chapters, the key characteristic of oligopolistic competition is interdependence. At the international as well as the domestic level, tourism firms may engage in 'Cournot competition', setting their output in the light of their assumptions about the other firms' output. Alternatively, they may engage in 'Bertrand competition', setting their prices according to their predictions of other firms' prices. The assumption that other firms will not change their output or prices is known as zero conjectural variation (Scherer and Ross, 1990). The behaviour of some international airlines may be explained by the reciprocal dumping model (Brander, 1981; Brander and Krugman, 1983), which examines the effects of Cournot competition in the international arena. As discussed in Chapter 5, the liberalization of the air transport market in Europe and elsewhere in the world has given airlines the opportunity of making further profits by selling to foreign consumers and distinguishing between full price and cheap fares. The outcome entails the provision of increased supply at lower prices. The extent to which supply rises and prices fall depends upon each airline's conjectural assumptions about the way in which the other airline will respond to its actions. If, on the other hand, the firms were to engage in Bertrand competition on the basis of prices, the outcome would be likely to be different; flight prices would remain high if each airline perceived that cuts in the prices of its flights would be fully matched by its rivals; otherwise, a price war could also emerge. Tourism firms' choice of competitive strategy is, therefore, as important in the international as in the domestic arena.

Moreover, the gradual liberalization of the international air transport market has encouraged a number of airlines to compete by differentiating their product from those of the other firms in the market. For example, some airlines provide a non-stop service to the destination at a relatively high price, as in the case of Air Mauritius, where the government prohibits charter flights (Seetaram, 2008), while others include stopovers and a longer flight time but charge a lower price for the flight. Although large airlines are concerned to exploit the economies of scale associated with large arrival and departure hubs in major cities, some smaller airlines, such as Trinidad and Tobago's British West Indian Airways (BWIA) and its successor Caribbean Airlines, can successfully supply transportation on specific routes (Melville, 1995), thereby obtaining particular niches of the market. Competition between oligopolies also takes the form of the provision of a range of qualities of a given product, known as vertical differentiation, or of products of comparable quality but different characteristics – horizontal differentiation. Thus, airlines such as Ryanair have specialized in the provision of low fare, no-frills flights, while other airlines have marketed flights with higher quality services at a higher price. A wide income range among consumers, in conjunction with low fixed costs but rapidly increasing variable costs of quality improvements, tends to be associated with a large number of both product qualities and firms. The opposite demand and supply conditions give rise to a small number of product qualities and firms, and trade is likely to force out of the market those firms whose products are of similar qualities at higher prices. This has occurred in the accommodation sector as some domestically owned hotels have been taken over by international hotel chains which can take advantage of economies of scale and scope as argued later in this chapter.

7.2.6 Competitive advantage and international destination competitiveness

The previously analysed theories provide a very useful insight into explaining tourism flows at an international level. Yet, they seem implicitly to assume that resource endowments, market size and/or product variety are sufficient to render a tourism destination successful in the international competitive environment. What seems to be missing from this discussion, however, is the role of resource deployment, effective management and distinctive competencies. For example, even if a country is well endowed with natural and built environmental resources and possesses abundant labour or other production factors, this will not necessarily result in the creation of a competitive tourism product. In fact, what really matters is how to convert a comparative advantage into a competitive one, with the ultimate goal to create value for the tourist and the other stakeholders.

More specifically and in addition to competing on the basis of price or quality, firms can compete by engaging in research and development in order to gain a technological lead over their domestic and foreign rivals.

Countries which gain such a lead are said to have a technology gap over other countries and if technological progress by the leading firms is ongoing the leaders are likely to retain their competitive advantage and boost their foreign and domestic sales (Posner, 1961). Examples from tourism sectors include the US and West European production of aircraft and of information technology such as CRS for travel, accommodation and destination facilities. Poon (1988) argues that Jamaica's SuperClub hotels are a further instance and that their success relies on innovation rather than imitation.

The open economy development of the product cycle theory (Vernon, 1966) takes explicit account of the role of factors of production in determining technology gaps and the location of supply. According to the theory, the development and supply of new products tends to occur in capital-abundant, high-wage, tourism generating countries, underpinned by domestic demand for technology-intensive products and a supply of relatively high-skilled and high-wage labour which is able to develop and produce them. Increasing standardization of the product, imitation by foreign producers and rising foreign demand result in a transfer of production from the initial innovating countries to medium-wage economies and subsequently to low-wage and therefore low-cost countries in the 'pleasant periphery' (Christaller, 1963). The theory is interesting in pointing to the role of both demand-side and supply-side factors in determining trade and to the process by which comparative advantage changes over time. It can help to explain the transfer of production from high-wage to lower-wage countries and it is consistent with the initial concentration of international tourism demand and supply in relatively high-income countries and the subsequent growth of foreign holiday tourism in lower-income countries, although other contributory factors such as rising incomes are also relevant. For example, it is clear that each country has a specific set of cultural and environmental endowments which preclude complete product standardization. That such endowments together with accommodation and entertainments are consumed on-site, in conjunction with demand for differentiated products, ensures that tourism production and consumption occur in both low- and high-income countries.

In his seminal contributions, Porter (1980, 1990) highlights the importance of attaining a competitive advantage by following appropriate strategies such as cost leadership, product differentiation or market segmentation (focus on a specific niche). Cost leadership may help a tourism service provider or a destination attain a cost advantage, whereas the other two strategies are essentially related to the achievement of a differentiation advantage in the marketplace. Porter identifies five groups of production factors which should be effectively deployed in this process, namely: natural resources; human resources; knowledge resources; capital resources;

and infrastructure. In the context of tourism, Crouch and Ritchie (2006) add cultural and historical resources and extend the infrastructure grouping to consider tourism superstructure (e.g. accommodation establishments).

The attainment of competitive advantage should enable a tourism supplier or destination to face competition and become attractive in the international market. In his five forces model, Porter (1979) determines competitive intensity and market attractiveness by referring to the corporate rivalry from the incumbents, the threat of potential new entrants, the risk posed by substitute products or services, the bargaining power of suppliers and the bargaining power of customers. Competitive advantage deals effectively with the impediments raised by these five forces and enhances the offering of a supplier in the value chain (Porter, 1990). By relying on both primary (such as inbound and outbound logistics, operations, marketing and sales and customer services) and support (administration, human resources management, technology development and procurement) activities, a tourist supplier should aim at maximizing value creation to enhance customer satisfaction. At a destination level, policy-makers should aim at achieving integrated quality management, which 'can be seen as a systematic quest for internal quality and external quality, i.e. economic improvement in the short term and local development in the long term' (EC, 2000: 15).

Based on the above, Ritchie and Crouch (2003) developed a generic conceptual model of destination competitiveness, based on five structural pillars. The first, and perhaps most important, is related to core resources, namely: physiography and climate; culture and history; mix of tourist activities; special events; entertainment; tourism superstructure; and market ties with origin countries and service providers in the tourism circuit (e.g. tour operators). The extent to which these core resources act as effective attractors depends on the second pillar of the model, related to supporting factors and resources consisting of: infrastructure; accessibility; facilitating resources; hospitality; entrepreneurial spirit; and political will in favour of tourism. The model then highlights the importance of the third pillar of destination management. This is associated with: organization; marketing; service quality and experience; provision of adequate information; human resources development; finance and venture capital; management of visitors; resource stewardship; and crisis management. Destination management is in many cases related to day-to-day operations. At a more strategic level, therefore, it is crucial to consider destination policy, planning and development, which jointly constitute the fourth pillar of the model. In this context, what really matters are issues of: tourism system definition; the overall philosophy; values and vision of the policy-makers; the positioning and branding of the destination; projections about development plans; the undertaking of a competitive and/or collaborative analysis; as well as the need to monitor, evaluate and audit the results of the pursued destination policy.

Finally, the fifth pillar consists of the so-called qualifying and amplifying determinants, which represent 'factors whose effect on the competitiveness of a tourist destination is to define its scale, limit or potential' (Crouch and Ritchie, 2006: 429). These 'situational conditioners' include: location; safety and security; the relation between cost and value; the existing inter-dependencies with other destinations and tourism suppliers; awareness and image; and carrying capacity. All these five pillars operate within an exogenously determined global macro environment but play a major role in affecting the competitive environment at a micro level. What they essentially do is to convert the comparative advantage of a tourist destination (based on resource endowments) into a sustainable competitive advantage (related to resource deployment).

7.2.7 Concluding comments on trade and tourism

In conclusion, this section (7.2) aimed at analysing alternative theories to highlight the relationship between trade and tourism at an international level. The theory of comparative advantage is supply-oriented, while the theory of overlapping tastes stresses the role of demand. The consideration of scale economies and product differentiation provides a synthetic approach, whereas the discussion on the importance of competitive advantage shows that tourism competitiveness also depends on how efficiently the various resources are employed. So far, all references to tourism have been made at the level of flows; corporate presence and location were given only limited attention. Nevertheless, these issues are of major importance as the emergence of transnational tourism corporations in the context of a globalized environment can also affect the direction of tourist flows to a significant extent. This topic is now examined extensively in the following section.

7.3 Globalization and transnational corporations in tourism

Globalization may be understood as 'a process in which the geographical distance between economic factors, producers and consumers becomes a factor of diminishing significance as a result of faster and more efficient forms of travel, communication and finance' (Robertson, 1992 cited in Fletcher and Westlake, 2006: 464). The WTO argues that globalization is essentially the result of liberalization, although the relationship seems to work in the converse direction too (WTO, 1999). In a globalized economy, goods and services may be produced using international factors characterized by extreme and instant mobility. Technological progress, economic changes in favour of market liberalization, cultural and demographic trends related to open and tolerant societies, as well as political stability and a pro-trade environment are among the most important factors affecting globalization (Fletcher and Westlake, 2006).

In what follows, the advantages and concerns regarding globalization will be initially discussed. Then, issues of corporate expansion will be examined in the context of the OLI framework. All the above will then set the foundations to study subsequently the relationship between tourism and globalization, with emphasis on the hotel sector and the international hotel chains.

7.3.1 Globalization: advantages and concerns

Proponents of globalization stress the convergence features of the process (Wolf, M. 2004). Based on classical equilibrium economics arguments and Samuelson's factor price equalization theory (discussed previously), globalization is supposed to assist poor states in raising their living standards by achieving higher growth rates than developed countries: in this way, existing disparities in quantitative and qualitative factors will eventually disappear. This convergence process is essentially based on three balancing mechanisms (known as spread effects). First, the liberalization of international trade enables developing countries to sell their products profitably to developed ones; this increase in demand will raise total output and employment in poor countries and improve prospects for future growth and investment projects. Moreover, this rise in aggregate demand will have only very modest (if any) impacts on the general price level to the extent that there are idle resources in the economy. Second, the liberalization of factor mobility will facilitate labour migration from poor countries to rich ones. This is because labour abundance in developing countries is associated with low wages, whereas labour scarcity in developed states is related to high remuneration; as expected, this wage disparity provides an excellent incentive for migration. As a result of this labour flow, total labour supply will shift leftwards in developing countries and rightwards in developed ones, improving wages in the former and reducing them in the latter. Third, free factor mobility will also favour capital outflows from rich to poor countries. This is because developing countries offer a higher return on capital due to its relative scarcity and the presumed diminishing returns prevailing in production. Consequently, capital relocation will lead to further investment in developing countries and raise their income and productivity. Similarly to the case of labour, capital returns will eventually be equalized across all countries.

Having the above in mind, globalization, alongside liberalization, may help developing countries exit the poverty trap and enter the world of prosperity. East Asian countries (with China being the most prominent) are often cited as successful examples of this process. In this context, international tourism may play an important role as it increases demand for the produce of developing countries and creates a positive business climate for investment in related infra- and superstructures. Low-skilled local labour finds it relatively easy to be employed in the tourism industry and living standards rise. As a result, and further to the discussion in Chapter 6, countries that choose to specialize in tourism may manage to attain a higher growth trajectory compared to others.

Still, this optimism is not shared by the globalization sceptics, who have developed an interesting argument to invalidate the balancing mechanisms discussed above. According to the proponents of divergence theory (heavily founded on dependency economics), the existing disparities between rich and poor countries will eventually increase (Amin, 1976). Among other factors, free trade is likely to affect developing countries adversely due to the low income elasticity of their products as opposed to the high income elasticity of goods and services produced by developed countries; the terms of trade for developing countries will deteriorate over time, whereas any effort to increase local production may lead to further problems and immiserizing growth, as argued in a previous section (7.2) of this chapter. Second, labour migration from developing to developed countries clearly has an 'eclectic' character in modern economies, as those who migrate are usually the younger and better qualified. As a result, rich countries become endowed with a skilled and highly productive labour force, whereas poor states are left with impoverished workers of diminished capability who suffer from exploitation. Third, relocation of capital to developing countries does not necessarily lead to better prospects, as corporate profits are usually repatriated to the developed countries where the headquarters and shareholders of transnational corporations reside. In this context, international tourism may play a negative role: developing countries usually offer a non-sophisticated, mass-tourism product of low income elasticity, whose value is expected to significantly depreciate over time; whereas, transnational hotel and other corporations may just take advantage of developing destinations to make a profit, which is subsequently returned to the origin countries (Britton, 1991). Some islands in the Caribbean seem to have occasionally suffered from such policies (Debbage, 1990).

In conclusion, and as also argued in Chapter 8 in the context of environmental and ecological issues, globalization may have both positive and negative implications for developing countries – international tourism is not an exception. Therefore, it is important not to over-simplify but holistically to address the emerging issues. As argued later in this chapter, protectionism may not offer a viable solution to the globalization problems: what matters primarily is to endow developing countries with a sustainable comparative and competitive advantage, which will allow them to realistically compete in a global marketplace gradually dominated by transnational corporations. In fact, the growth of the latter is inextricably linked to the advancement of globalization. In what follows, the reasons for 'going global' and the role played by transnational corporations in tourism will be discussed.

7.3.2 Corporate expansion and the OLI framework

A transnational corporation (TNC) may be defined as a company that operates in more than one country by engaging in production and/or service delivery. There are many reasons why a company may decide to expand its operations at an international level. The OLI theoretical framework suggested by Dunning (1981) provides a very useful background to explain foreign direct investment (FDI). In particular, a corporation may extend its presence abroad to enjoy ownership advantages; this is the 'O' in the OLI framework, which provides a rationale behind 'why' companies internationalize their operations; organizational structures, capital and human resource endowments, intellectual property rights and patents are examples of 'O' advantages at an enterprise level. Furthermore, by operating abroad a company may enhance its access into product and factor markets and exercise its oligopolistic and oligopsonistic power thanks to reduced unit costs, mass purchase of inputs and large production of outputs. In this way, a company can also spread its domestic market entry and setup sunk costs over a larger number of production plants at an international level. Furthermore, 'O' advantages are also associated with the mastering of operations in a business environment characterized by uncertainty and with the effort effectively to diversify risk.

With respect to location ('L') advantages, transnational operations enable a corporation to acquire access to specific foreign country resources and positive business conditions such as the existence of a high-quality and/ or low-cost labour force, adequate transport and communication infrastructure, tax concessions and government funding. Internationalization may also assist in overcoming trade barriers and/or other protectionist impediments imposed on imported goods and services and in reducing the cognitive and psychological distance (i.e. culture, language, etc.) between the origin country product and the host community. In this context, the 'L' features attempt to answer the 'where' questions in the process of transnationalization. Finally, internalization ('I') advantages allow a company drastically to reduce transaction costs in acquiring inputs and most importantly to minimize uncertainty and stochastic conditions by exercising effective and direct control over its intangible assets, such as logos, image and brand names. Internalization advantages explain the 'how' question in transnational business involvement and their accumulation is often related to mergers and acquisitions of local companies, which offer supply and/or demand synergies (complementarities) to the products of transnational corporations.

7.3.3 Tourism and globalization: the hotel sector as an illustrative example

The relationship between tourism and globalization has been examined by a number of researchers (Smeral, 1998; Knowles et al., 2001; Wahab and Cooper, 2001; Bianchi, 2002; Fayed and Fletcher, 2002; Cornelissen, 2005). McKinsey (2003) and Hjalager (2007) developed a four-stage model in examining the globalization of tourism. The first three stages are characterized by a low globalization profile as companies are primarily interested in international collaboration, e.g. at a marketing level (stage one); cross-border integration of business activities, e.g. through mergers and acquisitions (stage two); and gradual immersion into the logistics of the value chain, e.g. by outsourcing (stage three). On the other hand, the fourth stage provides the opportunity for high-profile globalization as tourism companies 'transcend into new value chains, adding value by integrating economic logics in other sectors' (Hjalager, 2007: 440). Brand extension and the creation of spin-offs is a typical strategy to be followed in this context. For example, the Virgin brand is not only apparent in the music and media industries worldwide (i.e. Virgin Records, Virgin Megastores, Virgin Radio, etc.) but is also involved in rail travel and six airlines (i.e. Virgin Atlantic in the UK; Virgin Express in Belgium, which subsequently merged with SN Brussels to establish Brussels Airlines; Virgin Blue in Australia; Virgin Nigeria in Nigeria; Virgin America in the United States; and finally the planned Virgin Galactic in space-tourism).

The globalization concepts and the OLI framework in tourism have been primarily associated with the international hotel sector, where chain affiliation at a global level is gradually becoming widespread (Dunning and McQueen, 1982a, 1982b; Rodriguez, 2002; Johnson and Vanetti, 2005; Anastassopoulos et al., 2007). According to the *Hotels Magazine* Giant Survey (2008), Inter-Continental Hotels Group is present in 100 countries, with 585,094 rooms in 3,949 hotels, followed by Starwood (present in 95 countries, with 274,535 rooms in 897 hotels), Accor (present in 90 countries, with 461,698 rooms in 3,871 hotels), Best Western International (present in 80 countries, with 308,636 rooms in 4,035 hotels) and Hilton Hotels (present in 78 countries, with 502,116 rooms in 3,000 hotels) in 2007. Chain affiliation has increased considerably in the US domestic market over the last decades. In South East Asia, transnational hotel corporations have an even more prominent role, given the explosive growth from zero since the early post-war years. On the other hand, the small family businesses still hold a major market share of the aggregate hotel room supply in Europe.

In many cases, this structural diversity is associated with some idiosyncratic elements of the non-standard hospitality sector, which are greatly appreciated by the wealthy sophisticated European clientele, as argued in Chapter 5. The increasing need for professionalism within the global environment, however, and the various advantages of hotel chains gradually change the scenery in favour of corporate association.

In particular, the first advantage of coordinating hotel services at a chain level is related to the unit cost reductions in terms of advertising, product development and training (Go and Pine, 1995). By internalizing these scope economies, hotel chains enhance their clientele, relying on otherwise unfeasible tools. For example, by using expensive database management and computerized global distribution systems, the hotel chains facilitate the booking process and acquire better knowledge of consumer preferences. In addition, hotel chains are more efficient in entering new markets and sustaining business, due to a set of intangible assets and logistical skills, mainly associated with their managerial and organizational expertise and technical superiority. Their know-how is augmented by the choice and training of high-quality personnel, who are attracted by the promising career prospects.

The sustainability and reinforcement of hotel chain activities may be also partly attributed to the creation and promotion of a specific trademark, through a standardization process of the accommodation product. Though the predetermination of hospitality characteristics can result in the loss of personal architectural and servicing flavour, branding is very efficient at reducing the inherent uncertainty features of travelling in unfamiliar environments. In fact, chains have a higher probability of providing a consistent product, including quality attributes such as design, comfort, performance, efficiency, degree of professionalism and attitude towards customers. All these qualifications may be strong enough to counteract the oligopsonistic power of tour operators and major clients, who usually play producers off against each other to obtain a large variety of high-quality accommodation for reduced prices (Papatheodorou, 2000).

The advantages of chain affiliation highlighted are expected to be magnified in the wider context of transnational hotel corporations. Because of their size, these hotel enterprises enjoy a greater range of scope benefits, which are subsequently enhanced by a number of marketing devices such as loyalty schemes (e.g. the Hilton HHonors programme). They also possess greater bargaining power against their intermediaries (e.g. tour operators), compared to national chains, and usually abstain from entering into detrimental commission payment games. Furthermore, by operating within a multicultural environment, they gain substantial experience and know-how in confronting complicated situations; their global management follows a portfolio approach to minimize specific country risk. In this way, by playing destinations off against each other and through internal transfer payments policies, they may increase profitability to a significant extent. Thanks to their international branding policy, transnational corporations enable the international traveller to pursue business or leisure, even in relatively unknown or politically unstable countries; for instance, Inter-Continental Hotels and Resorts was founded in 1946 by Pan American World Airways, after the encouragement of investment and network creation in Latin America by the US State Department (Inter-Continental Hotels and Resorts, 2008).

a. Contractual agreements in the international hotel sector

Following the OLI framework, international hotel chains should prefer equity control, in spite of the existence of quality standardization contracts or the extraction of trademark rents through franchise agreements. First, the world profit maximization strategy may be otherwise in conflict with local interests, endangering the achievement of scope economies at a global level. Moreover, moral hazard of the representatives may result in operational inefficiencies of local establishments, with global reputation side-effects for the corporation's trademark. Third, the service character of the hotel industry and the associated know-how induce equity control since the intermediaries may be prone to diffuse classified information and technological knowledge to competitors. Finally, internalization may be particularly important in the case of developing countries as it minimizes negotiation and transaction costs and exploits the monetary stability of the home country (McQueen, 1983; Dunning and Kundu, 1995).

Some evidence from the real world, however, supports the opposite view. In the early 1980s, only a third of the foreign-affiliated hotels were equity controlled by transnational corporations, whereas the majority of associations comprised some form of management and operational contracts (Dunning and McQueen, 1982a). This pattern in contractual arrangements is most prevalent in developing countries, where investment risk deters corporations from direct financial involvement (Chen and Dimou, 2005). On the other hand, the long duration of the management agreements (10–20 years) enables transnational corporations to delay the transfer of internalization advantages for a considerable time.

Several TNCs have had different types of integration with a given host country firm, for example a minority equity holding in a hotel combined with a management contract or franchise; majority equity holding and shared management with other TNCs; or leasing combined with equity holding, management or marketing arrangements (Dunning and McQueen, 1982b; Endo, 2006). In the early 1980s, non-equity integration was a characteristic of many French, Japanese and American companies' relationships with hotels in developing countries, the TNCs supplying technology, management and marketing expertise. Ownership was more common among firms based in other West European countries. In contrast, equity investment combined with franchising was fairly common in Asia and equity and leasing was characteristic of hotels in industrialized countries and the Caribbean. Hotels in some West African countries, where there is government opposition to majority ownership, engaged in management contracts with TNCs, as did hotels in some Middle East countries which had finance capital but lacked a skilled labour force. Franchising and marketing agreements tended to be more common in destinations with greater local expertise, notably Brazil and Mexico.

In a recent study, Chen and Dimou (2005) attempted to trace the country- and firm-specific reasons behind the choice of alternative modes of TNC involvement in the international hotel industry. They distinguished between hierarchical modes (with ownership being the most important) and market modes (with franchising as the most prominent example). According to their findings, 'the most influential factor on the development decision is the degree of proprietary content and idiosyncratic knowledge embedded in the service provided. The higher the market segment of operation, the higher the specialized skills and managerial expertise required for hotels to operate according to standards; therefore, the more likely a hierarchical mode will be used for their development' (Chen and Dimou, 2005: 1). Nevertheless, the significant heterogeneity of relationships between TNCs and host country firms in the international hotel sector demonstrates the inappropriateness of global generalizations about their nature.

In terms of integration practices, transnational corporations within the international hotel sector have engaged in both horizontal integration in the form of foreign direct investment, leasing, management contracts, franchising and marketing agreements and vertical integration between hotels, airlines, tour operators and travel agencies. For example, approximately 78 per cent of major hotels along the Kenyan coastline and around 66 per cent of those in Nairobi and National Parks and Reserves have had some foreign investment, although fewer than 20 per cent have been subject to total foreign ownership (Sinclair et al., 1992). Equity holdings have been acquired, for instance, by the tour operators African Safari Club, Hayes and Jarvis, Universal Safari Tours, Kuoni, Polmans, Franco Rosso and I Grandi Viaggi, as well as by BA and Lufthansa airlines. TUI, the world's largest travel conglomerate, is the most important leisure hotelier in Europe, with approximately 290 hotels and 173,000 beds in 29 countries; illustratively hotel chains such as Grecotel, Robinson and Riu are partly or fully owned by TUI (TUI, 2008). In his analysis of cross-border mergers and acquisitions in the tourism sector, Endo (2006) confirms that the majority of big deals take place in developed countries, with hotels and motels playing the most important role. New trends emerge, however, as investors from developing countries acquire tourism assets both in other developing as well as in developed countries.

A combination of foreign and domestic ownership has the advantage of spreading the risks associated with running tourism enterprises. It also provides the foreign participants, such as tour operators or airlines, with a strong incentive to maintain or increase the demand for tourism in the destination and can supply additional business expertise and increase investment in the area, as in the case of Australia (Dwyer and Forsyth, 1994). On the other hand, governments may spend large amounts on infrastructure for the tourism sector. Payments of profits and/or wages to foreign nationals are sometimes high and some foreign firms pass little expertise on to the recipient country. For instance, in hotels in the Virgin Islands and Grand Cayman, at the beginning of the 1970s, at least 43 per cent of wages and salaries were estimated as being paid to expatriates who predominantly filled the managerial and professional positions (Bryden, 1973). At the same time, many Caribbean governments also offered highly favourable fiscal incentives for hotel development, including the provision of credit on very advantageous terms, tax concessions and considerable infrastructure provision.

Contractual arrangements between tourism firms in different countries vary enormously from case to case, providing some countries with significantly more benefits than others (Dunning and McQueen, 1982a). Firms in many developing countries have little knowledge of the contractual terms which prevail elsewhere. In contrast, firms from developed countries have notable informational and first-mover negotiating advantages, as in the case of tour operators' contracts with hoteliers in developing countries. Over the longer term, a country's ability to move on to higher growth paths depends, to a great extent, on its ability to acquire additional knowledge, and foreign direct investment and contractual arrangements between firms provide possible means of doing so. However, the wide variations in the types and extent of foreign participation and the fact that foreign participants not only provide but also acquire local knowledge (Daneshkhu, 1996) mean that the outcome for the host country may be growth-enhancing or immiserizing (de Mello and Sinclair, 1995). In this context, some theorists argue that strategic tourism policy may make the difference, as now discussed in the following section.

7.4 Strategic tourism policy, protectionism and GATS

An important difference between imperfect competition in the domestic and international context is that, in the latter case, governments have an incentive to implement strategic trade policies (Krugman, 1989a). Strategic policies aim to help firms of their own nationality to achieve higher export earnings and/or to decrease outflows of foreign currency in payment for goods and services produced abroad. They include strategic commercial policy in the form of export credits or subsidies or import tariffs. For example, in the travel context, subsidies may be granted to legacy national carriers (even if they are no longer owned by the government) or taxes imposed on airport departures. Devaluation of the country's exchange rate and measures to restrain inflation increase the price competitiveness of tourism in the country. Strategic industrial policy includes, in the case of tourism, payments to domestic businesses, such as capital grants or low-interest loans for improvements in hotel and guesthouse accommodation, and state aid for research and development, for example in information technology relevant to travel agents, tour operators, car hire firms, airlines and hotels. The government can also threaten to alter the accessibility conditions or route allocation to foreign airlines, unless of course the market is fully liberalized. For example, competition on the high-density routes between high-income countries can be distorted by governments' allocation of favourable arrival and destination slots to their own national carriers and imposition of limitations on the numbers of flight arrivals by airlines from other countries. In some circumstances governments may not have to implement their proposed policy, as a well-publicized threat to restrict landing and take-off slots may be sufficient to deter foreign competitors from competing in the market. Thus, the range of strategies which are available to domestic firms in both tourism generating and host countries is widened by the government's actual or potential policies, so long as such policies are known and credible.

In practice it is difficult for developing countries to gain first-mover advantages, as competitive initiatives usually come from developed countries, which have acquired greater expertise in establishing competitive prices and supply. Relatively small country size may also impede a government's ability to engage in strategic policy formulation (Krugman, 1989b) and they may be subject to lobbying pressures from foreign investors or they may not possess sufficient information about the operation of specific sectors to intervene effectively (Alam, 1995). Developing countries' attempts to restrict the numbers of incoming flights from particular origin countries is a case in point, as tour operators sometimes successfully evade the restrictions. For example, tour operators avoided the Kenyan government's attempts to limit air traffic between London and Nairobi to scheduled flights by routing charter flights via Italy (Sinclair et al., 1992). In the case of Kenya, effective state control was maintained only over domestic flights, which were limited to the national airline, in which KLM subsequently obtained an equity holding (equal to 26 per cent in 2008).

Having the above in mind, strategic tourism policy may also provide some strong arguments in favour of protectionism. This may be defined as 'a protracted policy or attitude of a State, which for various reasons, prevents or impedes the internationalisation of its tourism market as a consequence of the application of instruments of administrative and economic intervention or of the lack of instruments and practices which could make such an internationalization possible' (UNWTO, 1988: 22). Protectionism may be implemented in a number of ways, including: the creation of an economic and administrative environment which is favourable to national enterprises; the erection of impediments and discriminatory commercial practices against foreign companies; the tolerance or even explicit support of market power abuse by a small number of national monopolies or oligopolies; and the existence of a vague institutional and legal framework, which can be interpreted in different ways (UNWTO, 1988).

Nonetheless, in deciding upon whether to propose or adopt a protectionist stance, the government must predict other governments' responses to the policies under consideration. The interrelationships between the strategies of governments, like those of firms, may be analysed by means of game theory, examined in Chapter 4. For example, the case of potential subsidies to firms is analogous to the well-known case in game theory of the prisoner's dilemma. Although all countries might achieve a pay-off in the form of higher net income in the absence of subsidies, each would be in a worse position if others were to subsidize their producers. It is clear that a range of outcomes may occur if governments engage in strategic policies to benefit their own producers, as well as to deter potential competition, and if firms formulate their strategies in the context of their governments' policies and other governments' and firms' likely responses to them. As shown in Chapter 4, the advantage of using game theory to examine optimal strategic tourism policy is that it makes explicit the range of possible courses of action and outcomes and the advantages and disadvantages which are associated with each. Unlike the comparative statics approach of comparing pre- and post-policy positions, it can shed light on the dynamic process in which both firms and governments are active players at the global level.

On these grounds and in order to avoid the coordination failure outcome depicted by the prisoner's dilemma, countries should best engage in an open dialogue to build confidence, remove mutual suspicion and set the fundamentals for trade conditions, which will prove beneficial to all. The GATS provides the ideal international forum for issues related to tourism. The GATS became effective in 1995 in the context of the WTO and intends 'to contribute to trade expansion under conditions of transparency and progressive liberalization and as a means of promoting the economic growth of all trading partners and the advancement of developing countries'

(WTO, 2006: 2). Progression is realized through 'Round Negotiations' undertaken on a regular basis; the first one started in 2000.

The GATS is based on two main principles. The first is market access, according to which 'each government shall accord services and services suppliers of other governments treatment no less favourable than that agreed and recorded in its schedule'. The second is national treatment, which requires that 'each government shall treat foreign services and services suppliers no less favourably than its own services and services suppliers' (UNWTO, 1995: v). Under GATS, international commercial services may be classified into four modes of supply (WTO, 2006; UNWTO, 1995), i.e.:

- cross-border, with tour operations and travel assistance being the most prominent examples in the context of tourism;
- consumption abroad, related, for example, to international visitors;
- commercial presence, when, for example, a tourism company establishes a branch office in another country; and
- movement of persons as in the case of consultants, tour guides and hotel executives.

The GATS identifies twelve core services sectors, two of which are of interest here, namely tourism and travel-related services and transport services. However, it should be noted that air transport services are partly exempted from GATS: the Annex of the agreement considers aircraft repair and maintenance operations, marketing and selling of air transport services and CRS bookings but excludes air traffic rights and other directly-related services (WTO, 2006). However, these exceptions are under review and may be lifted in due course, especially if multilateral or plurilateral air service agreements become the norm globally (WTO, 2007).

It seems, therefore, that GATS may provide an interesting forum to promote international tourism in a liberal context by removing restrictions associated with: tourism flows (e.g. visa requirements, air accessibility); work conditions of tourism employees; rights on real-estate ownership; management of tourism enterprises; regulatory and anti-competitive practices (WTO, 2000). From a pragmatic perspective, however, it should be always borne in mind that the evolution of GATS depends on a tough process of negotiation where each member-state tries to safeguard its interests based on its relative bargaining power. Unless, therefore, the rich countries realize the difficulties faced by the developing states and try to help them within a fairer trade framework (Fair Trade Federation, 2008), the GATS liberalization efforts may prove futile or solely to the benefit of developed countries. In fact, although the principles of fair trade were originally conceived in the context of trade in goods, tourism is a service par excellence where fair trade can be applied. This is because tourism can only flourish within an environment of peace and safety, which presupposes satisfaction rather than antagonism of communities involved in its production.

7.5 Conclusions

This chapter has considered a wide range of topics which relate to the analysis of tourism within a global framework. None of the topics could be examined with the depth each deserves and all merit further investigation, both as issues pertinent to economics and tourism in general and in the context of specific tourist origins and destinations. The discussion has shown that international tourism can be examined using economic theories relating to trade, industrial economics and growth. Building on the analysis of tourism supply in Chapters 4 and 5 and the regional economic framework of Chapter 6, this chapter has shown how market structure is important in determining patterns of specialization in production and trade at the international level. Within the competitive market structures which apply to some components of tourism as well as to other products, relative technological efficiency and/or factor endowments are key variables. The earlier discussion of the determinants of trade indicated that developing countries can use their natural and cultural resources as the basis for tourism supply. The ready availability of low-wage labour in them may also contribute to the growth of labour-intensive tourism production. Developed countries also have environmental and cultural resources which attract tourists and are able to organize many tourism activities with less labour-intensive production techniques. Thus, both developing and developed countries can use their specific factor endowments to supply tourism products with diverse factor intensities of production. Still, what also matters is to go beyond factor endowment and fruitfully engage in resource deployment; in other words, the challenging task is to convert a comparative into a competitive advantage and enhance destination competitiveness at a global level.

Under the imperfectly competitive market structures which relate to some components of tourism, product differentiation and market segmentation are of particular relevance. Some general equilibrium models of trade have attempted to take account of trade in both homogeneous products and products of different qualities (Helpman and Krugman, 1993). The pattern of inter-industry trade in homogeneous products can be explained by comparative advantage based on different factor endowments, while intra-industry trade between countries with similar factor endowments is based on product differentiation. Over the longer term, international tourism, like trade in goods, may contribute to changes in market structure, so that, for example, an airline which previously had a domestic monopoly becomes subject to oligopolistic competition in the global arena. Tourism firms and/or national governments may attempt to alter market structure in their favour by such means as inter-firm integration or strategic policy-making. Such active policies may be desirable to the extent that they do not entail protectionism and beggar-thy-neighbour strategies, as these are likely to prove futile when reciprocal actions from other competitors are considered. On these grounds, it is preferable to found international tourism relations in a framework of mutual understanding where market liberalization will be creatively amalgamated with the principles of fair trade.

PART IV
The economics of environmental issues in tourism and an appraisal of the economic analysis of tourism

8 GLOBAL ENVIRONMENTAL ISSUES AND TOURISM

8.1 Background

In this and the following two chapters, attention is turned to the environmental context of tourism. As with other forms of human and economic activity, tourism has both an impact on and is affected by the environment. As a major phenomenon and economic system almost wholly dependent on the environment, this interrelationship is highly significant, having widespread effects. On the one hand, tourism confers economic, environmental and social-cultural benefits but it also has detrimental impacts on these elements, with important structural consequences for employment, other industrial sectors and the style and quality of life of communities in host destinations. On the other hand, its long-term survival depends crucially on the preservation of the quality of the overall environment that is essentially its product. Three elements of the environment constitute this product: two are its physical capital and the third its human constituent. The first consists of natural resources, such as beaches, oceans and seas, mountains, lakes and forests. The second is comprised of human-made resources, for example the infrastructure, historic cities, heritage buildings, monuments and artefacts. The third is the cultural and social capital, for instance arts, language, mores, lifestyle and institutions.

This chapter examines environmental issues at a largely general level prior to considering their relationship to tourism. Except where considered necessary, it does not trace the development of environmental studies in economics (for example Hanley et al., 2001; Perman et al., 2003, or Tietenberg, 2006). In relation to tourism, the environmental issues have been appropriately covered elsewhere, for instance in Lew et al. (2004) Holden (2005; 2007) and Gössling and Hall (2006). After discussing definitions of the environment, its role and crucial importance to economic activity are explained. Next, the nature of global environmental issues is examined. As indicated in Chapter 6, two related factors are identified as being important determinants of changes in the environment and its quality. One is increasing globalization, affecting many aspects of individual nations' economies, financial institutions, sociocultural structures and even political systems. The other is the trend in international trade in goods and services, including the influence of worldwide agreements on it. In any examination of the environment, that concerning sustainability, both at a global and local level, must be included. Indeed, the identification and resolution of environmental problems is in essence embodied in the wider analytical framework of sustainability that covers the use of natural, productive and energy resources and waste and pollution, with the pursuit of sustainable development being seen as the ultimate goal. Thus, considerable emphasis is accorded to explaining this, also investigating tourism's position vis-à-vis global trends and how it relates to the pursuit of sustainability.

Economics has developed concepts, principles and methodologies with which to analyse environmental issues and there are many studies by the discipline, considered below, that can be applied to tourism. Accordingly, a substantial proportion of this chapter necessarily has to be devoted to the introduction and explanation of the economic perspective in general in an exposition of the subject's approach.

In Chapter 9 the analysis focuses more closely on the local environmental problems associated with tourism, showing that while the sector is a generator of them it is also adversely affected by activities outside its control. Central to the chapter is the exposition of the concept of market failure concerning what are termed in economics public goods, such as natural resources and externalities (social cost or benefit and distributional issues relating to intra- and inter-generational equity). An example at this point serves to illustrate the importance of examining market failure. Given that tourism's basic product is the natural and human-made environment, already referred to above, the first two aspects of market failure require extensive investigation. Effectively, much of the tourism product consists of public goods for which neither tourists nor the businesses serving them have to pay; in economics they are referred to as open access goods or resources. There is a danger that as they are free they may be overused and consequently degraded. This constitutes an externality as much as the increased pollution associated with tourists concentrated in an attractive and often fragile environment. A number of case studies are offered that demonstrate the difficulties of identifying and finding solutions to environmental problems associated with tourism activity.

Chapter 10 initially considers the valuation of resources, developing the argument of the problem of their non-priced nature and preservation in the face of market based economic activity. Then attention is concentrated more closely on solutions by looking at instruments currently proposed for mitigating the effects of environmental degradation and more general policy issues. Finally, the implications for tourism of the application of such instruments and the development of environmental policies are examined. An overall assessment of the role of economics is offered.

8.2 The environment and its role

8.2.1 Definition of the environment

Economists generally perceive the environment in terms of the total capital stock, which has three main elements:

1   natural capital;
2   human-made physical capital;
3   human capital.

The notation adopted by many environmental economists is: $Kt = Kn + Km + Kh$, where Kt is the total capital stock (the environment in its general sense); Kn is natural; Km is human-made; and Kh is human capital.

The first consists of the natural resources, which can be divided into two of economic significance. Base resources are land and water; from which the necessities of life, ostensibly renewable resources, are obtained: food, for instance cereals, animals and fish; productive materials, for example metals and timber; and energy, such as fossil fuels. The human-made consists of the physical stock of capital, such as the transport infrastructure, factories, offices, shops, housing, vehicles, equipment, household appliances. The human can be generically termed the 'cultural stock', for example political and social structures and institutions, education, training and skills.

Within economics there are two schools of thought as to the relationship between the three categories of capital that are relevant to environmental and sustainability issues. Mainstream, neoclassical economics conforms to the principle that market mechanism will resolve such issues and subscribes to the view that allows substitution between the three forms of capital as long as the total capital stock is maintained. Effectively, this means that natural resources can be degraded and reduced as long as human-made and human capital are substituted for it.

Increasingly, economists, especially those working in the ecological and environmental fields, explained later in the next chapter, are becoming more critical of the subject's neoclassical approach, recognizing that environmental problems are all-pervasive and persistent and not necessarily speedily responsive to market-based mitigation. Those who call themselves environmental economists, though largely adhering to most of the tenets of traditional economics, accept the need for some degree of market intervention.

Economists, displaying many shades of green, deepening as stronger environmental perspectives are held by those in the newer branches of the discipline, take the stance that no such substitution of natural capital by the other two categories should occur because it is the very foundation of all life and human activity (Turner et al 1994; Tietenberg, 2006). Such a stance begins to move towards the position held in other disciplines, for example cultural and ecological economics and geography, in which the environment is defined more widely and given a greater number of categories. Geographers, for example Cater and Cater (2007), have identified the following forms of capital or assets:

- physical natural resource stocks: ecological systems, on which people draw for their livelihoods;
- physical human-made: the basic enabling infrastructure, e.g. factories, shops, communications; transport systems, e.g. airports, ports, railways; utilities, e.g. electricity, gas, water;
- human: e.g. education, knowledge, skills;
- financial: e.g. banks, stock markets, credit facilities, investment companies;
- socio-political: e.g. institutions, kinships, networks and relationships;
- cultural: e.g. arts, heritage, customs and traditions.

Essentially the first two correspond to the economic forms of 1 and 2 above, while the remaining four can be subsumed under 3. The expansion of the forms of capital not only accords with developments in the more recent branches of economics that acknowledge the increasing importance of the human form, but also identifies elements that are relevant to the analysis of tourism.

The tourism environment can also be defined to include all three forms of capital that comprise its total capital. Resources, such as beaches, seas, mountains, lakes and forests, fauna and flora, constitute the natural resource base, while historic cities, heritage buildings and monuments are the human-made one. These forms are what might be called the primary tourism resource base and are essential components of the product. If they were to be degraded in a given destination, it is likely that tourism would decline. The third, human capital, either in total or one or more of its constituent elements in its expanded form, can also be affected by any change in the availability and quality of the base resources. Clearly, any degradation will eventually lead to a decline in the number of tourists and consequently adversely affect all environmental forms. An evaluation of what hereafter is referred to under the generic term 'environment', unless a distinction is required in considering issues specific to one of its components, is given within an economics framework to show the subject's particular relevance to the analysis of the interrelationship of tourism and the environment.

8.2.2 The role of the environment

The environment plays a crucial role in supporting the world's economies and takes several forms, which can be summarized as:

- a resource base for the production of renewable resources by agriculture, forestry and fishing;
- a source of all non-renewable energy and productive resources;
- a contributor to economic development and growth;
- a productive input to economic activity, for example leisure, recreation and tourism;
- a life support system for all forms of life;
- a sink for waste products and the assimilation of pollutants;
- a contributor to the quality of life through its aesthetic value.

Many early observations on the environment's role (Carson, 1962; Boulding, 1966; Daly, 1977; Lovelock, 1979; and Costanza, 1984), place it in the paramount position being vital to the continued existence of all life on earth. Such a notion presupposes that all human activities and structures, in particular economic ones, are subsystems of the environmental one (the biosphere). Daly (1991), for example, states that the basic thing to understand about the global economic system is that it is entirely dependent on the biosphere. The problem is that given the aspirations of the human race for greater levels of consumption, the economic system is geared for continuing growth but the size of the parent system is finite. Consequently, as economic development and growth accelerate, where development is the transition from primary to secondary and tertiary industrial sectors and growth is the increase in national income, they encroach on the biosphere, both displacing it and degrading it, as well as generating more waste and pollution. Given this, the issue is that continued economic development and growth and conservation of the biosphere are incompatible, so that a trade-off is necessary. However, increasingly Pearce (1991), Turner et al. (1994), Bowers (1997), and Oates (1999), take a stronger environmental stance, arguing that such trading off is not an option, given the grave consequences outlined above.

Reference to the environment in both the economic and tourism literature is somewhat ambiguous, especially if environmental problems are discussed, as it is not always made clear whether the term is being used in relation to its natural, physical, human-made or human elements; very often it is only the first that is under consideration. In trying to avoid confusion, the term 'environment' will be adhered to in the examination of global environmental issues and challenges in this chapter, but its elements will be referred to as 'capital', particularly where it is necessary to make the distinction between human-made and human.

8.3 The relevance of the concept of globalization to environmental issues

The phenomenon of globalization, related to development and growth, is increasingly being recognized as giving rise to economic and sociocultural change with profound implications for all forms of the environment, therefore presenting worldwide challenges. Due consideration must be given to the environmental impacts of current trends in globalization that act as a background to concerns about those emerging from similar trends in tourism.

In Chapter 7 tourism was examined within the context of the concept of globalization to establish the extent to which it parallels the processes, dominated by economic drivers and transcending national boundaries, that tend to lead to the interconnection and interdependence of all spheres of human existence worldwide. At one level these features are seen as creating uniformity, while at another globalization is perceived as a hegemony of developed countries' values and practices imposed on developing countries in relation to environmental standards and regulation, particularly sustainability (Rosenberg, 2000; Teo, 2002; Mowford and Munt, 2003). Accompanying globalization is the increase in international trade in goods and services, which brings its own environmental effects. However, it must be acknowledged that while globalization as a phenomenon appears to be accelerating, it can be viewed as more of a contributory factor to what are more general worldwide trends arising from growth and development in national economies and international trade. Here, global environmental challenges are examined at this more general level, where international trade and globalization are considered as inextricably entwined in their environmental effects.

Much attention has been paid to the interrelationship of globalization and the environment by academics (see, for example, Haas, 2003; Dauvergne, 2005). National governments have also considered the issues, and international bodies, of which two, under the umbrella of the United Nations Environmental Programme (UNEP) – the Global Ministerial Environmental Forum (GMEF) and the Global Environmental Facility (GEF), inaugurated in 1991 – were set up with the express aim of working to alleviate environmental problems arising from globalization. However, investigations have tended to concentrate on the ethical, governance through international agreements, political and social issues that arise, rather than the economic. Effectively, the literature is largely about global environmental problems surrounding sustainability rather than those associated with the globalization process. Indeed, there has been a debate over whether the original intention of considering globalization in relation to its links to environmental concerns has been severed (Dauvergne, 2005).

While the all-pervasiveness of globalization has been called into question (Mowforth and Munt, 2003), it can be argued here at the outset that there are few doubts about the global effects of environmental problems currently faced by all nations. Prior to considering tourism's role, issues arising from current globalization and trade trends concerning the environment, as widely defined above, are investigated and appraised from an economic perspective. The focus of economic analysis underpins the examination of tourism's environmental footprint, which is subsequently considered.

8.4 Global environmental issues

8.4.1 The natural and human-made capital stocks

There are no longer any boundaries to environmental problems, such as pollution and waste, that not many years ago were perceived as largely local. In the past the impact on the environment of the production of goods and services and consumption was not recognized. To an extent, therefore, current problems are a legacy of the unidentified consequences of human activity by earlier generations.

Initially concentrating on the first two forms of the environment, one of the early issues that came to light in the mid-twentieth century as exceeding national boundaries was the 'acid rain' that Scandinavian countries accused the United Kingdom (UK) of inflicting on their forests, resulting in their degradation, due to the UK's use of coal as a fuel in industrial processes and electricity generation. This was an example of what is termed 'positive feedback', whereby the primary actions cited above, industrial processes and generation of electricity, had the secondary reinforcing effect of creating acid rain. This was unforeseen and unintended, having detrimental effects with possibly further knock-on impacts on local environment and people, such as reducing timber production, polluting water bodies and courses, killing wildlife and possibly affecting drinking water and causing ill-health to inhabitants. Increasingly, the degradation of the ozone layer, through the use of CFCs in refrigerators, is another case in point. Loss of the ozone layer allows harmful radiation to enter the Earth's atmosphere, thus causing a potential hazard to the health of everyone in the world. Yet another, global warming, is of even greater importance, with far-reaching effects.

Thus, positive feedback is becoming more widespread and having global consequences by reinforcing any detrimental changes.

Conversely, negative feedback is self-correcting, any secondary effects of a primary impact being eventually diminished. For example, many eco-systems on the Earth have the capacity to assimilate any disturbance, perhaps by pollutants, to restore their former state; this is often referred to as the balance of nature. Nevertheless, most global environmental impacts tend to be those where detrimental feedbacks are generated, the principal ones being:

- climate change, giving rise to more extreme weather patterns, arising from global warming;
- rising sea levels, threatening low lying settlements and natural and human-made environments;
- pollution from air emissions, discharges into water, persistent residuals on and in land, noise and visual intrusion;
- generation of solid waste that is non biodegradable and often toxic;
- degradation and loss of wildlife habitats;
- loss of biodiversity in terms of the number of fauna and flora species;
- desertification;
- population growth;
- deforestation;
- land-use change, largely agricultural, arising from climate change;
- soil erosion and loss;
- water shortage.

8.4.2 Human capital

It is instructive to acknowledge the relevance of the last five categories of the environmental concerns identified above in considering the impact of globalization on the human form. It should be noted that population growth is particularly relevant in its impact on developing countries. Clearly, in itself, it is not strictly an environmental matter, but there are ramifications, because of the pressure put on scarce productive and energy resources and the last four challenges listed below it, which are the outcomes of human activity. Much attention has been paid by international organizations to these and also other attendant problems, such as natural disasters, poverty, food shortages, poor nutrition, disease and lack of education and training. For example there are programmes under the United Nations banner, the World Bank, the World Health Organization (WHO), developed countries' overseas aid agencies and non-government bodies, for example the Overseas Development Institute (ODI) and Oxfam in the UK.

Human capital is often more susceptible to external influences, especially where there is rapid change engendered by economic development and growth. The detrimental effects certainly have anthropological/cultural, political, psychological and social effects, but the economic impacts are pervasive in that they impinge on institutions and systems such as the financial, political and social ones. If these are put under stress and fail to function properly, or in extreme cases even break down, then the economy will be adversely affected. Also, there is no doubt that the quality of life can be affected as degradation of the natural and human-made physical capital occurs and results in greater waste and pollution. All the factors identified here affect the overall quality of life, or well-being, which comes within the purview of development, environmental and welfare economics. A seminal work is that by Dasgupta (1993) on population growth and poverty, but, increasingly, environmental economics studies and texts cover such challenges (for example Goodstein, 1999).

Brief reference has already been made in Chapter 6 to cultural and social structures that determine employment opportunities and income in examining gender roles in tourism. Quite a number of examples have been identified and investigated to indicate the interrelationship of economic activity and sociocultural factors that have environmental implications, especially in the use of natural resources. In addition to gender, social structures such as class, ethnicity and age give rise to differences in education, training and skills that impart advantages to some groups in society and disadvantage others. They also determine the power structures that prevail in a country's institutions locally and nationally, ultimately influencing its political structures. Cultural factors, ranging over beliefs, community cohesion, customs, heritage, inherited knowledge, mores, traditions and values, also exert similar influences that are instrumental in building and/or maintaining existing social capital and structures.

Sociocultural factors are particularly important with regard to access to and use of human-made and natural productive and energy resources and can have beneficial or detrimental effects on them, depending on the attitudes to conservation and therefore the way they are used. Of particular importance is the rate of current use as opposed to that in the future, and the consequent impact on the natural capital stock and its ability to sustain the present and future generations, notwithstanding the cultural and social significance over time of this form of capital for the way of life of communities. While traditional structures and practices have evolved to care for the natural capital stock and sustain life and culture, there are increasing pressures that threaten their continued existence. Where competition for resources occurs, then conflicts either between groups within a community or from external forces can arise. Well-known cases in the field of environmental economics are open access resources, such as common grazing and ocean fisheries, considered later.

There are many historical examples which affected access to and use of resources that graphically illustrate arrangements that were socioculturally determined within communities. Few cases have persisted, but those, mostly in isolated and rural communities suggest that the management regimes for open access resources have been environmentally benign. The Bush people in Botswana are a case in point. Regarding those communities that were or are now subject to change, the cause could have been both internal and/or external, often being the source of conflict that initiated the change or the outcome. In many instances all three forms of capital were degraded or collapsed, given the sociocultural structures, lifestyle and resource-use practices adopted. Other examples are the civilization on Easter Island and the Mayan settlements in Guatemala and Mexico abandoned before the invasions of the Spanish in the sixteenth century. External factors, the arrival of Europeans, undermined the Native American and Aboriginal civilizations in the United States and Australia respectively.

However, for current examples, given the importance of the human-made and natural capital stocks to their continued existence, tourism destinations are the most appropriate to illustrate the role of, and effect on, human capital with regard to resource use and environmental effects. This aspect is considered at the specific, local level in the following chapters. At this point, however, acknowledging that worldwide destinations have been subject to many external influences, they also exemplify the globalization and international trade trends effects, which will now be examined.

8.5 Tourism and global environmental challenges

As shown in Chapter 7 tourism is growing at a phenomenal rate. In 2006 (WTTC, 2007) the sector accounted for over 10 per cent of global GDP, the employment of nearly 235 million people, 9 per cent of the world total. It has been estimated that there were 800 million international tourists in 2006 and it has been forecast that there will be 1.6 billion by 2020.

The impact on global environments is already highly significant and clearly will be even greater in the future. As one of the most important activities internationally, which itself reflects globalization trends in other major industries, tourism has had a significant impact on both natural and human-made environments in the global–local nexus. The emphasis, however, has largely been on the detrimental effects of tourists' and suppliers' activities, which has given somewhat of an imbalance in the literature. It must be recognized that tourism has also conferred benefits that have increased the quality of natural and human-made resources and enhanced environmental quality.

Nearly all the environmental challenges identified earlier equally apply to tourism. It certainly adds to the first four: climate change; rising sea levels; pollution; and solid waste generation. As to the remainder, with the possible exception of desertification, population growth and, perhaps, deforestation, it contributes to: degradation of wildlife habitats; loss of bio-diversity; land use change; soil erosion; and water shortage in countries and locations that have selected tourism as a vehicle for economic growth and development. This is especially the case where existing natural and human-made environments are adapted to receive tourists and meet their needs. There are also many examples of sea, lake and riverside, forest and moun-tain resorts being newly developed to cater specifically for tourism. There are innumerable examples at the local level that illustrate the nature of the problems that are considered in the following chapter.

In a study of the influence of global environmental change on or by tour-ism by Gössling and Hall (2006), as editors of a number of case studies they identify climate change; rising sea levels; disruption of biological, chemical and physical cycles; transformation of land use, depletion of non-renewable resources; and loss of biodiversity as issues. Hall and Higham (2005), in a relatively recent text, focus more on the impact on recreation and tourism of climate change. However, global environmental problems associated with tourism are largely examined in the context of sustainability. This issue is investigated below, where a number of notable studies are cited relating to its conceptualization and the emphasis given to it at a general level, including the extent of initiatives that have been taken by the two principal world tourism bodies, UNWTO and the World Travel and Tourism Council (WTTC).

8.6 Solutions to global environmental problems

While the connection between human activity and global environmental problems was recognized more than forty years ago, over the last decade there have been more concerted initiatives to address it. The most pressing issues identified were the three that head the list given above. The examina-tion in detail of the actions proposed and implemented is not pursued here, but an outline is given of what are currently perceived as the most pressing problems and the policies being enacted.

The history, analysis and suggested solutions of the moves to deal with global environmental problems is extensively covered in the environmental economics literature; see, for example, Chapman (1999), Perman et al. (2003), Tietenberg, (1997, 2006), the last two references by Tietenberg being the most comprehensive. In the UK the report by the economist Stern (2006) was given official recognition by the Treasury. These texts consider the kinds of international actions needed and the extent of agreed approaches and their implementation regarding productive resources, acid rain, ozone depletion and pollution, already referred to. Despite a number of international agreements, action on climate change, caused by global warming through greenhouse gas emissions, has been extremely slow. Discussions were initiated in 1992 under the auspices of the UN Framework Convention on Climate Change (UNFCCC). Despite a path-breaking protocol concluded at Kyoto in Japan in 1997, subsequent meetings yielded no binding decisions. The principal stumbling block to progress was in setting emissions reduction targets, until a breakthrough in negotiations in late 2007, in Bali, with the refusal of Australia and the United States to agree on the proposals and the concessions to developing countries that included China and India.

However, somewhat paradoxically, the issue that has become paramount is the reduction of the emission of carbon dioxide, which is not as potent as two given much less attention, methane and nitrous oxide, as greenhouse gases, notwithstanding the fact that they are included in the protocols with hydrofluorocarbons (HFCs), perfluorinated compounds (PFCs) and sulphur hexachloride. Apart from the targets to reduce air emissions by 5 per cent per annum below a 1990 baseline between 2008 and 2012, the basis of international agreement has been emissions trading systems where countries in surplus will sell permits to those in deficit in meeting their targets. This has been a highly contentious element of the agreement, strongly criticized by environmental bodies.

That it has taken fifteen years to reach an agreement is indicative of a lack of political will, partly reflecting pressures from large industrial companies, particularly in the worst polluting countries, fearful of the impact of the detrimental effect on economic development and growth and whether such reductions in emissions would be cost-effective. An interesting aspect of this emerged in 2008 concerning vehicle emissions contributing to global warming, with individual countries of the European Union resisting the more stringent regulations proposed. Car manufacturers, in particular, are claiming that these will inflict higher costs and threaten employment in the industry. This aptly illustrates the narrow vested interests likely to undermine the urgent need for rapid action. The outlook is that real progress will remain painfully slow, with political considerations, reflecting heavy lobbing by large and powerful industries, especially those involved in the supply of fossil fuel energy, impeding action by governments.

8.7 The environmental implications of international tourism

International trade theory, embodying goods and services, tourism being an example of the latter, was examined in relation to tourism in Chapter 7, which underlined the benefits of trade where countries possess a comparative advantage in the production of particular goods or provision of services. With respect to the environmental effects, there are two levels at which international trade has been examined in the economics literature. The first is at what is essentially a microeconomic level in bilateral trade models. The second is at the more general macroeconomic or global level. The examination here essentially proceeds on the supposition that globalization largely manifests itself through the growth of and change in international trade. This supposition is justified on the grounds that trade, by increasing interrelationships between countries, facilitates environmental interchanges and agreements. For example, at its most direct level there can be trade in environmental goods and technologies to reduce pollution; agreements can be reached to set standards on the production methods and goods traded. The question is whether increased international trade enhances or degrades environments and increases or reduces pollution.

Bilateral models have been used, in effect defining trade in the sense of bargaining between two countries in an endeavour by each to secure economic gain. Game theory, covered in the context of market behaviour by producers in Chapter 4, has also been applied within international trade theory. Texts on environmental economics, for example by Chapman (1999), Hanley et al. (2001) and Perman et al. (2003), tend to follow similar lines by explaining the standard economic models of comparative advantage and gains from trade, then considering the effects of unilateral environmental policy actions. This theoretical analysis is subsequently broadened to suggest the effects of such policies at a global level before examining the empirical evidence of the effect of environmental policies. Here, the theoretical analysis is not considered in detail. Attention is concentrated on reviewing the predictions of the economic models and the empirical context.

8.7.1 Theoretical arguments

a. The environmental costs of international trade

The customary bilateral analysis adopted is the effect on trade between two countries in a situation where, say, country A enacts more stringent environmental policies domestically, perhaps by setting limits on air emissions and/or discharges into water courses of heavily polluting industries, such as chemicals, in which the country has a comparative advantage. The outcome is that production falls, so that the volume of trade declines. It has also been shown, theoretically, that comparative advantage can be adversely affected if production costs rise, for example if the industry is characterized by economies of scale, the inference being that higher costs may necessitate the raising of the prices of the traded good so its export to country B will fall. One supposition is that country B will seek to purchase the good, previously imported from country A, from another source that has not enacted environmental policies. Another is that international trade will simply fall if B develops the industry to obtain the good in question. Country A's action on its environmental policy can be interpreted as an example of the disadvantage a first mover may suffer. At an aggregate level, the current situation regarding international agreements on carbon emissions is that certain Asian and developing countries are exempted from controls, this giving them a competitive advantage vis-à-vis other countries. For example, the European Union has quite stringent regulations which have led to calls by Member States for restrictions on the import of goods produced more cheaply from Asian countries; such an action would be illegal under global trade agreements.

A variant of this case is where a country promotes economic growth domestically by investing in an industry in which it envisages it has a comparative advantage. The investment is likely to generate environmental degradation and pollution that represents a social cost, that is, one falling on everyone in the country if the industry is not required to internalize that cost. Should the country enact environmental policies, then the outcome is the same as previously argued. It should be noted that trade theory posits that reductions in the production of the good for which a country has a comparative advantage reduces the gains from trade.

Another scenario is to consider the international mobility of productive activity. If a firm producing a good that generates pollution is subject to stricter environmental policies in a country, it may decide to 'export' its pollution by locating in another country that has no such restrictive policies. This is known as the 'pollution haven' hypothesis. By this action the firm is likely to maintain its competitive position and market share, also increasing its profits, as it avoids the pollution regulation or charge, thus reducing its costs. Moreover, another incentive is that labour costs are almost certain to be less if relocation is to a developing country with much lower wage levels.

A disturbing feature of this, as posited in Perman et al. (2003) and Tietenberg (2006), is that the relocation from the developed to a developing country is welcomed, as the industry would contribute to its growth and improvement. The problem for the developing country is that the industry imported by it is likely to be a capital intensive one and so more polluting, as opposed to its domestic industries, which are most probably labour intensive. Notwithstanding this drawback, it may be reluctant to introduce environmental policies, as this might be a disincentive to the relocation that holds the prospect of increased employment. Another reason for its reluctance might be that it is competing with other developing countries to attract industrial investment. Competition between countries of this nature has been dubbed 'the race to the bottom', the theory being that countries progressively lower regulatory standards likely to disadvantage them economically, such as taxes on businesses, which raise revenue for the country. The end result, if all regulation is abandoned in order for developing countries to survive, is that poverty occurs; the World Bank (Dasgupta et al., 1995) has considered the relevance of the concept to environmental problems.

There are a number of environmental problems that arise from the phenomenon of relocation. The stimulus of this to growth and development in the developing country is accompanied by higher levels of environmental degradation, pollution and waste and the 'dirty' industry adds to this. The level of international trade increases because what was originally production within a domestic market is now exported back to that market in the developed country. This involves transportation; the necessary fuel used adding to the depletion of a non-renewable energy resource and increasing emissions and therefore global pollution.

Perman et al. (2003) and Tietenberg (2006) expand on the environmental impact of the 'haven' hypothesis by quoting Antweiller et al. (2001), who posited that there are three reasons for pollution to increase through the:

- composition effect: relating to the industrial structure whereby the ratio of dirty to clean industries increases, even if total output is the same, so that pollution increases,
- scale effect: trade liberalization, or any promotion of economic growth in an economy, will result in higher total output, irrespective of zero composition and technique effects, so that emissions/discharges increase,
- technique effect: the ratio of pollution generated increases for each unit of output if firms' production become dirtier because of their openness to trade.

There is intense competition to obtain products, such as those found in the resource bases of the oceans and seas in international waters, to which everyone has free access because no private property rights can be exercised over them, that has led to disputes between countries at both a bilateral and global level. In economics this is an example of market failure known as the 'public good' case, often referred to as the common property or open access resources issue. A notable example was the study by Hardin (1968) that he called 'the tragedy of the commons'. His was actually not quite an instance of a true public good as he took an example where there was an arrangement as to the responsibilities, as well as rights, of grazers of common land not to put more than an agreed number of animals to graze on a pasture. He showed that there was an incentive to cheat and put more stock on to the common land than agreed, leading to its degradation.

The issues associated with the use of such resources at specific locations are considered in more detail under the concept of market failure with respect to overuse, degradation and lack of property rights in the next chapter. However, there are issues of global consequence that should be identified at this point. At this level, some that parallel Hardin's example are agreements between countries on fishing rights. There are instances of these that have broken down where one country violates another's nominated territorial waters, either because the claimed territory is not recognized or because it is ignored. Two such cases were the cod wars that occurred between Canada and Spain and the UK and Iceland. Unlike these, there have been more widespread ocean fisheries issues where there are no such agreements, and where there are true open access resources. In international waters there is the case of tuna, for which the concern is that over-fishing will deplete stocks that may possibly not recover. Moreover, in the case of tuna fishing, environmental issues of concern arise globally because the techniques employed threaten wildlife; a case in point is the likely extinction of albatross and certain dolphin species. While disputes of this nature are not strictly concerned with international trade per se, they are indicative of wider and global issues where many countries are involved.

b. Counter arguments to the environmental costs of international trade

Several economists have disputed the allegation of the detrimental environmental effects of international trade. One of the most notable examples countering the allegation is what is known as the 'California effect' (Hanley et al., 2001), the origins of which were the vehicle emission controls enacted by the state of California that led to other states adopting them. The argument is that in order to sell goods to a country with stringent environmental product standards, the exporting countries need to adopt equivalent regulations. The adoption may well be voluntary and unconnected with trade if the exporting countries recognize the benefits to themselves, both in terms of improving their competitive position and the improvement in the quality of their domestic environment.

A similar proposition is the 'Porter' hypothesis (Porter, 1990), which asserts that there can be a competitive advantage in improving environmental standards, especially where doing so involves technical innovation that enhances production efficiency, such as lowering energy and materials inputs or allowing scale economies to be achieved. There are two reasons why Porter's notions seem to be feasible. One is that more stringent environmental regulations may be the trigger to introduce new production techniques. The second is that they may act in anticipation of environmental standards being improved in the markets served.

At an aggregate level the idea of countries voluntarily moving towards stricter environmental regulation has been associated with higher national income. While pollution in developed countries will at first rise as national income increases, at some point it will decline as new and cleaner production techniques are introduced, very likely reinforced by consumers' desire for better environmental quality. The inference of this hypothesis is that as developing countries' incomes increase they will follow the same path. This suggests that the liberalization of international trade that benefits all countries will contribute to enhanced environment quality.

8.7.2 Empirical evidence

Notwithstanding the fact that empirical investigations to test the validity of the theoretical predictions of the environmental effects of trade are increasing, the results so far do not unequivocally bear out some of the hypotheses. A particular problem is the paucity of available evidence in the face of no specific requirement for data to be collected in a form that would facilitate appropriate analysis. Moreover, while there are studies about individual countries, few have been conducted at an aggregate level and multilateral level.

In a relatively early investigation by Tobey (1990), using 1975 data and applying an index of environmental stringency, it was estimated that the impact on exports of heavily polluting industries was negligible. Jaffe et al. (1995), researching the impact on competitiveness in the United States, came to a similar conclusion. In a more recent study, Copeland and Taylor (2004) found that there was an adverse impact on trade flows, but it was quite small.

In research by Gallagher (2004), following the North American Free Trade Agreement (NAFTA) in 1994 between Canada, Mexico and the United States, there was no evidence of a downturn in trade. Mexico, a country well behind its northern NAFTA partners in industrial development and economic growth, undertook new investment that reduced production costs and pollution, with the exception of air quality, which was worse.

With respect to the pollution haven hypothesis, the evidence suggests that there was no exodus of industries to countries with less severe environmental regulation. Gallagher (2004), who examined NAFTA, found that there was no support for the hypothesis; neither Canada nor the United States was encouraged to export its dirty industries to Mexico, which wished to increase its rate of growth and had lower environmental standards. Likewise, Jaffe et al. (1995) and Levinson (1996) argued that there is little evidence for the pollution haven supposition because meeting environmental regulations is quite a small proportion of total production costs. However, Tietenberg (2006), in reviewing research within the United States, cited studies giving evidence of firms moving from high environmental standards states to those states with lower ones. Regarding the 'race to the bottom' thesis of countries continually lowering their environmental regulations to attract manufacturing firms, Tietenberg, reporting on research in the early 1990s, found that there was no empirical support for the assertion.

In the study by Antweiller et al. (2001), the model of composition, scale and technique effects of free trade and its environmental impact was developed and tested. It concluded that the composition effect, the one most likely to confirm the pollution haven hypothesis, was small in comparison with that for scale, which was significant in increasing sulphur dioxide emissions, offset by lower pollution as a result of the technique effect. The results largely confirmed the research quoted in Tietenberg (2006) and Gallagher (2004) regarding the scale effects of NAFTA.

Notwithstanding the fact that it has been disputed, there is supporting evidence from studies of the Porter hypothesis (1990), which parallels that of the 'California effect' concerning convergence by countries competing with each other for trade (Vogel, 1995). That firms do not suffer a competitive disadvantage if subject to more stringent environmental regulation is borne out by the research conducted by Tobey (1990) and Jaffe et al. (1995) already referred to. The enhanced competitiveness of innovative actions by firms, investing to improve environmental performance, was confirmed by Gabrynowicz (2003). This view was supported in a review of business initiatives in the United States conducted by Porter and van der Linde (1995). Later work on competitiveness by Estey and Porter (2002); Feichtinger et al. (2005) and Greaker (2006) have drawn similar conclusions.

At an aggregate level, the Environmental Kuznets Curve (EKC) and Ruttan trend, neither of which originally related rises in countries' environmental quality to the increase in their national incomes, have been claimed to exist and were first considered in empirical work by Grossman and Krueger (1995). The supposition has been taken up by most environmental economists, but notwithstanding the fact that one of the benefits of international trade will most likely lead to higher levels of national income, the link between it and an enhanced environmental quality has so far not been shown. The possible connection has been associated more with the pursuit of sustainability, so the Kuznet/Ruttan theses is investigated below.

There is a quite large body of evidence concerning the open access resources issue, outlined above, involving the process of international trade and its expansion. A number of examples indicate the magnitude of the problems. Further cases of impacts on open access resources relating to tourism are given later in this chapter. One of increasing importance is the environmental impact of the transportation of goods by air and sea alluded to earlier. While air transport, using the sky as an open access resource, has been indicted as contributing to global warming and pollution, considered in relation to tourism below, the role of freight transportation by sea has not been given so much attention. Virtually all shipping is fuelled by oil, which adds to the impacts relating to air transport. Vidal (2007) indicates that carbon emissions from the estimated 70,000 ships worldwide, increasing to 90,000 by 2020, are twice that of aviation. He reports on evidence that shipping freight (90 per cent of the global total) and passenger travel accounts for over 4 per cent of carbon emissions. Given its growth, emissions will most likely rise by 70 per cent and so it is potentially much more damaging than that of aviation, which has been accorded much more publicity.

However, a factor that tends to be overlooked is the environmental damage of shipwrecks, for example vessels carrying toxic substances, especially oil tankers; a number of high profile cases have occurred, such as the Torrey Canyon off the south coast of the UK in 1967, Exxon Valdes in Alaskan waters in 1989 and the Erika off the Brittany coast in 1999. Also, notwithstanding global agreements and conventions forbidding them, actions such as the illegal cleansing of oil tanks are commonplace. A related problem is where ballast water, taken on in a tropical region, is discharged in a temperate part of the globe, possibly introducing alien species or diseases.

Another example is the dumping of solid waste at sea, clearly indicated by the debris washed ashore worldwide. It is now evident that the litter problem is more widespread than thought. In a report in the *Independent*, a UK daily newspaper, Marks and Howden (2008), reported on the 'plastic soup' of solid waste, twice the size of the United States, that is swirling around in the Gyre currents in the North Pacific ocean. It has been estimated that this mass of plastic items amounts to 100 million tons. According to UNEP, plastic waste is responsible for the deaths, when eaten, of 100,000 birds and marine mammals. Moreover, it is known that much of this waste is toxic. In degrading it can absorb concentrations of toxins, such as polychlorinated biphenyls (PCBs). Also, minute amounts are digested by small organisms, entering the food chain of marine fauna, and even humans.

Recent evidence indicates that there are much more widespread and serious environmental effects of human activity on the world's oceans. Related to the problem of global warming and the higher levels of carbon dioxide is the acidification of the oceans. This is threatening the balance of marine ecosystems, such as coral reefs, and the ability of marine organisms to create their protective shells of calcium carbonate (Connor, 2008a), and this could lead to their collapse. At the annual meeting American Association for the Advancement of Science in 2008, Connor (2008b) reported on the ecological state of the oceans and indicated that over 40 per cent had suffered environmental degradation from 17 different sources. In addition to shipping, land-based development, agriculture, pollution and over-fishing were cited as the principal culprits. The worst affected waters were the Atlantic coast of North America, Caribbean, China Seas, Mediterranean, North Sea, Persian Gulf, Red Sea and western Pacific. Only 4 per cent of the oceans appeared to be unaffected. Another study reported at the meeting of the American Association for the Advancement of Science (Connor, 2008c) found that in Antarctica a micro-organism, the pteropod snail, which constitutes a key part of the food chain for fish, penguins, seals and whales, is under threat of extinction. This parallels the case of sand eels which is referred to below in giving examples of marine environmental problems with relevance to tourism.

Finally, attention must be drawn to how far environmental issues have been incorporated into international agreements on trade and services. The original post-Second World War umbrella organization on trade was the General Agreement on Tariffs and Trade (GATT), extended later to include services (GATS), which eventually became the World Trade Organization (WTO) in 1995. The main aim of the WTO is to engender multilateral trade, eliminate trade barriers and lessen unilateral restrictions.

There are no specific clauses in the WTO concerning environmental issues. What have arisen in trade between countries are disputes where a country restricts imports from another whose goods do not conform to the domestic economy's environmental standards or whose production processes are environmentally damaging. There have been specific unilateral actions restricting trade of this nature. For instance, for a period in the 1990s, the United States banned imports of tuna that was not harvested using dolphin friendly methods, until it was ruled as against the then General Agreement on Tariffs and Trade (Tietenberg, 2006). The succeeding body, the WTO, is increasingly being compelled to consider such issues, especially if the question of the possible risk to the lives, health and safety of humans and animals arises, such as in the case of imported toys that were painted with a lead content paint. Another aspect of legitimate restriction on trade stems from 1973 concerning the trading of endangered fauna and flora species; CITES (Convention on International Trade in Endangered Species) governs this. Examination of the effects of international trade agreements is covered in the environmental economics texts, for example Hanley et al. (2001), Perman et al. (2003) and Tietenberg (2006).

It can be concluded that the appeal of the hypotheses on the relationship between international trade and environmental effects is their attractiveness to politicians and environmentalists. The former wish to implement both trade and environmental policies that appear to promote economic growth and development; the latter, tending on balance to be negatively predisposed, oppose the pursuit of economic goals and perceive increases in international trade as being environmentally damaging.

8.8 Global environmental issues, international trade and tourism

8.8.1 Background

As indicated above, the environmental economics analysis of the interrelationship between trade and the environment is quite extensive. This is not the case with respect to the attention paid to tourism. Certainly there are studies taking an economic perspective on tourism's role regarding environmental issues at a local level (Mathieson and Wall, 1982; Bull, 1999; Tribe, 1999; Holden, 2007). However, the subject has contributed little at the global level regarding either its interrelationship with globalization and trade and the consequent environmental effects (Wanhill, 2007). The study of these elements has been largely left to other social scientists, particularly geographers, for example Mowforth and Munt (2003), Lew et al. (2004) and Gössling and Hall (2006). Therefore, to an extent, recourse has had to be made to the non-economic literature to support the arguments put forward below, notwithstanding the fact that the majority of studies tend to be concerned with tourism's relationship to sustainability, the consideration of which follows this section.

Although the boundaries between global and local environmental issues related to tourism are blurred, it is possible to offer some observations at a general level that have an international dimension. Students, when introduced to tourism at a macroeconomic level, are very often surprised that it is treated in national accounts in the same way as the international trade in goods, where expenditure on holiday travel abroad is treated as an import and conversely tourist expenditure by foreign tourists is recorded as an export. The point of identifying this here is that tourism outside the national boundaries is basically the same as international trade; it is simply classed as foreign trade in services. Given this, it is clearly part of the apparent trend towards globalization and the resulting rise in international trade. It is thus examined within such an analytical framework. Nevertheless, it must be remembered that tourism is unique as a traded service in that it can only be consumed by travel away from home requiring accommodation and purchases, while abroad, of other goods and services. In a sense, therefore, the tourist, as a consumer, has an environmental effect, as well as the suppliers of the goods and services bought. On the supply side the essentially fragmented and small scale structure of tourism must be borne in mind. Other aspects of the environmental impact of tourism, seldom referred to in the environmental economics studies, is the impact of trade on the sociocultural environment and the extent to which an economic sector such as tourism is crucially dependent on the quality of the natural and human-made environments.

8.8.2 The relevance to tourism of globalization and trade and their environmental impact

All the hypotheses introduced earlier regarding trade and the environment are to varying degrees relevant to tourism, as they indicate the likely negative and positive impacts. There are two possible exceptions, the first, as shown below, relating to the enactment of more stringent environmental regulation that, prima facie, would be a deterrent to tourism by making countries relatively more expensive to visit. The second case is the open access resources issue, also examined below, for which the connection is more tenuous because it is difficult to pinpoint the extent to which tourism contributes to such an environmental impact; statistics seldom differentiate between the sources of such effects.

Given the paucity of empirical testing of some theoretical studies on trade and the environment in the context of tourism, partly exacerbated by the unavailability of appropriate data, consideration here is to an extent to infer what their relevance is likely to be, in the absence of hard evidence. With respect to this assertion, the reader is reminded that while there is a literature on tourism and globalization, referred to in Chapter 6, which examines such aspects as its competitiveness and marketing, there is little concerning environmental issues. One study that does both is that by Wahab and Cooper (2001), in which there are three chapters on globalization and sustainability.

The factors of most significance, on which some research has been conducted, relate to the engendered growth and development of tourism that changes the economic structure; type and stringency of the environmental regulation; nature and level of pollution and impact on competitiveness, efficiency, innovation and productivity of both countries and businesses. Much attention has been paid to the effects of trade and globalization on the economic and physical environments, but there are also marked influences on the human one in terms of sociocultural and political effects.

International tourism makes a large contribution to the growth and development of both developed (high income) and developing (low income) countries that itself generates trade in goods and services, simply to satisfy the needs of tourists in destinations. The theory of comparative advantage is relevant here because some commodities originate from specific countries and thus they will have an advantage. For instance, tradition, expertise and climate tend to favour wine production in certain countries. Likewise, horticultural production, a labour intensive activity, gives countries with a plentiful supply of workers and lower wages, in conjunction with a climate conducive to rapid cultivation cycles, a trade advantage. Therefore, the volume of international trade increases as exports rise from countries with a comparative advantage to supply tourism destinations. Clearly, for the two examples given, the necessity to transport these commodities increases what has become the environmental issue of 'food miles'.

An interesting point made by economists, noted in Chapter 6, is the link between international trade and tourism that facilitates commercial relations, foreign direct investment and enlargement of markets, stimulating growth in host countries. This may trigger countries to select tourism as a means of diversifying their economies. This is certainly understandable, given the theory of trade. Many countries possess resources that are potentially very valuable tourism assets. These might be natural environments, such as their climate, beaches, lakes, mountains and often abundant and exotic wildlife. Also, it could be their human-made and human environments, for example historic buildings, sites and artefacts, and their cultural and social attributes. In essence, they have a tourism comparative advantage over countries that are less well endowed. Cases in point are the sand, sea and sun resorts developed in Mediterranean countries with the advent of cheap air travel.

The environmental implications of increased global trade arising from the growth in international tourism are likely to be greater for low-income countries that indeed decide that tourism is an appropriate means of promoting development. There have been a number of studies of the effects of tourism on such countries' economies that touch on the environmental effects. Pearce (1989), Coccossis (1996), Wahab and Pigram (1997), Hall and Lew (1998), Wahab and Cooper (2001) and Southgate and Sharpley (2002) have investigated this, but they tend to do so in relation to the impact on sustainable development, considered below, and the sociocultural components of the human environment. Pearce (1989) gives the most comprehensive account of the environmental effects of tourism development, in coverage reflecting that by Mathieson and Wall (1982). However, the literature virtually ignores any reference to how the consequent increase in the international trade of goods and services affects the environmental quality of countries' pursuit of growth and development through tourism.

Developing their tourism sector, even if they endeavour to do so domestically, will almost certainly necessitate the improvement of the infrastructure, the manufacture of materials for this and the erection of buildings, such as hotels, restaurants, shops, and the need to offer a wide variety of services. If the industries to supply the resources required are domestically founded and are labour intensive, because of lower wage costs compared with capital intensive ones, they will almost certainly be of a lower quality and perhaps raise pollution levels. The structure of the economy will change markedly if every effort is made to concentrate on domestic production sources and workers to serve tourism. This will have widespread consequences for cultural, economic and social environments. Also, in the short run at least, it is unlikely that countries can achieve self-sufficiency. Accordingly, they will have to import productive and consumer goods and services, encouraging businesses from higher income countries to locate there, both during the development and the subsequent operational stages. Again, the increase of such trade will lead to a rise in the volume of freight transportation, adding to air emissions, thus adding to global environmental problems.

The Kuznets/Ruttan supposition runs counter to the projected negative environmental effects of tourism as a generator of economic development and growth. If over time tourism does succeed in increasing countries' national incomes, then a positive outcome can occur. While in the early stages of rising incomes, environmental degradation and pollution increase, in due course countries can afford to improve the quality of their environments. Also, it should be acknowledged that although tourism has detrimental environmental effects, it is better than many other forms of economic activity that might be introduced to stimulate development and growth. Moreover, given its need to preserve what is in effect its product, it has both an incentive and the potential to improve the quality of the environment. Thus the quality of the experience for tourists is enhanced and the demand for a better environment by inhabitants go hand in hand. Over time destinations can differentiate their tourism product and attract higher paying tourists, enabling businesses to improve their performance, which then reinforces the upward trend in environmental quality. However, a caveat is that should the tourism strategy fail there may be a temptation to divert into a less sustainable use of natural resources, a market value for them having been conferred by the sector.

It can be posited that a country with a significant tourism sector that enacts more stringent environmental regulations than others with which it competes for visitors will suffer a comparative disadvantage; for example, offering beach holidays in the already over-supplied mass market. Owing to this, and as there are many substitute destinations that will now be relatively cheaper, tour operators in generating countries offering such holidays will probably remove that country from their brochures. On the other hand, tourists who are environmentally aware may prefer to go to countries that have tighter regulations.

The pollution haven hypothesis undoubtedly applies, given that developing countries, as indicated above in the section on solutions to global environmental challenges, are currently exempt, under international agreements, from meeting targets for the reduction of CO_2 emissions. They are also likely to have lower environmental standards simply because they cannot afford to invest in the requisite technologies that offer cost reductions. Lower standards act as an incentive for firms in pollution intensive industries from developed countries, such as those manufacturing materials serving the construction industry undertaking tourism projects, which are invited to locate to low-income destinations. Conversely, countering the haven hypothesis are the circumstances in which cleaner industries relocate to destinations and demonstrate the cost reduction advantages of their production methods and techniques, to which attention is now turned.

The California effect and Porter hypotheses, principally concerned with competitiveness, are also associated with the efficiency, innovation, productivity and profitability aspects of countries and businesses. These aspects have attracted researchers in tourism because they have implications for the performance of destinations as a whole and of businesses, in both generating and host countries. Examples are travel agencies, tour operators and transport companies in the former, and the management of attractions, hotels, restaurants and services in the latter. It is recognized that the aspect identified above have an international dimension that therefore relates them to trade theory. However, hardly any tourism studies acknowledge this, as is also the case with regard to the environmental issues that arise.

The essence and significance of the two hypotheses is the benefit of the voluntary adoption of techniques that reduce pollution and enhance environmental quality. However, their vital feature is the demonstration effect, which encourages countries and businesses to follow the same course of action because they posit that production costs will be lower and/or profits higher if more up to date and cleaner techniques are adopted. This adoption of cleaner technologies is crucial to tourism that is reliant on the quality of its environment. An excellent paper by Razumova et al. (2007) has reviewed the Porter hypothesis, pioneering its application to tourism and environmental issues, but it does not consider the possible diffusion via trade links. The paper cites studies on competitiveness, innovation and technical change with reference to their relevance to the management of tourism enterprises relating to environmental quality and ecotourism, for example those by Lindberg and Huber (1993), Hjalager (1996; 1998) and Huybers and Bennett (2003). Currently, how far the Porter hypothesis applies to environmental issues in tourism is uncertain for there is little evidence to make a direct connection. In the investigations quoted above, the connection is only implied for the demonstration effect.

The argument that the composition, scale and techniques of economic sectors are changed by international trade, giving rise to environmental effects, has implications for tourism. To an extent the detrimental impact has been shown above in the section on the economic exposition and in the cases of the promotion of tourism to stimulate economic growth and development and the pollution havens thesis. The quality of the environment in destinations will be impaired by both domestic and imported industries introduced to support the development of tourism if their operations use unclean production methods. This will constitute a threat to the continued viability of tourism, especially if the quality of tourists' holiday experience is adversely affected. One notable example of this is the over-development of coastal resorts in Spain between the 1960s and 1980s and the desertion of them as more attractive destinations were served by airlines, such as eastern bloc countries after the demise of communist regimes in the 1990s.

A more direct effect on tourism, however, is apparent because of its structure. On the supply side, as a fragmented activity, consisting mostly of small enterprises, often with a limited season, it very likely suffers from a lack of capital to invest or reinvest frequently to meet higher environmental standards. This is probably the situation in low-income countries, in which such standards to qualify for certification and accreditation are difficult or impossible to attain. Furthermore, tourism companies originating from developed countries that are based in the destination, aware of possibly less stringent environmental regulations, may well adopt lower standards that aggravate the situation, resulting in a negative demonstration effect.

Conversely, there is potential for the composition, scale and technique effect to confer a benefit on tourism enterprises directly that in some respects is a reflection of the life cycle hypothesis proposed by Butler (1980), on which there is an extensive literature. For example, one way is to change the composition of the tourism product, perhaps by segmenting or upgrading it into forms that are more environmentally friendly. Ecotourism is one such approach. The scale and technique of operations can be increased by collective action, such as the formation of cooperatives or associations in destinations that enable individual businesses to adopt cleaner technologies and practices. The demonstration effect of the Porter hypothesis may well contribute to this outcome.

To conclude the physical environmental interrelationship of tourism and globalization and trade, attention is now turned to issues concerning open access resources worldwide. The two most important aspects in this context are air and sea travel, the atmosphere and oceans and seas being the respective open access resources, although, in Holden (2007), some reference is made to those that are land-based, such as lakes, rivers, mountains and wetlands. It also should be recognized that the availability of water can be perceived as a natural resource with open access characteristics where its supply crosses national boundaries, for example rivers and underground aquifers. Tourism is in competition with other economic activities for this resource, especially where a developing country is diversifying its agriculture, such as Kenya with respect to its supply of vegetables and flowering plants to Europe, or industrializing to facilitate economic development that will increase the water use; in particular, the generation of energy uses large amounts of water.

Of the 2.5 per cent globally air transport contributes to CO_2 emissions, a large proportion relates to tourism; even business trips can be included as a majority of them possess an element of leisure while travelling and in the destination. Tourism is set to double by 2020 and it has been esti-mated that air travel will increase by 3.5 per cent per annum, that is, by more than 40 per cent over the same period, with nearly 3 billion passengers (Hickman, 2007). This suggests that the airline industry's CO_2 and other emissions, irrespective of technical advances likely to improve the effi-ciency and fuel economy of aircraft, will rise proportionally. Hickman paints a gloomy picture with respect to the emissions by aircraft, which run into almost countless millions of tons.

The impact of cruising as a form of tourism is also growing rapidly, at over 10 per cent each year, and will certainly increase (UNWTO, 2003a). In 2005 there were 150 ships in service and new vessels being introduced are getting larger, the latest coming into service in 2008–9 being over 150,000 tons, with a capacity to carry 4,500 passengers. It is being pro-jected that ships of 500,000 tons, carrying 7,500 crew and passengers, will be constructed within five years (Dowling, 2006b). These will travel further and for longer periods than currently, thus adding to air emissions, often in more fragile environments, such as the Arctic and Antarctic. The litter problem, already referred to, will be further exacerbated by the growth of cruising, notwithstanding the up-dating in 1998 of the international regula-tions on maritime pollution (known as MARPOL) and independent actions by the cruise industry itself to improve its own environmental performance (Sweeting and Wayne, 2006). A continuing problem is the existence of the registration of ships under flags of convenience where environmental stand-ards are often ignored (Dowling, 2006c). Moreover, the effect of the trans-port of freight by both air and sea to meet tourists' needs in destinations, also covered above, will also increase in line with the greater numbers of tourists.

There are other environmental impacts on open access resources arising from this explosion of tourism. One that may be construed as a local prob-lem, but increasingly is being perceived as a global one, is the discharge of sewage into lakes, rivers and above all the sea. It has long been an issue in the Mediterranean, almost an enclosed sea, into which 70 per cent of sewage from resorts is discharged; in developing countries of the world the propor-tion is nearer 90 per cent. A similar situation exists in the Americas, as identified by Anderson (2003) in a considered report to the US House of Representatives, focusing on the Gulf of Mexico, but which has a world-wide significance. Whilst recognizing that it is difficult to establish what proportion of such discharges is attributable to residents or tourists, there is no doubt that where built environments are primarily tourism destinations the volume generated by the sector is substantial. This hazard no doubt contributes to the spread of diseases. The WHO has estimated that 80 per cent of illnesses are waterborne in developing countries.

Discharges of sewage have created a further threat. If untreated they give rise to algae blooms (see Anderson, 2003; Gidwitz, 2005) that themselves are toxic to humans and wildlife. Moreover, intensified by global warming, they increase eutrophication and plant density, the decomposition of which de-oxygenates water, which kills animal life. While initially local in their effects, such blooms do eventually spread out into seas and oceans. As indicated in Cater and Cater (2007), there are detrimental effects on marine organisms such as phyto and zooplankton, which are important food sources for fish, marine mammals and sea birds. This example represents a positive feedback effect on both land and marine based wildlife of pollution and climate change.

Notwithstanding the fact that tourism contributes to environmental problems, it is also adversely affected by them from other activities. The watching of many forms of marine wildlife, such as birds, dolphins, seals and whales, a growing form of leisure activity by tourists, is threatened. For instance, Cater and Cater (2007) cite the example of sand eels, the staple diet of many seabirds. Both the sand eels and plankton thrive in relatively cold water, within which they feed. Thus the reduction in their food source and global warming, which raises sea temperatures, reduces their numbers in locations where sea fauna and flora exist. The reduction of sand eels has also been caused by over-harvesting them, in international open access resources waters, for animal feed products and as a fertilizer. First noticed in the 1980s in the UK, particularly in the off-shore islands of northern England and Scotland, this collapse in the number of sand eels is accelerating, in turn resulting in a number of species of seabirds failing to breed. The RSPB (Royal Society for the Protection of Birds) has campaigned vigorously against the fishing of sand eels because of the danger of the extinction in the UK of seabirds if the populations of their prey fall below a replacement level.

Many other examples of pollution emanating from largely tourism locations and the overuse of marine open access resources have been identified in Cater and Cater (2007). They were not specifically studying the effect, on tourism in general, of open access marine resources environmental issues, as they were primarily concerned with the prospects for marine ecotourism. However, in the cases given they indicate an awareness of the wider global issues, such as the impact of commercial fishing and aquiculture. For example, they refer to coral reefs, mangrove forests and sea grass beds as being under threat, with adverse ramifications for indigenous populations.

In this review of the interrelationship of global trends and tourism the emphasis has been on the physical environments, tending to concentrate on the negative aspects. Clearly this is because they are more visible and possibly reflect the relatively short-term and incremental and fragmented nature of much tourism development and the objectives of businesses in pursuit of profits in an activity that experiences a large measure of uncertainty.

However, it should be acknowledged that the development and growth of tourism, already considered above, has had a positive effect on both the physical and human environments. While the process has been slow and does not constitute a very significant proportion of total tourism, its alternative forms, particularly ecotourism, have been given much prominence in the literature. However, as the impacts are overwhelmingly local, they are examined in the following two chapters.

8.9 Sustainability

8.9.1 An outline of the meaning of sustainability

Throughout the exposition above of the interrelationship of globalization, international trade and tourism, the linkage to sustainability has been acknowledged. It was also emphasized, in delineating the capital forms constituting the environment, that there is a connection with sustainability. Referring back to the definitions of the environment given above, the interface of the natural with human-made and human elements indicated that there might be a trade-off between economic development and growth and sustainability. Similarly, it has been implied that there is a trade-off between tourism development and environmental quality and sustainability.

However, there is much confusion and consequently debate as to what sustainability means. Over many years there have been many definitions (see WCED, 1987; Pearce et al., 1989) depending on what viewpoint was taken and a vast literature now exists. A number of key references are identified below, including those of relevance and applicability to tourism. Currently, the issue continues to exercise the minds of academics, business people, politicians and domestic and international NGOs (non-government bodies). Tracing the history of the sustainability issue and actions advocated to achieve it is not dealt with here; it has been extensively covered in both the economics and tourism literature (see for example Sharpley and Telfer 2002; Perman et al., 2003; and Holden 2007). However, it is necessary to clarify the different terminologies that have been used. Also, an exposition of the economic approach to it gives more rigour to its meaning.

A degree of consensus as to what sustainability is, generically, is emerging as researchers recognize that a more precise definition is required. It is increasingly being perceived as a term embracing both economic activity and environmental matters, particularly the conservation of natural resources. There are also implications for the cultural, social and political sectors, having acquired a wider meaning than the term 'sustainable development' (SD), with which it has often been mistakenly used interchangeably. Similarly, within tourism studies frequent reference is made to 'sustainable tourism', 'sustainable tourism development' and more recently simply 'sustainability'. An understanding of the word 'development' is crucial here. In economics it refers to the process of transition from the primary sector, such as agriculture, to the secondary, for example manufacturing, and tertiary, that is the service sector, of which tourism is an example. It includes increases in the standard of living, a more equitable distribution of income and wealth, improved educational and literacy attainment, the extension of work skills, better health and enhancement of the general well being of the population.

Development should be distinguished from economic growth, which is an increase in productive capacity and output of goods and services. It is normally indicated by an increase in national income or product GDP (gross domestic product) or, by adjusting for capital (Km) depreciation, NDP (net domestic product) in real terms. It is often measured by estimating per capita income, allowing for changes in population size. Estimates in national accounts of GDP do not include the benefits of development cited in the previous paragraph, nor the environmental costs falling on Kn and Kh that are considered below in examining sustainability in more detail. In many spheres of economic activity, for instance as applied in many tourism studies in referring to sustainable development, it is apparent that the term refers to both development and growth, thus blurring its precision. Indeed, tourism development is largely perceived as being the physical process of providing such features as the infrastructure, hotels, bars, restaurants, attractions and services. At a more specific level sustainability is concerned with what perhaps should be more correctly referred to as 'financial viability', in that a particular locality, industry or business will be sustained in the long run by maintaining or increasing revenues and profits. If it is understood in this sense there is likely to be a direct conflict with environmental goals, which use of the term sustainability should imply, as it embodies economic, social and environmental goals.

The perception of the key aspects of sustainability – the issues; its principles; conditions required to attain it; its measurement; indicators of it; policies on it and the instruments or mechanisms applied to attain it – depends on the stance taken. However, there is widespread consensus internationally that the overriding issue, as reflected in the much quoted WCED (1987) definition, is intra- and intergenerational equity, which respectively relate to the current and future generations.

With regard to the current generation, there are two concerns with what is essentially poverty. The first is where there are differences in per capita incomes within a country, often referred to as vertical inequity. The second is horizontal or spatial differences in income levels per capita that occur both within and between countries. Nevertheless, it is mostly considered as the difference in incomes between developed and developing countries. The issue of intergenerational equity relates more to the productive and energy resources available in the future and the state of the environment.

Again, reflecting the stance taken on sustainability, other intra- and intergenerational issues encompass to a greater or lesser extent those identified above in the section on global challenges.

The principles of sustainability in the definition below consist of both economic and environmental goals and range from the fundamental concerns of the preservation of the resource base and the availability and efficiency with which non-renewable and renewable resources are used, to the quality of life and community cooperation, conservation of all forms of life, maintaining biodiversity and enhancing the quality of natural environments. Given these principles, a reasonable working definition that satisfies the desire for economic progress, conservation of the resources base, optimal use of productive and energy resources, intra- and intergenerational equity and natural environmental goals is:

> the management of the resources base and its products in order to secure economic development and growth, while safeguarding the quality of the human and natural capital stocks for the current generation, but also to pass on to future generations the opportunity and means to maintain the same economic, environmental and quality of life benefit.

To consider the conditions for sustainability to be achieved, measure the process, outline the policies and identify the instruments or mechanisms required, attention is turned to examining economic perspectives. These impart more rigour to what has been over thirty years a rather vague and ambiguous concept, reflecting many viewpoints that have offered little guidance on making it operational.

8.9.2 The economic approach to sustainability

There is a continuum of stances in economics ranging from what has been termed the techno-centric, at one extreme, reflecting mainstream economic concepts, principles and analysis in which efficiency is the byword, and the eco-centric at the other, which acknowledges the relevance of the natural sciences, particularly ecology (Turner et al., 1994; Holden, 2007). The former is sometimes referred to as the 'optimistic' perspective. It is largely representative of mainstream economic analysis that emphasizes the need for efficiency, tending to be laissez faire in its approach, taking the line that the market mechanisms of price, supply cost, technical change and substitution are capable of attaining sustainability. The latter standpoint, a 'pessimistic' one, argues that market forces work too slowly and that their imperfections and failure to take account of salient factors, examined in the following chapter, are ignored by consumers and producers, giving rise to environmental problems that impair the pursuit of sustainability.

a. Sustainability conditions

To indicate the continuum of stances on sustainability, and enlarge on this aspect, the principal economic approaches are summarized, roughly in the order from weak perceptions to increasingly stronger ones. Most of the models adopted conform to the notion that the environment in general consists of the three forms of capital identified above: Kn, Km and Kh, which add up to total capital Kt. The models, briefly considered, which state the conditions or rules for sustainability are:

1   non-declining utility over time;
2   non-declining consumption over time;
3   management of resources to maintain future production opportunities;
4   non-declining natural capital over time;
5   the sustaining of the yield of renewable resources and services over time;
6   to satisfy the conditions of ecosystem stability over time.

- Utility is the satisfaction gained from the purchase of a good or service, synonymous with welfare, that an individual obtains. It is, however, difficult to identify and measure over time. Consequently, in the context of sustainability, consumption is used as an indicator of it.
- The basis for maintaining non-declining consumption over time, known as the Hartwick rule (Hartwick, 1977; Hamilton and Hartwick, 2005), is that the rents (surplus of revenue over costs) derived from the use of non-renewable resources should be invested in Km and Kh. It presupposes that the use of such resources should be efficient; it does allow for substitution between resources and the components of the total capital stock, particularly an increase in Km by expending and degrading Kn.
- The corollary of non-declining consumption is maintaining future production opportunities. This is achieved if the production potential of Kt is non-declining. In a weak sustainability situation, this again implies substitution between the elements of Kt, most likely that Km and Kh increase while Kn is likely to decline. If no substitution is envisaged then all three elements should be non-declining.
- A stronger condition is that Kn should be non-declining over time, given its unique functions set out above in the section on the role of the environment. This suggests, in a strong stance on sustainability, that if Kn declines then Km and Kh would as well . Conversely, reflecting the Kuznets/Ruttan hypothesis, if the economy is growing and incomes increase, the value of Kn rises and steps are taken to conserve or even

enhance it. Under this condition there would be the imperative of valuing Kn. Since many elements of Kn, such as open access resources, examined earlier, are not traded in the market and are therefore in danger of being overused, prices need to be derived for them that reflect their value to society. The issue of the valuation of Kn is analysed in the following two chapters.

- This approach largely accords with the principles put forward by Daly (1977). Essentially, it embodies conditions 2, 3, and 4 expounded above, but with the added proviso that the scale of economic activity should be bounded. It is a stronger stance in positing that a sustainable state exists where the product base is held constant and is managed to maintain the yield of renewal resources indefinitely. Furthermore, a maximum sustainable yield (MSY), for example cereals from agricultural land or fish from water resources, is one where the highest feasible flow of renewable resources is obtained. For non-renewable resources (NRRs), Daly's rule reflects the Hartwick one by positing that a proportion of the returns gained from NRRs should be reinvested to expand the yields of renewable resources (RRs) that over time would be a complete substitute. Additionally, a further condition is that the assimilative capacity of the natural environment should not be breached by pollution. This condition is examined at a microeconomic level with regard to tourism in the following two chapters.

- This condition reinforces the previous three in the movement towards a very strong sustainability stance by emphasizing a state in which ecosystems, essentially Kn, are stable and resilient. Stability means the ecosystem is in a state akin to the economic concept of equilibrium; that is, no tendency to change. Resilience is the ability of an ecosystem to retain or restore itself to an organizational structure and function following disturbance, for instance the ability of water ecosystems to assimilate pollutants.

Currently it is not being claimed by economists that the six conditions are operational. They are still largely hypotheses because testing them empirically is in its infancy. Theoretical economists would contend that this is not their purpose; they are benchmarks for deriving principles, policies and instruments or mechanisms to implement the transition to sustainability and identify indicators or measures of progress. The discipline also underlines very firmly that the form the derivation takes depends on the relative weakness or strength of the stance adopted, reflecting the underlying tenet of economics that choices have to be made that involve opportunity costs. Moreover, it recognizes that the goal of absolute sustainability, advocated by strict environmentalists, is unobtainable; in reality it should be seen as a transition or process.

A reasonably fair conclusion of the review of the economics literature on environmental issues (see for example Field and Field, 1998; Gilpin, 2000; Hanley et al., 2001; Tietenberg, 2006 and Holden, 2007) over the last three decades or so, is of an increasingly stronger stance on sustainability. While its examination has tended to be at a conceptual and theoretical level, there has been some investigation of methodological matters to consider how it can become operational. Most attention has been paid on the means of measuring, or identification of indicators as to, whether or not economies are becoming more sustainable.

b. Sustainability measures or indicators

Macroeconomic approaches are to modify national accounts to incorporate environmental aspects, either on a physical or valuation basis. In using income per head to give an indication of welfare, conventional accounts do not include a number of factors that influence general well being and the quality of life and Kn. For example, they do not measure the distribution of income within the population, nor changes in people's health and education, that is, enhancement of Kh. While additions to Km are accounted for, the effects of economic activity on the Kn, stock and its quality, are ignored. A particular case in point is pollution, which, if expenditure on its mitigation is incurred, is actually recorded as increasing national income, a somewhat perverse action.

Physical measures are essentially ad hoc deductions from NDP or Net National Product (NNP) relating to the depletion of the Kt or its components. Of especial importance is the Kn stock, particularly the loss of the resource base, such as agricultural and forest land, mineral sources and oil fields from which productive and energy resources are derived. Many countries, for instance those in the European Union, have partially recognized the shortcomings of conventional accounts by introducing environmental accounts. In the UK they are contained in the publication of the national accounts each year; the most recent (United Kingdom National Income, 2007) indicates the energy reserves and consumption, greenhouse gas emissions, materials flows, environmental protection expenditure and taxes. As such they are a very rudimentary indication of actions to secure environmental improvements and progress towards sustainability.

A physical index of sustainability of international origin, that is more concerned with poverty in developing countries, was the United Nations initiative in 1990 called the Human Development Index (HDI), under the development programme. It is actually about economic and social development and its basis is the level, but more importantly the distribution, of income. As such it is an issue of intragenerational equity, although to an extent it is similar to the following two measures of sustainability.

The ecological footprint index first came to be widely recognized in a publication by Wackernagel and Rees (1996). It is one measure of condition 6, given above. It considers the impact of the demand on the land required to support life and assimilate pollution and waste products. It is expressed in the hectares needed to sustain one person and relates these for the world population to available productive land.

Variants of ad hoc approaches translated into monetary terms are concerned with what is known as green accounting. These identify the adjustments that need to be made to national income accounts to indicate the policies and instruments required to pursue sustainability. The genuine savings or adjusted net savings approach proposed by the World Bank (Hamilton, 2000) posits that the excess of savings (net national savings) over capital (Km) depreciation should have education expenditure (an increase in Kh) added as an investment. From this, depletion of natural resources, based on the calculation of resource rents, and the damage from pollution should be deducted to arrive at net genuine savings. A negative figure denotes a decline in wealth and thus sustainability; in effect, this result can act as a policy imperative to increase savings.

The Genuine Progress Indicator (GPI), a variant of the Index of Sustainable Economic Welfare (ISEW), was first mooted by Daly and Cobb (1989), but not inaugurated until 1995 by the body Redefining Progress and is updated on a year to year basis for the United States (see for example Talberth et al., 2006). Its rationale is that if governments adjusted GDP statistics by calculating indicators of economic and social progress that improved well being and the quality of life, a truer picture of sustainability would be identified. This would act as the foundation for policies to pursue sustainable development and planning. The indicator is based on physical indicators translated into monetary costs or values to arrive at the adjusted estimate of GDP. The kinds of adjustments made to the consumption data of GDP in the GPI are on: income distribution; resource depletion; long-term environmental damage; life span of the public infrastructure and consumer durables; changes in leisure time; housework; volunteering; higher education; crime; pollution; and reliance on foreign assets. Originally the base year was 1950 and the results into the twenty-first century show a marked decline in the index in the 1970s and therefore a reversal of any sustainability trend.

The application of TSAs, the framework that indicates the contribution of the sector to the economy introduced in Chapter 6, has the potential to address the measurement of environmental issues. It is increasingly being applied to sustainability issues in tourism as an adjunct to official national accounts. A recent example is the comprehensive exposition by Costantino and Tudini (2005) on how to incorporate into the framework ecologically sustainable tourism. Increasingly, also, CGE models, considered in

Chapter 6, are including environmental variables and incorporating cost–benefit analysis (CBA), an exposition of which is given in relation to tourism in the following chapter.

However, the drawback to the measures that are based on the extension of national accounts, expressed in value terms, is that as yet they are unable to encompass more intangible indicators. Examples are the quality of life and the environment, air and water quality, the aesthetic benefit of the natural environment and community cohesion. This issue is considered in the next two chapters.

8.9.3 Global institutions for sustainability

There are many institutions concerned with sustainability issues. Key organizations with a truly global span are those under the auspices of the UN, such as the UNDP and the UNEP. One organization of significance is GEF, inaugurated in 1991 and made permanent in 1994. Three agencies are involved: the World Bank, UNDP and UNEP. Another is the International Institute for Sustainable Development (IISD). There are a number that have a specific role, such as the Forest Stewardship Council (FSC), which monitors the provenance of timber and certifies that it has come from sustainable forests. A similar body is the International Tropical Timber Organization (ITTO). The World Wildlife Fund (WWF), the RSPB, Birdlife, the International Union for the Conservation of Nature (IUCN) and the Whaling Commission are concerned with wildlife conservation. A recent initiative has been the GMEF of UNEP, which in 2007 selected as its focus globalization and the environment.

8.10 Sustainability and tourism

8.10.1 Background

It is evident from the exposition above of economic approaches at the macroeconomic or aggregative level, given the increasingly stronger stance advocated, that analytically at least, they are rigorous, precise and prescriptive with respect to the conditions that should prevail, and thus what the principles and policies and their implementation should be to pursue sustainability. However, two principal questions arise in considering where tourism stands on the issue. In which direction is it moving and what are the implications for it as a global activity if sustainability is becoming a more urgent imperative? Seeking answers to these questions necessitates a review of thinking in the study of tourism and attitudes in the sector, particularly its global institutions. At the outset it should be noted that most research on sustainability and tourism in economics and other disciplines has been at the local or micro level, covered in the following chapter. Notwithstanding this, it is possible to offer an assessment of the relationship at a general or macro level.

Examination of sustainability and tourism has occurred over as many years as in the wider context of economic growth and development, and an extensive literature exists that traces its history, the different perspectives taken on its meaning and the principles, conditions, measures, policies and instruments to make it operational (see for example, Bramwell et al., 1996; Stabler, 1997; Wahab and Pigram, 1997; Sharpley and Telfer, 2002; Williams 2004; Saarinen, 2006; Bramwell, 2007 and Holden, 2007).

As outlined earlier, 'sustainable tourism' (ST) and 'sustainable tourism development' (STD) are the terms most often being used. These have been referred to in both weak and strong stances, representing the two extremes on a continuum very similar to the economic one. The weak stance, which interprets ST as maintaining the long-term commercial viability of tourism and its constituent businesses and destinations, is hardly one that squares with sustainability when ST is related to a stronger stance, equated to one where ecological, environmental and sociocultural ones, as well as economic, are dominant factors.

The past direction and nature of the tourism sector is likely to continue to determine its growth globally and its attitudes to sustainability. As shown in Chapters 6 and 7, it has been perceived by many governments, particularly in developing countries, as having the potential to earn foreign currencies, contribute to the balance of payments and generate income, employment and economic development and growth. The term STD reflects this possible function. Therefore, a dilemma arises because of what are essentially opposing aims where economic goals are incompatible with environmental ones. Thus a trade-off or balance has to be struck. Another aspect that muddies the waters is that sustainability, when considered in relation to tourism, is very largely concerned with micro issues at a local and managerial level. In effect the focus is a tourism business centred one. Accordingly, its contribution to sustainability on a global scale is limited. Whether in relation to selling holidays via travel agencies, tour operators and promotions by national tourism boards, or the provision of travel, accommodation, attractions and services in destinations, there is virtually no coherent sustainability strategy. Any action is mainly confined to improving environmental performance by reducing energy and materials consumption and the generation of solid waste and pollution. Reference is made to some examples of these actions, at the local level, in the following two chapters.

8.10.2 The global tourism bodies' perspective

Bodies representing tourism interests, such as the UNWTO and WTTC, have advocated some sustainable practices worldwide. The UNWTO has published several documents of a general nature relating to ST, ranging from a bibliography, a guide for policy-makers, what amounts to manuals of ST good practice, indicators and environmental protection, to more specific aspects on eco-labelling, ecotourism practice, ST in destinations and poverty alleviation. However, a number of them are more concerned with management and promotion than sustainability, such as guides on tourism planning and the development of resort, coastal and rural tourism and dealing with congestion on natural and cultural sites. The WTTC is a global business leaders' forum, consisting of chief executives from all tourism sectors. Its primary purpose is to promote tourism by working largely with national governments. Although it has formed links with environmental initiatives, for example Agenda 21 and Green Globe, and cooperated with the United Nations in the past, its sustainability credentials are modest. To an extent, therefore, these two bodies are pulling in two different directions. The UNWTO recognizes to a degree the need for stronger policies and more effective measures to improve tourism's sustainability performance and response to global warming (UNWTO, 2003b), whereas the WTTC (2007) largely ignores it. Thus, it is no wonder that tourism as a whole simply reflects the support given to sustainability by politicians, but in practice continues to promote its development and growth, hardly heeding warnings by natural and social sciences and environmental bodies, for example Friends of the Earth (FOE), Greenpeace, ODI and Tourism Concern.

8.10.3 The academic perspective

It is in academic circles that most attention has been devoted to tourism development and its implications for sustainability, increasingly from a conceptual perspective, arguing that this is necessary in order to be able to posit how these seemingly irreconcilable objects can be made operational and compatible. Its origins in the social sciences generally were very much founded on economic concepts and principles regarding development and growth. Spatial economists, working in the urban and regional field (for early examples see Christaller, 1966; and Richardson, 1969, 1979), introduced the theoretical bases for the core–periphery concept in tourism and the adaptation of regional multipliers to income and employment tourism multipliers considered in Chapter 6. Rostow (1960) modelled the stages of economic growth that was paralleled in regional economic models, charting the rise, flourishing and decline of manufacturing (McCann, 2001), echoed in the holiday resort life cycle hypothesis (Butler, 1980). It was the burgeoning of environmental economics in the early 1970s, largely, initiated by Boulding (1966), which established the foundation from which tourism studies on sustainability and the beginnings of interdisciplinary approaches have been developed. This was reinforced by the advent of ecological economics in the late 1970s, in which Costanza and Daly played a central role (Costanza et al, 1997). The pluralistic methodology of this sub-discipline of economics, by breaking down the boundaries between the social and natural sciences, has moved the debate on by relating economic development models to environmental policies and management and concern for intergenerational equity. It has linked theoretical and spatial economics, anthropology and sociology to different branches of ecological systems analysis and biology. The concepts and methodologies of ecological and environmental economics are revisited in the following two chapters in looking at specific tourism environmental and sustainability issues.

Notwithstanding economics' original influence over the last three decades, the environmental problems that conventional economic models of the subject did not resolve have led to the derivation of new approaches. Nevertheless, currently, the contribution by economics to research on tourism sustainability remains problematic. It is still largely grounded in the single discipline of microeconomics and tending to reflect the long-term viability of tourism, rather than a real concern for the natural environment. This position is exemplified in Lanza et al. (2005), who present examples of the more recent technical economic approaches with policy implications, including a few cases at a macroeconomic level. The focus of the contributions is on the development and growth of tourism as an export, mainly in developing countries. Consequently, this emphasis raises the fundamental issue of the need for a trade-off of tourism and environmental goals. The book covers such aspects of tourism as forecasting, demand, hotel pricing, growth performance, development, expansion, financial returns and management. However, it does also consider such sustainability issues as the use of environmental taxes and extension of TSAs to embody ecologically sustainable tourism indicators.

In contrast to the clear direction economic conceptualization and methodologies have taken with respect to tourism sustainability, other social sciences have struggled to develop their own approach which nevertheless has evolved over the last two decades, now being much more holistic. Geography, in particular, with its emphasis on spatial and time dimensions, central to an understanding of tourism, has made a significant contribution to the issue; many references given in the examination that follows originated from authors rooted in the subject. It has increasingly related sustainability to the need to re-conceptualize tourism that was, referred to in Chapter 1, in which the dissatisfaction of viewing tourism as an industry and the recognition of its complexity were acknowledged. One aspect in particular that has assumed ascendancy is the impact of tourism development on destinations and their communities, especially in developing countries.

It was at the beginning of the 1990s that serious concerns were expressed as to what exactly ST and STD stood for and whether it was possible to attain them, given the acceleration in the pace of international tourism. An early debate centred on the incompatibility of tourism's growth and the maintenance of environmental quality. Hunter (1995) comments on this in an outspoken critique of the prevailing opinion that the preservation of the tourism resource base was paramount to secure continued development of the sector to meet the needs of tourists, suppliers and communities in destinations. His perception was of two interpretations of ST that indicated the degree to which sustainability was given priority, one where tourism should be totally encompassed in a sustainability regime and the other partially incorporated by it. His ideas were expanded in Hunter and Green (1995) and in Hunter (1997). The criticism was even more vociferous about the lack of a theoretical base on sustainability and that attitudes towards it were skewed in favour of ongoing tourism development, arguing that ST was almost entirely isolated from SD generally. He identified four scenarios representing weak to strong stances on SD, the weakest being a tourism imperative and the strongest an ecological one. Coccossis (1996) took a similar stance, also recognizing seven dimensions: cultural; economic; environmental; governmental; managerial; political; and social. Harrison (1996) gave a sociological slant to the debate in pressing for a more holistic conceptualization that emphasized the importance of the interrelationship of social groups in the process of achieving sustainability. Other contributors to re-conceptualizing and redefining sustainability in the 1990s were Clarke (1997), suggesting a framework for operationalizing it akin to those identified by Hunter and Coccossis, Butler (1999), who produced a very perceptive and succinct review of the state of the art at that time, and Hall and Lew (1998), who edited a book of guides to contemporary approaches.

The inauguration of the *Journal of Sustainable Tourism* (JOST) gave an impetus to the investigation of sustainability issues. Since the beginning of this century, the re-conceptualization of tourism and sustainability has become a central theme in the more general tourism literature. Bramwell and Lane (2000) recognized the need to consult those in destinations affected by tourism development, citing the pressures, often from developed countries, to adopt more sustainable practices. Hardy et al. (2002) emphasized the complexity of tourism as a phenomenon and called for approaches to ST to be interdisciplinary, also contrasting the proactive methodologies of economics with the reactive ones of tourism. Shaw and Williams (2002), in a concluding chapter, began to both widen and move discussion towards the application of the principles of ST and the techniques required to analyze and assess its attainment. Liu (2003), in a specific critique of the state of the study of sustainability, while reiterating problems raised in earlier studies, alleged that research was stuck in a groove of continuing to simply formulate and discuss principles and assumptions at a micro level in what is essentially a macro issue. Fundamental weaknesses in the literature were highlighted, such as in overlooking the intergenerational equity issues, the nature of tourism demand in destinations, the complex nature of tourism resources, the negative cultural and social effects, the pace of development and growth of tourism, and the means and instruments needed to control the adverse environmental elements of tourism activity. A paper that linked globalization to sustainability (Teo, 2002) continued to subscribe to the tourism development and SD trade-off thesis, identifying issues regarding the balance that could be struck in an economic framework, taking account of the global–local nexus which Milne (1998) considered. Teo perceived the problem as the responsiveness of tourism, as a composite structure, to its multiple inputs and the number of stakeholders involved, where there is the tendency for a managerial approach to prevail in pursuing ST. Farrell and Twining-Ward (2004, 2005; Twining-Ward and Farrell, 2005) developed the argument for re-conceptualizing tourism in order to understand more fully the nature of ST and STD. Their approach reflected a stance that tourism is inherently subject to unpredictability, operating in a world that is a complex, yet adaptable system, and embracing the concept of the sustainability transition as described above. Consequently, as indicated in Chapter 1, tourism is not susceptible to orthodox, reductionist, positivistic, linear, deterministic scientific analysis. They used the term 'sustainable science' as the framework in which global and environmental change can be examined, also suggesting that such a paradigm allowed for a better understanding of the links between human and natural systems to be developed, in which ST is seen as a transition. In their 2005 paper, the authors spelled out more clearly the steps required, effectively conditions, for progress towards ST to be made and for the co-existence of human and natural systems. Essentially they proposed that a new philosophy of interdisciplinary research giving full recognition to systems reflecting an ecological methodology to secure a sustainability transition is needed. Miller and Twining-Ward (2005) expanded on ideas for operationalizing the ST concept by considering indicators. A contribution by Bramwell (2007) also suggested new methodological directions.

Tribe (1997, 2000) had effectively anticipated new methodologies, particularly his notion of extra-disciplinarity, for what is now being termed trans-disciplinarity. Although as an approach it is not new, indeed the term has hitherto been attached to much tourism research that strictly is not, it is now rapidly gaining recognition in many spheres of academic endeavour. Nowotny (2003) and Dickens (2003) give an overview in the context of the

wider social sciences. As it is concerned largely with communities in destinations, the perception of trans-disciplinarity as a joint problem-solving
investigation between researchers, providing the expertise, and those
stakeholders affected, offering their practical experiences, appears to be a
suitable methodology for tourism researchers to adopt. The advocacy of
trans-disciplinary research points to the extension of or, perhaps more correctly, the virtual obliteration of the boundaries of separate disciplines,
including economics, especially its ecological sub-discipline.

8.11 Conclusions on the state of the analysis of sustainable tourism

This fairly extensive review of the sustainability issue over the last decade
and a half is justified on the grounds that it demonstrates that effectively
nothing has changed. The debate in the tourism studies arena is still raging
over its conceptualization; principles; key issues; operationalization; appropriate policies; measurement and monitoring, as the most recent examina-
tions of the concept above have underlined. With respect to ST practice,
investigations have tended to be at a specific case study level (see for example Bramwell et al., 1996; Wahab and Pigram, 1997; Bramwell and Lane,
2000; and Hall and Richards, 2002), rather than attempting to delineate
more general guidelines, as Stabler (1997) argues. Recent studies still tend
to emphasize ST in relation to the growth and development and long-term
survival of tourism and the interaction between it and communities in destinations (see for example Sharpley and Telfer, 2002; and Butler, 2005).
The latter amounts to a review of the complexities of defining ST, such as
identifying indicators, setting targets, and consequently attaining it. Butler
relates such problems to resorts in the declining stage of mass tourism in
their life cycle when they wish to undertake re-investment to market a
higher quality form. Likewise, Moscardo (2008) hardly refers to the environmental issues arising from tourism development in calling for an innovative approach in which the initiative and control are taken by the host
stakeholders to ensure their cultural, economic and social needs are met.

In economics, some progress has been made since the initiation of its
ecological discipline and the development of more sophisticated and appropriate models, such as CGEs, referred to in Chapter 6, and as indicated above
in the exposition of the modifications to national accounts, using TSAs, to
include environmental issues. However, the contribution of the subject to
sustainability at an aggregate level in its own journal, *Tourism Economics*
(Wanhill, 2007), and in books is slight. Recent contributions have begun to
address this lacuna, but there are still shortcomings in the analyses. One is

that of continuing to view sustainability as the achievement of the long-term viability of tourism. Another is adherence to the notion of the trade-off between tourism development and the conservation and quality of the natural environment. Currently, models do not incorporate a sufficient number or range of appropriate environmental variables. Costantino and Tudini (2005) in their application of TSA accounts exemplify the existing state of the art from an economic perspective. However, Johnston and Tyrrell (2005; 2008), acknowledging the ecologically based and theoretical approach by Casagrandi and Rinaldi (2002), have begun to address the shortcomings. In their 2008 paper they incorporate resiliency in the dynamics between ecological, environmental and social factors and tourism.

In defence of the economic modelling of sustainability, it is still developing its approach within the context of environmental economics so that attention to the application to tourism is still in its infancy. Notwithstanding this, with the subject's emphasis on the policy implications of the investigation of sustainability, it has the potential to make a valuable contribution to the debate and the difficulties of reconciling tourism development and environmental goals. However, of more concern is that economic approaches are seldom referred to in the wider tourism literature. Possible reasons for this are discussed in the final chapter of this book.

8.12 Final observations on global environmental issues and tourism

This chapter, recognizing the paramount position of the biosphere, has identified within an economic analytical framework the major global environmental issues, with an emphasis on the implications for the use and conservation of natural resources of globalization and international trade trends. The environmental challenges these pose have been examined to indicate the extent to which tourism is affected by, or affects, the quality of the global environment, as defined in its widest sense within the chapter. The centrality of the concept of sustainability to the resolution of global environmental problems and the implications of its pursuit, particularly with regard to its operationalization, were investigated and related to its perception in the context of tourism.

Given the urgency to resolve the many global environmental problems, it is dispiriting that political complacency over several decades has meant that little progress has been made to reach a consensus on the derivation of coherent and effective strategies, appropriate policies and the instruments by which to implement them. Tourism development and growth and their effects are illustrative of this lack of progress. To an extent the academic study of tourism and its role in the emergence of environmental issues has, in general, echoed the political inaction. While acknowledging the complexity of tourism as a phenomenon, researchers have still not agreed on such fundamentals as an acceptable conceptualization, definition and theory of it. This has consequently hampered the development of an analytical framework and methodologies for considering the associated environmental issues.

Economics, because of its well-developed theory, principles and methodologies and its policy orientation, has a firmer foundation for making progress in meeting global environmental challenges and the derivation of the necessary policies and instruments to deal with them. However, its application to tourism at a macroeconomic level, particularly in relation to environmental issues, has not been very extensive, as indicated above with regard to sustainability. Most economics research has involved microeconomic modelling that is now being refined and applied in a more general context. The contribution of the subject to resolve more specific and local environmental problems that are relevant to those connected with tourism is investigated in the following two chapters, to which attention is now turned.

The overall conclusion to be drawn is that currently the prospect for the resolution of global environmental issues associated with tourism is bleak. It is partly the result of the analytical shortcomings identified above but is more strongly related to the continuing tendency for the study of tourism to be conducted within single disciplinary boundaries. Despite pleas evident from the literature cited for inter-, multi- and trans-disciplinary research to be undertaken, the move towards this is painfully slow.

9 THE ANALYSIS OF TOURISM ENVIRONMENTAL ISSUES AT THE LOCAL LEVEL

9.1 The content and scope of environmental and ecological economics

This and the following chapter pick up on the reference made at a number of points in the previous one on the environmental effects of tourism that arise at the local level. In economics such issues fall largely within the purview of microeconomic analysis. The last two to three decades have witnessed the emergence of a number of economic fields considering environmental issues, such as accounting/auditing, ethical, green, neo-political, natural capital, and resource valuation economics. However, two, ecological and environmental economics, have become the two principal sub-fields of the mainstream subject. This has had two effects on the analytical approach of the discipline on environmental matters. While still rooted in the general principles of economics, the two fields have derived their own concepts, theories and methodologies specific to the issues to be addressed. As a consequence they have taken economics in new directions, especially with respect to their scope and approach. In particular, ecological economics has established itself as a trans-disciplinary field of research, explained below.

Initially, in this chapter the elements of the two fields are outlined and the difference between them identified, before setting them within an analytical framework that demonstrates the investigative processes and the methods employed, also embodying those that are examined in Chapter 10. Subsequently, the nature of local environmental problems is considered, including those associated with tourism, and an exposition is given of the economic approach to their analysis, drawing out the likely operational implications for the sector and wider policy. Because of its relevance to understanding the causes of most environmental issues, the concept of market failure is given due attention, which additionally informs the analysis in Chapter 10. Illustrative case studies are presented that indicate the impact of tourism's current activity and its expected future development and growth on local environments, as defined in its widest sense in Chapter 8, to include the Kn, Km and Kh elements.

9.1.1 Environmental economics

Environmental economics is considered first as it predates and maintains the chronological time frame of the development of the fundamental concepts, theories and methodological approach in which ecological economics has its roots. The origins of this field of economics go back to the nineteenth century conservation concerns in North America, particularly the management of the resource bases of agricultural, fishery and forestry products to perpetuate their capacity over time. Then it was simply referred to as resource conservation. However, it was Hotelling (1931) who laid the foundations of current analysis with respect to the optimal utilization of non-renewable energy and productive resources, such as fossil fuels and metal ores. The scope of the field was widened when pollution and the allocation of resources not traded in the market were perceived as issues. Both of these aspects were subsequently absorbed into the broader perspective that acknowledged the contribution to and therefore value of natural resources when sustainability became the ultimate objective of environmental economic investigation, as indicated at the aggregate level in the previous chapter. However, the principal purpose of the field, as in conventional economic analysis, remains the identification and analysis of the policy implications, including the instruments to mitigate the adverse environmental effects of economic activity. Pearce and Turner (1990), Gilpin (2000) and Holden (2007) trace the historical antecedents of the field, but most texts on the subject indicate how its scope has continued to broaden; see for example, Perman et al. (2003), Tietenberg (2006).

The field conforms largely to tenets of conventional economic analysis based on rational choice theory and the resolution of environmental problems within a market context, although it does acknowledge the relevance of market failure, examined below, as a factor that needs to be taken into account. It recognizes the interrelationship of the environment and economic systems. With respect to its analytical framework, environmental economics has applied cost–benefit analysis, incorporating methods for valuing non-traded goods and services, to appraise the extent and impact of environmental issues, especially the income and consumption distributional effects, based on welfare economics principles. To resolve environmental problems it advocates the use of price-based instruments and the workings of market mechanisms, such as price, supply cost, technical advances and substitution. To facilitate the greater efficiency of these market forces, it argues for a clearer definition of property rights as a means of negotiating over disputes concerning the allocation and use of resources, including the 'free rider' problem, and the beneficial or detrimental environmental effects generated. These concepts are explained later. Increasingly, environmental economics is moving towards a position akin to ecological economics by recognizing the value of biodiversity to economic and human activity, particularly with regard to the market benefits of improving the quality of life.

An outcome of the development of environmental economics, undoubtedly a sign of a separate and maturing discipline, was the inauguration in 1979 of the Association of Environmental and Resource Economics (AERE). Two journals are now devoted to the field. A fundamentally technical one, founded in 1974, is the *Journal of Environmental Economics and Management* (JEEM). A more general one, the *Review of Environmental Economics and Policy* (REEP), was first published in 2007.

9.1.2 Ecological economics

As an offshoot of environmental economics, the origins of ecological economics lie in the concerns of economists, such as Boulding (1966), Daly (1977) and Schumacher (1973), and natural scientists, for example Carson (1962), Lovelock (1979), Costanza (1984; 1989), and Ayres (1998). It emerged as a separate field because it was considered that mainstream economic analysis was inappropriate for investigating environmental issues and because of a growing acceptance that all human activity is a subsystem of the biosphere. A key feature of the field, emphasizing the role of the environment identified in Chapter 8, is that the biosphere places crucial environmental constraints on population increases and the pursuit of economic development and growth. It also considers that negative feedbacks, often unforeseen or indeed ignored, exacerbate environmental problems, particularly as markets work inefficiently and too slowly to resolve them. Consequently, a greater degree of intervention in economic processes is necessary (Common and Stagl, 2005).

Ecological economics acknowledges the role of natural sciences, especially biology, ecology and physics. The laws of thermodynamics and physical limits are relevant in the analysis of economic activity as it interacts with ecological and environmental systems. Of primary importance is the first law of thermodynamics, which states that in a closed system, matter and energy cannot be created or destroyed, only transformed. The implications of this for economics are that growth in the productive process, involving both materials and energy inputs, will generate solid waste and pollution, necessitating action to reduce or, preferably, eliminate them. The second law has been invoked by ecological economics (Georgescu-Roegen, 1971), which relates the entropy law to the economic process to posit that finite energy and materials resources would limit consumption and production and consequently economic growth. From this Georgescu-Roegen suggested a conceptual framework for ecological economics. The inference of this concept is that only renewable resources should be utilized, rather than simply transforming natural materials, such as metal ores and fossils fuels, into waste residuals that contribute to pollution and global warming. The applicability of this notion to economic activity has been contested by Ayres (1998). Nevertheless, Georgescu-Roegen's contribution to the development of ecological economics has been very influential.

Ecological economics constitutes a more pluralistic approach than environmental economics, embracing environmental and natural sciences and behavioural, cultural, political and social aspects of the social sciences, to research into environmental issues and suggested policies. The field is essentially concerned with the long-run sustainability of the environment and its principal focus relates to the limits of economic growth determined by the finite capacity of ecological and environmental systems, thus requiring a move towards a steady state that is analogous to climax or stable ecological systems. Two other aspects of the field are the concerns over the problems associated with the scale of economic activity and its distributional inequities, reflecting the view taken by Schumacher (1973). As with environmental economics, the ultimate goal is seen as sustainability that recognizes the dependence of Km and Kh on Kn. As explained in Chapter 8, this represents a strong environmental stance that does not allow the substitution of Kn by Km and Kh.

The methodological position of ecological economics parallels that which is occurring in tourism studies on its conceptualization, referred to in Chapter 1, and its advocacy of trans-disciplinarity to sustainability, reviewed in section 8.10.3 of Chapter 8. This certainly marks it off as a separate field of economic enquiry. This is underlined by the setting up of the International Society for Ecological Economics (ISEE) in 1989 with the aim of advancing an understanding of the relationship between ecological, economic and social systems for the mutual well being of nature and people. In the same year, the introduction of the journal *Ecological Economics* (EE) confirmed the establishment of the field as a discipline.

The research areas of the field, as revealed in the journal, reflect its aim to integrate ecological and economic modelling, while acknowledging the need to assess critically the basic assumptions of current ecological and economic paradigms. Issues covered, in the local–regional–global nexus, include: valuation of natural resources; the sustainability of agriculture and its development; renewable resources; conservation and management; the implementation of environmental policies; examination of the ecological and economic effects of genetically modified organisms; the identification of case studies of ecological–economic conflicts or complementarity; developing an ecologically integrated technology; incorporating natural resources and environmental services into national accounts.

9.2 The context of the microeconomic analysis of environmental issues

The exposition in Chapter 8 of the fundamental and vital role of the world's natural resources, and the extent and magnitude of global environmental issues, demonstrated the growing acceptance that human activity is crucially dependent on conserving the environment and its quality. The elements of ecological and environmental economics, outlined above, indicate a growing acceptance by economists that what can be termed the economic system is closely interactive with the overall ecological one. Indeed, as suggested earlier it is increasingly being perceived as a subsystem of the biosphere. This notion is pursued at this point by initially putting into a wider context the suggested position of the economic system in relation to human and natural systems.

Rather than showing this position as a fully embodied subset of the biosphere, Figure 9.1 is consciously structured in a hierarchical form to illustrate the interrelationship and role of the many factors that have a bearing on the economic perspective on environmental issues; its purpose is not to be a paradigm for economic modelling. It is intended that the figure should be interpreted from the bottom up as this reveals the linkages of the forms of human activity and, to an extent, the stages in a transition to global sustainability to safeguard the biosphere. In this respect it complements Figure 9.2, considered below, which should be read in the same way regarding the methods and process by which economic analysis of environmental issues proceeds. However, it is acknowledged that there is no clear distinction between each level and that their linkages and interaction are both upward and downward.

The left side of Figure 9.1 relates the three capital components of total capital or the environment defined in Chapter 8 to particular levels. The five bottom levels, containing the cultural, socio-economic, legal and political aspects of human activity and moving from the individual through the local, regional and national to the global level, are an adaptation of those proposed by Harrison (2007) to develop an analytical framework for tourism. His discussion is apposite for it parallels the approach adopted here in applying economic analysis to tourism later in this chapter. The matter in these boxes in the figure should be largely self-evident. However, the content of those above, pertaining to environmental factors, especially what is meant by the biosphere, ecological systems and the stances on sustainability, requires an explanation.

In Figure 9.1, the box labelled the biosphere has two elements. The first, the biosphere as a whole, is best understood as the total collection of all organisms on the Earth and the space on it they occupy. In effect,

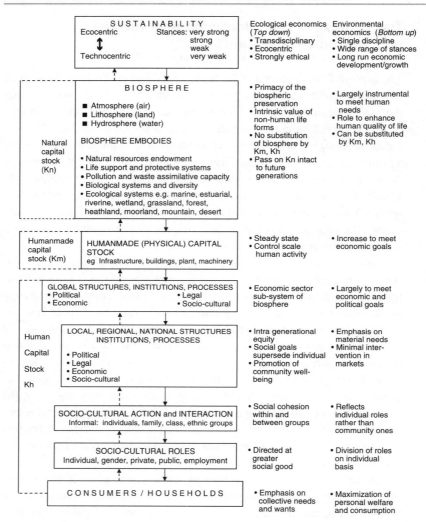

Figure 9.1 The context of the socio-economic system and the economic analysis of environmental issues

what earth and natural scientists refer to as the biosphere is simply the planet itself. Biospheric space consists of the air (atmosphere), water – lakes, rivers, seas and oceans (hydrosphere) – and land (lithosphere). The biosphere contains all the Earth's ecological systems that interact with the environment and each other.

The second element embodies examples of the ecological systems that maintain biological communities of living organisms found in similar physical environments that are influenced by ecological processes, such as fire, rainfall, sunlight, temperature and wind. In a broad classification, the types of systems found on the planet are assigned respectively to the atmosphere, hydrosphere and lithosphere. Some examples are shown of biosspheric systems in the kn box; the last two are related to tourism in particular. In the natural sciences, systems analysis as a formal enquiry began to be applied to ecology in the mid-twentieth century in order to facilitate the implementation of more informative management practices, especially to conserve and enhance ecological systems and their processes. Their effective management has a direct bearing on the viability of tourism. For instance, as already suggested in Chapter 8, given that a significant proportion of the tourism product consists of natural resources, a lack of or poor management will very likely result in their degradation and so adversely affect tourist demand for them. A notable example is ecotourism, which has been extensively promoted in recent years. Cater and Cater (2007) aptly illustrate the issues with respect to marine environments and Fennell (2007) covers a number of cases of this form of tourism and thus ecological systems. This aspect of the interaction between ecological systems and tourism is considered below.

9.2.1 The stances on sustainability

The extent of the relevance of this brief explanation of the biosphere, and its constituent ecological systems to the economic analysis of environmental issues, arises from the different stances taken on the importance accorded to the role and value of Kn and sustainability. Stances can be conceived of as lying on a spectrum from a very weak one at one extreme to a very strong one at the other. Four positions are identified in Figure 9.1, based on those posited by Turner et al. (1994): very weak, weak, strong, very strong. In identifying their categorization, the author consider the type of economies, management strategies and ethical positions in attaching green and sustainability labels to each.

a. Very weak sustainability stance

This is signified by a concentration on economic growth, exploitation of non-renewable energy and material resources through unrestrained markets to serve consumer demand and the producers of goods and services and the ethical supposition of only the instrumental value to humans of all natural resources and other forms of life. The inference is that the market forces of consumption price, supply cost, substitution of Kn by Km and Kh and technical advances will resolve environmental problems.

b. Weak sustainability stance

There is recognition of the need for conservation in the management of resources, the provision of incentives to move towards a green economy and markets, the modification of economic growth, rejection of the substitution principle of Kn by Kh and Km, although this is allowed to take place between each of these elements of Kt. The ethical position adopted is of concern for intra- and intergenerational equity, but the instrumental value argument of the weak stance is maintained.

c. Strong sustainability stance

The preservation of resources, rather than simply their conservation, is advocated in an economy akin to Daly's (1977) and Lovelock's (1979) posited 'steady state' concept, in which environmental standards and limits are set in macroeconomic policies. Economic and population growth is assumed to be zero, implying no increase in scale, in a system that emphasizes the pivotal role of ecological systems. The interests of society at large supersede those of individuals.

d. Very strong sustainability stance

In this deep green economy, in addition to the policies identified for the weak ecocentric one above, very stringent policies are implemented so that non-renewable resource use, pollution and waste are minimized. Such a position involves a trade-off between socio-economic goals and environmental ones. The ethical position recognizes the intrinsic value of non-human life forms in what is termed a 'bioethics' regime.

A very similar categorization, undoubtedly based on Turner et al. (1994), is identified in Holden (2007), discussing Baker et al. (1997), in relation to the position of sustainable tourism on the spectrum. However, the reason for introducing the notion of the technocentric–ecocentric spectrum is less concerned with tourism's stance at this point. The interest, initially, is with suggesting the position of environmental and ecological economics on it, shown on the right side of the figure, in order to focus on the relative importance they attach to the role of the economic system and the biosphere and its ecological systems. A second reason is to suggest the extent to which the respective stances are likely to contribute to or inhibit a transition to sustainability.

Sustainability is placed above the biosphere in the hierarchy in Figure 9.1, as it is perceived as the ultimate goal of all human activity, in which the biosphere and ecological systems play a crucial role. Nevertheless, it is acknowledged that the influence of all sectors identified in the figure is a two-way one, irrespective of the place of each in the hierarchy.

9.2.2 The sustainability stances of environmental and ecological economics

It is instructive to enlarge on the natures of environmental and ecological economics as it enables their respective positions on the environmentalism stances spectrum in Figure 9.1 to be justified. It also helps to address the issues raised above regarding the interrelationship between the economic system and biosphere, the impact of the environmentalism stances on sustainability, their feedback effects on the biosphere and the methods and processes of economics with respect to environmental issues given in Figure 9.2.

It is not easy to reach a consensus on the distinction between environmental and ecological economics that unequivocally signifies their respective stances on the technocentric–ecocentric spectrum. Daly and Farley (2004) and Common and Stagl (2005) indicate that, methodologically, their approaches to environmental and sustainability issues are similar. Both fields acknowledge the interdependence of the economic system and environment and the tension between two essentially conflicting goals. They acknowledge that economic development and growth are central tenets of all nations' policies that conflict with the desire to conserve natural resources, promote biodiversity and the maintenance of environmental quality. Their stances on environmental policy objectives and the instruments required to achieve them and to mitigate the adverse effects of economic and human activity generally coincide. There is also a measure of agreement on the limits of markets to resolve the problems of the degradation of natural resources, pollution and waste at a global and local level.

The key difference between the two fields lies in the foundations of their methodologies, already outlined in the previous section but worth rehearsing at this point. Ecological economics' pluralistic/trans-disciplinary approach, embracing such subjects as anthropology, biology, ecology, ethics, geography, physics, politics and sociology, is in marked contrast to the essentially single disciplinary one of environmental economics. Accordingly, the former was originally located at the deep-green ecocentric end of the environmentalism spectrum, using the systems perspective of the natural sciences, simultaneously pursuing both socio-economic and ecological sustainability. Its economic stance is in line with the stationary or steady state economy position of Daly (1977), also taking a political economic slant in which a number of aspects of human existence are not susceptible to economic analysis.

Conversely, environmental economics started out from a conventional economic analytical base at the weak, technocentric, extreme of the environmentalism spectrum. Increasingly, it has moved towards a more ecocentric position as demonstrated in the current book and journal literature (see for instance Turner et al., 1994; Perlman et al., 2001; and Tietenberg, 2006), and the focus of articles in the journals of *Environmental Economics and Management* and the *Review of Environmental Economics and Policy*. A feasible inference to make is that environmental economics continues to have a wider range of environmentalism stances than ecological economics. Nevertheless, it is a reasonable conclusion to draw that the two fields overlap significantly and that the latter borrows many of the methods of analysis from the former.

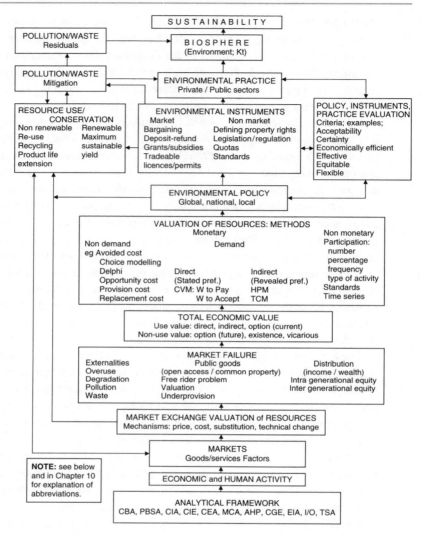

Figure 9.2 The scope, content and process of the economic analysis of environmental issues

With respect to Figure 9.1, a final observation is that ecological economics' regard for the preservation of the biosphere, in considering that it encompasses the economic system, is adhering to the notion of the ultimate goal of ecological sustainability. It works on a top-down basis to the economic and social level to ensure that this goal is attained. On the other hand, environmental economics can be perceived as working in the opposite way, bottom-up.

The interrelationship of the economic system and the biosphere has been touched on at the beginning of Chapter 8, where the role of the environment and the global challenges facing humans were identified. The examples illustrated quite forcibly the current technocentric oriented structure and level of economic activity at the expense of the conservation of the biosphere.

9.2.3 The scope, content and process of the economic analysis of environmental issues

It is against the background of the development of environmental and ecological economics given in the previous section, and their applicability to tourism, that Figure 9.2 was compiled. In the figure an attempt has been made to reflect the contribution of both these fields by identifying their principal areas of study and the methods adopted. It largely reflects the commonality of their approach, but as far as is possible it indicates how they differ. No claim is made that the figure is comprehensive. Indeed some areas of environmental and ecological economics, for example the analysis of such concepts and issues as the precautionary principle, safe minimum standard, the uncertainty of outcomes of particular actions or the effect of policy initiatives, cannot easily be depicted in it. Nor is the figure meant to show the interrelationship between consumption, production and the environment as presented in economics texts, for example Turner et al. 1994.

While Figure 9.1 puts economic activity into the wider context of systems analysis, relating it to the two principal economic fields' stances on sustainability and environmental issues, Figure 9.2 looks in more detail at the scope, content and process of economic analysis. The figure is again based on the premise of pursuing the transition to sustainability of human activity in which the economic system is entirely embodied within the biosphere. It has been constructed to indicate not only the economic analytical approach to and process with regard to sustainability and environmental issues, but also the use of non-renewable and renewable resources and the mitigation of pollution and waste. The figure endeavours to reflect the two extremes of stances on sustainability. The left side, from the markets and market exchange valuation boxes directly to the resource use/conservation and two pollution/waste boxes, indicates the very weak, 'technocratic', essentially traditional economic analytical stance. The centre column largely relates to the very strong, 'ecocentric' one. Identifying the distinction between ecological and environmental economic approaches is not an issue here.

As is made evident from the examination above of their representative bodies' objectives and journal papers, there is no fundamental difference between them in the analytical methods employed. It is more a question of the nature of their disciplinary bases and focus of research. In general, the two fields are increasingly in agreement on the need to intervene in the operation of markets, the definition of economic value, the valuation methods adopted, the policies needed and instruments employed to attain environmental goals and the transition to sustainability. Thus the centre route through Figure 9.2 is now considered common to both fields. It is worth emphasizing in this respect that irrespective of the different shades of opinion within economics, its principal rationale is to contribute to the derivation of economic policy and instruments and to investigate their implications for society at large regarding the distribution of income and wealth, allocation of resources and economic efficiency.

Figure 9.2 should be read from the bottom upwards. The second box, concerning human activity, corresponds to the bottom six in Figure 9.1 that are embraced by Km and Kh. The two boxes that head the figure, sustainability and biosphere, are common with those in Figure 9.1 and so do not require further explanation. The content within the boxes labelled markets, market exchange valuation of resources, market failure, resource use/conservation and pollution/waste (two boxes) is explained later in this chapter. The remaining boxes concerning total economic value, valuation methods, environmental policy, policy instruments, environmental practice and their evaluation are examined in the following chapter. As appropriate, the interpretation of Figure 9.2 from the economic perspective is illustrated by examples from tourism, concentrating on issues arising from its operation and performance at a local level.

9.3 Capital appraisal frameworks

Figure 9.2 indicates the breadth and complexity of the requirements of an economic analysis of the use of resources to meet human needs, but simultaneously caring for the environment to sustain both over time. The figure also represents the elements of the appraisal of large-scale capital projects with long time spans that have significant environmental impacts. It is necessary to devise analytical frameworks as a basis for making decisions on such investments. There are several grounds for conducting capital appraisals. A choice has to be made in the face of competing demands for scarce resources. Effectively this ascertains the opportunity cost of resources used in a specified project, by which is meant the establishment of the benefits and costs of the alternative uses to which they might have been put. Projects in both the private/commercial and public sectors should be subject to appraisal to ascertain the expected returns or benefits and costs. The returns should exceed the costs, i.e. there should be a net benefit; otherwise the project is not worth undertaking. Typical cases for the selection of one project in CBA analysis are to accept/reject a single project, or choose one of several discrete or mutually exclusive contenders. Where a number of projects are under consideration, of which not all can be undertaken, the selection should be based on a ranking system that places them in the order of the highest net returns to the lowest. This ensures that the most profitable, or those yielding the highest net benefits, are adopted.

In economics, a distinction is drawn between a social and commercial appraisal as different criteria are used. The former is founded in welfare economics as it considers the gains and losses for society at large, including any redistribution of income and wealth. Moreover, the appraisal endeavours to correct for the effects of market failure, examined below. In the latter sector the concern, over the life of the project, is the maximization of returns over costs to ensure an adequate rate of return on the financial capital invested in the project. The internal rate of return should be at least equal to the external rate of return, i.e. the market interest rate that represents the opportunity cost of funds invested within the enterprise or project. However, projects in the private sector do have social consequences and also implications for the environment and sustainability that are not taken into account in any strictly commercial appraisal. Examples related to tourism to illustrate this will be offered below.

The box at the foot of Figure 9.2 identifies a number of analytical frameworks that might be adopted in different circumstances. The most appropriate and comprehensive are given first, working from left to right in the box. Also included are techniques that can be used independently or contribute to widening the scope of the more extensive frameworks.

9.4 Cost–benefit analysis and its variants

Cost–benefit analysis (CBA) originated in the 1930s in the United States, where there was a need to show not only the direct benefits of public expenditure on flood prevention to agriculture, but also the wider indirect benefits to others affected by floodwater. The method is favoured by economists because of its wide-ranging scope and adherence to the subject's principles. Since its first introduction, it has been used to appraise electricity generation, airport construction and expansion, ports, roads, railways, health and education programmes; see, for example, Dasgupta (1972), Little and Mirlees (1974), Pearce and Nash (1981). It has also been applied to environmental projects, such as the preservation of forests, wetlands, wildlife reserves and coastal erosion programmes. In many instances, projects of this nature have included recreational activities (Hanley and Spash, 1993) but few that would relate specifically to tourism (Stabler 1999).

Closest to CBA, the PBSA (Planning Balance Sheet Analysis) was first devised in the 1950s and extensively developed over the next thirty years by Lichfield (1988) by extending it into CIA (Community Impact Analysis) and CIE (Community Impact Evaluation). As a land-use planner, his interest was essentially focused on urban development and its effects. Lichfield argued that the PBSA method represents a useful compromise between theoretical acceptability and practice because it is simpler, more easily understood and overcomes the problem that many benefits and costs are not measurable in monetary terms. His view was that if these can at least be identified they should be listed on a balance sheet either as assets or liabilities. This was considered important, and in his later CIA method he developed this to indicate which sections of a community were likely to gain or lose from the project, so taking into account, on social grounds, the distributional impacts as well as the efficiency effects emphasized theoretically. It should be noted, however, that Lichfield's contribution to project appraisal was undertaken prior to the development of the valuation methods now incorporated into CBA, examined later in this chapter.

9.4.1 Cost-effective analysis

A narrower project appraisal method, CEA (Cost-Effectiveness Analysis) has been widely applied in both the private and public sectors. It considers different ways of achieving a given objective at least cost, to satisfy what is known as the equi-marginal principle. It has not been considered feasible to include environmental effects in this method and thus it ignores the non-market, social benefits and costs that CBA does. Nevertheless, CEA has the merit of being simple and cheap to apply.

9.4.2 Multiple criteria analysis

The MCA (Multiple Criteria Analysis) developed in the 1970s and 1980s by Nijkamp (1975, 1988), Paelinck (1976), Voogd (1988), and more recently MCDA (Multiple Criteria Decision Analysis) (DTLR, 2001), extends CEA to embody a number of feasible options, often weighted to reflect their relative economic and environmental outcomes, not susceptible to monetary evaluation. It expresses the preferences of decision-makers where the options are ranked differently by various criteria. The purpose of the analysis is to combine and evaluate the various options to produce a single choice option. The framework allows for those affected by the project to adjudicate on the options and criteria in making the choice. A recent example of its application to environmental issues by Park et al. (2004) considered the effectiveness of mechanisms to secure landscape and habitat enhancement that included amenity value.

9.4.3 Analytic hierarchy process

A more mathematical approach to decision-making between alternatives is the analytic hierarchy process developed by Saaty (1987; see Zahedi, 1986), which has been suggested for the evaluation of alternative approaches to conservation and restoration by Lombardi and Sirchia (1990) and Roscelli and Zorzi (1990). This, too, is designed to formalize the process of choosing between alternatives in the absence of full information. Increasingly, risk analysis is being incorporated into such evaluatory and decision techniques.

9.4.4 Other frameworks

The remaining items in the box at the foot of Figure 9.2, CGE, I–O and TSA analyses, have already been explained in Chapter 6. As shown there, they are essentially models constructed to measure, for example, the contribution of tourism to economies, acting as databases for other applications and to facilitate comparisons within and between countries. Both CGE and TSA possess the potential to include environmental variables, particularly CGE, which can model at the industry, regional and national level the impact of tourism. Work on simulating the means of implementing environmental policies in general began in the 1990s (see, for example, the relatively early research by Xie, 1996). The incorporation of environmental variables into tourism CGE modelling, either directly or by combining it with CBA, is quite recent, the analysis by Dwyer et al. (2007) being especially notable.

The employment of EIA (Environmental Impact Assessment) and its associated method, SEA (Strategic Environmental Assessment) has largely been confined to the European Union countries, although there have been similar approaches, RIA (Resource Impact Assessment) and DA (Damage Assessment), in the United States. These are designed to assist decision-making in land-use planning and development, regarding large-scale business projects, such as chemical plants, industrial sites, distribution centres, retail parks and leisure facilities. In the public sector, cases have considered port, airport, power generation, reservoir construction and renewable energy schemes. Impact assessments are basically policy and regulatory tools in which the effects of projects are expressed in physical terms only, being conducted primarily by engineers and scientists. Effectively, they cover only the first few stages of a CBA to the point where benefits and costs are identified.

This summary of the principal analytical frameworks and adjuncts to them has been given to indicate that the appraisal of large-scale projects is essentially tailored to meet specific objectives. Their emphasis on environmental and sustainability issues does not necessarily mean that all have been applied in practice to the appraisal of the effects of tourism projects. However, they are feasible methods that the sector could and should employ given that the human-made and human environments constituting the tourism product are vital to the survival of tourism destinations.

Further reference will be made to them later in this chapter and the following, where examples of the environmental issues raised by tourism are identified.

9.5 Cost–benefit analysis: its scope and content

In economics CBA remains the most favoured framework. Hitherto, there had been little study or application of the method to the proposed development and ongoing effects of tourism. However, as stated above by reference to the representative study by Dwyer et al. (2007), tourism has revived interest in CBA as an appropriate analytical method. Similarly, the derivation of more sophisticated and reliable methods of attaching values to benefits, such as those provided by natural resources that have no market price, has strengthened its feasibility and relevance in practice. Thus, a closer examination of it is warranted. Given its attention in the public sector and the consequences for communities, the ultimate aim of CBA is the maximization of social welfare, meaning an outcome that gives the greatest benefits to the greatest number of people. The process of conducting a CBA given here largely conforms to that identified in Stabler (1999) for tourism.

Most tourism policies and development do not include environmental objectives and consequently the sector is not required to take account of their impact on local natural, human-made and human capital. They are more likely to emphasize the generation of income and employment that contribute to the diversification of local and regional economies, as demonstrated in Chapter 6. This is the starting point for CBA where, for example, there is a proposal to construct a holiday complex in a rural area or designated urban historic conservation quarter in a city, with the aim of developing tourism as an integral part of a projected regeneration. Irrespective of whether the project is in the private or public sector, the appraisal must go beyond the immediate commercial objectives to consider the ramifications for the wider environment and community.

There are implications for the allocation of land, capital and human resources from existing uses, say agriculture in the rural area or housing or open space for residents in the city, to the tourism project. Furthermore, distributional issues arise regarding the effect on the income and wealth of landowners or occupiers, employed persons and businesses. The boundaries of spatial impact have to be delineated in the sense that resources and inhabitants in both the immediate area and further afield are affected, for example upgrading the infrastructure and provision of services, such as energy and water supply, drainage and waste disposal facilities. The impact can of course be beneficial, especially where these services have been inadequate or did not even exist. This would certainly be the case in developing countries. Pollution, for instance emissions to air, discharges of effluents into water courses, noise and visual intrusion, are almost wholly detrimental and impose increased costs in the area. These aspects are important initial considerations of CBA, as indicated in the first two stages of the sequence of the process given below outlined together in section 9.5.2.

9.5.1 The CBA process

1   Establish what is being appraised and what and who is affected.
2   Define the scope of and objectives of the project and its physical impact.
3   Identify the environmental, financial, political, social and time constraints.
4   Consider other analytical methods for an appraisal.
5   Identify and evaluate the benefits and costs:
    -   direct/primary (monetary);
    -   indirect/secondary (measurable in physical if not monetary terms);
    -   intangible (identifiable but not necessarily measurable).
    Indirect and intangible items are generally referred to as social benefits and costs.
6   Discount the benefits and costs to present value.
7   Conduct a sensitivity analysis.
8   Investigate allocative and distributive effects and apply weighting to reflect the desired objectives of the project.
9   Apply the decision rules in order to establish the viability of a given project or select one or more projects from a number proposed.

9.5.2 An explanation of key stages in the CBA process

a. What is being appraised: the scope, objectives and impact

Steps 1 and 2 identify the physical and socio-economic impacts on the locality, namely the human-made (Km) and natural (Kn) elements of the total capital stock or environment (Kt) and the impact the project will almost certainly have on the allocation of resources. The key aspect with respect to who is affected, the human capital (Kh) of Kt, is to ascertain the effect on the distribution of income and wealth. Some sections of the community will gain and others will lose. In theory Pareto optimality would posit that the project should yield benefits for some without anyone suffering a loss; in practice this is unlikely. The criterion is whether the benefits accruing to some people outweigh the losses suffered by others. Regarding what and who is affected, a distinction should be made between additionality and displacement effects. The former concerns the establishment of the net impact both phsically and socio-economically, for instance that the creation of an urban park would enhance the well-being of the community. The latter is whether the project simply replaces a similar activity or facility with no net gain.

b. Identify the constraints

The third stage concerning the constraints relates to both the conduct of a CBA and the project. Undertaking an appraisal is time-consuming, requiring consultation with affected parties and the collection of vast amounts of data; the process is therefore expensive and may be such a large proportion of the total cost of a project as to render it unfeasible. With respect to the project itself, there may be a financial capital constraint and there is always the possibility of competing projects being given priority, especially in the public sector where political acceptability may be a factor.

c. Consider other analytical methods

The fourth stage of the process is relevant where the project is a comparatively small one, for example the addition of a new attraction, say a leisure centre, to an existing resort. A full CBA may be unnecessary because the effect on natural and human-made environments is minimal and the principal objective may be to generate employment, suggesting that a multiplier analysis would be sufficient. Where there are various options from which only one is to be chosen an MCA might be appropriate.

d. Identify and evaluate the benefits and costs

The fifth stage, particularly the indirect and intangible (social) elements, where their magnitude in a large-scale project is a significant proportion of the total benefits and costs with wide-ranging spatial effects, constitutes the core of CBA. In such a case the commercial, environmental, sociocultural impacts need to be given due consideration. The objective at this stage is to ascertain the net effects of the project and therefore amounts to an inventory of the value of the impacts to be compiled. This should cover the initial construction and the ongoing operation of the development.

The proposal to build a golf course, with its associated facilities, and a holiday complex of a hotel and self-catering accommodation and appropriate services on sand dunes of high wildlife value on the Aberdeenshire coast in 2007 represents a suitable example. Its impact is likely to be far-reaching both in terms of benefits and costs and spatially. The direct benefits of the project are the net revenues (after deduction of running costs) to yield an acceptable rate of return on the financial capital invested and to maintain the viability of the enterprise in the long run. The direct costs will be the acquisition price of the land, the cost of the construction of the building and facilities and subsequently the ongoing inputs of goods and services, staff salaries and wages, insurance, maintenance, energy, etc. when the development becomes operational. Multiplier analysis, as shown in Chapter 6, by estimating the direct, indirect and induced income and employment effects indicates the economic impact in the locality.

This approach does not embrace the wider impact of the development in the area identified in the introduction to this section above, relating to the allocation of resources and distribution of income and wealth and pollution generated. Pearce (1989) has admirably analysed the principal effects, both beneficial and detrimental, of tourism development that includes sociocultural factors (Kh). However, the effects of this project are even more widespread and diffuse when its impact on the Kn is considered; they are almost entirely costs and they will degrade it. There are likely to be adverse effects on the fauna and flora and biodiversity as a result of disturbance and the inevitable loss of the sand dunes, a fragile habitat, when development of the golf course and complex occurs. Once operational, the resort will subject the environment to further degradation, especially if it engenders the development of supporting facilities and services, such as retail outlets, restaurants, bars, banks and food stores for both tourists and day visitors and staff who settle in the area. Infrastructural improvements will probably accompany this expansion. Notwithstanding the fact that these may confer material benefits on the original inhabitants and so enhance their living standards, it is a debatable point that it will yield a better quality of life if traditional customs and mores suffer and the influx of tourists increases congestion, noise, crime rates and litter, which are a feature of increased tourism in developing countries. This aspect of the illustrative case study is not pursued here as it is referred to later in examining the effects of market failure and policies to mitigate the adverse effects of tourism, as well as the efforts by the sector to improve its environmental and sociocultural performance.

These examples of intangible impacts are difficult to identify and measure, and are ignored in a commercial project approach that is not required to take account of them unless obliged to under regulations. In this particular case, because of its size and potential impact, an EIA would probably be mandatory, but this would only consider the physical aspects so the costs remain implicit. Ideally, at the evaluation stage, the CBA method would express all the benefits and costs in monetary terms. Progress has been made in doing this for the intangible ones identified above with the development and refinement of the valuation methods shown in the box in the centre of Figure 9.2, which will be explained in the next chapter.

Another illustration of the relevance of application of CBA to a tourism project, unusual in the literature on the phenomenon, is one conducted by

Dixon et al. (1993) that considered the means of simultaneously attaining economic and environmental objectives in the Bonaire Marine Park in the Caribbean. It was essentially a multiplier approach to estimating the benefits of the park, but contained elements of a CBA by considering the costs of its development that included reference to those that were intangible and also identified who benefited or who paid.

Dixon's study gives some observations that are relevant to the examination below in the section on resource use and conservation, especially renewable resources, for which a number of case studies are drawn from the ecotourism literature that convey the impacts on natural resources and communities. The use of CBA in these cases to identify and evaluate the benefits and costs of this particular form of tourism would have established its wider implications and ascertained the net effects. There is virtually no reference to the application of the CBA method in the examples cited.

With respect to the monetary evaluation of benefits and costs, it should be noted that even for those that are direct, market prices are not necessarily appropriate. In CBA adjustments are made or what are called 'shadow prices' are derived. First, they are distorted by imperfect competition, particularly monopoly, where control of supply of goods and services results in higher prices than would prevail in a more competitive market. Second, future prices should be adjusted for inflation or deflation so that the estimates for the life of the project are in real terms. Third, the prices of goods and services subject to taxes and subsidies also need to be adjusted to reflect their supply cost. For the constructed values of intangibles referred to in the previous paragraph, the concept of total economic value (see box in Figure 9.2) has a bearing on the price to be attached to them; an exposition of this concept is also given in the next chapter.

e. Discounting the benefits and costs

The sixth stage in the CBA appraisal process is to discount all the benefits and costs over the life of a project back to their present value (PV) to establish whether the net present value (NPV) is positive, therefore indicating that it should proceed. Clearly a negative NPV would result in rejection of the project. The rationale for discounting is that positive interest rates and inflation underpin time preferences by the current generation, which values present monetary returns more highly than future ones. Also, on the same grounds, the opportunity cost of the financial capital in the project is a consideration. A simple example demonstrates the argument for discounting. One million dollars put into a financial investment at, say, the current market interest rate of 10 per cent, would yield 1.1 million dollars in a year's time. Working backwards, where there is an expectation of a million dollars in a year's time, with an interest rate of 10 per cent, the PV of that sum would be 0.90909 million dollars. To illustrate the effect of discounting that will reinforce the impact of the discount rate in hypothetical cases given below, with a 10 percent interest rate, a million dollars in five year's time would have a PV of 0.62092 million dollars and in 10 year's time a PV of 0.38554 million dollars. A positive inflation rate, say 5 percent, would reduce the real present value of that sum even more. Effectively, deferring the benefits that can be obtained from funds to some future period reduces current welfare, which justifies the discounting argument. On the other side of the coin, also discounting the costs to their present value lessens their burden in the future, based on the assumption that continued economic growth means that income will be higher from which to fund the project.

The NPV of benefits and costs and the viability of projects are sensitive to the rate of interest chosen in the discounting procedure, arising from their life span and magnitude and timing of the capital and ongoing costs, likewise the benefits. This sensitivity is particularly relevant where environmental elements are significant, and the same may also be the case with tourism development. Two hypothetical examples indicate the effects that are especially relevant to projects with environmental elements.

A HIGH DISCOUNT RATE

This will discriminate against a project with a long life span, a large initial capital outlay and ongoing costs in the early years, with benefits not beginning to accrue until much later. In brief the PV of the costs will be high as their PV will be relatively larger. The benefits will have a much lower PV so that very likely the costs will outweigh the benefits, leading to a negative NPV and therefore rejection of the project. Conversely, where the capital outlay and ongoing costs are spread over many years and the benefits are large and occur in the early years, the project is more likely to have a positive NPV.

A LOW DISCOUNT RATE

This will yield higher NPVs for benefits and costs. In general, projects with long life spans, if the timing and size of the benefits and costs are similar to those given above for the converse high discount rate case, are more likely to be viable where the discount rate applied is lower.

9.5.3 Environmental projects and the discount rate

Prima facie, projects with a substantial environmental content are more likely to be long term, with a considerable lag before benefits occur. Moreover, capital and ongoing costs can be substantial, given that such projects involve constant and prolonged management, for example reforestation, coastal erosion protection, flood control and wetlands creation and maintenance. The inference to be drawn is that the discounting of the benefits and costs will result in a negative NPV and rejection of such projects. The decision then becomes largely political or social in that only by applying a very low or virtually zero discount rate does a positive NPV occur. This has certainly been the case in the past, a case in point being land reclamation in the Netherlands.

Accordingly, environmental projects tend to fall within the purview of the public sector, being the responsibility of central government or its agencies.

9.5.4 Differences in discounting approaches in the private and public sectors

The issue of the choice of the discount rate, alluded to immediately above, has been the subject of an extensive and prolonged debate in economics, which is not pursued here. However, what should be emphasized is the generally different approach to the technique of discounting in the private and public sectors. It is worth reiterating what was stated earlier about the internal rate of return (IRR) that commercial enterprises normally employ in project appraisal, which is compared with the rates of interest prevailing in the market, i.e. the external rate of return (ERR). The IRR indicates the return on the financial capital invested expressed as a percentage of the sum involved. The criterion is that the IRR should exceed the ERR for the investment to be viable. This is in contrast to the NPV approach used in the public sector. Although the techniques are different, for most appraisals the IRR and NPV techniques yield the same results, but there are circumstances where the IRR does not produce a single result and this is why economists prefer the NPV approach. The reader is referred to the literature for an examination of the two techniques; for example a clear exposition can be found in Hawkins and Pearce (1971).

9.5.5 Concluding observations on CBA

In the sequence of CBA traced above a number of aspects have been omitted. Furthermore, it should be noted that examination of the remaining stages, of a CBA listed above 7–9, is not central to appreciating the relevance of the method to tourism. Therefore, readers interested in learning more about CBA are recommended to consult Dasgupta (1972) and Pearce (1983), while Hanley and Spash (1993) have demonstrated its application in an environmental context. Stabler (1999), in a critical assessment, has indicated that such is tourism's use of and impact on the environment that CBA should be mandatory for all large-scale developments by the sector.

9.6 A critique of cost–benefit analysis

It is not denied that CBA has been subject to criticism within and outside economics, both in principle and as applied to environmental issues; see for example Pearce and Nash (1981), Porter (1982), Hanley and Spash (1993) Layard and Glaister (1994), and Foster (1997). Currently, the economic approach is underpinned by the fundamental economic market principles. There is a presumption that individuals act rationally (i.e. consistently) to maximize their self-interest so that the value of goods and services is expressed through prices in the market. Choices have to be made because resources are scarce in relation to unlimited wants. The method, originating from conventional economic analysis, is perceived as technocentric with too much emphasis on economic efficiency. Therefore, it can be argued that it is unable to embody the sociocultural, political factors of human exist- ence and the complexities of ecological systems, especially the greater risk and uncertainty surrounding the outcomes of human actions on them. Moreover, it is not seen as a test of sustainability. Certainly from an eco- logical standpoint, echoed to an extent by economists in the field, it has been regarded as inappropriate for dealing with environmental factors because of their inherent intangibility. In particular many fauna and flora, because of their uniqueness, are considered to be priceless by ecologists and society at large so their value cannot be measured in monetary terms.

In a practical context, it is an expensive exercise in terms of the data acquisition and analysis and often time consuming. It is not easily under- stood by practitioners and is possibly subject to institutional capture by the entity under scrutiny, such as a property developer, who can manipulate the appraisal to meet his/her own objectives.

Notwithstanding these misgivings about CBA, it should be emphasized that the method is conceptually simple, wide-ranging in its scope, well founded in economic theory, where the projected outcomes are expressed objectively in monetary terms that facilitates aggregation and the compari- son of projects on a similar basis. Many problems encountered with CBA largely stem from forcing specific studies to conform to the standard method rather than clearly defining and delineating the project and then adapting or deriving a version of CBA within which to conduct the appraisal. Lichfield (1988), referred to above, has shown that urban conservation is a case in point because of the characteristics of the built environment in terms of its function, physical structure, location, surrounding environment and legal status, as well as the many cultural, ecological, economic, environmental and social effects of an intangible nature which are generated.

Within the field of tourism economics, CBA is increasingly being favour- ably received as contributing to the enhancement of research. As indicated earlier, innovative work in modelling the tourism sector has started to incorporate elements of CBA. Also, the refinement of the methods of valuing non-market benefits of the environment promises to meet objections to it for project appraisal.

In the regular tourism literature there are expositions of it as a project appraisal method, for instance Curry (1994), Burgan and Mules (2001) and Archer et al. (2005). However, examples of the application of CBA to tourism are actually travesties of its conception and methodology in economics. The term is grossly misused in many cases where it is more akin to a multiplier analysis (for example Samples and Bishop (1984), Goldman et al. (1994), Fleischer and Felsenstein (2000)) or is simply related to tax revenues from tourism (Florida TaxWatch, 2000).

9.7 Tourism development and cost–benefit analysis

The record of tourism development in the past is unsatisfactory. There are many instances of an almost haphazard and unplanned growth of coastal resorts, particularly in countries bordering the Mediterranean. The Bravas in Spain, and the Balearic and Canary Islands, catering for mass tourism specializing in sand, sea and sun holidays, are cases in point. The quality of their environments has been degraded and the natural beauty of their coasts impaired. In addition, so great has been the clamour of new destinations in Asia, and more recently in the former Eastern Bloc countries, to develop tourism, that they are in danger of going the same way. One large-scale tourism project that was planned, as part of a regional development strategy, was the Languedoc-Roussillon area in southern France. Regrettably, the environment was not a very significant factor in the project appraisal, economic development and growth being paramount (Ashworth and Stabler, 1988). The outcome has been that the natural, human-made and human capital of the area has suffered.

It should be noted that unless required as a condition for the granting of rights of development, it is clear that there is generally no incentive to include, voluntarily, environmental elements in any tourism sector project appraisal. Undertaking these kinds of works in the public sector benefits tourism since such resources constitute the sector's product base, effectively being an instance of the free rider problem encountered in market failure, examined later. In the face of the far-reaching and often unintended detrimental environmental effects of tourism, the sector has not adopted the potentially powerful tool of CBA, even in a simplified and practical form as advocated by Lichfield (1988) as a means of making more informed decisions and avoiding the pitfalls into which tourism development has hitherto fallen. There is not an extensive environmental literature of case studies that have assessed the effects of tourism development projects on natural resources. To be pertinent to new developments, ex ante appraisals should be conducted, particularly if the sector is committed to operating more sustainably. In reality most CBA appraisals, whatever the nature of the projects, have been ex post. Such examples are of limited value in establishing whether, a priori, a specific project is feasible. The increasing awareness by tourists of the adverse environmental impacts of continued tourism development on destinations and their communities is certainly putting greater pressure on the sector's businesses to accept greater responsibility for their actions by identifying and incorporating environmental protection measures into any investment proposals. Some attention is paid to the sector's initiatives in this respect in the following chapter.

9.8 Market valuation of resources and market failure

In following the analytical sequence of ecological and environmental economics in Figure 9.2, it is not necessary to add to what is given in the boxes concerning 'economic and human activity' and 'markets' as they are self evident. In following the sequence, continuing to move up the figure, the boxes 'market exchange valuation of resources' and 'market failure' are considered in the same section as both have a bearing on the use of resources and generation of pollution and waste, indicated in the three boxes in the top left part of Figure 9.2, which are subsequently examined. In giving an explanation of the determination of the market exchange value of resources and market failure, attention is drawn to the opening section of the chapter on ecological and environmental economics and the two columns on the right of Figure 9.1.

9.8.1 Market exchange valuation of resources

The user price of resources, the cost of supplying them, the possibilities of substitution if their relative prices change and the opportunities for technical change to facilitate the more efficient use of them, or increase supply, or the employment of substitutes, are the principal mechanisms by which markets execute their allocation. Taking the case of oil became a pressing issue in 2008, as the demand for such a finite resource increased. Price and forces were set in motion to decrease demand, increase supply, seek substitute forms of fuel and introduce techniques for greater efficiency in its use and supply.

These mechanisms reflect the stance of what was referred to earlier as the 'Optimists' or weak sustainability case in environmental economics, summarized in the right-hand column of Figure 9.1. This stance posits that markets possess the ability to achieve the conservation of scarce resources and resolve problems relating to their depletion, the generation of pollution and waste and the degradation of the environment, provided that property rights can be exercised. This stance is countered by those who are more pessimistic about the ability of largely unfettered markets to achieve such a resolution, summarized in the column headed ecological economics in Figure 9.1. They point to the imperfections, inefficiencies, the difficulties of defining property rights and the inability of markets' mechanisms to work quickly enough. Above all, they cite the existence of the failure of markets.

9.8.2 Market failure

In order to identify the inefficiencies of markets and draw comparisons economics constructs a benchmark of the ideal conditions for the efficient allocation of resources and the distribution of the goods and services; the analysis of perfect competition in Chapter 4 indicated this. In reality, these conditions are not met in several ways and therefore allocative efficiency is not achieved. It is within the field of welfare economics that circumstances of market inefficiency, perhaps misleadingly referred to as 'market failure', have been investigated. The so-called failure is not that markets do not function but that they do not take account of factors that give rise to welfare problems. Effectively, the rationale for ecological and environmental economics largely concerns the identification and mitigation of market failure in connection with the role played by the environment, especially its natural capital (Kn) component, already considered in Chapter 8. As the box 'market failure' indicates in Figure 9.2, the three principal welfare issues arising in the operation of markets are concerned respectively with externalities, public goods and equity within the current and future generations. There is a measure of interaction between all three.

Although it has critics (see for example Randall (1993), who argues that market failure would not persist if imperfections and transaction costs could be eliminated and property rights clearly defined) it is a firmly established concept in an environmental context. It is perceived as the rationale for land-use planning and other forms of government intervention because ameliorative measures, examined in the next chapter, are required in the management of natural resources and environmental quality. The existence of goods and services not traded in the market, such as public goods, necessitates the estimation of demand for them and their valuation, also covered in the same chapter.

9.8.3 Externalities

An externality occurs if the consumption or production decision of one party unintentionally affects the utility of another consumer or the output, revenue and profit of another producer. The decision can have either a beneficial or detrimental effect. No payment is made by the party benefiting from the externality to the party conferring it and, conversely, no compensation is made to the party suffering from the detrimental externality by the party imposing it. Externalities can be categorized as:

- Consumer on consumer
- Producer on consumer
- Producer on producer
- Consumer on producer

The emphasis given in environmental economics tends to dwell on the detrimental effects of externalities, particularly relating to pollution through air emissions, discharges into water bodies, noise and visual intrusion, where the impact relates mostly to the second and third categories. With regard to beneficial externalities, the major one, enjoyed by both tourists and tourism providers, was implied in mentioning the free rider issue in the concluding section on CBA and tourism. The tourism product of natural resources and the built environment is made available at the expense of the communities and the destinations. They do not receive any direct payments for this provision, although it is acknowledged that indirect ones are secured through the expenditures by tourist and tourism businesses. The nature of the externalities and the impacts they have are best explained by illustrative examples of each that will have been or might be associated with tourism. More specific evidence of them emerges in considering the environmental performance of the sector in the next chapter.

a. Consumer on consumer

The externalities tourists as consumers are subject to are most likely to take place while travelling to and from and staying in a destination. For example, at peak times, tourism times for summer and winter, especially when there are public holidays, airports become very busy and check-in times take longer and likewise security checks. Travellers are very likely to have to allow more time. Likewise, at a destination, delays in clearing immigration procedures will occur at peak arrival times. Many destinations, especially those that offer a different experience, say, ranging over beach, cultural, gastronomic and wildlife interests, may attract tourists who are mutually incompatible with each other. For instance the Balearic islands and many Asian countries have found that conflicts have arisen when noisy young people who wish to go clubbing and drinking all night, who disrupt and impair the enjoyment of older tourists wanting a more tranquil experience. Residents in the destination will suffer similarly and may also be subject to boorish behaviour and higher rates of crime.

b. Producer on consumer

Producer on consumer externalities associated with tourism can occur both from the commercial and industrial economic activity unrelated to the sector and providers within it. There are cases of many cities and towns, which are also tourism destinations, which possess economies that create pollution and degrade their environments. One very twenty-first century form of pollution is the photochemical fog generated by road travel within these urban areas. Those immediately coming to mind are Bangkok, Beijing, Los Angeles, Mexico City and Rio de Janeiro. In some previous Eastern Bloc countries polluting agricultural and industrial activity and economic development projects sit side by side with tourist attractions, the environment of which is degraded. Examples are the pollution of lakes, rivers and loss of natural habitats; a notable case is the River Danube and its delta. With respect to tourism providers imposing externalities on their consumers, there are those associated with the visual intrusion, such as inappropriate modern hotels and leisure facilities in or adjacent to conservation areas in historic cities, caravan parks sited in areas of high scenic value. Both tourists and local residents can suffer detrimental effects, for instance where holiday complexes catering for large numbers of tourists cause congestion at peak times and the holding of noisy entertainment and sporting events in peaceful rural areas. As alluded to in Chapter 8, in coastal resorts over-development of tourism accommodation and facilities generates pollution that creates health hazards for both inhabitants and tourists.

c. Producer on producer

As indicated in the examples for the externalities engendered by general economic activity on consumers above, tourism sector providers can suffer the same detrimental effects. They can also generate externalities that confer benefits and impose costs on each other. As argued in the field of spatial economics, businesses have complementarity effects, which increases their turnover and profits, even though they may be in competition with one another. While he has not explicitly stated it as such, Pearce (1989), in referring to the detrimental effects of tourism development in local destinations, such as anti-social behaviour, higher crime rates, incidence of disease, congestion, noise, pressure on local services, was effectively identifying externalities that tourism providers contributed to.

d. Consumer on producer

In economics, this category is perceived as being inconsequential but the congestion, air and water pollution examples can be conceived as costs that producers, including those in the tourism sector, can suffer. The congestion caused by holiday traffic, even away from destinations, can interfere with road freight transportation of a whole range of products that may be en route to tourism businesses.

Thus externalities are an important determinant of market failure, about which more will be added in considering the social benefits and costs of tourism and the policy instruments employed and their effects in Chapter 10. In conclusion, it is worth noting that if an external benefit is being conferred the market is providing minimal goods or services. Conversely, if a detrimental externality, such as pollution, is generated, then the market is producing too much.

9.8.4 Public goods

The term is somewhat confusing as it implies that it is the public sector, central or local government, which is making provision of a good or service that markets do not. The confusion is compounded for there are instances where a good is provided publicly which falls within the economic definition of 'public'. A public good is stated to exist when / where property rights cannot be exercised or it is physically or financially difficult to do so. Examples were given in Chapter 8 in discussing global environmental issues relating to the pollution of the atmosphere and oceans and seas, where the terms 'common property' and 'open access' were used and the distinction between them made clear. The reader is reminded that in the case of the former, following Hardin (1986), where rights of access and use are accorded through customary or legal means, informal rules or regulation are exercised to control over-exploitation. In practice this does not always occur, as individuals are tempted to cheat to gain a commercial or personal advantage. In the latter case there are no restrictive rights on use; access is open to all and exploitation of the resource is uncontrolled. As shown in Chapter 8 with respect to fisheries and dumping of waste, the oceans have suffered depletion of stocks and degradation from pollution.

a. The private–public goods spectrum

A way of delineating these two categories of public goods is to perceive them lying on a spectrum of pure private goods at one extreme and fully public at the other. A private good is one over which complete property rights can be exercised, so there is rivalry for its consumption and so it is exclusive to a single consumer. An example in tourism would be the reservation of a hotel room or self-catering accommodation. Economists posit that a pure public good exists where, if it is provided, it is available to all, i.e. is inclusive, and consumption by one does not deny it to another.

The classic cases cited are defence and street lighting. Many tourism resources, such as beaches, the sea and mountains, are ostensibly public goods but there are circumstances where they are not, such as when they become congested in peak seasons; they then become subject to consumption rivalry.

Assuming that pure private goods are located on the left and full public goods on the right of the spectrum, then, moving from left to right, the order of the location of intermediate goods is: those subject to congestion; common property resources; open access resources. There can be variations in the extent to which each of these three possesses the characteristics of private and public goods.

9.8.5 Intra- and intergenerational equity

Less obvious and not always cited as an aspect of market failure are the issues associated with the distribution of income and wealth. Markets do not take into account the inevitable inequities that arise between the rich and poor and are considered both politically and socially unacceptable. Consequently, a case is made for intervention by governments and their agencies to correct for the maldistribution of welfare. Intra- and intergenerational equity were identified as the principal objectives of sustainability in the previous chapter and a key feature of CBA, referred to above in this chapter, where a judgement has to be made about the weighting of benefits and costs of a project to favour particular sections of society in current and future generations.

a. Intragenerational equity

The two dimensions of intragenerational equity issues are vertical and spatial both within and between countries. The emphasis within a country, via subsidies or taxes, is to achieve a more even distribution of income, but economic development, as already shown in Chapter 8 with respect to tourism's role, is a factor that may or may not have an effect on reducing income inequality. There is a spatial dimension within a country because of income variations regionally, as well as between rural and urban areas. A factor that is relevant but not given as much attention is the difference in living costs that affect consumers' welfare. However, a growing spatial issue that is becoming more apparent is that of wide differences in global incomes that indicate that one in four of the world's population is living in poverty. The Rio Declaration in 1992 identified concern for the poorer nations of the world and the differences in opportunities and lifestyles of the rich and poor. This intragenerational inequality has an environmental dimension with respect to the rate of use of resources, the impact on the natural and human resource base and the attainment of sustainability. It is also an essential element in valuing resources and devising policy instruments.

b. Intergenerational equity

Without underrating the claims of the present generation, environmental economists, especially where sustainability issues are concerned, argue that consideration of the intergenerational effects of current human activity are paramount. Thus the natural assets rule, equated to the strong approach to sustainability, acknowledges that the present generation holds in trust the natural environment that should be passed on undegraded to future generations. The issues identified above regarding intragenerational effects are equally relevant to intergenerational ones. They similarly reflect market failure as, on the grounds of equity, they are perceived as important but are not taken into account in the operation of the market, simply because the time horizons of the present generation are relatively short, so that future benefits and costs are discounted more heavily. They also occur because of the public good and externality problems.

9.8.6 Market failure in the context of tourism

In practice, there have been few attempts to relate the activities of tourists and the operation of tourism firms to policies necessary to deal with environmental issues, which the economic analysis of market failure prescribes. That the full implications of the market failure concept have not been fully grasped by practitioners in general and in tourism in particular is not surprising. The term itself is a source of puzzlement and confusion since the business sector does not perceive that markets actually fail, and it certainly does not fully comprehend that they do not function in accordance with an economic ideal. This is, in part, a reflection of the lack of fully developed operational techniques within economies for dealing with market failure issues, but it is also the result of the lack of political will by governments to take concerted action.

Moreover, the issues are considered too fundamental and far-reaching to be within the empowerment of tourism firms. This is exemplified in developed countries but is even more apparent with respect to many tourist-emergent developing countries, where access and property rights regarding resources are obscure or non-existent, government legislation and regulations are absent, ill-considered or badly enforced and there are no appropriate institutional structures.

In considering the issues in tourism relating to externalities, public goods and intra- and intergenerational equity below, an attempt is made to identify the areas of tourism activity where these occur and to indicate the relevance of their analysis in economics to the sector. Some evidence is offered of initiatives by tourism businesses to mitigate the detrimental effects of their operations, particularly with regard to intragenerational inequities. Actions by the sector regarding externalities and use of public goods are referred to below on the use of resources in the following section of this chapter.

9.8.7 Externalities and public goods interactions

Some examples relating to tourism have already been given above to illustrate in general terms those of a largely detrimental nature. Those inflicted by tourism businesses on each other and destination communities and, similarly, by tourists on other tourists are the most significant. The issue of the sector's businesses and tourists as free riders in the use of public goods has been quite extensively covered at the macroeconomic level in Chapter 8 and brief reference has been made to it above. Effectively, tourism businesses and tourists are not paying the full cost of the provision of the resources that will tend to lead to excess demand for them. It is not putting it too strongly to reiterate that this is exploiative, with the result that Km and Kn resources, especially the latter, are likely to be overused and thus degraded, with host communities bearing the cost. Therefore, in common with other spheres of human economic activity, the sector's use of public goods and generation of externalities are inextricably interrelated.

While there is this linkage of these two aspects of market failure, they are not synonymous. They are conceptually distinct and economists argue that the policy implications are different. As has been shown, the open access cases raise an allocation issue, in the absence of a market price for them, by which to establish their opportunity cost in an alternative use. Taking rural land resources as an example, in a competitive market the demand for land for commercial development, including tourism, say a hotel or leisure complex, will nearly always outbid demand for agricultural land as the returns will far exceed those of farming. Where land has traditionally been used for informal recreational purposes, it carries no price as an indicator of its value, yet it is likely to have a high social one. In short, a comparison of the commercial and social value of amenity land is not conducted on a comparable basis. Thus, there is a need to devise ways of assessing the demand for and value of such non-priced resources. Conversely, to an extent, there is a market for assessing the social benefits and costs of externalities and appropriate policy instruments to enhance their benefits or mitigate their costs. The methods advocated by economists to estimate the values of public goods and the instruments to deal with externalities are examined in the next chapter, using illustrative examples relating to tourism.

9.8.8 The intra- and intergenerational issues

This aspect of market failure is certainly of importance in the tourism sector and a brief consideration is accorded here. However, it should be immediately acknowledged that the issue of intragenerational and intergenerational equity and tourism is a wide one that covers not only socio-economic and political but also ethical and moral issues, as Lea (1993), Wheeler (1994), Page and Dowling (2002), Holden (2005) and Fennell (2007) have indicated. In Chapters 7 and 8, in discussing international aspects of tourism, with the emphasis on the impacts of the development of the sector, the environment was defined in its widest sense, covering the Kn, Km and Kh elements. Reference was made to the inequalities that occur relating to the hegemony imposed by developed countries on developing ones, particularly with respect to economic power. However, another dimension is the tendency to dictate how developing countries should conserve their natural environments. This attitude is presumptuous, because what is often overlooked is the link between poverty and unavoidable environmental degradation through over-cultivation of land, over-grazing and reliance on forests for fuel. The issue is aggravated by the hegemony within tourism destinations that can give rise to inequalities that contribute to poverty because of the socio-economic structures of communities, for example gender roles, social class and the distribution of wealth and income. Poverty has thus become a key issue that is examined in the section on intragenerational equity below.

a. Intragenerational equity

It is not too sweeping an assertion to make that in order to address such inequities most initiatives have come from developed countries that have tended to concentrate on the problems of the current generation. Many tourism businesses have adopted practices that involve working with destination communities for their mutual interests; see for example Wearing and Neil (1999) and Fennell (2007) with respect to ecotourism, examined further below. However, it is organizations such as the ODI, not directly operating within the sector which have advocated and undertaken programmes that have used tourism as a component in specifically combating poverty. Recent examples by the organization are contained in two briefing papers (2007a and 2007b) that have considered methods for measuring the impact of tourism initiatives on communities in Central Africa, Ethiopia, the Gambia, Laos, Mozambique, Namibia, South Africa, Sri Lanka and Vietnam. A government funded body, the Department for International Development (DFID) in the UK, has wide-ranging poverty alleviation programmes for assisting countries, mainly in Africa and Asia, but it is not centrally relating these to tourism projects. Internationally, the United Nations (2006) and UNWTO (2006; 2007), while recognizing that the larger share of tourism receipts received by developing countries increased over the fifteen years to 2005 and has helped to relieve poverty, have suggested additional ways of alleviating it through tourism. Their principal motives are indeed to reduce intragenerational inequities with such an emphasis. However, it should be noted that these initiatives are quite a small proportion of total tourism. Moreover, the programmes are mostly at a very local level; and generally the case studies do not show whether the impact diffuses into the country as a whole.

Another aspect of the pro-poor issue has been the concept of debt for nature that received some attention in the 1990s. Its basis is that governments in developed countries, or wildlife organizations such as Friends of the Earth and the World Wide Fund for Nature, or possibly international financial entities, would purchase a proportion of a developing country's government debt in exchange for an undertaking that programmes to conserve wildlife habitats and indigenous fauna and flora would be introduced and maintained sustainably. Such schemes if related to tourism would generate income and employment and would be enhanced if local communities participated in the planning process, decision-making and ongoing management of them. Holden (2007) cites examples in Ghana, Madagascar, the Philippines and Zambia. A new twist to the concept has been the idea of what amounts to a debt for carbon-trading arrangement. In 2008, Guatemala made an offer to the UK to hand over control of a proportion of its rain forest, relating to the UK's offset of carbon dioxide emissions, to be left as it is in exchange for financial compensation to prevent its exploitation for logging to contribute to economic development. At the time of writing the UK government has not responded to the offer.

There are numerous examples of wider intragenerational inequities connected with tourism, particularly in the discrepancies in the quality of life of visitors and residents of host communities. The organization Tourism Concern was set up with the central aim of campaigning against the exploitation of, and inequalities suffered by, indigenous communities in developing countries by those from developed nations; for instance, the displacement of lower income groups from areas where tourist demand is high has led to soaring property and land prices, as has occurred in Africa, Burma, Egypt, the Gambia, Morocco and the Philippines (Tourism Concern, 1995), the destruction of food growing land and fishing waters in Hawaii (Puhipan, 1994), the corruption of culture in Fiji (Helu Thaman, 1992) and the need for ethical tourism (Wheat, 1999). In issues devoted to particular themes, Tourism Concern has covered charity challenges (2000), racism (2001), fair trade tourism (2002), tourism clashes, covering villagers versus tourists (2004), and tourism and globalization that included who controls tourism, the improvement of the living conditions of workers and whether locals benefit from tourism.

b. Intergenerational equity

Essentially, the connection between tourism and intergenerational equity arises in the issues surrounding sustainability and the environment that were covered in Chapter 8, where it was posited that the long-term viability of the sector's businesses is dependent on maintaining environmental quality. It was concluded that the sector has given insufficient support to sustainability and actions to pursue it to the benefit of destination communities were much less in evidence. It is not the intention at this point to revisit the debate but merely to suggest that tourism businesses are not really in a position to act unilaterally to promote intergenerational equity. As shown, it is the governments in destination countries that choose tourism as the route to economic development and to relieve poverty and it is they that need to recognize that conservation of all three environmental elements, natural, human-made and human, should fundamentally underpin those goals. What also is involved in the intergenerational equity objective are macroeconomic factors, for instance investment in education, training, health and the development of the infrastructure and financial institutions identified in Chapter 8, where an indication of the bodies involved was given; these are not rehearsed here, where the focus is on the microeconomic perspective.

Notwithstanding the increasing interest in the ethics of tourism referred to above, the picture painted in the academic literature is a gloomy one, as it is alleged from the evidence of many case studies that tourism firms and tourists have a relatively short-term interaction with destinations. Wheeller (1994, 2004), a long-standing and scathing critic of the tourism sector, argues that it is characterized by avarice, greed and self-interest. Mowforth and Munt (2003) are sceptical about the long-term prospects for developing countries of emerging tourism patterns; Page and Dowling (2002) outline how the complex structure and objectives of the supply side hamper the derivation of long-term strategies to benefit destinations in the future; Fennell (2007), in examining sustainability in tourism, is also pessimistic about trends in the future. Nevertheless, as shown in Chapter 8 with respect to the international organizations, there are examples of programmes where the tourism sector is involved and is taking the longer view regarding the overall development of destinations. Buckley (2003a) presents a collection of worldwide case studies, many of which indicate cooperative interaction.

An interesting example of a tourism programme having implications for intergenerational equity originated indigenously in Zimbabwe, rather than externally by the tourism sector. Inaugurated in 1989, the CAMPFIRE programme (Communal Area Management Programme for Indigenous Resources) in Zimbabwe (Barbier, 1992) was an early initiative hailed internationally as a model for tourism that benefited local communities, which had the potential to be self-sustaining over the long run. Ownership of wildlife on communal lands was conferred on local communities, giving them the right to sell quotas for hunting and photographic safaris. The revenue from these was used to finance recurrent expenditure, including salaries and vehicle costs, generating considerable employment. It was also possible to support local community social projects, such as clinics, crèches, footbridges and housing. In addition, there were sufficient funds to pay compensation for animal damage to people and crops. It was immensely successful for a number of years (Weaver, 1998), at one time involving thousands of people and constituting as much as 5 percent of GDP. Unfortunately, the programme has virtually ceased to exist, given the current political situation and economic decline in Zimbabwe.

Most tourism firms, particularly those in the nature-based field, are quite small and aware of the precarious position they are in, knowing of the high failure rate of businesses within two years of commencement. Their survival depends vitally on their long-term commercial viability and preservation of the environment and maintenance of the social structure of communities. What is increasingly being recognized, concerning long-term sustainability in destinations, is the need to reconcile the partially conflicting objectives of the simultaneous attainment of environmental, economic or financial and social goals to benefit communities in destinations. The tourism literature has considered the extent to which it makes a contribution to such an attainment; see for example Buckley (2002, 2003b), Page and Dowling (2002) and Cater and Cater (2007). To this extent there is some evidence that, implicitly, intergenerational equity is beginning to be taken account of, notwithstanding the fact that, in comparison with the mass market, the proportion of such tourism is small. References and illustrative case studies related to the issue of equity are given in the section below on tourism's use of renewable resources under the heading of responsible tourism.

The role of economics with respect to intra- and intergenerational equity issues in tourism lies in methods such as CBA, as demonstrated above, regarding the identification of which sector of society is affected and the technique of weighting benefits and costs to attain specified social goals. Approaches also of relevance are models, also cited in the explanation of Figure 9.2 and CBA, for example CGE and TSA analyses, that indicate the significance of tourism in economies and the distributional effects. Other techniques being advocated, especially by the ODI (2007a, 2007b), that are potentially relevant are local expenditure tracking, analysis of destination communities' dependency on tourism and values chain assessment that traces the expenditure linkages between generating and destination countries.

9.9 Resources use and conservation

In completing the exposition of the scope, content and process in Figure 9.2 of the economic analysis of environmental issues in this chapter, an examination of the use and conservation of resources is undertaken and its relevance to tourism indicated. It should be stated at the outset that the analysis of this area of environmental economics has become extensive and extremely technical so that only a brief exposition can be offered here. An elementary introduction is provided by Turner et al. (1994); for an accessible explanation consult Tietenberg (2006) and for a more advanced analysis Perman et al. (2003).

9.9.1 The elements of conservation economics

A suggested classification of resources that facilitates an analysis of the issues concerning their use and conservation, including policy implications, is:

Non-renewable (NRR)	Renewable (RR)	Unlimited
Productive: Minerals, e.g. copper, lead, zinc	Agricultural	Aggregates
	Fisheries	
	Forests	
Energy: Fossil fuels, e.g. oil, coal, gas	Hydro-electricity	Tidal
	Bio-fuels	Solar
		Wind

As well as concern over the possible depletion of NRRs and the conservation of the resource base of RRs, this area of economics also considers the issues associated with the generation of waste and pollution and their consequent impact on the environment, particularly its natural component. Tourism, as a major global economic activity, consumes a sizeable proportion of the world's resources and, as already shown, has a significant effect on the environment. It is thus relevant to examine the elements of conservation economics and to relate them to the operation of the tourism sector at the local level.

9.9.2 The economics of resources conservation

a. Non-renewable resources

The rate of depletion and possible exhaustion of non-renewable key productive and energy resources is a central economic issue. The problem of maintaining their supply is likely to become more acute in the face of growing demand by Asian countries, such as China and India, experiencing rapid economic development and growth, and the increasing scarcity of new exploitable sources.

The economic analysis of NRRs is founded in what is known as the 'Hotelling Rule' (Hotelling, 1931), which sets out the conditions for the optimal use of such resources over time. Basically, the rule for the supply of a resource is to maximize its present value by estimating the future returns and costs over the expected time period that it is economically exploitable; that is, to the point where the net returns equal zero. The rule makes a number of conditional assumptions as to the competitive structure of the market that it is not essential to pursue here. Fuller explanation can be found in environmental economics texts such as Tietenberg (2006).

It is, however, instructive to draw attention to the wider view of resource conservation that demonstrates the consequences for economic growth of the possible solutions to the threat of depletion and environmental degradation that are equally relevant in the more specific case of tourism. Economic analysis starts with expounding the role of the market mechanisms of costs and prices in inducing conservation through reduced demand and substitution of scarcer resources by cheaper alternatives. Also, technological developments lead to new methods for the exploration and extraction of resources and improve the efficiency of their use, such as the extension of product life that reduces the demand for the constituent resources and facilitates recovery (collection), re-use and recycling.

The issues of re-use and recycling (the former using a product unchanged, for example drinks bottles, whereas the latter involves processing, for instance scrap metal or paper) have become more important in recent years on two counts. The inference is that re-use only involves the recovery cost, but recycling is costlier because labour, machinery and energy are required in the process. First, there is the problem of disposing of products that are not recyclable as the capacity to use landfill sites is rapidly being exhausted. The problem of waste generation raises the issue of the full costs of its disposal. Ignoring its externalities effectively means waste disposal is underpriced. For example, materials not recovered and recycled if disposed of by combustion or landfill create detrimental externalities, such as the production of harmful emissions, toxic chemicals and sterilization of amenity and developable land. In addition, in the interests of public health, central and local government intervene to set standards or regulate or levy charges, involving administrative costs. Intervention distorts market prices and costs, while administrative expenses are often not fully assigned to the service, so that the full costs are not passed on in user charges. The inference is that more recycling would occur if the full costs of disposal were to be identified and met by users of such services.

Second, there is an economic issue over the opportunity cost of obtaining resources from primary as opposed to the secondary sources of re-use and recycling. While the cost of exploiting primary resources can be estimated reasonably accurately, this is not the case for secondary sources. The extent to which recycling takes place is influenced by the nature of the material, at what stage in the product cycle recovery occurs, who uses it, what proportion is unusable residual waste and how it is disposed of, and the ease or difficulty with which it can be recovered, one crucial determinant being the scale of the operation. These physical factors determine the costs of recycling, which may be considerable, but an important additional variable is the price of the material that is governed by demand in relation to supply.

In general, economists argue that recycling becomes feasible if the costs of recovery, including re-processing of material, are lower than those of exploring for and exploiting primary sources. However, market prices and costs do not reflect the true benefits and costs as while they take account of those that are private they do not embody the additional non-priced benefits and costs, i.e. externalities. On the other hand, in the case of, for instance, discarded cars, whether the separation and recycling of their component materials is worth it might depend on incorporating a larger percentage of recyclable material when they are first manufactured, so reducing unusable waste. Consequently, the social costs generated by primary resource utilization are lessened, such as the despoliation of scenic areas where metal ores are mined, the generation of noise and dust and associated health hazards. The collection, sorting and reprocessing costs are not easily identified. Generally, it is only high value materials, such as aluminium, copper and lead, that are worth recycling, notwithstanding the fact, that there may be environmental benefits of recovering materials like cardboard, glass, paper and plastics so as to reduce landfill waste disposal. Moreover, whether the reprocessing of recovered materials creates fewer externalities than the extraction of primary materials is not at all certain. It is an area of environmental economic analysis that has not been fully investigated empirically, although the theoretical arguments for doing so are clear.

The two issues raised in the foregoing discussion can be illustrated diagrammatically to indicate the theoretical analysis. The relationship between the use of a resource from primary sources and its recovery for re-use or recycling as a secondary source is examined in Figure 9.3.

Figure 9.3 depicts the supply curves for the primary and secondary sources of a resource facing each other. The total quantity demanded and supplied is shown on the horizontal axis. The supply cost from primary sources is the vertical axis on the left of the figure, so that its supply cost (SP) and quantity supplied rises moving from left to right. The secondary sources cost axis is on the right of the figure and its supply cost (SS) and quantity supplied rises from right to left. Where the solid lines of the respective costs SP and SS intersect indicates the proportion of the total resource

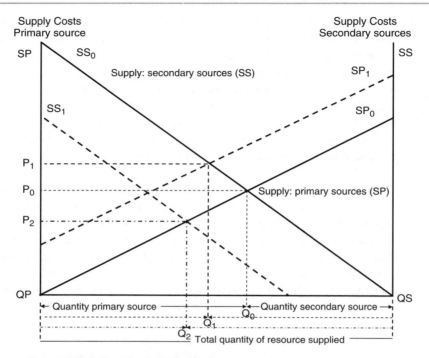

Figure 9.3 Optimal level of recycling

marketed on the horizontal axis; quantity QPQ_0 is from primary sources and QSQ_0 from secondary sources. Should the supply cost from primary sources rise, shifting the supply upwards to the left as shown by the broken line SP_1, perhaps because of their greater scarcity of economically exploitable quantities, then the proportion supplied from it will decrease and that from secondary sources will increase, shown as QPQ_1 and QSQ_1 respectively. Starting from the initial position, should the supply cost from secondary sources decrease, shifting the supply downwards to the left as shown by the broken SS_1 line, possibly because of technical advances that reduce the recovery and recycling costs, the proportions of total supply would be QPQ_2 from primary sources and QSQ_2 from secondary sources. In effect the relative supply costs have changed so it is feasible to recycle. As depicted in the figure, the inference could be drawn that the shift downwards of the secondary source supply line implies that a positive quantity would be supplied at a zero price. However, this would be feasible if a policy were to be implemented to encourage more recycling by subsidizing supply from secondary sources.

One of the fundamental concepts of environmental economics is the identification and evaluation of both the private and social benefits and costs.

It is assumed in Figure 9.4 that any economic activity, including in tourism destinations, generates both social benefits (MSB) and social costs
(MSC), which exceed private benefits (MPB) and private costs (MPC). The
private optimum is at P where MPB = MPC, whereas the social optimum is
at S, suggesting that a reduction in the level of economic activity from P to
S is necessary. However, this social optimum does not equate to what can
be called the ecological optimum indicated by E, beyond which the assimilative capacity of the environment (EAC), shown by the horizontal broken
line, is exceeded, generating an unassimilated detrimental externality, abc.
Should this be cumulative, environmental degradation increases in each
succeeding period. The conceptual distinctions between private, social and
ecological optima are easily identified, but deciding which should be pursued is crucial in terms of minimizing the wider detrimental effects of the
activity and the attainment of long-term sustainability. The issue is also a
business, socio-economic and political one as to where between E and P the
level of activity should be; it is revisited in Chapter 10.

It is possible to combine the analyses given in Figures 9.3 and 9.4 to
consider the relative merits of obtaining resources from primary and secondary sources with regard to the social benefits and costs associated with
each. For example, the net social benefit or cost of an activity can be ascertained by establishing whether the MSB outweighs the MSC, or vice versa.
In the case of both primary and secondary sources of the procurement of a
resource, the net MSB or MSC can be incorporated into the SP and SS

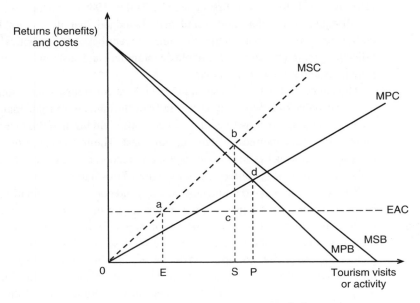

Figure 9.4 The concept of an economic optimum as applied to tourism

supply lines shown in Figure 9.3. A net MSB would move SP supply downwards to the right and the SS supply downwards to the left. Conversely, a net MSC would have the opposite effect by moving SP supply upwards to the left and the SS upwards to the right. As stated above, it is currently an unsubstantiated empirical question as to what the position is on the case for increasing recycling on social as opposed to economic grounds. A CBA of individual materials is required. Organizations such as the New Economics Foundation (NEF) and the Friends of the Earth (FOE) argue that recycling can make a significant contribution to achieving greater efficiency in the production of goods and in pursuing sustainability.

b. Renewable resources

In defining the environment in Chapter 8, a distinction was made between base resources and the consumable products they provide. The key issue that environmental economics has highlighted with respect to renewable resources is the need to conserve the base resources from which agricultural, forestry and fishery products are obtained. An associated and dominant issue is that relating to open access resources, which has also been examined in Chapter 8 and above in this chapter concerning market failure.

Maximum sustainable yield, an essentially biological concept and a long-standing basis for the modelling of renewable resources in economics, has surprisingly not figured strongly in the more strategically oriented focus of sustainable development (SD), except for the much discussed resort life-cycle hypothesis first posited by Butler (1980), who has reviewed it subsequently (Butler, 2005) and acknowledged that it has not been embraced by the tourism sector. Yet, this concept has strong practical significance in implementing sustainable development and sustainable tourism principles and policy in particular.

The concept of maximum sustainable yield is concerned with resources that are capable of renewal either naturally or through appropriate management. A key issue is how to achieve maximum yield but maintain sustainability from the economic use of the resource base. The concept is particularly relevant to open access and common property resources, on which attention is focused since many tourism resources, are of this nature.

The economics of maximum sustainable yield considers the relationship between the price of the product, the cost of exploiting it, the yield in terms of the physical quantity and the stock or population. The yield is determined by the exploitation effort and the total stock of the product. The greater the effort, for example increasing the number of hours worked or the efficiency of the equipment, the greater the yield. However, the size of the product stock governs its rate of renewal or recruitment and thus the quantity available.

The problem becomes an economic issue by considering the revenue generated in relation to the cost of the effort. Normal profit-maximizing conditions can be applied which will determine the quantity of the total stock of the product taken, which may be below or above that necessary for the resource to renew itself. If the cost of exploitation is reduced or the price of the product rises, or both, then profits will increase which will, in turn, attract more entrants. Since access is open and thus free apart from the costs of entry, it is likely that exploitation is beyond the point of maximum yield. Should the total stock fall below a given threshold (a biological or ecological issue) then the population will crash, leading, in the case of animal or plant species, to extinction. In the conventional economic exposition of renewable resources, such as fishing and forestry, the shape of the total revenue curve is dictated by the underlying biological relationship between the resource stock, its rate of reproduction or renewal and the harvest rate which is determined by the harvesting effort (see Norton, 1984; or Conrad, 1995 for a full explanation). It is posited that there is an inverse relationship between the stock and effort. The total cost curve corresponds to that usually constructed in economics, in which increasing returns prevail over a range of output, so flattening the curve before decreasing returns set in, giving rise to a steeper curve. It is also assumed that normal profit, i.e. that required to keep firms in business, is included in the cost curve.

The concept of maximum sustainable yield is applied to tourism below in section 9.10.5 after considering the sector's use of resources. Non-renewable resources are examined first, with particular attention being devoted to the issues of irreproducibility and irreversibility that arise where they are considered to be unique and under threat of being lost completely. In the section on renewable resources use in tourism, open access resources and their possible overuse and degradation are issues given due emphasis.

9.10 Tourism and the use of non-renewable and renewable resources

9.10.1 Non-renewable resource issues in tourism

Given that it is one of the major economic activities in the world, tourism has a marked impact on the demand for non-renewable resources. It also generates significant wastes which, although not as hazardous as pollution from extractive industries, manufacturing and chemicals production, can create acute disposal problems (Stabler and Goodall, 1996) as well as major environmental problems. The operation of tourism firms reflects the market-driven characteristics of other economic sectors where the environment is treated as a free good. For the sector, the environment is an essential input as well as an element of the final product, so that problems occur, especially over-exploitation of the natural resource base and the generation of non-priced adverse effects. Much tourism expansion, particularly its concentration in certain areas, has neglected the long-term dependence of the industry upon the environment (Cater and Goodall, 1992). The difficulties of taking coordinated and concerted action, at least to mitigate if not eliminate environmental problems, are formidable.

9.10.2 Tourism's use of non-renewable resources

There were many initiatives in the 1990s to improve the environmental performance of the supply side regarding the use of resources and reduction of waste, when attention concentrated on the causes and effects of firms' operations. In practice, combating problems was largely confined to limited initiatives by certain sectors of tourism, notably the hospitality sector. These were mainly concerned with holiday accommodation. Tourism associations recognized this contribution and urged businesses to institute programmes to avoid the use of materials likely to be environmentally harmful, substitute purchases of recycled products for those from primary sources and to reduce waste by an absolute cut in the consumption of materials and energy and by recycling materials as far as is practicable (Troyer, 1992; International Hotels Environment Initiative, 1993; Middleton and Hawkins, 1993; Scottish Tourist Board, 1993; WTTC, 1994; Beioley, 1995; Dingle, 1995).

Rather more ambitious are decisions only to use suppliers who meet the environmental standards expected by purchasers and action to educate and encourage tourists to adopt environmentally responsible behaviour. It is not unduly critical to assert that most utterances by other sectors paid little heed to developing environmentally responsible behaviour (Brierton, 1991; Hunter and Green, 1995). This, in part, was a reflection of the failure by governments to commit themselves fully to pursue environmental policies, despite being signatories to international agreements, as indicated in Chapter 8.

However, within tourism, as in other industrial fields, as well as a lack of comprehension by businesses of environmental issues and objectives (Stabler and Goodall, 1997), an element of complacency was engendered in that in some areas of activity, it was felt that the potential to attain sustainability was limited. Furthermore, firms lacked knowledge of the implications, in monetary terms, of investment in, operation, monitoring and management of systems to achieve specified objectives (Institute of Business Ethics, 1994; Forsyth, 1995). Understandably, the private sector perceived the introduction of environmental practices as adding to costs. Currently, this perception has tended to persist, the sector bowing to pressure by tourists or laws and regulations by governments in destinations, rather than investigating the likely competitive advantage of using resources more efficiently and implementing their own schemes. Recent work on the relevance of the Porter hypothesis by Huybers and Bennett (2003) and Razumova (2007) with respect to improving the environmental performance of tourism businesses was discussed in Chapter 8. In a study that considered the role of tourists, tour operators and governments in reducing waste, Kuniyal (2005) undertook a study of its management in the Himalayan trails and another for expeditions. It was advocated that there should be an administrative scheme introduced to reduce the quantity of articles, constituting two-thirds of articles taken into the area, which cannot be reused, recycled and composted.

It was also acknowledged by the tourism sector that there were many destinations, including islands or small states, often set in fragile environments, where reductions in the use of primary sources of materials combined with recycling could make a significant contribution to achieving sustainable tourism (Stabler and Goodall, 1996). The emphasis on the hospitality sector has tended to wane as the wider environmental effects, as defined in Chapter 8, to include renewable natural resources and sociocultural elements, considered below, are increasingly perceived as a bigger problem of tourism activity.

9.10.3 The irreproducibility and irreversibility of resources use in tourism

Ostensibly the use by the tourism sector of natural environments or resources, and possibly some human-made and human resources, should be examined in the subsection below on renewable resources, but there are circumstances where these can be perceived as non-renewable. This is particularly the case where they are unique. Some natural resources that attract large numbers of tourists are irreproducible, such as the Giant's Causeway, Grand Canyon, Great Barrier Reef and the rain forests of Asia and South America, if they were to be severely degraded, or even destroyed, by tourism development or overuse. Flora and fauna, especially those indigenous and unique to particular locations, are also irreproducible. Indeed it has been estimated by international conservation bodies concerned with bird, insect, marine, mammal and plant life that with the loss of their habitats through human activity or improper management, as many as 50 species are becoming extinct each year.

An aspect of this threat is that irreversible trends may occur, for example the fragmentation of or reduction in the size of habitats that render them incapable of maintaining the population. There are many such designated resources of this nature that are vital to tourism, for example forests, game reserves, marine conservation areas, national parks, nature reserves and wetlands that are being encroached upon and subject to overuse. In addition, there are numerous and varied landscapes, sometimes influenced by human activity, which, although not actively utilized, form a backdrop to tourism, for instance coastlines, lakes, rivers and mountains, and are a key component of the product for destinations. Overcrowding and over-development often occur in relatively small locations. At peak times, visitors can outnumber the resident population by a factor as high as three or more. Tourism firms and tourists in both generating countries and destinations are seldom aware of the damage being caused; indeed it is usually unintentional. For example, there is compaction of snow on ski runs, which damages plants and the eco-system (Tyler, 1989).

Other effects are more deliberate, for instance off-road vehicular use (Sindiyo and Pertet, 1984). Excess numbers also increase the demand for secondary resources, water and energy, that may be scarce in developing countries and islands (Romeril, 1989a, 1989b). Loss of flora and fauna occur where tourism expansion (Andronikov, 1987), climbing (Pawson et al., 1984) and hunting (Smith and Jenner, 1989) have taken place.

Except where they are easily reproduced, many human-made resources, especially heritage artefacts, are unique and so are at risk in the same way as natural resources. Heritage buildings and monuments can be damaged by pollution and gardens and parks degraded by inappropriate and excessive use. Paintings and murals in galleries and artefacts in museums can deteriorate because of greater exposure to strong light and body moisture (Goodall, 1992). The influx of tourists with different life-styles, large financial resources and demand for non-indigenous services not only disturbs existing economic systems, but can also destroy traditional cultures (Pearce, 1989). These problems have been recognized by many involved in tourism and have become issues of concern (Cater and Goodall, 1992; Eber, 1992; Jenner and Smith, 1992).

It should be appreciated, in the examples given, that it is difficult to separate entirely tourism's use of non-renewable and renewable resources. Further and more recent case studies of tourism's use of resources are given below in the section on those that are renewable, which also examines the sector's endeavours to develop responsible tourism to safeguard natural, human-made and human capital. These examples include reference to tourism's consumption of non-renewable resources.

9.10.4 Concluding observations on tourism's use of non-renewable resources

In this short review of the use of non-renewable resources, pollution and waste by the tourism sector, it is apparent that the 'bottom up' approach to conservation has not been entirely analyzed so that the total benefits and costs have not been taken into account. Firms have seized on the conservation element of the sustainability issue because of the possible expectation of higher revenues, such as from demand by tourists seeking a higher quality experience, or financial benefit, for example the cost savings from better energy and water management and waste minimization effected in hotels (D'Amore, 1992; Hill et al., 1994). If cost savings are passed on in lower prices and consequently attract more tourists, then, in total, energy and materials use and waste generation may be no lower in a particular establishment or destination. Moreover, other forms of environmental damage may occur, such as over-use of natural resources and pressure on local services in the area in which a hotel is located. These externalities are perceived by individual businesses as remaining outside their responsibility. Thus while moves by tourism firms towards the conservation of materials and energy reflect the market oriented approach, their conservation has not been fully considered in the wider context of sustainability.

9.10.5 Renewable resource issues in tourism: the relevance of the concept of maximum sustainable yield

The elements of maximum sustainable yield, explained in general terms above, are applied to tourism in Figure 9.5. They are also applicable to land-based open access resources and common property resources should the owners not adopt, or should they abandon, sustainable management practices.

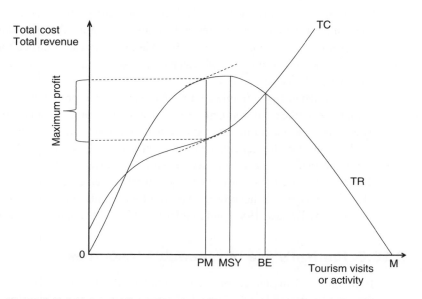

Figure 9.5 A bioeconomic model of maximum sustainable yield applied to the use of open acess resources

Suppose tour operators and tourism firms in destinations make use of the natural resources base that can be degraded by over-use. In Figure 9.5 it is assumed that profits are maximized at PM, where there is the greatest vertical distance between the total revenue curve and the total cost curve. The resource, however, is capable of sustaining a larger number of users or visitors and generating greater total revenue but at a declining rate of profit to the right of PM, given the shape of the cost curve. The point MSY indicates the maximum revenue possible, which declines to zero at point M because of a reduction in the resource's capabilities to support visits. As posited by Butler (1980) in his life-cycle hypothesis, this could occur because of visitors' perceptions of a decline in its quality or as a result of deterioration in its physical properties, for example a ski resort where the ability of the snow cover to support the activity declines with increasing use. Another possible instance is the case of safari tourism, where increasing tourist numbers disturb the breeding patterns to such an extent that the fauna reproduction rate declines.

The level of BE shows the break-even point for providers meeting tourists' demand. This presupposes that the resource is one characterized by free access so that apart from the costs of marketing holidays (which the shape of the cost curve suggests is subject to eventually decreasing returns) there is freedom of entry. Thus, suppliers will continue to exploit the resource even though abnormal profits are competed away, for example guides in wildlife habitats where over-supply diminishes the earnings of individuals. Should the marketing costs be lower, then the nearer to point M the use of the resource will be, threatening its renewability. On the other hand, infrastructural and market constraints on continued exploitation of the resource base sometimes occur, for example because of difficult air or road access, lack of energy and water supply and limited accommodation, facilities and services.

As applied in bioeconomic circumstances, the reinterpretation of models within the context of tourism shows that over-exploitation can occur. Therefore, the economic analysis of maximum sustainable yield holds implications for the management of open access resources used by the sector, particularly the need to secure workable agreements on their use in destinations. The issues of classifying natural resources and the problems of their management have been explored by a number of writers and are reviewed by Berkes (1989) and Stabler (1996b) in the context of tourism.

9.10.6 Tourism's use of renewable resources

Both tourism firms and governments often fail to consider the total nature of the product offered and the impact of tourism on the natural resources component, prioritizing the short-term benefits which they obtain from tourism.

Businesses and residents of destination areas, although accepting the demands made on them in order to enjoy the benefits arising from tourism, may be unwilling or unable to become involved in decision-making concerning their economic and sociocultural environments, as previously indicated in earlier chapters and above in this one. Consequently, the conservation of these environments is compromised. Those who do become actively engaged in protecting them are more likely to oppose the expansion of tourism *per se*. Such issues have been discussed in the literature under the community approach to tourism development and sustainability as espoused by, for example, Murphy (1985), Keogh (1990) and Hall and Richards (2000), which echoed the debate in the natural resources management field by Berkes (1989).

However, to ascertain the response of tourism at the operational level in destinations to the use of ostensibly renewable resources, it is to the examination of the non-mass market sector that attention must be turned. Over the last decade or so there has been an explosion in the study of what might be termed, generically, 'responsible tourism' that has generated a debate as heated as the one concerning sustainability, to which it is related. There is now a vast literature on this niche market that is referred to under seemingly innumerable headings, identified below. Also, there is a wealth of case studies that illustrate tourism's response to the challenges it faces to be seen to be acting responsibly with respect to the economic, natural resources and sociocultural environments on which it has an impact.

9.10.7 Responsible tourism

To appreciate the scope of responsible tourism and its applicability to renewable resources use, it is instructive to trace briefly its development. A relatively early call for tourists to be more sensitive to the effects of their behaviour while on holiday was made by Krippendorf (1987), both with regard to the environment and hosts' life-style and culture. In the tourism literature there has been much confusion as to how to describe what has been given various labels, the most frequently quoted being: adapted tourism; adventure; alternative; appropriate; community-based; conservation; cultural; ecological; ethical; fair-trade; green; high value; natural area; nature-aware; nature-based; popular ecotourism; pro-poor; responsible; self-reliant; small group; soft; social; sustainable; and wildlife tourism. Three terms have assumed prominence, 'ecotourism', 'geotourism' and 'sustainable tourism', of which the first has been widely adopted, and under which the others listed here have been largely subsumed, notwithstanding the fact that many scholars contributing to this aspect of tourism perceive it as essentially nature-based. Acknowledging its almost universal usage in the literature, the term ecotourism is adopted here. Representative studies of ecotourism are Cater and Lowman (1994), Wearing and Neil (1999), Newsome et al. (2001), Weaver (2001), Page and Dowling (2002), Fennell and Dowling (2003), Buckley (2003a), Hall and Boyd (2005) and Fennell (2007). The concept has been exhaustively discussed regarding its definition, in order to delineate its boundaries, and is threatened with getting bogged down in the same conceptual and definitional morass that sustainable tourism has.

Ecotourism has been accorded much importance as the fastest growing tourism sector, albeit from a low base. There are several bodies active in this field such as Conservation International, the Ecotourism Society, Eco-source Network, Green Globe 21, United Nations Commission on Sustainable Development (UNCSD) and WTTC. The United Nations declared 2002 to be the International Year of Ecotourism and there was also the Quebec City Declaration on Ecotourism that emerged from the World Ecotourism Summit in the same year. There are also two journals that devote articles to the subject, the *Ecotourism Journal* and *Journal of Sustainable Tourism* (JOST).

Page and Dowling (2002) argue that it is necessary to define ecotourism in order to show that it is an appropriate medium to bring about a greater environmentally benign understanding that leads to an appreciation to inspire action for the environment. A recently published new edition by Fennell (2007: 24) is essentially a review of the current position. After discussing the many definitions of ecotourism, he put forward his own:

> Ecotourism is a sustainable, non-invasive form of nature-based tourism that focuses primarily on learning about nature first-hand and which is ethically managed to be low impact, non-consumptive and locally orientated (control, benefits and scale). It typically occurs in natural areas, and should contribute to the conservation of such areas.

With the many labels given above describing ecotourism, Fennell's definition identifies some of its elements and aptly summarizes what it should aim to achieve, including prescriptions for tourism firms on their operations and the behaviour of tourists respectively. The adaptancy thesis emphasized that tourism should take a form that was responsive to the effects on communities' environments in destinations and was an early perspective taken. Much attention has been concentrated on developing countries, but a significant proportion of case studies are centred on developed countries. The key aspects of ecotourism, as an appropriate form of tourism, continue to be the impact in destinations on the use of renewable resources and environments, especially open-access and common property ones, and that it should avoid as far as possible conflicts over their availability to stakeholders; conserve natural and sociocultural capital; minimize the impact on them; be small-scale; be largely non-consumptive of resources; be ethical in relation to local communities; be locally oriented.

In developing countries, an important feature of ecotourism has been on the relief of poverty and the preservation of the sociocultural capital. The pro-poor initiatives and programmes have already been examined in the section above on intragenerational equity. However, in connection with these actions reference needs to be made to what has become known as the 'triple bottom line'. This is the notion that ecotourism should endeavour to satisfy economic or financial, environmental and sociocultural objectives. The first is actually concerned with tourism businesses' necessity of earning profits to ensure ongoing viability, the second relates essentially to the tourism resource-base, with the emphasis on its natural elements, and the third to the conservation of human and cultural capital. The feasibility of attaining the triple bottom line has been questioned as part of the wider debate as to whether ecotourism is simply yet another tourism oxymoron in the sense that it is not possible to achieve all three objectives simultaneously.

The issue has been succinctly summarized by Buckley (2002; 2003b), who has argued that ecotourism does not necessarily need to be defined. He asserts that its meaning is clarified if it is considered purely in terms of inputs and outputs. His primary objective was to establish whether or not a given form of tourism has eco credentials. Buckley's approach is interpreted here by identifying the inputs as the natural, human-made and human capital that constitutes tourism's base resources and the outputs as the benefits derived from and the costs borne by these resources. This effectively sidesteps the need to consider the service or operational element that all forms of tourism have; in any event this aspect is concerned with the consumption of non-renewable resources, which is not an issue at this juncture. Ecotourism is an acceptable type of tourism if the condition is met, under the triple bottom line objective, that there are net benefits. Buckley, gives sound reasons as to why its pursuit means that, strictly, all three of its elements ought individually to yield net benefits, rejecting the notion that it should allow costs to outweigh the benefits in one or two elements but that the overall outcome gives a net benefit. In other words, each of the three elements should have a positive conservation effect. An accounting procedure is required to measure the benefits and costs of the process of achieving the triple bottom line that should encompass all those identified in CBA, examined earlier in this chapter. Furthermore, adaptations of the techniques of TSA and CGE would be relevant in estimating the outcomes.

Buckley (2003b) asserts that businesses in most spheres of economic activity have understandably paid more attention to, and taken initiatives on, the financial bottom line that meets their own objectives, but perhaps only incidentally yield economic benefits to the area in which they are located. They have not given so much regard to the environmental and sociocultural elements. He sees ecotourism, if it conforms to the triple bottom line as he posits it, as contributing to sustainability, but suggests that it is the responsibility of those alive now to ensure that the interests of future generations are met. Therefore, unlike most other economic sectors, tourism has the potential to deliver net gains for all three components of the triple bottom line. However, the inference is that the burden, which additionally will include the effects of their customers' behaviour and actions while participating in this form of tourism, will fall on tourism businesses.

Whether the tourism sector is meeting the triple bottom line criteria, and is indeed attaining the desired effects of ecotourism identified above in Fennell's (2007) definition and the paragraph immediately below, as well as the pro-poor objective, is an empirical matter. Additionally, there are issues connected with the operation of businesses engaged in the ecotourism sector, such as working within the laws and regulations in place in destinations, adhering to any codes of conduct, certification and eco-labelling and abiding by the standards set by bodies of which they may be members. It should be noted that there is very often a dichotomy between what is intended by these legal and regulatory measures and what happens in practice. Important features of operations are to ensure the involvement of the local community in decision-making over tourism planning and development, control over resources, the management of facilities, services, visitors and activities, balancing the interests of all stakeholders in the distribution of benefits. With respect to resources, tourism operators need to be mindful of whether a contribution is made in cash or kind to environmental and sociocultural projects and their management, including new techniques. Projects might involve the acquisition of land for conservation, rehabilitation or reallocation. Other possible initiatives are to offer information, interpretation and education to tourists.

Case studies are now presented that illustrate, as far as space permits, tourism's performance with respect to most of these objectives and issues. They are also intended to indicate the effects on renewable resources that constitute the tourism product base and the core of this section of the chapter on the economics of resource use and conservation.

9.11 Case studies of tourism's performance in the use of renewable resources

Studies have been conducted of various aspects of the impact of tourists and businesses on environments in localities and communities in destinations, both in developed and developing countries. With an emphasis on the latter, the cases are drawn from an extensive literature, with the exception of examples relating to Australia, New Zealand and Belize that reflect the first named author's own experience and as yet unpublished research on the locations. Particular reference is made to Buckley (2003b), an excellent source of a large number of cases in global regions loosely corresponding to the continents. The value of his compilation lies in the impacts of tourism development on natural and human-made resources and sociocultural structures and the role of international and local tour operators and organizations. Doan (2000) in a review of ecotourism articles in the tourism literature, has also analysed the phenomenon's effects. Most of the texts on ecotourism, a selection of which have been cited above, contain case studies, as have the reports of the international bodies referred to in the section on intra- and intergenerational equity.

9.11.1 Marine tourism

Marine tourism covers numerous activities ranging over beach-based ones, bird watching, boating, diving, fishing, geology, sea watching, surfing and wildlife study. These activities, with a number of case studies commentating on the operations of international and local tour companies and the impact on natural resources and communities, have been admirably covered by Cater and Cater (2007). They also investigate the consumption and non-consumption, pro-poor and triple bottom line issues in the marine tourism context. An example that reflects their focus was a study by Holland et al. (1998) on recreational bill-fish activity on the US Atlantic and Central American coasts that raised the issue of whether or not it was a consumptive activity. However, the term consumptive was not fully explained in that it did not indicate whether it was confined to the 'product', in this instance fish. Consumption could include tourists' transit and accommodation requirements while in the destination, disturbance of wildlife, erosion and pollution.

An interesting finding of this study was that the economic benefits from sports fishing were greater than the food value of commercial catches. Moreover, the observation was made that consumptive recreational hunting and fishing generated more benefit than the non-consumptive form because they were conservation oriented and constrained exploitative commercial activity that depleted stocks. Other relevant texts on marine tourism are Halfpenny (2002), and Garrod and Wilson (2003), Luck (2007). The two cases of marine tourism given here concern the impact of sea watching on the natural and human environments. One is whale-watching in the developed country of New Zealand, the other is of a less developed area in Mexico.

Kaikoura is a small town on the east coast in the South Island, north of Christchurch in New Zealand. The example is an interesting one as it relates to a business run and staffed entirely by Maoris, a hitherto under-privileged section of the country's society. What started as a sideline activity to fishing in the 1980s now forms a significant proportion of the economic base of the locality. Kaikoura Whale Watch is a virtual monopoly, having absorbed smaller operations since its inauguration; this has led to comments by tourists, after the tour, about the relatively high charges, given the uncertainty over the quality of the experience. The control of the business by the Maoris has been the source of some friction locally, as has concern over its environmental performance, notwithstanding the fact that it has benefited the community by generating income and employment and improved the quality of life of all inhabitants of the locality. It has also been instrumental in triggering the provision of accommodation and an improvement in local services. A detrimental impact is alleged to have been the hiking of land and property values deterring settlement by incomers seeking work. Because there are few tourism facilities, those that exist are rather basic, so the area has tended to attract low-cost tourists who spend little in the town. Notwithstanding the claim by the company that it has minimized its environmental impact, there is some evidence that its activities have had a negative effect because of the disturbance of the cetaceans and resident species of fur seals, white-fronted terns and red-billed gulls.

The Mexican case has drawn on the work of Young (2003) and concerns conflicts over the use of open access resources of regions on the Baja California peninsula in Mexico. It is also a good example of consumptive versus non-consumptive use of natural resources and a form of tourism that does not relieve poverty in the destination. The regions are prime fishing areas, 2,000 kilometres long, and constitute 22 per cent of Mexico's coastline. The main problem is to reconcile this activity, which is the main livelihood of the residents that are relatively poor, and the conservation of the natural environments that are increasingly appealing to tourists. Fishing disturbs the cetaceans which migrate to the area to breed and is a threat to the growth of whale-watching tourism. The waters in the regions have been over fished, reflecting Hardin's (1968) common property resources case, irrespective of the competition emerging between the interests of local people and tour operators promoting coastal tourism and whale watching. There are also issues concerning the local power structures and the political processes that influence both the local economy and tourism development. Added problems are that both international and domestic tourism is very seasonal and most holidays are marketed by external tour operators so that revenues accruing locally are minimal; also because expenditure locally by tourists is low.

9.11.2 Island tourism

Tourism in small islands been has been given special attention in recent years; see for example Briguglio et al. (1996), which has covered issues concerning sustainability. The following two cases, the first on a long established resort in a developed country and the second where tourism has been adopted for economic development, illustrate some of the problems that arise.

Green Island, a small sandy cay, part of the Great Barrier Reef marine park, is close to Cairns, Queensland, Australia. It is extensively used by day visitors and has residential accommodation and some basic facilities. Its natural environment has been quite badly degraded through unregulated landings by independent sea-borne visitors, overcrowding, noise from boats, erosion as a result of trampling, visual intrusion, introduction of alien plant species and pollution from inadequate sewage disposal. It is a typical example of many cases of resorts in developed countries that were customarily used by local people and subsequently became a destination for international tourists. Until recently its development was unplanned and uncontrolled. With greater awareness of its limited capacity to cater for nature-based tourism and the adverse environmental impact, redevelopment was undertaken in the early 1990s to deal with the problems cited here and to control access. This new regime has been only partially successful.

The next case represents the situation in which tourism is promoted as the major activity (Robinson, 2001), where there is a discrepancy between the intention to conserve natural environments and what actually occurs in practice. The Maldives in the Indian Ocean are low-lying islands whose economies are based on fishing and tourism. The government instituted a policy of designating and developing specific resorts for high value tourism on selected islands to minimize the natural environmental and sociocultural effects by stringent controls on the size and location of the resorts, water management and consumption of and waste disposal of non-renewable resources and the use of renewable resources. In practice, restrictions on visitor arrivals have been exceeded, over-development has occurred in inappropriate places, the living conditions provided for tourism workers have been poor, landfill of solid waste has been condoned. Residents have received limited benefits as they are not permitted on to the tourist resort islands and the distribution of the wealth and income to them has been uneven.

9.11.3 Community initiated tourism

An example where a local private initiative, initially concerned with the conservation of wildlife, has successfully combined its primary objective with tourism to help fund its operations and generate income for the local community is the Bermuda Landing Howler and Spider Monkey Project in Belize. The monkeys were under threat of extinction from forest clearance and farming that fragmented their habitat. With the help of some external guidance, village elders persuaded local farmers and inhabitants to leave corridors of trees for the monkeys to move around as they rarely come down to ground level. Basic accommodation is provided by outsiders who have settled in the area, or by residents in the village, and there is a guide and some interpretation of the programme. The project has spread into the proximity of the original site and economically supports a significant proportion of the community. A feature of the programme has been how it has educated local people to appreciate the value of the wildlife and given them a vested interest in its conservation. However, there is some discontent in the wider area as those in other villages consider that the distribution of the rewards is uneven and that Bermuda Landing receives most of them. Also, an issue that did occur, which was of concern to the local community, was the attempt by outside operators to provide tours; this was eventually resolved by the compromise of using local guides and services. Other similar initiatives related to the conservation of wildlife, the Toledo and Coxcomb Basin sanctuaries and the Manatee conservation project, exist in Belize. Being in less accessible areas of the country they are relatively low-key operations.

Costa Rica in Central America is renowned for its wealth of wildlife, particularly its plant and bird species. It has some notable ecotourism credentials through its tour operators and private nature reserves, but with economic development and the rapid growth of tourism many of its natural resources are likely to be degraded. Lumsden and Swift (1998) argued that the country faces the prospect of becoming a mass tourism market and that stronger controls over access to its national parks and wildlife reserves is necessary, adding that it is at a crossroads with respect to the future of its tourism sector. In a later comparative case study by Stem et al. (2003), it was asserted that the scale of tourism in Costa Rica has been detrimentally affecting its ecotourism reputation and that this form of tourism is no longer delivering socioeconomic and environmental benefits.

Concentrating on developing countries, Munt and Mowforth (2003) give a number of examples of how the growth of what they term 'new tourism', which threatens to be simply another form of mass tourism, has had an effect on their environments and the response by communities to the challenges they face with a desire to achieve economic development through tourism. With respect to community initiated tourism in developing countries, Hall and Richards (2000) consider tourism and sustainable community development and their study contains case studies mostly concerned with rural areas in Europe, covering Albania, Hungary, Ireland, the Netherlands, Scotland and Portugal.

9.12 Some wider implications of studies of tourism's use of renewable resources

The case studies given above offer a number of insights into both the beneficial and detrimental effects of tourism's role in the use of renewable resources, with the emphasis on natural ones, as well as those that are human-made and sociocultural. The review of a considerable number of other ecotourism studies yields several additional features of the tourism business sector's and tourists' activities that have implications for the use and conservation of resources.

There are indications that tour operators in both developed and developing countries exert considerable influence on destinations. In Asia, there has been a rapid increase in mountaineering and trekking in the Himalayas, particularly the Annapurna area of Nepal, where tour companies such as World Expeditions have had an influential presence. It endeavours to employ local people, source supplies locally and assist community run businesses. It also prescribes how tourists should behave and advises them on their use of consumables. There is a similar situation in Africa, where, for example, Conservation Corporation Africa (CCA) operates more than twenty game reserves and lodges. While it has a self-imposed strict code of practice, the intention and reality have not always coincided. The CCA has initiated triple bottom line reporting and does make contributions to conservation and community projects. However, these examples are rare and are a tiny proportion of their revenues.

There are cases of tourism development on a very large scale by single entrepreneurs that operate in destinations with fragile environments that are virtual monopolists. This is not necessarily an adverse phenomenon as this can have beneficial conservation outcomes. The exposition in environmental economics of the optimal use of resources, outlined earlier in this chapter, considers the effect of the market competitive structure. By exercising control over the price and quantity of supply, covered in Chapter 4, a monopoly reduces the quantity demanded that slows the exploitation of non-renewable resources (Tietenberg, 2006). Thus, in the context of tourism it can have a similar effect. A notable example is the Maho Bay enterprise in the Virgin Islands developed by Stanley Selengut, examined by Honey (1999). He was able to demonstrate that close control over the construction and operation by a single entity that incorporated environmentally friendly materials and practice could be commercially successful.

Many international and local conservation and environmental organizations are involved in tourism destinations that work in conjunction with central governments, local authorities, NGOs and commercial enterprises. There is however, a difference in the circumstances between developed and developing countries and in regions globally that affects of the form of conservation and the degree to which it takes place. An important factor is the customary and legal position regarding the ownership, occupation and use of land. This dimension is not pursued here as environmental issues concerning tourism and its relation to common property and open access resources have been covered in the section earlier in this chapter on market failure pertaining to public goods. There is the possibility of conflict, confusion as to responsibility and duplication over land use and conservation. In developed countries designated national park and protected areas are under public or quasi-public control. There can be conflict between commercial use, leisure activities and conservation, particularly where users, business entities and management agencies have an interest in the resources. Buckley (2003a), in a concluding chapter of his book containing case studies on ecotourism and its role, discusses many of the issues identified here regarding its contribution to conservation.

A good example of the involvement of several organizations in a developing country's tourism sector is the Annapurna area of Nepal, noted for its culture, landscape, and wide variety of fauna and flora in tropical, temperate and Alpine climate zones. Gurung and De Coursey (1994) set out the situation, indicating that the economy is a subsistence one. The area has suffered loss of forest, destabilization of its hillsides, the accumulation of litter, pollution from inadequate sanitation and the undermining of its culture as a result of haphazard and rapid tourism development. It badly needs a structure that meets the region's human development aspirations, involves the community, attains conservation targets and moves towards a more sustainable form of tourism. The central government tour operators (as noted above), local tour companies and indigenous and international NGOs are all active, giving rise to the dangers of a fragmented and uncoordinated management of tourism and the economy. A particular problem is the separating of revenues generated locally by the central government. Buckley (2003b) gives a quite detailed account of the more recent circumstances and the activities of the Annapurna Conservation Area Project to control the adverse effects of tourists' behaviour and businesses operating in the region and to generate benefits for the local communities.

9.12.1 Concluding observations on tourism's use of renewable resources

An appraisal of responsible tourism, under the ecotourism banner, with respect to its impact on the natural, human-made resources and communities in destinations is an appropriate way of indicating the tourism sector's use of renewable resources. There is a continuing debate in the tourism literature about the impact of ecotourism, the estimates of its size ranging from 8 to 20 per cent, depending on the definition adopted. The subsection on responsible tourism above has given an exposition of it and an indication of both its intended and unintended outcomes. Its potential positive and negative effects, drawing on commentaries, largely reflect those for tourism generally and will not be rehearsed here.

On balance, the recent weight of opinion, even of those who basically take a neutral or positive stance on ecotourism (see for example, Weaver 2001;

Halfpenny, 2002; Page and Dowling, 2002; Cater and Cater, 2007; and those who are sceptical or strongly critical, such as Wall (1994; 1997), Honey (1999), Ross and Wall (1999), Buckley (2003b), Wheeller, (2004), Fennell (2007), suggest that there are dangers that it will have the same impact as mass tourism, albeit on a smaller scale. A somewhat more cynical view of it is that it is merely a marketing ploy by tourism businesses and that its potential to conserve the three environments – natural, human-made and human – particularly to benefit local communities, is very much in doubt. As for attaining sustainable tourism and economic development, the prevailing view, as indicated in Chapter 8, is that this is very unlikely. Thus, from the case studies examined and the trends charted by organizations concerned with environmental issues globally, tourism's use of renewable resources appears to be as exploitive as other economic activities. This possibly means that environments will be degraded and polluted in many destinations and does not augur well for the future of the sector's ability to give tourists a rewarding holiday experience, notwithstanding the fact that tourists themselves bear some responsibility for the problems.

9.13 A postscript on the economics of environmental issues relating to the use of resources in tourism destinations

From the explanations of recycling in Figure 9.3, maximum sustainable yield in Figure 9.5 and social benefits and costs, covering the externality of pollution and degradation, especially concerning open access resources, in Figure 9.4, the links between non-renewable and renewable resources and the mitigation of pollution and waste identified in the top left of Figure 9.2 have been traced. These figures also indicate the relevance of economic analysis to sustainable development and tourism, moving environmental action away from the as yet largely ineffective energy and materials conservation and waste management policies on which firms, including some in tourism, have focused their efforts. The notion of sustainable yield and the revenue depicted in Figure 9.5 can be reinterpreted as sustainable tourism capacity in the context of open-access environments, both cultural and physical. The prescription economics offers is that the primary tourism resource base, or the parts which comprise open access resources, should be subject to management regimes that ensure that the maximum contribution to tourism and the local economy is achieved while allowing constant regeneration. Economic analysis of the sustainability of open access resources tends to reach the pessimistic conclusion that over-exploitation and degradation will occur in the context of costless access. Thus, in addition to the need for detailed information of a physical, biological and economic nature to ascertain maximum sustainability, the management of such resources entails control measures which may need to be regulatory, in the case of resources to which access cannot be controlled, rather than the monetary measures normally advocated by economists. Further consideration is given to the policy instruments that can be applied to achieve resource conservation and pollution and waste abatement in Chapter 10.

9.14 Conclusions

As indicated throughout this chapter, in examining elements of the process of economics' analysis of environmental issues depicted in Figure 9.2, the subject can play a central role at both a strategic and operational level by informing practitioners of the issues and the consequences of current economic trends and their environmental impact. At a more applied level, the subject can indicate, particularly through its analytical framework, methods of appraising tourism projects, as demonstrated above in the sections on CBA, the issues raised by market failure and the use of non-renewable and renewable resources and resulting pollution to attain environmental objectives, including its quality (Stabler, 1996a; Keane, 1997).

The subject is in a position to provide the necessary data and analysis for tourism organizations and businesses to make informed decisions, although the potential has not yet been fully realized. Economics can also make a contribution to the largely applied tourism literature, which contains few texts with rigorous economic explanations of the aspects of environmental problems and issues identified in this chapter. The dearth of texts in the general tourism literature of such analytical developments occurring in economics is due in part to the lack of communication by economists of what are often quite esoteric approaches. Notwithstanding the progress made in disseminating the economic perspective on tourism, discussed in the concluding chapter of this book, the subject has tended to focus on specific areas largely found in research reports and journal articles, although some economists have contributed general introductions to tourism economics, such as Lundgren et al. (1995), Bull (1999), and Tribe (1999) which do not consider environmental issues in detail. Another reason is that environmental concerns in tourism are largely examined by non-economists, who seldom are aware of economic theory and its applications, as argued by Burns and Holden (1995) and Holden (2005, 2007).

One key aspect of tourism which economists have analyzed and which has implications for environmental change and damage is economic development and its impact. The economic literature, examined in Chapters 6 and 7, has concentrated on estimating income and employment generation, and foreign currency earnings, which developing countries gain from international tourism, and their implications for their balance of payments (for example, Archer, 1977b, 1989; Lee, 1987; Baretje, 1988; Pearce, 1989). The studies have not considered the impact of environmental deterioration on demand and thus income, employment and currency receipts, nor have they estimated the social costs of tourism's role in economic development, either environmental or other. It is in this area that CBA can make a valuable contribution; hence the more detailed attention given to it in this chapter. Notwithstanding the emergence and development of ecological and tourism economics, there have been few economic studies directly related to the possible beneficial effects of tourism on the environment in some areas. For example, conservation of fragile ecological areas and their designation and, more importantly, maintenance of wildlife reserves, and the wider concerns relating to the sociocultural effects in destination communities have been addressed. In this context, it is clear that some conservation areas, such as African game parks, rain forests and marine environments, are sustained because they are seen as valuable tourism resources, not because of their intrinsic value.

Most economic studies of relevance have arisen from the amenity demand for or conservation of natural, mainly rural, environments and the need to attach a value to them in making resource allocation decisions. The other areas of economic analysis are the techniques for measuring costs and benefits and the policy instruments for effecting environmental improvement and their evaluation. They, too, have been developed in economics with virtually no reference to tourism and will be considered in the next chapter.

10 THE VALUATION OF RESOURCES AND ENVIRONMENTAL POLICY INSTRUMENTS

10.1 Introduction

Chapter 8 was concerned largely with the wider global environmental issues and economic impact of tourism with particular reference to sustainability. After introducing the scope and content of the microeconomic analysis of the environment, the core of Chapter 9 was to demonstrate the relevance to tourism of capital appraisal methods, the elements of market valuation, the consequences for the environment of market failure and the use of productive and energy non-renewable and renewable resources and their conservation. Of particular importance is the effect of the sector on the natural resource base.

This chapter starts from the point in Figure 9.2, presented in Chapter 9, which concerns the valuation of resources following on from the problems associated with market failure. There is almost a sequential relationship between the three main forms of market failure identified in Figure 9.2. The remainder of the figure concerns the concept of economic value, valuation methods and environmental policy instruments and their evaluation which determine environmental practice as it contributes to sustainability.

Prior to giving an exposition of the methods of valuing resources not traded in the market, the concept of total economic value is explained. An exposition and appraisal of the valuation methods are undertaken and their application to tourism assessed. Attention is then turned to the achievement of environmental goals and targets, including the mitigation of pollution, the use of resources and the associated benefits and costs. The relative advantages and disadvantages of price or market-based as opposed to regulatory instruments, are then considered. The chapter is concluded by offering some observations on the state of economic methods and environmental action in relation to tourism.

10.2 The valuation of resources

The problems arising from the existence of collective consumption or public goods, the generation of externalities and the need to consider distributional issues are founded in the need to evaluate demand in order to make decisions on the allocation and optimal use of resources. Issues associated with these problems, analysed in Chapter 9, are equally applicable in the valuation of resources considered here; in particular, the assessment of the demand for the use of resources that are un-priced, for example open access lands, should be on a basis that allows for comparison with prices in a market context. Demand evaluation, or perhaps more correctly the estimation of consumer benefits, is necessary, irrespective of whether or not demand is expressed through the market. Except under specific conditions, economics accepts, as was indicated in the examination of CBA in Chapter 9 concerning shadow prices (which will be revisited again below), that the market price does not necessarily reflect the value of a good or service. Moreover, for non-traded commodities, for which there is no market price, it is not suggested that they have a zero value. Accordingly, means have to be devised to attach value to them. There are thus two elements to estimating consumer benefits. The first is to establish what is meant by value in use and non-use. The second is to employ methods to ascertain that value.

10.2.1 Total economic value

Although the non-priced elements of the environment have been emphasized, it must not be forgotten that a number of environmental goods and services are traded in the market, where excludability, i.e. private consumption rights, can normally be exercised. In the market, the exchange value is indicated by the price at which it is traded. Nevertheless, in economics it is recognized that the use value can be greater than this price for all but the marginal consumer, on the assumption of a downward sloping demand curve from left to right, as there are many purchasers who are willing to pay a price above that which prevails in the market. This is known as consumers' surplus, which yields an aggregate use value above the price paid by individual consumers and the quantity each purchases. In practice, the concept of consumers' surplus is rarely invoked for priced goods to ascertain total user value, but it is of interest because it forms the theoretical foundation of valuation methods to establish the willingness to pay for non-priced goods.

Although environmental benefits are often unmeasured and un-priced, the true value of many collective consumption or public goods can be considered as being much greater because they are unique, so that if they are over-used an irreversible trend may be initiated which leads to their destruction, as indicated in Chapter 9. Moreover, as unique resources they cannot be reproduced. This often imparts a non-use or passive value in addition to the use benefits. Thus they have a value that transcends their exchange and use value, i.e. respectively, any price paid and the consumers' surplus. In the context of environmental and collective consumption goods, a number of non-use benefits or values have been identified which should be added to the use values to yield what is known as total economic value (TEV).

With respect to static benefits, which are generated at one point in time and arise from existing resources, such as heritage sites and public spaces, in effect from the stock of resources in their present state, it is possible to perceive the natural and human-made environment as possessing total economic value consisting of use value and non-use value.

10.2.2 Use value

Use value can be both direct and indirect. For human-made resources traded in the market, such as the occupation of an historic building, whether purchased or rented, use value would represent the direct exchange value, which may include consumers' surplus, as indicated in the first paragraph of 10.2.1. The appearance of the historic building almost certainly also confers social benefits, giving pleasure to the local community and tourists, very likely enhancing their quality of life. These social benefits constitute indirect value. Likewise, natural resources not traded in the market, for example a lake, river or amenity open space, would confer similar benefits on residents and visitors, as an un-priced positive externality of wider socio-cultural benefit. This social value does not necessarily arise from their utility in the market, as with most human-made resources in an instrumental sense, i.e. their direct use to humans. Many resources possess indirect value irrespective of that which is instrumental. These should be identified and their value estimated under a different category of use; this is examined in the next section.

10.2.3 Non-use value

Economists suggest that in addition to use value, whether direct or indirect, there are two other forms of value: option and existence or intrinsic. Here there is some confusion, as in the economics literature option value has been considered by some as relating to potential demand should people's circumstances change, for instance a higher level of income or the increased leisure time that would enable them actually to use resources that hitherto they were prevented from gaining access to. However, other commentators on total use value perceive option value as relating to a desire by the current generation to ensure that resources are passed on to future generations; this view is examined below.

Existence or intrinsic value is that which suggests resources, particularly natural ones, such as other living organisms and the physical environment, for example forests, mountains and wetlands, which are not necessarily of instrumental value to humans, have a right to exist. There is also an inference that existence value is one that relates to the wish to pass on resources intact for future generations. This is also discussed below.

a. Option value

This is the potential benefit that consumers might derive from resources. It is an expression of a willingness to pay for their preservation so that they retain the possibility of using them in the future. In this sense, option demand is a quasi-use value. It may be extended to include an option for others to enjoy the consumption of certain resources; a kind of vicarious demand, a term that is occasionally used as a form of demand in its own right. Some economists distinguish between demand by the current and future generation by designating the latter as 'bequest value'. This term has been coined to suggest the value that the present generation places on resources, where it expresses a willingness to pay for their preservation for the benefit of future generations. This, however, can be construed as a form of option demand and is viewed as such here. For example, people who may have no intention of using specific natural or human-made resources in their own lifetime might be willing to pay to prevent a forest being felled or a beautiful coastline from being developed or an historic building from being demolished.

b. Existence or intrinsic value

This is a more complex and unclear form of value in that it can be considered as unrelated to demand in that it has no instrumental value. People may have preferences for and therefore place value on the continued existence of resources, with no intention of ever using them. Therefore, the preservation of both natural and human-made resources may be advocated because it is recognized that they have intrinsic value. This is especially relevant in the case of unique human-made resources, such as historic artefacts, buildings and cities, and natural resources, for instance rare fauna and flora, and World Heritage Sites, for example the Grand Canyon, Great Barrier Reef and the Jurassic Coast in the UK. Thus, total economic value (TEV) can be summarized as:

TEV = exchange value + consumers' surplus + option value (use and non-use) + existence value.

So far, the analysis has been based on the implicit assumption that there is a given stock of resources and environments, so that the question is what value do people attach to these. However, in reality their values are continually changing, particularly as over time humans have different attitudes as to their worth. Therefore, ideally it is necessary to endeavour to observe and estimate changes in the value of resources and environments, for example how can the possibility that future generations may have different tastes be taken into account? This is an issue that has been addressed at a conceptual level, as shown in the examination of CBA in Chapter 9, but it poses acute practical problems that so far have not been resolved. In effect it is necessary to move on from the estimation of static environmental benefits and adopt methods that measure dynamic ones. Even so, these are likely to be inadequately measured using market prices because, as shown in the assessment of CBA, such prices do not reflect the true value.

Furthermore, it is no longer just a question of estimating the size of a given stock of resources or environments; it is the possibility that action may be taken which acts as a catalyst in making the stock larger. For instance, many human-made resources, such as the creation of lakes, enhancement of the landscape and improvement of the built environment, also produce indirect benefits if people are attracted to live in or visit such areas. This argument derives ultimately from a seminal contribution of Davis and Whinston (1961). While they were concerned with strategies by householders to under-invest in their properties, acting as free riders and so leading to the creation of slums, their argument can be considered in reverse. Intervention by public agencies to conserve and improve the environment will result in some owners being willing to pay to improve the quality of their properties, resulting in a rise in the values of all property and rents and land values in an area, referred to as gentrification in the UK. This argument will be considered below in the context of tourism.

10.3 The total economic value of resources

While the notion of TEV is the foundation of the estimation of the benefits derived from resources in environmental economics, its relevance to evaluating the tourism resource base has hardly crept into the literature. However, given tourism's reliance on the natural environment it is not difficult to appreciate its implications and application. With regard to its implications, TEV suggests that the resource base, which is treated largely as a free good by tourism firms, has a much greater value than has been envisaged. This should increase the importance of the environment as an input into the tourism product by raising its opportunity cost in terms of alternative uses in, for example, agriculture, industry or urban property development. It should also emphasize to governments, tourism bodies and the business sector the necessity of safeguarding this resource base.

However, it must be recognized that virtually all resources used by tourists and tourism businesses have other uses. Whether tourism complements or conflicts with these is an empirical matter on a case-by-case basis. Indeed, it is very likely that there are multiple uses for resources so that a number of competing demands need to be reconciled. Tourism may complement other uses where, for example, the development of a resort contributes to an improvement in the infrastructure, facilities and services. Conversely, in a rural area, such as a national park or nature reserve, it might conflict with environmental goals by causing erosion, trampling or disturbance of wildlife, notwithstanding the fact that it might contribute to conservation management through entry fees and on-site expenditure by visitors.

Where tourism can co-exist with other uses, the question arises as to what the level of usage ought to be to optimize the use and non-use value gained. The decision should rest on the relative value of tourism as against other uses and whether, as the level of a specified use is increased, the value derived from it starts to decline; in effect, diminishing returns set in. Basically, the choice is akin to a problem met in the real property market on the allocation of land, (see Evans, 2004) regarding the existing use value of an urban site and its alternative use value; in short the opportunity cost of the current use of the site.

In the explanation below, for simplicity, it is assumed that the total quantity of the resource available is exogenously determined and that the choice is between the expansion of tourism and one other alternative use of a resource. For land-based uses the choice could be between tourism and agriculture; this is not necessarily suggesting that each is a sole user. Figure 9.5 in the previous chapter can be adapted to demonstrate in Figure 10.1 the point where the value generated by tourism and the alternative use of a resource is optimized. The horizontal axis of the figure indicates the total availability and capacity of a resource. Here, as an example, it is instructive to consider a marine resource, reflecting the case study given in Chapter 9 on the seas adjoining the Baja California peninsular in Mexico. The figure shows the tourism marginal economic value (TMEV), denoted by OPM, moving from left to right, consisting of such activities as catch and release, non-consumptive fishing and other water-based recreational activities; the shape of the curve largely being determined by the circumstance explained in relation to Figure 9.5. For the alternative use, the commercial marginal economic value (CMEV) curve MPN, similarly shaped to that for tourism, runs from right to left. The use of the resource is optimized at PQ, where tourism's use is OQ and commercial fishing's is MQ. This is signified by the fact that at point PQ the value of tourism is greater than commercial fishing. Clearly, from PQ towards M the value of commercial fishing is greater.

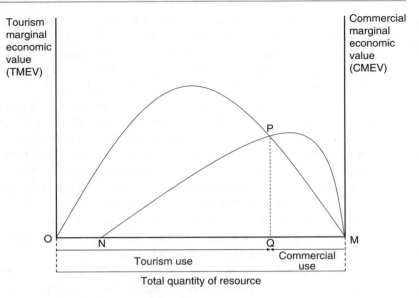

Figure 10.1 The optimal use of a resource

The decline in the value of tourism can occur because of a number of factors, such as overcrowding, disturbance, pollution and visual intrusion, that lead to a decline in the number of visits and thus the tourism value generated by the resource. The reason for the decline of the value of commercial fishing is explained in the next paragraph.

The total value of tourism is the area under the curve OPM and for commercial use it is the area under the curve MPN. The curve for tourism is based on the assumption that tourism is likely to be compatible with other uses, including in the example commercial fishing; it may also possibly contribute to the conservation of the resource and its wildlife. Hence its total economic value is shown as being larger. Conversely, the commercial total economic value, on the supposition that it is solely fishing, is lower, the value becoming zero at N before it is the only user of the resource. This accords with the maximum sustainable yield thesis, given in Chapter 9, that the stock is unable to reproduce itself when it falls below a critical mass and thus a viable recruitment level. What total value is for each use is an empirical question; Young (2003) indicates in the Baja California case that the value generated by tourism is greater than commercial use.

It can be posited that the respective curves are the net marginal economic values (NMEV) after taking account of their use (exchange and consumers' surplus) and non-use (option, embodying existence and bequest) values, examined above in the exposition of TEV, which is akin to marginal private benefit (MPB). Also, it is possible to conceive that marginal social benefit (MSB) has been added and the marginal private cost (MPC) and marginal social cost (MSC) deducted, as shown in Figure 9.4.

The application of TEV to tourism, bringing out the distinction between static and dynamic benefits, is demonstrated by an example of the role of heritage conservation and tourism in urban economies, particularly the regeneration of post-industrial towns and cities. Attention is concentrated on the built environment. Cities, it is argued, have ceased to have an important function in the direct production and distribution of goods. Rather they are reverting to the functions that they had before the industrial revolution as sites for administrative services, businesses and commercial activity, as generators of cultural services and, increasingly, as the providers of urban amenities, both enhancing the quality of life of their residents and providing the economic base for urban tourism and other leisure-oriented service activities. The key difference with the pre-industrial era is that the set of activities identified above now contributes a growing part of urban economic output and welfare.

The likelihood of an urban area succeeding in changing its function in this way depends, in part, on, for example, its location, culture, legacy of historic buildings, conservation areas and open spaces. If the city or town can both make the best use of its inherited assets and change its function, intervention will assist the regeneration of the local economy. By investing in conservation policies for architecturally significant buildings or areas, public agencies may encourage owners of adjoining properties to upgrade them. People with new skills can be encouraged to move into those areas or skilled people persuaded to remain there. This could lead to new businesses being located in the conserved buildings/areas and, in turn, facilitate the emergence of new business opportunities, additional spending and engender yet more new business creation. In effect, the implication is that public spending fills a gap that the market, left to itself, would not; see for example Davis and Whinston's (1961) 'neighbourhood effect'. The initial public investment also acts to give leverage to further private investment, both in upgrading other buildings in the area and in the creation of new businesses or the retention of businesses that would otherwise have failed or left. However, this is only the medium-term effect. The argument can be extended to the longer-term process of cumulative upgrading as the import of new skills reinforces the base of local entrepreneurial activity and improves the image of the area. Such a process is akin to multiplier-accelerator analysis, considered in Chapter 6, which has been extensively considered in relation to regional development (Richardson, 1972) and investigated with respect to the role of specific economic activities, particularly tourism (for example, Sinclair and Sutcliffe, 1988a; Archer, 1989; and Johnson and Thomas, 1990).

Stabler (1995, 1996b, 1998) and Allinson et al. (1996), in a report for English Heritage, make a distinction between static and dynamic benefits in urban environmental improvements, that may well be prompted by the desire to expand tourism, or introduce it in a process of local economic development. The static benefits of the impact of an improved urban environment can be measured by means of 'one-off' studies, employing CBA as an appraisal method, reflecting the TEV concept by estimating property values and/or willingness to pay to live or work in or visit a city or town (the valuation methods are reviewed in the next section). Conversely, the assessment of the dynamic benefits has to be attempted by observing and evaluating the process of change, which is more difficult given the techniques derived so far in environmental economics. It therefore constitutes an issue that needs to be addressed if the current static benefit analysis is to be developed to accommodate the dynamic nature of economic activity and the consequent variation in benefits, over time. Moreover, it is imperative that in addition to private net benefits, net social benefits are measured and included, i.e. the gross social benefits minus the social costs. Initially, a case-study approach seems feasible to derive general principles that would include the role of tourism as an economic activity, in which its net social benefits are identified and assessed. Of course, it is conceivable that the sector's social costs outweigh its social benefits.

Clearly, the identification and enhancement of static and dynamic benefits are not confined to an urban infrastructure. They apply equally in rural, mountain and marine contexts where tourism may be the trigger for the improvement of the physical, economic and sociocultural environments. The measurement of static and dynamic benefits will be discussed further in this chapter after the appraisal of environmental economic valuation methods.

10.4 Valuation methods

In the development of the analysis of the amenity and pollution aspects of environmental economics, much attention has been paid to their non-traded nature. The economics literature addressing the issue of how to value environmental deterioration or improvement of the amount consumers would be willing to pay to preserve, say, an unspoilt beach, forest, historic building or undeveloped landscape of high scenic value, is vast. The main methods which have been developed to place a value on environmental attributes and which could be applied to the valuation of non-priced tourism resources are:

- contingent valuation (CVM)
- choice modelling (CM)
- hedonic pricing (HPM)
- travel cost (TCM)
- combinations of these

Other useful methods, where detrimental impacts occur, are those rooted in household production function analysis such as avoided costs and estimation of the opportunity cost, provision or replacement cost.

The methods identified above can be classified according to whether they seek to place a value on the good or attribute, by directly asking respondents their willingness to pay for an improvement or their willingness to accept a degradation, or indirectly, by using prices from a related market which does exist. Contingent valuation is an example of the former, while hedonic pricing and travel cost methods are representative of the latter. In addition, there are more qualitative approaches, such as the Delphi technique, a direct method, which might be employed. The box in Figure 9.2 in the previous chapter contains a summary on valuation methods and approaches that simply measure participation as a basis for possible valuation. Only the methods that yield monetary valuations are examined in this chapter and reproduced here. It should be noted that choice modelling, given its affinity with CVM and its increasing use as a component of the latter, would be considered as a stated preference approach. The HPM and TCM are revealed preference approaches, as indicated in the summary below.

The methods are eminently suitable for valuing the benefits, and sometimes costs, generated in proposed projects, particularly where they are indirect and intangible ones. Thus, the methods now constitute the means of valuing these in the CBA capital appraisal approach, as indicated in its examination in Chapter 9.

10.4.1 Monetary valuation of resources methods

Non-demand	Demand	
	Direct	*Indirect*
Avoided cost	**Direct**	**Indirect**
Delphi	(Stated preference)	(Revealed preference)
Opportunity cost	CVM	HPM
Preservation cost	CM	TCM
Replacement cost		

The four recognized in environmental economics as most suitable and therefore the commonly applied methods, the CVM, CM, HPM and TCM, are appraised first as being most relevant to the valuation of tourism resources. As shown in the summary above, the CVM and CM are direct approaches; the latter has been used in conjunction with CVM but increasingly is now applied independently. Indirect techniques are based on the concept of consumers' surplus in that, whether or not a price is paid for a specific resource, a measure of the net benefit can be estimated by proxy for what are goods and services for which there are no markets. Here, as indicated above, attention is concentrated on the HPM and TCM as the principal methods applied in the economics literature.

10.4.2 The contingent valuation method

This method has been continuously developing since the 1970s and there is an extensive literature on the method and its application in the environmental field. There are innumerable illustrative case studies concerning the conservation and the leisure use of natural and human-made resources that are of relevance to the impact tourists and tourism businesses have on such resources. It is only possible to give a broad overview and evaluation of the method here; several references are identified below that offer detailed and quite rigorous expositions of it.

The CVM directly questions consumers on their stated preferences in two possible approaches. The first seeks their willingness to pay (WTP) for the creation of an environment or an improvement in its quality. The second is to ascertain the willingness to accept (WTA) compensation for a fall in the quality of the environment or to be prevented from gaining access to it or suffer its complete loss. Since respondents are questioned directly, it is possible to ask them whether they would be willing to pay, for example, to preserve a rural informal recreational site which they might use or even a tropical rain forest of which they are not users, and not likely to be. Thus an advantage of the method, giving higher values than others examined below, is that it is possible to obtain, at least in principle, non-use option and existence valuations as well as use values, i.e. the TEV as identified above. A relatively early standard text explaining the method is that by Mitchell and Carson (1989), but more recent ones are Hanley and Spash (1993) and Perman et al. (2003). Significant contributions to the development in the method were made in the 1980s by Brookshire et al. (1983), Desvouges et al. (1983), Hanley (1988) and Heberlein and Bishop (1986).

- The CVM is a survey-based methodology with responses being obtained either by a postal questionnaire or by interview on the site of the project or in the home or by recruiting a panel of respondents. Sometimes such a panel is used as a pre-test for the main sample survey. The first stage is to establish the size and nature of the survey, especially if it is to be by sampling a specified section of society off-site, to ascertain who is affected, and how, in terms of their utility.
- The second stage is to ensure that respondents have sufficient knowledge of the proposed action and its likely impact on the environment and themselves. A hypothetical reason for payment or compensation in the eyes of consumers is set up. For example, respondents may be told that the government is considering clearing and improving a nearby

former industrial site for a leisure development, but only if additional funds are raised.

- The third stage is to inform respondents of how much additional revenue is required and how the scheme would be paid for. This is referred to as the 'bid vehicle', for example a local income tax or an entry fee.

- The fourth stage is to question individuals, by the various means which have been developed within the method, on their maximum WTP to ensure the scheme goes ahead, or on their minimum WTA compensation for loss of the resource or artefact. This can be done by open ended questions, or an initial suggested sum or bid card that contains a number of payments. Increasingly, however, dichotomous bidding, also called a referendum model, is being applied, whereby respondents are asked to answer yes or no to a given amount; such questioning can be repeated until a given WTP is agreed.

- The fifth stage is to analyse the responses to estimate the average and/ or the maximum WTP and WTA, also to ascertain how feasible the estimates are. The conduct of the exercise should be evaluated by examining the socio-economic profile of respondents, such as their educational level and income, whether they fully understood the purpose of the proposal, what problems and biases (see below) were identified and how were they resolved. Also, a comparison could be made with the evidence from other similar investigations.

- Finally, an estimation is made of a bid curve involving a regression of the WTP/WTA on a range of explanatory variables which are thought to affect the bid and on which information has been elicited in the survey, for example: incomes that might bear on their travel experiences; visits to and knowledge of specific sites; educational levels that might determine their knowledge of environmental resources and issues connected to them; social class; age; gender and residential location. The estimate in the sample is then aggregated by multiplying the mean or median by the population (N) representative of the sample (n) in the survey.

10.4.3 An assessment of the CVM

Dwelling a little longer on the explanation of CVM, it is instructive to consider a number of features that have been analysed in its development that could give rise to potential discrepancies in estimates of WTP and WTA and the validity of the method. In the literature a distinction is often made between those that are often referred to as biases and other issues with the method that could occur. An extensive review of the many biases or errors which may be inherent in CVM studies is given in Garrod and Willis (1990).

Biases

- Starting point bias relates to the base price or level of compensation respondents may be given respectively in WTP and WTA bids. This can occur where the investigator offers an opening bid to the respondent. As indicated above in stage four, there are alternative techniques for avoiding this problem, such as the dichotomous bid approach.

- Strategic bid bias is possible where a respondent overstates the WTP in the belief that the proposal is simply hypothetical or will go ahead if the bids are high enough and there is no suggestion of payment having to be made. Conversely, an under bid is likely to be made if it is perceived that the project will be undertaken and that payment will have to be made; in this case, the respondent is essentially a free rider. The bias can be overcome if it is posited that all respondents pay the average bid.

- Payment vehicle bias can occur over the form of payment; for example, the respondents' attitude to a direct one, such as an entry fee, or indirect one like a tax or contribution to a capital fund.

- Hypothetical bias is possible where respondents' valuations in the survey differ from those in a real situation. This may be because they have difficulty in appreciating accurately the choices being presented to them. The bias may be compounded by insufficient information being given on the proposal or that it influences the bids made. A misvaluation may occur if the respondent finds the circumstances difficult to relate to or simply fails to take the survey seriously, therefore giving unconsidered answers. The researchers thus need to make the issue and means of dealing with it as realistic as possible.

- Misspecification bias is where, in the first two stages, the perceptions of the investigator and the respondents are dynamically opposed.

- Mental account bias relates to the proportion of income or wealth that respondents aim to devote to the consumption of environmental goods and services and bids for individual projects that when aggregated exceed their total planned allocation. It is conceivable for a single bid in a CVM investigation to exhaust their entire budget where they feel very strongly about a given environmental issue. Willis and Garrod (1990, 1991b, 1991c) have extensively examined this bias.

- When a respondent appears unable to distinguish between, on the one hand, a WTA payment to be denied one visit to a site and, on the other hand, a WTA payment for permanent removal of the possibility of ever visiting the site, then what is termed the temporal embedding problem arises. Discrepancies between WTP and WTA are likely because of respondents' perception of the implications of each. It is possible that the WTP is seen as a payment for each occasion when a resource is used over a given time period, whereas WTA is perceived as forgoing for ever the existence of a resource, in which case the compensation for its loss is much higher. It is therefore necessary to try to establish the total environmental perception of individuals.

- The WTP and WTA may also be affected by a number of socio-economic variables. For example, wealthy, high income and educational attainment respondents, who are more environmentally aware, are likely to make higher bids for projects or the preservation of environmental resources.

10.4.4 Other issues associated with CVM and an appraisal

These are mostly connected with the conduct, analysis and interpretation of CVM investigations and thus their reliability and content, construct and criterion validity. It is a fundamental consideration of all surveys that they should be both reliable, in that they should yield consistent results when replicated, and valid, meaning that they should be representative of circumstances in the real world. Attention here is concentrated on the validity of CVM, largely reflecting the exposition in Hanley and Spash (1993) and Perman et al. (2003).

The validity of CVM depends crucially on the construction of the hypothetical market being credible and realistic, in which sufficient information is given to respondents and they have knowledge of the resource and the issues relating to it. The design of the survey should be such that the sample is truly representative of the relevant population and its size is large enough to be statistically robust. Two key elements of the method that are viewed as central are to place emphasis on WTP, rather than WTA, and the means of payment. Every effort should be made to reduce or eliminate the biases identified above. The estimates emerging should meet the expected results when the study is instigated.

The CVM is considered by administrators and politicians as a democratic approach to establishing people's attitudes and views on environmental and resources use issues; the findings of studies have certainly influenced action, particularly in the US. However, economists have been rather more cautious about it because of concerns over its inherent hypothetical nature. They tend to favour revealed preference methods. A review in the US by the NOAA (1993), albeit in the relatively early development of CVM, identified a number of reservations, but these were essentially positive in making recommendation on its use.

10.4.5 Applications of the CVM

A rapid growth in the number of applications of some form of CVM has taken place since the mid-1980s. This may be partly attributed to the potential ability of the method to value option and existence values in ascertaining TEV and the feasibility of it in forming policy. The majority of studies have been on the use and need for the conservation of natural environments, the existence and quality of which are under threat from commercial development, or overuse and degradation if they are amenity resources. It can be argued that few such resources can be considered as exclusively for tourism. Effectively, all tourists are indulging in recreational activities once they are visitors in a destination. Moreover, publications that appear in the tourism literature are as much about the relevance of the CVM method as the resources examined as case studies. An illustrative example is that by Choong-Ki Lee and Sang-Yoel Han (2002).

In the UK, the option value to preserve existing landscape was found to be between 10 and 20 per cent of the total site valuation (Willis, 1989; Bateman et al. 1994). Lockwood et al. (1993) used a CVM survey in order to assess WTP to preserve national parks in Victoria, Australia. The survey highlighted the relative importance of existence and bequest values, which constituted 35 per cent and 36 per cent of the total valuation respectively. Non-use value was found to be over three times that of use value in a survey of the Somerset Levels and Moors Environmentally Sensitive Area (ESA) scheme (Garrod et al, 1994). Other related areas of application of contingent valuation have included improved park facilities (Combs et al., 1993), a ban on the burning of straw (Hanley, 1988), the benefits of canals (Willis and Garrod, 1993a), forestry characteristics (Hanley and Ruffell, 1992; 1993), wildlife (Willis and Garrod, 1993b), the value of elephants (Brown and Henry, 1989) and tourism-related traffic congestion (Lindberg and Johnson, 1997). In a paper related to the evaluation of urban resources in the UK, Willis et al. (1993) assessed the usefulness of CVM for estimating willingness to pay to gain access to historic buildings. They considered CVM to be a constant, robust and efficient estimator of WTP, although non-use value was not measured in their particular study.

10.4.6 Choice modelling

This method is related to CVM in that it is a direct approach that elicits stated preferences of WTP or WTA. It can be incorporated into CVM, but perhaps can be more correctly viewed as a development of it (see Bennett and Blamey, 2001) that contains an extensive representative list of references. Whereas CVM gives one alternative to the status quo, CM is designed to allow for a range of choice responses. It is related to the dichotomous choice technique in CVM and is also similar to the characteristics demand theory (Lancaster, 1966), an exposition of which has been given in Chapters 2 and 3. It is gaining favour because it allows for many characteristics of the environmental issue being examined to be explored. As yet it has not been widely applied in the tourism field, although Hanley et al. (2001), in a review of the method, includes illustrative cases. It has the same drawbacks as CVM in survey design and operation, a particular problem being the difficulty of respondents to decide on a choice of the several offered.

10.4.7 Hedonic pricing method

The hedonic pricing method, hereafter referred to as HPM, is an indirect one of eliciting valuations from consumers by considering their revealed consumption preferences in related markets. It was developed by Rosen (1974), and is based on the characteristics consumer theory of Lancaster (1966), referred to in explaining the CM approach immediately above and earlier in Chapters 2 and 3 with respect to tourism demand. The method can be used to estimate the value of un-priced characteristics of goods and services. It aims to determine the relationship between the attributes of a good and its price and is arguably the most theoretically rigorous of the valuation methods.

It takes as its starting point that any differentiated product can be viewed as a bundle of characteristics, each with its own implicit or shadow price, for example in the case of residential property, where it has been widely applied. One early area of research has been the relationship between property values and air quality (Anderson and Crocker, 1971). Another example is that by Cheshire and Sheppard (1995), who investigated how the value of location-specific attributes is capitalized into land prices if they are not included as independent variables. Among the variables they included were local amenities provided through the land-use planning system. The characteristics may be structural, such as the number of bedrooms, size of plot, presence or absence of a garage. They can also include environmental characteristics, for instance air quality, the presence of views, noise levels, crime rate and proximity to shops or schools. Accordingly, the difference in price between two houses, identical in every respect except for one, should accurately reflect consumers' valuation of that characteristic.

This holds implications for tourism as it might explain how the facilities and services related to it attract people to move into an area. Likewise, it is possible to attribute the impact of the designation of sites as amenity or conservation areas, which attract tourists to historic cities and towns, by observing the difference in value between two identical sites, one of which is regarded as a conservation area and the other which is not. Conversely, negative features of the sector, such as congestion, noise, pollution, crime rates and pressure on facilities, could act as a deterrent to location in a tourism destination and depress property and land values. Thus, the price of a given resource, for example the destination environment, can be viewed as the sum of the shadow prices of its characteristics whether they are positive or negative.

Hanley and Spash (1993), following Freeman (1979) on the valuation of non-market goods, Hanley et al. (2001) and Perlman et al. (2003) offer reasonably accessible explanations of the HPM. The feasibility of the method depends on being able to relate variables reflecting environmental quality to the price of a good, which empirically may in turn rest on the availability of the requisite data.

The HPM involves two stages, the first being to estimate the price/value of a good, say hotels in a seaside resort, by applying explanatory variables such as the property and site characteristics, e.g. number and size of bedrooms; facilities (bars, restaurants, leisure services, meeting rooms); car park and garden. Neighbourhood characteristics reflecting environmental quality would also influence the value, e.g proximity to the sea, parks and attractions, air, pollution, noise and crime levels. This hedonic price equation allows for implicit prices for each characteristic to be estimated. One or more environmental explanatory variables from stage one can be incorporated into stage two of the HPM procedure to estimate a demand curve for the value of environmental quality or resources or services. The area under the demand curve can be measured as an estimate of their total value.

10.4.8 An assessment of HPM

Unlike the hypothetical basis of CVM, the HPM as a revealed preference approach is founded in actual consumer behaviour and is considered as credible by economists, given its close relationship with the widely acknowledged contribution of the Gorman–Lancaster theory to the analysis of demand reviewed in Chapters 2 and 3. Useful explanations of the method can be found in Hanley and Spash (1993) and Palmquist (1999); the latter contains a review of it. However, HPM does rely heavily on the weak separability utility function of consumers and the assumption that the demand for goods embodying environmental attributes are complementary, meaning that any change in environmental factors will have an impact on property values. It is also assumed that the property and land markets are in equilibrium and that consumers aim to maximize utility, possessing perfect information on environmental attributes in all locations. Another assumption is that existing environmental attributes determine property values, but it is likely that expected changes could exert an influence. It is possible that environmental variables that are interrelated with others in their effect on property values are omitted and so lead to inaccurate estimates. The interrelationship of variables creates the statistical problem of multicollinearity. Lastly, the functional form selected will determine the shape of the demand curve and the area under it; accordingly, different functions will result in variations in estimates of the values obtained. The method does not capture non-use values, so that it is not possible to estimate TEV.

10.4.9 Applications of HPM related to tourism

Most studies using the HPM have been concerned with the impact of amenity resources, such as those which have considered the effect of the proximity of environmental and neighbourhood variables on residential property prices, for example a forest or nature reserve or water courses. They are relevant to tourism in that the methodology may also be used to estimate the effect of such variables on holiday prices (see Chapter 3) as well as identifying and measuring the environmental impact of tourism. Davis (1964) was a pioneer in applying the method to big game hunting in a forest. Garrod and Willis (1991a, 1991b, 1991c) found that the presence of forestry might have a large positive impact. A similar result is observed in the case of location near waterways. Willis and Garrod (1993a) estimated that the presence of a canal or river raises the value of a property by an average of 4.9 per cent, while the proximity of at least 20 per cent woodland cover raises it by 7.1 per cent above that of an identical property without these features. More recent examples of the application are by Pendleton (1999) and Pearson et al. (2002).

There also exists a significant body of research, with possible relevance to tourism in historic urban areas, into the impact of architectural style and historic zone designation on property valuation. Asabere et al. (1989), for example, showed that architectural style has a strong impact on residential property valuation in their sample of 500 properties sold in Newport, Massachusetts, USA, between 1983 and 1985, with older styles of architecture commanding premiums of around 20 per cent. Moorhouse and Smith (1994), in their study of nineteenth-century terraced houses in Boston, also found that individuality of any style commanded a higher price. Hough and Kratz (1983) argued that architecture has certain public good characteristics which may be undervalued in the market, while Ford (1989) evaluated the effect of Historic District designation on the prices of properties sold in Baltimore, Maryland, USA, between 1980 and 1985. Designation was found to have a positive but insignificant impact. This result was corroborated by Asabere et al. (1989) and also by Schaeffer and Millerick (1991) in their study of the prices of 252 properties, prior to and after designation in Chicago.

A related area of study in which the HPM has been used is in research on the pricing of package holidays where environmental characteristics are part of the tourism package. The price competitiveness of package holidays cannot be compared directly because of differences in the characteristics of the packages supplied. However, the HPM can be used to estimate the price differences, which are due to variations in the mixes of characteristics. For instance, in the case of holidays in the Spanish province of Malaga offered by UK tour operators, the characteristics included the categories of hotels in which accommodation was supplied, the hotel facilities and locations and the tourist resorts themselves, all of which affected the prices of the holidays (Sinclair et al. 1990). The HPM has also been used to compare the price competitiveness of package holidays in cities supplied by tour operators from different countries (Clewer et al. 1992).

10.4.10 Travel cost method

The travel cost method (TCM) was developed by Clawson and Knetsch (1966) following an initial suggestion of the plausibility of the technique by Hotelling (1949). The method is based on the premise that the costs of using recreational sites or tourist areas can be used as a proxy measure of visitors' willingness to pay and thus their valuation of those sites. It has been applied to the demand for boating, fishing, forest visits, hunting and even international tourism. Even if visitors do not pay to gain entry to a site, they have incurred expenditure either implicitly or explicitly in travelling to it, which could be used as a measure of (or at least a lower bound to) their valuation of that site. Travel and on-site time can be perceived as an implicit cost, while explicit costs include travel, petrol or public transport fares, entry charges and expenditure on site. Whether on-site and travelling time should be incorporated into the estimate of total cost is a point of debate in the literature (Smith and Desvouges, 1986; Chevas et al. 1989). For sites to which the majority of visitors walk, valuing the time they take is the only measure which can feasibly be used (Harrison and Stabler, 1981). This might suggest that on-site time should be the true focus of the debate. If it is decided to include on-site travelling time it may be difficult to assign a value to it. The opportunity cost, in terms of forgone earnings or leisure time which could have been spent doing something else, might be estimated. This aspect has been addressed in Chapters 2 and 3 in considering the income leisure trade-off in the tourism demand.

Studies employing the travel cost method almost exclusively consist of those for visits to rural recreational sites such as important areas of high scenic value, forests, lakes, mountains or watercourses. These applications can be categorized according to whether they use an individual, zonal or hedonic formulation of the travel cost model, or a combination of all three.

10.4.11 Stages in the TCM process

- Conduct a survey on site of visitors, or in the home if investigating potential visits to given sites. An initial survey as a pre-test is desirable as it informs stage two, particularly the identification of the catchment area as a basis for distance travelled, which also contributes to the derivation of the trip generation function.

- Establishing the catchment area of the site is important, as it is the basis for establishing the furthest distance visitors are likely to travel to it; an issue that can arise is the point where outlier observations are omitted from the analysis, as they have an impact on the estimation of the demand curve considered below. Outliers also are a factor in the calculation of visitor rates for a given population in the zonal TCM, explained below.

- Derive a trip generation function (TGF). Its purpose is to estimate how many visits will occur, but it also is the stage when likely explanatory variables can be considered, in addition to those identified in the introductory paragraph above, such as mode of transport, size and composition of the visitors' party; resident or tourist; wealth; income; occupation; educational level; age; class; nature of the visit, i.e. sole purpose, or in transit, or meandering on a pleasure outing. The TGF, where the dependent variable is the number or rate of visits, can take two forms:
  - trips generated by individuals;
  - trips on a zonal basis; as a visit rate for a given population size, e.g. rate per 1,000 population in each zone, covering the catchment area around the site. Zones can be concentric rings (Cheshire and Stabler, 1976), or postal districts or local authority areas, or rectangular in the case of linear sites, for example a canal (Harrison and Stabler, 1981); a problem with the last is the many access points it has. This is also the case with very large sites, for example national parks.

  There is no clear indication that either of the TGFs is superior to the other, nor is there any widely agreed approach; for a review of them see Willis and Garrod (1991a).

- Estimate the demand curve for the site. The demand curve derived indicates that as travel costs increase (the vertical axis), the numbers of individuals, or visit rate in the case of the zonal approach, will fall (horizontal axis). The functional form of the demand curve should be such as to ensure an intersection of the curve with the travel costs, visit numbers or rate axes.

- The final stage is to calculate the total value of the site by measuring the area under the demand curve. This yields the consumers' surplus for a single visit. The estimate so obtained can be aggregated to provide a total value for a given period. This, however, is problematic as use of a site almost certainly varies over the week, higher visit rates occurring at weekends and seasonally. This suggests that a series of surveys need to be considered to take account of these factors.

10.4.12 An assessment of TCM

The use of TCM raises many conceptual, analytical and operational issues. Here it has only been possible to give an overview. There are, however, several publications that give useful introductory reviews and explanations of the method. Hanley and Spash (1993), offer an accessible and clear exposition related to its incorporation into CBA in the environmental field. Perman et al. (2003) is a more rigorous approach in which appendices in the relevant chapter deal with the technicalities of the method. Bockstael (1995) and Kling and Crocker (1999) are reviews of the application of TCM.

- Multipurpose trips raise the issue as to what value should be assigned to the specific site being examined. Cheshire and Stabler (1976) made the distinction between sole, transit and meandering visitors identified above. For the first, the entire cost of the visit was attached to the site, while for meanderers information on the number of sites visited on the day was sought and a portion of the total travel costs allocated between them. The issue is whether the value placed on each site by respondents should be equal; if not, one solution is to ask them to offer a score on the value attached to each site, so that a weighting scheme can be applied. The transit visitors, who were in a minority, presented a problem and eventually it was decided to remove their responses from the estimation of the value of the site.

- The value of time has been a source of dispute in the literature, not so much as to whether or not it should be included; that has been widely accepted, particularly where visitors incur no monetary travel costs, as Harrison and Stabler (1981) found with regard to informal recreational visits to canals in the UK. The problem is what value should be included for time; currently the issue has not been fully resolved. In the canal study, a sensitivity analysis was applied taking a number of fractions of an hourly average earnings rate secured from income information sought from respondents.

- The mathematical function chosen to derive the TCM demand curve is a similar issue to that found with HPM. The need to obtain an intersection with the axes and the position and shape of the curve determined by the function will in turn affect the area under the curve and thus the estimate of consumers' surplus. The truncation to achieve an intersection may be arbitrarily decided on when the curve suggested by the data is asymptotic to the axes. The functional forms most frequently employed are linear and semi-log and log-log. The characteristics and appraisal of the functional forms have been examined by Common (1973) and Hanley (1989).

- Quality changes can occur where a site is degraded, which reduces the number of visitors, or is improved, for example the provision of new facilities, or enhancement of its natural environment, which increases the number of visitors. Such changes imply that the site should be re-surveyed and the change in demand estimated.
- Competing sites can have an influence on the number of visitors, especially if they are in close proximity to the survey site and their quality changes. Also, if the other sites give rise to a meandering type of visitor, necessitating a greater allocation of their travel costs across each site, the estimates of value will be lower.
- The visits of tourists are perceived by some researchers as a distorting factor on the distance travelled and therefore travel costs. This can be resolved by obtaining the location of departure on the day of the visit, treating them as temporary residents. Another recourse is to apply the adjustment for multiple trips and competing sites by apportioning their travel costs from their usual residence over all the sites in the location where they are staying. However, overseas visitors present a more acute problem on which there is no consensus as to how to treat their travel costs; one solution is to remove them from the survey.
- Whether to apply marginal or total costs of travel is an issue where visitors arrive by car. The problem is analogous to the tourists' case above. The majority of researchers have used marginal costs as they are perceived as reflecting the marginal utility theory, in which the marginal cost of consumption is taken into account.

The TCM effectively only measures the consumption values of visitors at one point in time; it does not capture the option or non-use values, so does not estimate TEV. With respect to the value of time, it is possible that the income–leisure trade-off is not conceptually acceptable, it being argued that the opportunity cost is that relating to another leisure activity.

10.4.13 The TCM and CVM and HPM approaches

The CVM and HPM methods have an affinity with the TCM in certain circumstances. For example, where it is planned to improve the environmental quality of a site for which there is an entry fee that is likely to adversely affect the visit rate of a similar un-priced one. In such a case it is of interest to the management of the latter, by means of a CVM approach, what the WTP for both sites will be after the improvement has been completed. While the linkage of CVM and TCM has not been pursued, that of HPM/TCM certainly has. There are situations in which the characteristics of two sites are a factor in determining the number of visits. Where two sites are under consideration that are identical except for one important environmental attribute, the difference in visit rates and therefore the estimate of the total value of each can be explained by the value of this one attribute. Brown and Mendelsohn (1984) developed the approach in a study of the demand for freshwater fishing where characteristics such as landscape, level of visits and the quality of the fishing were perceived as significant. The drawback of the combination of HPM and TCM is that it suffers from the problems associated with both.

10.4.14 Applications of TCM

There were many applications of the TCM over the last quarter of the twentieth century. Englin and Mendelsohn (1991), for example, estimated the value of alterations in the quality of forest sites using a hedonic pricing travel cost model (HPTCM). They found that some site attributes, such as dirt tracks and alpine fir trees, had saturation levels below which they are an economic good but above which they are a bad. Hanley and Ruffell (1992) used the HPTCM to evaluate consumer surplus across different types of forestry, each with different physical characteristics. The study showed a strong relationship between visits per year and the mean height of trees, reason for visit, length of stay and importance of visit, while forest characteristics were insignificant. In a study of canals (Willis et al. 1990) and another of botanic gardens (Garrod, et al. 1991), consumer surplus valuations were estimated to be significantly lower than the financial operating loss. Applications to the valuation of the quality of fishing (Smith et al. 1991) and deer-hunting sites (Loomis et al. 1991) have also been considered. Very few studies using TCM have been conducted to evaluate urban resources, two principal reasons being the problems of separately identifying the benefits generated by specific resources where there are many, and the number of different reasons those travelling to cities and towns have for their visits.

An interesting departure from most TCM applications has been that by Maddison (2001), who developed the method to predict the possible effects of climate and changes in it on the visitation rates of British tourists to beach holiday resorts abroad. In doing so, he extended the variables describing the attributes of destinations to include beach length, hours of sunshine, precipitation and temperature, using a pooled travel cost model (PTCM). He discussed the shortcomings of both the model, concerning the handling of price and quality change, and data, suggesting that the inclusion of socio-economic variables, such as age, class, education, income and security at resorts, should be incorporated. Predictions were made of the changes in visit rates and consumers' surplus of the changes in climate variables. The results, which Maddison considered were probationary because of the shortcomings he identified in the study, confirmed that the pattern of visit rates would alter for selected resorts, suggesting that climate variables were indeed significant determinants of destination choice.

10.5 Other methods of valuation

The CV, HP and TC methods account for the vast majority of practical applications. However, reference should be made to other methods of valuation as they appear in the economics, leisure and tourism literature. In the environmental field, the methods outlined below have been applied mostly to investigate the effect of pollution mitigation, but they are equally relevant to measuring the effect on environmental quality of economic activity. In this respect they can be applied to the tourism product base, particularly natural resources, by indicating the costs of the adverse impact of the sector's operations. Production function approaches are indirect methods of valuing non-priced goods and services.

10.5.1 Production function methods

The basis of the production function methods (PFM) is that firms, households and individuals combine factors of production and commodities respectively with environmental services in order to produce other goods and services, for example their holidays. The approach is based on the Cobb–Douglas production function: $Q = f(L+K)$, where L is labour and K is capital, the function being adapted to include an environmental variable (E), perhaps related to its quality, so that if L and K are held constant, the effect of a change in E can be estimated. A complication, reflecting the complexity of demand for leisure pursuits and tourism which involves several components, some of which may have to be combined sequentially and others simultaneously, is the incorporation of the value of time in household production functions (HPFs). This acknowledges, as indicated above, the trade-off between work and leisure and also different forms of leisure. Within an HPF the value of time can be formalized as a basis for estimates of the value of time in the TCM approach to resources valuation.

The method is potentially useful as it allows 'before' and 'after' situations to be assessed so that if the quality or quantity of an environmental attribute changes, agents change their expenditure patterns in order to take account of this change in the environmental attribute. The problem with PFM is that it has been difficult to make it operational in what might be termed its pure form. Many authors (for example, Bateman et al. 1992) argue that PFM underpins other approaches, such as dose response and avoided cost (sometimes referred to as averted expenditure), as well as TCM.

10.5.2 Dose-response function

This method is another based on the Cobb–Douglas production function. It has generally been applied in the context of pollution to estimate the effect of changes in environmental quality on the quantity of output; it has been widely applied in agriculture. Their relevance to tourism can be indicated by examples. In the context of human-made resources, air pollution damage to buildings might affect their utility to occupiers or impair their visual appearance and so deter visits by tourists. The tourism sector itself can cause pollution and damage to natural resources, such as fragile land and marine based wildlife sites, which impairs their quality. This is equivalent to the dose-response in the production of goods. The method seeks to measure and quantify financially the effect of changes in the quality of environmental services. It is closely related to avoided cost and averting expenditure methods.

10.5.3 Avoided cost and averting expenditure methods

The basis of avoided cost and averting expenditure methods is that economic agents may be able to undertake expenditure to minimize the effect of a fall in environmental quality. For example, if the quality of drinking water is considered to have fallen below an acceptable standard in a tourist destination, the local authorities may decide to install a water filter system, the cost to be borne by local suppliers. The sum of these expenditures across all affected parties could be viewed as an implicit valuation of (a lower bound to) the fall in water quality. In a study of direct expenditure by consumers, Hansen and Hallam (1991), for example, use a production function approach in the valuation of recreational fishing benefits of an improvement in freshwater stream-flow relative to 'consumptive' river uses, such as agricultural irrigation. They found that marginal valuations of stream-flow changes are higher for recreational fishing than for their corresponding use in agriculture. Clearly, however, it is likely that the averting expenditure and the environmental service forgone are imperfect substitutes. For example, many tourists experience a loss of welfare as a consequence of the environmental degradation in general, as well as of unique sites of high scenic value, but do not consider the averting expenditure, which may have a high threshold, for instance the use of public transport at an immediately higher personal cost rather than using a car to reach a holiday destination, or a site for which the visual intrusion of parked vehicles adversely affects its attractiveness. Thus, valuations using avoided cost methods may produce underestimates and this has been found to be the case in many human life valuations (Dardin, 1980; Dickie and Gerking, 1991). There is, however, no specific evidence of tourists being directly unwilling to pay to avoid being exposed to risky environments that might endanger health and life.

The relevance of these methods to conservation and natural resources is that they potentially can measure the impact on welfare of ongoing changes in the environment, i.e. dynamic benefits and costs. The examples cited from the literature largely refer to degradation of the environment and being applied in this way they could indicate the impact of a 'without' conservation situation. However, in principle there is no reason why they cannot measure the impact of an improvement. Their main shortcoming is the difficulty of making them operational.

10.5.4 Opportunity cost

Estimating the value of resources used for a specific activity in terms of the opportunity cost is a standard concept in economics but is seldom applied in practice because of the difficulty of establishing an acceptable measure. It is possible to ascertain the difference in value between the uses of resources without any constraints as opposed to their use with them, because of environmental objectives. For example, the obligations placed on the ownership of an historic building by being listed, and its subsequent maintenance, are likely to reduce the market value or rents derived from it (Allinson et al. 1996). The difference can be taken as an indication of its heritage value. The problem is that this is almost certainly an under-valuation as it regards conservation in purely market terms because non-use benefits and externalities are ignored, such as the benefits accruing to tourists wishing to look at conserved buildings. A possible resolution of this omission is referred to in discussing replacement cost.

10.5.5 Replacement cost

Replacement cost is a variant of opportunity cost as an accepted economic approach to ascertaining the value of resources. It differs from the latter because it does not consider alternative uses of a specific resource but the cost of providing a similar resource as a substitute. This cost can represent the value of the original resource. There are two problems with this method of valuation. The first is that it is suitable where resources are easily reproducible, for instance an artificial boating lake, but is inappropriate for natural environments where the resources are unique, such as an ancient woodland containing species of wildlife only found in it and that has taken many centuries to evolve to its present ecological system. Second, it suffers from the same shortcomings as the opportunity cost approach in that non-use benefits and externalities would normally not be included, if market-based replacement cost is considered. This problem can be overcome by applying one of the methods for estimating total economic value, such as CVM, in which case the whole valuation exercise might as well be conducted in this way, thus making the calculation of replacement cost redundant.

10.5.6 Provision cost

Early studies by suppliers of the benefits derived from leisure resources considered that their provision expenditure could be taken as a measure of the value of such resources. While such expenditure may reflect society's willingness to pay for amenity and environmental goods, the approach is patently illogical. Providers, by merely spending more on such projects, can claim that they have greater value. This strategy might be adopted to achieve the objectives of the providing organization. It is very unlikely that valuation undertaken on this basis will correspond to that which would be stated or revealed by consumers and, as with the opportunity cost and replacement cost methods, it would certainly not measure TEV.

10.5.7 Delphi technique

Instead of attempting to ascertain the WTP for an environmental improvement or WTA for a degradation of it by individuals who may be irrational, impulsive or poorly informed, it is possible to employ a panel of 'experts' and to elicit their views on the valuation of environmental changes. The technique was developed by the RAND Corporation in the 1950s (Dalkey and Helmer, 1963) and has been found to be particularly useful in cases where historical data are unavailable or where significant levels of subjective judgement are necessary (Smith, 1989). The method consists of the researcher assembling a panel of those who are believed to have some knowledge concerning the issue. It is important that the panel be from diverse fields, with different approaches and therefore different viewpoints and subjective valuations. The numbers in panels have ranged anywhere from four (Brockhoff, 1975) to over 900 (Shafer et al. 1974).

In the next stage the panel is given information on the study and asked for individual valuations. The responses are circulated to all panel members, who are asked if they wish to revise their own valuation in the light of the other responses. This process continues until some convergence of views between the experts, in the light of discussions with their colleagues, has occurred. An advantage of the method is that it requires little specialist statistical knowledge and is relatively simple to conduct, although the selection of an appropriate panel, and the design and wording of the questionnaire, are likely to have a crucial influence on the final outcome. Some may also view the process of valuation by so-called 'experts' as undemocratic and artificial.

In discussing the environmental impact of tourism and means of assessing it, Hunter and Green (1995), in addition to canvassing expert opinions, emphasize the necessity for the local community to be involved in the decision-making process and the valuable contribution it can make to directing or assisting the investigation of potentially significant impacts. Korca (1991) acknowledged the role of experts and local residents in using the Delphi technique in a study of tourism development in the Mediterranean, in which a panel composed of representatives of both groups was assembled to identify the impact of continuing to expand tourism. Green and Hunter (1992) incorporated local public opinion when using the technique to assess redevelopment at a site in northern Britain. Green et al. (1990a 1990b) also considered the more general application of the Delphi technique to tourism.

10.6 A concluding overview of the application to tourism of the main methods of valuing non-priced resources

A number of possible methods of valuing non-market goods have been examined, almost all of which have been applied in a leisure or environmental quality context. They are equally relevant to tourism, given the role of the environment as its resource base. Their principal relative strengths and weaknesses have been identified and discussed above. Nevertheless, it appears that there may be some trade-off between economic rigour and breadth of applicability.

10.6.1 The contingent valuation method

The CVM is possibly the least rigorous and technically developed of the three methods, but arguably now the most popular. It has become widely accepted by politicians, particularly in the US, as a valid basis for compensation in cases of environmental degradation, since it is compatible with the concept of democracy and valuation by the population at large. It has the merit of being, not only simple, flexible and easily understood, but also the broadest in covering use and non-use values and both residents and visitors. Its main drawback is that, to date, it has investigated virtually only hypothetical situations. However, it could be applied to real-life circumstances where a decision has to be made on specific resources when WTP and/or WTA need to be ascertained and payment and/or compensation actually made. However, despite a substantial body of academic work on its use with respect to environmental quality and pollution, it has as yet been widely employed neither as an appropriate method of establishing society's evaluation of natural and built environments, nor as a feasible means for making decisions on resource allocation. This applies equally to its use in tourism, where although it has been applied in ecological and environmental research fields, it has hardly engaged the attention of tourism economics (Wanhill, 2007).

10.6.2 The hedonic pricing method

The hedonic pricing method (HPM) takes account of the value that consumers place on environmental attributes only insofaras they enter into product prices or are capitalized into land and property values, i.e. captured internalized benefits. Its validity in the context of environmental issues in tourism therefore depends on the extent to which the price of the tourism product is determined by the attributes of its environmental components, for example the quality of the beaches or facilities or attractiveness and cultural associations of historic buildings. As shown in Chapters 2 and 3, and the references given there, it has been incorporated into the determinants of the demand for tourism. Even if such environmental attributes are embodied in market prices, the only ones HPM captures, the technique is neither suitable for the consideration of option, existence or bequest values, nor does it incorporate non-market benefits enjoyed by visitors. Thus, this feature indicates that the estimates emerging from the method give rise to gross under-valuations of the environmental or historic conservation elements of the tourism product. Nevertheless, its advocates assert that it is acceptable from an economic standpoint because of its rigour, reliability and robustness.

The presence of market segmentation, which is widely ignored or overlooked in the application of HPM to property values, is a significant aspect of tourism. For example, even within seemingly homogeneous tourism markets, such as those for packaged summer sun holidays, there is segmentation by such variables as price, quality, destination and accessibility. The HPM must acknowledge such segmentation and estimate separate equations for each segment. These are viewed as serious drawbacks at a technical level and impair the confidence that can be placed in the results. The HPM requires data on a wide range of variables. Much of such data is not available in a suitable secondary form, thus necessitating its collection from primary research, making it a costly method.

10.6.3 The travel cost method

Willis et al. (1993) argue that the travel cost method (TCM) works best when the majority of the visitors live at a long distance from the sites, which are also rural. It is apparent that it is more reliable if visitors travel by car.

By making adjustments to their travel costs, as discussed above, TCM models are capable of measuring the benefits of tourists' visits. However, for the valuation of urban recreational sites, the distance travelled by permanent residents and tourists staying in the location, which is likely to be very short, tends to nullify the employment of the method. However, the TCM may be relevant in these cases provided it is based on the value of travel time technique, so that even those living or staying nearby, and who walk to the recreational site, incur such a cost. Clearly, the TCM remains appropriate for valuing visits to urban sites in circumstances where both residents and tourists in the locality travel some distance. One interesting and not entirely academic implication of the estimates of the value of sites using the TCM is when visitors are so attracted to an area that they move to live nearer to sites in order to visit them more often. Consequently, for most visitors, the TCM will undervalue benefits because travel costs will be lower.

The TCM is perhaps the least preferred of the three main methods technically, since it has neither the strong economic theoretical underpinning (the consumers' surplus concept is contested) nor the well-developed economic theory of the HPM. It does not possess the attributes to measure the wider non-use values to measure TEV. It is also difficult to aggregate values obtained for individual sites. Since, by definition, an on-site survey would not capture the value placed on it by non-users or by those who do not make a visit during the sample period, an under-estimation of consumer surplus would occur (Smith and Desvouges, 1986). Like the HPM, the TCM suffers from requiring large resources in terms of time and funds to collect, process and analyse data.

10.7 Achieving environmental targets: policy instruments

The economics literature tends to be dominated by the consideration of two aspects of the pursuit of environmental goals. The first is the advocacy of the price mechanism. The second is the tendency to concentrate on mitigating the effects of environmental degradation, particularly pollution. In the political and social consciousness of environmental problems, the words 'polluter must pay' have become a catchphrase for action. This is misleading for not only does it imply that the sole environmental issue is pollution but it infers that those economic agents, such as primary and secondary producers, are the principal sources of environmental problems, whereas in reality consumers also cause them. It also suggests that the costs of dealing with these problems should fall entirely on the perpetrators.

More holistic thinking, as demonstrated in ecological and environmental economics, does not take this somewhat over-simplistic view.

While recognizing that pollution and environmental degradation are prominent issues giving rise to social costs, there are many instances of benefits which are bestowed on society at large, for example by owners of natural resources and by businesses, public bodies and individuals in their normal day-to-day activities, for which no payment is made by users. These branches of the subject also accept that any intervention to correct market failure or imperfections may itself distort the distribution of welfare and allocation of resources. Indeed, the form of intervention, i.e. the instruments employed, can have unintended effects. Therefore, it seems appropriate to consider both the nature of the instruments and their evaluation, which have been identified as separate issues in Figure 9.2 in the previous chapter.

Economists largely favour the price mechanism because it tends to lower the costs of dealing with environmental problems, to make the cost of production and consumption more visible and to offer incentives and disincentives in the market place, potentially minimizing distortions. They also acknowledge that some non-monetary instruments play a role in circumstances where price-based instruments might prove inappropriate or where they are complementary, or have an impact on prices and costs. This explains the presence in Figure 9.2 of non-market instruments, which in practice tend to predominate in developed countries. It is appropriate to make a further distinction between once-and-for-all instruments, for example those concerning capital investment, and those that are recurrent.

In examining the various instruments that have been proposed in the environmental economics literature, attention is concentrated on the issues each raises rather than simply describing their purpose and how they are applied. A fundamental debate concerns the relative merits and demerits of price-based and regulatory (sometimes designated command and control) methods; see, for example, Turner et al. (1994) and Tietenberg (2006). It must be appreciated that the economic analysis of the mitigation of environmental problems and the instruments employed tends to be through static or partial equilibrium analysis. Ideally, more dynamic analyses should be applied; when further developed, CGE analysis, considered in Chapter 6, would be appropriate.

10.8 Priced-based instruments

It needs to be stressed at the outset that the economic analysis of the instruments advocated by economists is quite technical. It is not possible to give comprehensive coverage of the ramifications of their effects on economic activity; the exposition here is very much an overview. Tietenberg (2006) gives a quite wide-ranging and accessible account of instruments applied in the environmental field and Perman et al. (2003) offer a more advanced approach, with quantitative analyses in appendices.

Price-based instruments for dealing with environmental issues take two forms. One, for example a tax or subsidy, attempts to modify the operation of existing markets, whereas the other, such as tradable permits and quotas, tries to create markets that did not previously exist. The economic argument for using the price mechanism stems from the recognition that the cost structures of different activities vary and so, in contrast to regulation which imposes a standard compliance, there can be more flexibility in the response by those subject to price-based environmental policies. For instance, in the case of a tax on air emissions, a firm can elect to pay the tax or alternatively fit control equipment, the response being determined by the relative costs of the tax instrument and the cost of investment in equipment and/or procedures to reduce or eliminate the environmentally damaging activity. The reverse could apply with respect to conferring un-priced benefits, whereby a subsidy would be preferred by the supplier if it exceeded the cost of capturing formerly un-priced benefits via the market. For example, as stated earlier, many landowners of natural resources incur costs in making them available to tourists, yet cannot exclude visitors, perhaps because the cost of fencing and/or employing people to collect an entry fee would outweigh the revenue generated; a local authority may pay a subsidy to ensure the resource remains available. The subsidy might relate to management of the resource as an amenity or compensation for damage or loss of revenue from an alternative productive use, such as crops or timber. Another factor of some importance is the effectiveness of price-based instruments with respect to the burden on the public purse. Regulation through the price mechanism may lower enforcement costs; in effect, once the price instruments are implemented and accepted, there can be virtual self-regulation by agents active in the market. The principle underpinning price-based instruments is that the full production or consumption costs or benefits should be estimated in order that the appropriate level of charge, tax, subsidy or grant can be ascertained to achieve a socially optimal position where the marginal social benefits and costs are equal, i.e. the full costs or benefits are reflected in market prices, as shown in Figure 10.2. In practice, in addition to effectiveness in the sense of achieving given targets, a priori the choice will depend on such considerations as the degree of uncertainty in administrative costs and long-term effects, projected efficiency and flexibility as economic conditions change, possible spin-off benefits, and the estimated distributional impact on consumers and producers.

Concentrating on marginal social costs, Figure 10.2 is essentially a simplification of Figure 9.4 in Chapter 9 and indicates why an economic optimum position is gained at OS. The figure shows the private optimum at OP where marginal private cost (MPC) equals the marginal private and marginal social benefit, where MPB and MSB have been aggregated to depict a composite marginal benefit curve. At the activity level OP and price PP in

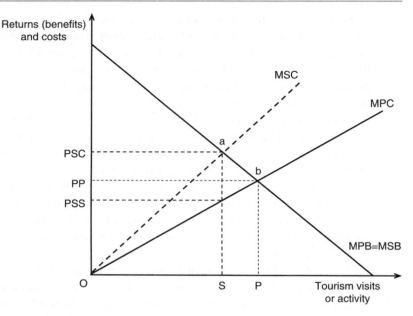

Figure 10.2 The concept of an economic environmental optimum

a market economy, which does not take account of environmental costs, the marginal social costs (MSC) are ignored. Consequently the price does not reflect the full costs to society of the production of the good or service and its consumption. By levying a charge or tax equal to the difference between the MPC and MSC at OS an economic optimum is achieved, i.e. where MSB equals MSC. Reduction of the level of activity to the left of OS would be economically inefficient as the loss in benefit outweighs the reduction in MSC, i.e. MSB is greater than MSC. It is worth noting that consumers as well as suppliers, bear some of the burden of a charge or tax on suppliers as market price increases to PSC after its imposition; suppliers also bear part, as their net of tax price is PSS as opposed to PP previously.

In the context of tourism, a landing fee or bed tax has the effect explained above and in Figure 10.2 if it is levied on suppliers in the first instance. A proportion of the charge is passed on to tourists, which deters some from making a visit; demand falls from OP to OS. Thus, the charge or tax not only raises revenue that can be used to meet the cost of environmental protection but also reduces the pressure of demand.

10.8.1 Payments to improve environmental quality and business performance

Long-term environmental improvements can be encouraged by incentives in the form of one-off capital grants to encourage investment in more efficient and up-to-date technology to reduce materials and energy use and the generation of waste, or even to eliminate some activities entirely. Subsidies are essentially a negative tax and a recurrent form of investment grant. Both are made on the grounds of enhancing social benefits by encouraging those conferring them to continue to do so and/or increase the quantity supplied. The problem with the use of grants and subsidies is that they can be extremely costly for governments if open-ended. Moreover, there is no certainty that there will be a sufficiently large take-up where schemes offer less than the total cost. Such intervention distorts the cost structure of particular industries if the take-up is uneven. It can also create international differentials, giving a competitive advantage to industries in countries where such instruments are applied in comparison with others in which they are not.

Economists' attitudes to subsidies are somewhat ambivalent because they have been conditioned to be concerned with identifying and evaluating the unintended outcomes of encouraging new entrants and existing, often inefficient, producers. For example, government support of farmers financially, to ensure they remain in business, has given rise to over-production, overuse of resources and environmental degradation. Nevertheless, there are circumstances, particularly with respect to situations involving public goods and externalities, where grants and subsidies could be beneficial. For example, in an urban context, conservation of heritage buildings is supported by public sector capital grants and subsidies or tax breaks because it is recognized that the maintenance costs of such artefacts to occupiers and owners are much higher than for modern properties. In this sense the level of grant, subsidy or tax break is an implicit recognition of the social benefits generated by the built and natural environment.

The payments of grants or subsidies to do this can be demonstrated by referring back to the diagrammatic analysis in Figure 9.4 in the previous chapter. Ignoring the MSC curve, if a vertical line is drawn from the intersection of the MPC and MSB curves to the quantity axis (level of activity in Figure 9.4), it is clear that the quantity supplied is increased because it is to the right of OP, the private optimum where MR = MPC. To encourage this increase a subsidy, equal to the vertical distance between the MPC and MSC curves above the intersection of the MSB and MPC curves, is required as this meets the MPC of the increase in supply. In the context of tourism, such a subsidy might be paid where occupiers or owners require an incentive to maintain or restore historic buildings or conservation areas that are attractions in tourism destinations' cities and towns. In some cases historic buildings may be suitable for accommodation or facilities, such as bars and restaurants. Thus the level of both private and social benefits can be enhanced. Another important example is the change in agricultural policy in the European Union since the mid-1980s, whereby grants and subsidies are paid to farmers who farm in more traditional ways and enter into environmental schemes, thus enhancing the appearance of the landscape and rural environment for wildlife and, where feasible, visitors.

10.8.2 Charges

The distinction between charges and taxes is not always clear and they are occasionally considered together, suggesting that they have similar characteristics. Charges can be applied to emissions of pollutants, supply of products or services and consumers' use of environmental facilities and so are relevant to both the supply and demand side of tourism. Charges can raise prices to consumers and almost certainly increase producers' costs, and to this extent have the same effect as taxes. However, there is a technical difference in that economists perceive charges as being levied to deter specific activities or the use of environmentally harmful products or to meet the full costs of output and/or services and facilities. Taxes, on the other hand, are conceived as meeting identified levels of consumption or production, especially in relation to maximizing net social benefits or (minimizing net social costs) by attaining an economic optimum (see Figure 10.2). The distinction can best be explained by illustrating briefly the various forms of charges. In the first two, the charge is essentially a selective indirect tax, but unless related directly to the environmental costs generated by the product, is not designed to affect consumption or production to achieve a social optimum.

a. Air emission, effluent discharges and solid waste charges

These can be levied to deter the generation of particular wastes and are acceptable to polluters if the charges are less than the cost of the required investment, where possible, to eliminate emissions and discharges entirely. They can also be employed to cover the cost of disposing of waste or the necessary remedial action to offset the effects of air emissions and discharges into watercourses. User and waste disposal charges are similar to emission charges where specific facilities or services are publicly provided to collect, treat, dispose of or recover waste materials. These charges should cover the net cost of providing the facilities or services. It is possible that the providing body can secure income from recycling some materials and therefore offset the gross cost. Commercial organizations are normally charged, which can result in illegal fly-tipping to avoid such costs and thus exacerbate environmental problems and render the charges counter-effective. In the UK, where a landfill tax has been introduced, the charges levied by local authorities have risen, thus tending to increase the level of such illegal disposal. A more recent issue in the same country has been the proposal to charge the way households are charged for waste disposal, which could increase the incidence of illegal practices, the outcome being environmental degradation, the cost falling on society at large. There is no evidence that tourism businesses have indulged in fly-tipping but it is conceivable that their contractors might have done. To this extent, other charges, which do not encourage such adverse behaviour, are preferred.

b. Product charges

In order to deter the demand for or supply of goods and services that have a detrimental environmental impact, product charges can be imposed. The charge could be set at a level that effectively makes the price of the product prohibitive, so that demand and supply are entirely curtailed or at a lower level, which decreases demand or supply. It may also induce substitution of more benign materials or services. The use of product charges is less effective where demand is inelastic or the volume of sales is small. Such charges can be applied at the output stage if they lead producers to modify their operations to make use of certain materials, or at the point of sale to reduce consumption. In both cases, a charge is essentially a selective indirect tax, but unless related directly to the environmental costs generated by the product, is not designed to affect consumption or production to achieve a social optimum.

c. Deposit and refund schemes

These might be considered as a variant of charges instruments, but increasingly they are being perceived as separate because they concern the conservation of material resources through re-use and recycling as opposed to environmental quality. For example, the hospitality sector purchases many commodities that are supplied in packaging and/or containers. To encourage and fund recycling of these, a number of countries have initiated deposit-refund schemes for metal and plastic beverage containers (Turner et al. 1994).

10.8.3 Taxes

Levying a tax is the foremost economic instrument for achieving environmental improvements. Unlike charges a tax should, ideally, be set at a level which takes account of the social costs generated rather than solely the private costs of the activity. If correctly estimated, it induces a move to an economic optimal position which maximizes net social benefits and reduces environmental degradation, as shown above in the section on payments to improve environmental quality and business performance, in Figure 10.2 and the textual reinterpretation of Figure 9.4 in section 10.8.1. In theory, taxes can be applied on capital, for example, to deter the use of inefficient and polluting technologies, and/or products or activities that pollute.

Much debate in the economics literature is concerned with the merits of taxes (see for example Turner, 1988) in comparison with regulation. The principal argument is that with regulation producers have no incentive to reduce pollution below the standard imposed, which, in any case, may be arbitrary and bear no relation to the costs, and therefore remedial expenditure falls on society. However, notwithstanding other advantages of taxes, such as the lower cost of implementation, difficulty in avoiding them, the possibilities of inducing investment in new, cleaner technologies or substitution to less polluting products or processes, there are some disadvantages. These are mostly concerned with the practical difficulties of applying them, in particular the problem of identifying the source, measuring the impact physically and estimating accurately the environmental costs on the one hand and the social benefits generated by production of the good or service on the other. There are even wider issues, for example regarding the eventual tax burden and acceptability, but since these are common to any form of intervention which affects costs and prices, observations on them are deferred at this point.

10.8.4 Tax-subsidy devices

The simultaneous use of a tax and subsidy approach to deal with environmental problems is a relatively recent phenomenon in environmental economics and is exemplified by the deposit-refund, recycling credit and disposal schemes designed to facilitate the reclamation of materials. Their effectiveness depends on the relative costs of primary and secondary sources, examined in Figure 9.3 in Chapter 9, and the ability to develop the necessary structures to run the schemes, which connect consumers with manufacturers through collection operators. A number of countries in Europe and states in Canada and the USA promote recycling schemes, most using either a levy at the production stage, which may be passed on to the consumer, or a charge on the product at the point of sale. Refunds are made as the materials are reclaimed, either at public sector recycling locations or by suppliers. There is little evidence to indicate how efficient such schemes are and unless supported by regulations restricting disposal of recyclable materials, they are unlikely to be adopted by consumers and manufacturers.

10.8.5 Quasi-price instruments: marketable/tradable licences, permits and quotas

Licences, permits and quotas are a form of regulation and ostensibly should be considered in the following section. However, increasingly they are being implemented with the greater flexibility of making them tradable so will be examined here. Licences are essentially official or formal non-transferable permissions whereby an individual or group of firms or persons is given the legal authority to produce or use a good, service or resource, or permitted to do a specific thing. An example is the right to produce a good or service or use a trade name over which another has property rights. A permit is less formal and can take a tradable form; it does not necessarily entail a legal agreement between the granter and recipient. In practice, the terms licence and permit are largely used interchangeably. A quota is the allocation of permission to produce or use a given quantity of a tangible good.

Issuing licences, permits or setting quotas as a means of controlling output or the use of a resource has an impact on both those subject to them and the operation of the goods and services markets. It is but a short step also to control directly the residuals of any production or activity, particularly pollutants, by granting licences or permits or quotas to emit or discharge them. More contentious is how to control the over exploitative extraction of consumable products from open access resources that threatens the long-term yield of such products; see Chapter 8 concerning issues regarding open access resources. Two recent cases to illustrate the problems in the context of environmental policy instruments are the limited revival of the ivory trade and the hunting of whales. Control of the quantity of ivory yielded or whales caught is intended to preserve the species and the diversity of their gene pool, recognizing also that they have a value in the wildlife watching market. However, if the conditions governing the issue of the licences/permits to cull the species are not met, the quotas set are very likely to be flouted and over-exploitation to occur, so that there is a danger that the species will become extinct, given the impossibility of fully policing the nature of the resources in which elephants and whales live.

A similar major example is the long experience of applying quotas in the fishing industry. These have been imposed internationally and unilaterally by countries, to restrict the tonnage and/or species landed because of the depletion of fish stocks. Their purpose is to preserve a population at such a level that it is able to reproduce at a rate to sustain the industry in the long term. The outcome, not necessarily intended, has led to much higher prices for consumers and the exodus of fishermen from the industry, where their livelihood was no longer economically viable, notwithstanding the fact that schemes have been initiated in some countries to give them incentives to retire, such as severance payments. Anderson (1995) and Conrad (1995) have examined the devices used to deal with the depletion of stocks and the restructuring of the industry. There is a parallel of this in tourism, examples of which are given below in the section on the relevance of the policy instruments to the sector.

It is also of interest to examine licences, permits and quotas when they become tradable and how this might work in a market context if they can be freely bought or sold. Making them tradable allows the market to determine their value and producers to assess the best course of action in order to comply with them. In the case of pollution, the outcome is much the same as with other pricing mechanisms. If a particular producer finds that the cost of purchasing a licence/permit is lower than undertaking investment to reduce emissions, then a licence/permit will be chosen. Conversely, a business which finds the cost of abatement lower than that of the licence/permit it has been granted will sell it. Of course it is conceivable, depending on the basis on which the initial allocation was made, for firms to sell a part rather than all of their rights. For example, if the licence/permit allows the emission of 100 tons of a residual and the actual amount is only 80 tons, then the remaining tonnage can be traded.

The economic analysis of tradable pollution instruments considers their efficiency in terms of an optimal position that depends on how they are offered when issued. They could be free, or issued at a given price, or auctioned. In essence, it is assumed that the initial quantity of licences issued is based on the estimates of targets set; with perfect knowledge the target for the allowed emissions or discharges will be the assimilative capacity of the environment (see Figure 9.4), i.e. the supply curve will be vertical. The demand curve is dependent on the perception of would-be purchasers, as suggested in the previous paragraph, and will be downward sloping from left to right, reflecting the marginal abatement cost to the purchaser. The intersection of the demand curve with the supply of licences/permits will be the optimum position. Once they are issued the market will be like that for any good, the demand curve as already indicated and the supply curve upward sloping from left to right up to total capacity.

Thus, the advantage of licences, permits and quotas as environmental policy instruments, for instance in relation to pollution, is that they potentially lower the cost of compliance and can reflect the variations in cost structure of businesses and their dynamics and in terms of entry into and departure from an industry or market. The drawback of such instruments is ascertaining in practice of the aggregate permitted residual, which raises the fundamental issue of ascertaining the capacity of the environment to assimilate it. Thus regulation ultimately has to be exercised in the light of experience, raising or lowering the permitted level when appropriate and monitoring whether firms comply. This may impose costs on society at large. There is some evidence of how the instruments work with respect to air emissions in the USA (Tietenberg, 2006). There are moves in other countries to extend the idea to the use of resources, along the lines of those already implemented in agriculture and fishing. At an international level, the allocation of quotas on the emissions of carbon dioxide has already been referred to in Chapter 8. Countries which emit less CO_2 than their quota can sell the balance to countries that are exceeding theirs.

10.9 Non-monetary instruments: regulation and standards

As with both permits and quotas, regulation and standards, often referred to as command and control, are of interest in economics because of the effect they have on consumer prices and production costs and therefore markets. They have also been widely applied in many countries with respect to air emissions, effluent discharges and noise, so that it is possible to assess the effectiveness of regulation in achieving environmental goals. In economic terms the monetary assessment of the cost of standards and targets in comparison with the benefits of achieving a reduction in pollution and/or an environmental improvement can be considered at two levels. The attainment of maximum economic efficiency requires that regulation should be judged by whether net social benefits are maximized. At a lower level it can be appraised on a cost-effective basis, the criterion being that benefits should exceed costs for a given environmental objective. Hartwick and Olewiler (1998) have published an accessible review of regulatory instruments and Perman et al. (2003) provide a comprehensive and quantitative exposition.

The appeal of regulation is that it may obviate the need to estimate the required level of, for example, pollution or resource use, its certainty in attaining specified targets, its transparency, its acceptance by the main sectors of society, especially industry, its relative simplicity in monitoring and enforcing compliance and the placing of responsibility for meeting identified standards on those causing problems. On the other hand, in most countries there is no coherent regulatory strategy. Most policies have been introduced to deal with particular problems. Promotion of national, let alone international, standards and regulations is still in its infancy. For example, the environmental mediums of air, water and land have been considered individually and it is only recently that a more integrated approach has been advocated to reach an agreed standard, for example parts per million of pollutants. These standards tend to be technologically based, with the cost of the investment required to meet them taken into account. For instance, the EU has adopted what is termed the best practicable environmental option (BPEO). Other suggested approaches are best practical means (BPM), best available technology (BAT) and best available technology not involving excessive cost (BATNIEC). Nevertheless, with increasingly rigorous standards and consumer pressure to effect environmental improvements, the burden on producers is likely to increase, with cost consequences which may be passed on in the form of higher product prices. Moreover, greater inflexibility may be introduced the more aggregative the regulation, so that variations in local and industry level conditions are not reflected. It is possible that the effectiveness of regulation may diminish over time because those subject to it will influence the body responsible for administering it. This occurs because both parties need to confer to implement regulations. An analogous example is the procurement of planning permission by property developers, which has been termed 'rent seeking' (Evans, 1982), particularly when there are opportunities for mutual benefits to both parties, for example 'planning gain', whereby the community secures, say, leisure facilities within an allowed development.

These observations point up the necessity of there being an effective institutional structure involving central and local government and NGOs to administer, monitor (as in the operation of non-transferable licences on pollution levels) and enforce the regulatory instruments. This is underlined by the fact that in practice there are multiple environmental objectives, embracing economic efficiency, health and safety, equity, technical feasibility, sustainability and political objectives to meet national and international policies. The benchmark for institutional action is the existence of uncertainty, very often arising from a lack of information on the source, level and impact of environmental problems. Regulation practice proceeds by the adoption of such principles as the polluter pays (if identified), safe minimum standards (SMS), embodying approaches like BPEO and BATNIEC given in the previous paragraph, and the precautionary principle that endeavours to satisfy the multiple objectives identified above.

Basically, the institutional role can be effective by working towards the internalization of the externalities of inequity, use of open access resources (public goods cases), pollution and environmental degradation, by facilitating the defining of property rights so that the legal system can determine responsibility for mitigation of environmental problems. The role can also be performed by simply banning the use of inputs or production of outputs that give rise to hazardous wastes, land-use zoning through the planning system to separate more industrial activity and transport systems generating pollution from residential locations, and the dissemination of information and education. Moreover, regulation can work in conjunction with priced instruments; for example through non-compliance charges for exceeding landfill quotas or failing to recycle waste. Financial redistributions can be made to environmental schemes, such as revenues from mineral extraction regulations and compensation to those who suffer from preventable actions. For example, the pollution of potable water supplies by the utility company responsible for its delivery in a state fit for human consumption can be legally forced to make compensation payments to those adversely affected by, for instance, suspension of supply to deal with the problem.

10.10 Some general observations on policy instruments

Although it has not been exhaustive, this review of policy instruments and their respective advantages and disadvantages has highlighted some of the more important issues arising from the means for achieving environmental goals. Some common economic evaluation threads are discernible in the examination of the instruments, whether priced or regulatory.

First, a key factor from an economic viewpoint is whether the measures are cost-effective and/or efficient. These criteria will not be rehearsed at this point except to state that although conceptually it is possible to appraise the effect of the instruments, in practice it is extremely difficult, especially with those which are regulatory. In general, economists conclude that price instruments are more cost-effective and efficient.

A second and equally crucial issue is that for nearly all the instruments considered, certainly for taxes, charges and tradable devices, there is an implicit assumption of an identifiable limit to the marginal social cost (or benefit where grants and subsidies are being offered), which the instruments will prevent being exceeded. Two observations are apposite. The first is the practical difficulty of ascertaining the allowable environmental damage, which is the foundation of the concept of limits to acceptable change as an environmental target. The second is that the economic optimum, towards which the instruments are meant to guide activity, does not necessarily accord with optimal positions propounded by other disciplines, such as ecology; see the relative positions regarding the level of activity of the private sector (OP), the economic optimum (OS) and the ecological optimum (OE) in Figure 9.4. For example, the European Union Bathing Waters Directive identified standards (environmental indices) which are relevant to human health and consumer satisfaction but which are not relevant to the marine environment and its biodiversity. Regulation may be more arbitrary if it is decided that a reduction in environmental damage is desirable per se without any reference to an economic or any other optimum or assimilative capacity of the environment, as shown in Figure 9.4.

Another aspect which emerges is the equity issue; in particular whether the polluter pays principle is actually effective. A further dimension of this issue is the degree of regressiveness (i.e. lower income groups pay proportionately more than higher income groups) of the instruments, irrespective of whether the burden falls on consumers rather than producers. The basic economic concept of elasticity of demand and supply gives insights as to the possible burden, while knowledge of variations in demand elasticity over different incomes offers evidence on the degree of regressiveness of specific instruments. Clearly, the more inelastic demand is for products in relation to supply, the more the burden of an environmental charge or tax falls on the consumer. Priced instruments, especially if they are specific rather than ad valorem, i.e. levied at a flat rate rather than on a percentage basis, in essence being in their effect like indirect taxes, are usually regressive. Their impact can be offset by governments compensating lower-income groups, but using revenue from charges and taxes in this way may reduce the funds available for environmental improvements where the public sector is responsible for implementing them. Much the same arguments apply to the differential effect on specific firms or industries where the instruments distort the allocation of resources.

Passing reference has been made to the implementation of the instruments but this aspect needs to be spelt out more fully. There are a number of evaluative factors for judging whether instruments are both effective and workable. Effectiveness concerns the achievement of particular aims or objectives, whereas the extent to which instruments are workable is more a function of cost and feasibility. Costs arise when there is a requirement to administer an instrument. It becomes necessary to ask whether such administration can be undertaken through existing structures – such as the fiscal authorities for the collection of charges, taxes and payment of subsidies or tax relief, the land-use planning system for regulations, and specialist environmental bodies for licences or permits or quotas – as opposed to setting up specific organizations and structures.

There may also be a requirement for monitoring activities and enforcing compliance. In addition there are the operating costs incurred by those subject to instruments, especially the ones more firmly set in a market context, for example the transaction costs of securing a tradable permit or quota. Other factors determining whether particular instruments are feasible or not are the ease or difficulty with which the nature, sources and possible sufferers of environmental problems can be identified, the possibilities of ascertaining and measuring the extent to which their impact has unintended and detrimental effects and how far they are considered acceptable by society in general and the business sector in particular.

10.11 Environmental policy instruments in the context of tourism

Tourism, like other economic activities, is a consumer of substantial quantities of materials and energy and a generator of wastes. For example, burnt aircraft fuel has a significant polluting effect on the stratosphere. Tourism activity also tends to be concentrated in specific areas, often over-exploiting scarce resources such as water and creating waste disposal problems and pollution. Accordingly, many tourism activities should be subject to the same environmental policies and instruments that are applied to productive industries. Yet, tourism possesses characteristics which mark it off as different and therefore potentially requiring a different means of dealing with environmental problems.

- First, the generally fragmented nature of tourism supply results in the identification of the source of its environmental effects being more difficult, while periodicity and seasonality give rise to an uneven temporal impact with consequently more intense physical effects (Sutcliffe and Sinclair, 1980).
- Second, that impact tends to be physical in terms of sheer numbers of people (therefore exceeding the capacity of transport modes and tourism resources): a visually intrusive built environment and evidence of tourists (such as cars parked in scenic areas); over-use of natural environments leading to degradation; disturbance to wildlife and reduced biodiversity; waste generation.
- Third, perhaps reflecting an implied criticism of mainstream environmental economics, tourism, in common with a number of other activities, has the potential to generate considerable benefits. This aspect tends to be submerged in the economic analysis of policy instruments, which dwells almost exclusively on the social costs of pollution and the direct costs of its mitigation.

With respect to combating detrimental environmental effects concerning tourism's use of materials and energy, the instruments examined are clearly relevant insofar as the sector consumes substantial quantities of natural resources-based inputs. Current policies are virtually specific to manufacturing, where there are tangible products and consequent chemical effluents or emissions are generated, so that the blanket imposition of emission and product charges, permits and quotas would, prima facie, not appear to be applicable to tourism. This would suggest that taxes and subsidies, deposits and refunds and regulation seem to be the most appropriate instruments for dealing with the environmental problems generated by tourism, reflecting its specific characteristics. However, as already suggested earlier, the use of permits and quotas is indeed relevant to the sector. Emissions by cruise ships and jet aircraft, referred to in Chapter 8 as the primary modes of tourism travel, are instances of an emission whose source is obviously identifiable. This issue has not been solved by any of the currently advocated instruments, as there is no international consensus on action. Notwithstanding the downturn in international tourism in 2008, as a result of inflationary rises in fuel prices and a global recession, the long-term increase in air travel will continue, particularly in the long-haul market segment, which includes a significant proportion of the growth in ecotourism. Thus the contribution of tourism to global environmental problems may be increasing in relation to that by manufacturing industries, some of which have a wider choice in adopting innovative and less polluting technologies. It is extremely difficult to combat this problem of emissions into the atmosphere while there is a reluctance to raise the supply costs to consumers by imposing taxes on aviation and shipping fuel, or more draconian measures such as the imposition of permits or quotas limiting the numbers of flights or cruise ships. In the short run it is unlikely that the number of flights and cruises would be drastically reduced; aircraft and ships would possibly continue to operate with lower payloads, which is both technically and economically inefficient. The solution is more likely to be a very long-term restructuring of the two sectors by taking aircraft and cruise ships out of commission altogether, as has occurred in the fishing industry. Similarly, the reduction of emissions is dependent on technological developments, which involve the use of less polluting fuels, for example biological ones, or even a completely new means of propulsion.

Tourism concentration in certain locations suggests that instruments designed to deal with local problems, such as waste, litter, noise, degradation of physical resources, visual intrusion and cultural and social deterioration, are particularly suitable. In a sense the hospitality activities involved in tourism are much the same as those of households but on a larger scale, with some spatial concentration, such as in seaside or skiing resorts or certain districts in historic urban areas. In this context, the issue is largely one of materials and energy use and waste disposal, requiring instruments aimed at encouraging reductions in the use of packaged materials or energy saving. User charges to meet the full cost of waste disposal, combined with product charges levied on manufacturers of products used, perhaps related to the amount of packaging, would provide an incentive to cut consumption. Similarly, deposit and refund schemes would encourage recycling, where feasible. Reductions in energy use can be stimulated by capital grants to use more efficient heating and lighting systems and to improve insulation. Materials and energy use and waste generation are the aspects of environment issues which have been targeted by firms themselves, as shown by initiatives for which references were given in Chapter 9 in the section on the use of non-renewable and renewable resources. In the longer term, the potential commercial pay-offs, in terms of enhanced competitiveness, were identified in Chapter 8 in examining the Porter hypothesis, concerning the adoption of new technologies and management systems that confer savings in materials and energy use. The impetus for buttressing such market actions with publicly applied instruments is often local, suggesting the need for administration at the local government level.

It is, however, the physical impact of tourism which is more crucial and intractable. The provision of the needs of tourists threatens to destroy the resource base, particularly where natural and fragile environments are involved.

This pressure is aggravated in certain locations, which are subject to the temporal characteristic of their tourism. Some, for instance, possess environments that are often at their most vulnerable in bird and animal breeding seasons, when visitor numbers are at their highest. In other cases, attempts to reduce numbers of tourists at weekends or in the high season by encouraging off-peak and off-season tourism are not efficacious; sustained use over the year can be worse for some resources; for instance, hiking can erode paths more severely in wet and/or winter periods.

The problem of concentrations of large numbers of tourists suggests that controls are required in some destinations (Wanhill, 1980). One obvious approach to the problem, in addition to the use of the price mechanism, is to use regulatory instruments on both the demand and supply side to restrict visits. Some reference has already been made in Chapter 9 to means of restricting access in earlier discussions of problems concerning renewable resources.

With respect to the supply side of tourism, transport operators and hoteliers could receive severance payments to leave the industry, or they could be compensated to reduce their capacity. The land-use planning system can act on the supply side by restricting the accommodation, facilities and services made available. This, however, would be very difficult to implement for existing destinations since it might involve demolishing a proportion of the built environment.

Although there are some moves in tourism destinations that possess much sought after but fragile environments to restrict entry into them (Buckley, 2003a), the use of permits and quotas has not been widely adopted to exercise control over businesses within them. Examples of areas in the world whose environments might benefit from their imposition are Central America, Indonesia, Nepal, the Philippines and Antarctica.

In developed countries in Europe, there are acute pressures in coastal resorts in the Mediterranean and winter sports locations in Switzerland and Austria. In the UK, one of the most densely populated countries in the world, the idea of permits and quotas has been mooted for access to the countryside, especially to environmentally sensitive areas subject to congestion at peak periods. The use of permits which restrict entry into the national parks in North America and the 'take' in hunting or fishing is a case in point. The example of Bermuda is slightly different in that by controlling the amount of hotel accommodation available it restricts the numbers of tourists entering the island. This has had the effect of raising its holiday prices and curtailing demand by lower income tourists and thus making it exclusively the preserve of the wealthier tourist. While the restriction of the numbers of tourists has had a beneficial impact on the island's human-made and natural environments, it may have had detrimental effects, for instance residents who have been displaced or suffered more limited economic opportunities.

Whether these licences or permits would be tradable might depend on the circumstances. A more firmly market-based method would be to levy taxes on suppliers who exceed set limits, or entry taxes, perhaps at a punitive rate, to control numbers of tourists. Charges and taxes, such as the fairly widespread imposition of departure and entry taxes at airports, are already in operation but not at a level that is likely to deter entry. This form of tax is used as a revenue-raising device but is unlikely to be applied to mitigate the environmental impact of tourism travel by air. There are exceptions; for instance, in Vancouver the exit tax on air passengers is specifically stated as being for the improvement of the airport and the environment adjacent to it.

The problem with physical or financial means of controlling numbers is that inefficiency and inequity are likely to result. Economies of scale in providing travel, accommodation, facilities and services may be lost on the supply side, while if increased costs are generated and passed on to tourists, or a direct charge made on them, those in lower-income groups may be priced out of the market. Moreover, those working in tourism sectors in host communities may be disadvantaged if the level of activity is reduced. There are also other factors which may create significant costs, for example the administration of the instruments or simply whether they are cost-effective. Other issues raised are their compatibility with environmental and regulatory objectives, their acceptability and what kind of incentives they offer to tourists and firms to find their own solutions, which, on the one hand, might circumvent the restrictions but, on the other, obviate the need for publicly applied instruments. Rather than attempt to restrict numbers, an alternative approach, which might resolve the problem of overcrowding in existing destinations, would be to levy user charges and allocate the revenue raised to institute environmental protection or reparation. This strategy is not favoured by many environmental economists for it suggests attempting to cure the affliction as opposed to preventing it from occurring.

Thus, reducing the use of materials and energy and the generation of waste, and controlling numbers of tourists, can involve the application of virtually all the policy instruments examined and evaluated. Current policy thinking has not advanced so far as to perceive the operation of service activities in the same light as manufacturing activity. Consequently it has not considered the full range of instruments as being applicable as indicated here. Given the potential and actual detrimental effects of tourism, the time is approaching when restrictions on its development and operation almost certainly have to be more severe.

To this point, the evaluation of policy instruments in the tourism context has echoed that given in the literature of emphasizing the environmental costs of economic activity. However, the many possible beneficial effects of tourism mean that the instruments presently proposed might counteract such effects. The general consensus in economics is that beneficial externalities should be enhanced, the instruments being to compensate or subsidize those making such provision whether or not they are in the private or public sector. Prime examples, already given, are the occupiers or owners of listed heritage buildings or sites, who incur additional costs to conserve them. Both grants for capital costs and subsidies for recurrent expenditure are seen as the principal instruments. However, many tourism resources are natural and open access environments so that property rights do not exist or cannot be exercised. With respect to unique resources of global significance, such as Antarctica, the oceans, rare flora and fauna, whose importance transcends national boundaries, the possibilities of allotting grants or subsidies are, at the very least, problematic. It is necessary to establish international bodies and agreements to oversee and manage such resources. This means instituting international regulations and administrative structures; there is little evidence of such a holistic approach, as demonstrated in the case studies examined in Chapter 9.

Past experience with conventions on whaling and fishing give rise to pessimism that agreements can be made binding and adequate funding made available for enhancement and/or policing of resources, although the position in Antarctica gives cause for a greater degree of optimism. There is a strong case for, and indeed likelihood of, the introduction of some kind of ownership structure, and thus of property rights being enforced through international law. These kinds of issues go beyond mere implementation of policies and instruments, which are applied at a specific and operational level. Overall, it would appear that policy instruments advocated in environmental economics are equally relevant to service activities such as tourism. However, their application needs to be assessed within the wider context of the global environment and strategic policy-making if the instruments are to be effective in both a restrictive and positive way.

10.12 An overall appraisal of the application of environmental economics to tourism

In this chapter and the previous two, the review of environmental analysis, in economics in general and tourism in particular, prompts three main observations.

- The first is that conceptual and theoretical developments in economics have not been fully incorporated into the analysis of tourism.
- The second is that there are some disparities between the policy desiderata proposed in economics to achieve environmental goals and those recognized as important in tourism.

- The third is the continuing debate in economics on the extent to which market intervention is necessary to pursue the ultimate goal of sustainable development, which extends to create uncertainty of commitment in practice.

There is a need for economists to ensure that the substantive prescriptions of environmental economics and its policy implications are brought to the attention of those involved in applied work. Nevertheless, it would be improper to suggest that environmental economics is a fully mature and comprehensive system of thought, and that its methods have been sufficiently developed to be applied in all contexts. Some elements of the subject are already well advanced and applicable. For example, the concept of TEV, the CVM, HPM and TCM methods for ascertaining the level of non-priced demand for such objectives as environmental conservation and the policy instruments for attaining it, are well developed and operational. On the other hand, the analytical framework in which to appraise environmental policies is still being debated in economics. This is certainly so with respect to the boundaries of CBA in terms of what benefits and costs to include and their magnitude; in brief, when are the impacts so minuscule as not to warrant measurement? A further subject of debate is the rationale for and means of taking account of intra- and intergenerational distributional effects.

A more fundamental issue, the cause of considerable controversy, is the extent to which it is necessary to intervene. The market failure concept, as suggested earlier, is not accepted universally in economics. There are those (for example, Coase, 1960; Demsetz, 1969; Randall 1993) who suggest that the existence of such factors as externalities and public goods are merely imperfections which can be resolved by the market if property rights are properly defined and internalization can occur.

Yet another area where environmental economics is still in the process of being developed and made operational is where its concepts and analysis rest on a scientific base. For example, the rate of consumption, which allows renewable resources to replenish themselves, and the establishment of the levels of emission or effluent discharge, which enable the environment to assimilate pollution, are scientific issues. Thus, the concept of maximum sustainable yield, notwithstanding the economic limits of exploitation being determined by benefits and costs, is founded on the physical/biological imperative of the ecological processes involved. Hence the arguments put forward within environmental economics must be made in conjunction with those emanating from the natural sciences (Hohl and Tisdell, 1993).

The issue of environmental protection and sustainability is also political; leaving matters to market forces has been called into question. Some environmental economists assert that the discipline should be more persuasive in showing the environmental implications of the current emphasis on economic growth and, in turn, tourism expansion (Pearce et al. 1989; Cater and Goodall, 1992; and Goodall and Stabler, 1994). The organization Redefining Progress (1995) reflects these concerns in advocating, as a start, that national income measurement should account for the impact of economic activity on the environment. The evidence of global warming, ozone layer depletion, acidification, deforestation, desertification and de-biodiversification is cited as pointing to the need to move from the implicitly accepted weak, or at best intermediate, sustainable development position, to the strong one (Turner et al. 1994). The weak stance allows for the substitution of human-made for natural capital and the reliance on innovation via the market, whereas an intermediate position, although taking account of the constraints imposed by not allowing renewable inputs to be degraded, still considers that technical progress will resolve problems. The strong position argues for the non-substitution of natural capital for that which is human-made but goes further in advocating the precautionary and safe minimum standards principles (Pearce et al. 1989) where uncertainty prevails, particularly when decisions on resource use might introduce irreversible trends. It may imply lower economic growth and possibly no enlargement in the scale of activity.

With respect to tourism, reflecting these prescriptions, there is a requirement to resolve the extent to which sustainable tourism is consistent with maximizing its present value over time, i.e. maintaining its future commercial viability. In addition, the conventional economic analysis concerning market failure, which advocates an optimal position where marginal social costs equal marginal social benefits, may be incompatible with an ecological optimum and so lead to the long-term degradation of the environment. Furthermore, ignoring the differential spatial impact of any environmental policies applied may increase materials and energy use; for instance, some areas may have a greater capacity to absorb pollution than others and therefore the imposition of common policy instruments may be unnecessary and wasteful. A more selective approach is required. Sustainable development and tourism concepts and policies have originated in developed economies where there is already a high proportion of human-made capital. The pursuit of sustainability is not only more likely but has a greater chance of being achieved in these countries. To impose the same regime on developing countries whose environmental problems may be very different not only extends to a measure of arrogance but also may be inappropriate (Curry and Morvaridi, 1992).

Research, both in general and specifically in relation to tourism, which needs to be undertaken to move towards sustainability can be summarized as:

- systematic investigation of the scale and nature of the environmental impact of tourism activity at the local, regional, national and global scale;

- establishment of a real understanding of consumers' preferences for environmental protection and sustainability rather than the views represented by tourism firms for commercial gain;
- examination of the ecological and economic concepts of and prescriptions for sustainability and the means of reconciling them, recognizing that the holistic and stochastic interrelationship of ecosystems is unlikely to yield deterministic perspectives and solutions;
- development of environmental policies based on the notion of safe minimum standards.

At present the focus has been limited to defining sustainable development, considering how best to measure it, together with its weak, intermediate and strong scenarios and their policy implications. A major obstacle to progress in conserving the environment continues to be the establishment of its economic value. It is crucial that environmentalists convince governments and businesses that the environment is not a disposable commodity but an integral part of the resource base, as are materials and different forms of energy. Moreover, its conservation is not always an additional cost, but can in fact be the reverse.

Attitudes in tourism firms reflect what can only be described as the complacent political stance. Currently, woefully inadequate support is being paid to sustainable tourism and environmental conservation. The underlying philosophy remains that any environmental action should not undermine the viability of businesses. The response to requests for environmental responsibility has been piecemeal and muted. Incentives to pursue environmental goals more vigorously will need to be offered to encourage tourism firms to safeguard their resource base and reduce the generation of detrimental externalities. They also need to be convinced that some regulatory measures may be needed. Economics, through its tenets and analytical approach, can demonstrate both the costs and benefits of environmental action and where and on whom these fall. It is a social and political, and perhaps even ethical, matter to decide on the distribution of such costs and benefits in society.

11 WHITHER THE ECONOMICS OF TOURISM?

11.1 Background

At about the time that the contract for a new edition of this book had been agreed, ten years after it was first published in 1997, and the remaining original author was undertaking a preliminary search of the literature since that date, he was invited to give a keynote paper at the Tourism and Travel Research Institute in Nottingham at its annual conference in December 2007. The suggested focus was to give the impressions gained of the changes in the research of tourism over the period from the preparation of the first edition to the present day. In the event the paper presented consisted more of an examination, with a measure of critical comment, of a limited number of illustrative themes and issues perceived as important in the economic study of tourism, or at least susceptible to analysis by the discipline. However, revisiting the literature on tourism studies, after more than a decade away from it, led to the realization that the subject was not only developing rapidly, but there was quite a sharp dichotomy between the nature of research into it in the wider tourism context in comparison to that having an economic perspective. Nevertheless, there were indications that those working firmly within an economic theoretical framework were beginning to question where the discipline stood in relation to other approaches and its relevance to tourism practice. The conclusion was that a much closer look at the nature and tenor of tourism economics research should be undertaken. Thus this chapter attempts to avoid the temptation to simply summarize the content of the previous chapters as a conclusion to the book, to show the contribution economics has made to the study of tourism. Instead it offers a brief exploration of the extent of the progress made in economic research over the last decade or so, its strengths and weaknesses and likely future direction. The chapter is effectively a drawing together of a number of significant strands of developments in the economics of tourism identified and examined throughout the book.

The early part of the chapter to some extent reflects the initial literature review and the approach adopted in the keynote paper, but also draws on a number of observations made by Wanhill (2007) relating to his perception of the trends in, and need for further research on, certain areas of tourism in his paper on submissions to *Tourism Economics*, an overview of which was given in Chapter 1. Where apposite, in contributing to a commentary on its performance and development, a number of issues regarding economic research into tourism identified in Chapter 1, and others relating to specific topics throughout the book, are briefly rehearsed before an appraisal and evaluation are offered. Subsequently, the scope of the appraisal is broadened to set economic analysis into the wider context of tourism studies. In particular, the subject's theoretical approaches are critically examined and comment is made on the possible necessity of it breaking out of its boundaries in collaborative research.

11.2 Perceptions of tourism economics research

Re-acquaintance with the tourism studies literature, with particular reference to that having an economic perspective, yielded the impression that there are three features of the subject's attention to research. In positing the scope and content of the three features no hard and fast boundaries are drawn; it is recognized that there are possible overlaps between them. Furthermore, it is acknowledged that to an extent the categorization of research areas can be construed as somewhat arbitrary, reflecting the authors' perceptions, and is therefore open to dispute. The interests presented do not constitute an exhaustive identification. They are illustrative of the range of studies that either are, or have the potential to be, relevant to tourism, also acting to highlight the principal economic issues considered in previous chapters.

Where considered appropriate, appended to each feature categorization is a commentary for selected items that includes where in the book they are examined, a reiteration of some important shortcomings, or identification of those not discussed in the requisite chapter or chapters. Shortcomings that may have subsequently been addressed or remain as outstanding are noted. Furthermore, the commentaries identify areas of research, considered from an economic perspective, that have not been extensively acknowledged in the wider tourism literature. This aspect is discussed below in an overall evaluation of the apparent status of the economic study of tourism.

The first feature examined is perceived as consisting of established interests of some years standing that are ongoing areas of research very much within the single disciplinary mainstream economic methodological framework. The second feature is referred to as emerging themes and developing methods of analysis that also largely follow the traditional economic approach. These are areas of research that were either non-existent before the 1990s or were in their infancy. The third of the features is essentially one containing what can be described as representing two almost interwoven threads of tourism research. One concerns both the long established and more recent research interests relevant to the wider context of tourism studies that economics has virtually ignored, or has given very little attention to. Conversely there are economic concepts and methods that are appropriate to the analysis of tourism that have not been acknowledged by those in other disciplines in the broader research field. The other relates to themes and issues that have been recognized in other fields of the subject, particularly ecological and environmental economics that have developed appropriate methodologies for their analysis, but which have hardly been adopted in the study of tourism, where their application would be apposite. The research areas contained in all three categories of features, especially those that are under-researched in the third, include those either identified specifically by Wanhill (2007) or considered so by an inspection of his paper of the low submission rates on certain topics. The citation of studies in support of the discussion on the state of economics research into tourism in what follows is kept to a minimum as the relevant studies are largely contained in the chapters identified in examining each interest.

11.2.1 Established research interests

- modelling of demand and forecasting
- structure of tourism supply
  - firms, markets, pricing and employment
  - structure–conduct–performance (SCP) and game theory
- effects of the activities of transnational companies
- impact of tourism activity at the local, regional and national level: I–O and multiplier analysis
- tourism as a form of international trade contributing to foreign currency earnings and the balance of payments
- productive and energy resources use and efficiency, waste and pollution
- the valuation of unpriced amenity and tourism resources

The modelling of demand and forecasting is one of the most developed and rigorous areas of the economic analysis of tourism, hence its quite extensive and detailed examination in Chapters 2 and 3. It has benefited from the availability of large databases in the two principal approaches of time series and econometric analyses, although there are discrepancies in it. This has given rise to problems of unreliable estimations in more advanced and complex econometric models, as identified in Chapter 3. There it was suggested that error correction models have gone some way to resolve the problems. However, as Wanhill (2007) points out, the shifts in consumer behaviour regarding the type of tourism demanded, and the effect of tourism policies, do influence the estimates.

The economic analysis of the structure of tourism supply in Chapters 4 and 5 is founded in industrial economics. The traditional approach of market competitive structures and pricing has not figured strongly in tourism research until recently, particularly the concepts of oligopoly, duopoly and contestable markets that characterize certain tourism sectors. The inclusion of the SCP framework and game theory in the first edition of the book, which has been retained, was novel in positing that they were relevant to the analysis of tourism supply; there was little research on their application at the time. There has been research recently on tourism sectors within the SCP paradigm, in particular the travel trade, but this is not the case with game theory.

Examination of the activities of tourism transnational companies, considered in Chapter 7, has been less concerned with their competitive structure; rather the emphasis has been on the effects of companies based in generating countries of their control over the distribution of revenues and profits. Considerable attention is given to the proportion of expenditure by tourists on the elements of their holidays, especially on travel and accommodation, that is retained in the generating country as opposed to that accruing to the destination, mostly in developing countries.

The I–O and multiplier methods for measuring the economic impact of tourism, mainly at a local level, are explained in Chapter 6, where some reference is made to their weaknesses. The principal problems with I–O analysis are the restrictive assumptions made, such as constant proportions of inputs and outputs as output changes, and the failure to embody productive input constraints, flexible prices and industry interactive effects. They continue to be applied but are tending to be superseded by CGE modelling that is more firmly set in an empirical context, considered under the next features category. A major shortcoming of these long-standing methods is that they measure only income and employment generation; the costs of these are not incorporated. Also, they are mostly applied using fixed prices that limit the possibility of them reflecting dynamic conditions in the real world. They also tend to overestimate the effects of expansion in economic activity.

Tourism as a form of international trade and its impact on the economies of destinations, particularly developing countries and their regions (also related to the activities of the transnational companies considered in Chapter 7), is analysed in Chapter 6 in the section on the instability of earnings. It is an area of research that has waned in economics somewhat in recent years, yet remains an important issue in tourism. This is especially so as responsible tourism (see Chapter 9) entails travel to more remote locations, in which communities are likely to be very poor, often reliant on companies based in developed countries, and raises all the issues of the instability and smallness of earnings remaining in the locality, and of foreign currency and the contribution of tourism to the balance of payments. The wider tourism literature, largely outside economics, on the expansion of ecotourism and its impact on destinations has recognized the issue and has been instrumental in giving it a more visible profile.

The last two items, well-established research interests in the wider leisure field, are analysed in Chapters 9 and 10 respectively. Curiously, the use of resources and the generation of waste and pollution was extensively studied in the mid-1990s, but the interest was virtually confined to the initiatives by the hospitality, attractions and facilities sectors of tourism, as indicated in Chapter 9. It is an area relating to environmental issues that has largely fallen out of the purview of researchers. There has been an upsurge in the study and advocacy, particularly in the wider literature, of alternative or responsible tourism, also considered in Chapter 9. Given the argument for greater care to be taken of natural environments, and studies that suggest that ecotourism represents the acceptable face of tourism, it is surprising that suggest the well-established analyses developed in environmental economics have not been applied to assess the impact of the growth in this form of tourism.

Similarly, the valuation of non-priced resources, the public goods case explained in Chapter 10 and examined in the next category of features, has hardly entered into tourism studies; yet this aspect of tourism's use of resources is crucial in deciding on development of the sector and its scale. With respect to environmental issues in tourism and the contribution of economics to analysis and solutions to the problems that arise, the attention of the reader is drawn to the concluding section of Chapter 10.

11.2.2 Emerging themes and the development of methods

- demand analysis developments;
- measurement of the contribution of tourism to economies deriving TSAs;
- impact modelling using CGE;
- role of tourism in generating economic development and growth;
- application of IT to the transport, intermediary and hospitality sectors;
- competitiveness between countries and destinations at a local level;
- innovation–largely in the hospitality sector;
- efficiency and productivity in the intermediary and hospitality sectors;
- tourism governance;
- poverty alleviation;
- globalization.

Demand analysis has recently taken new directions, with greater attention being increasingly paid to the characteristics framework of demand, considered in Chapters 2 and 3 and first mooted by Lancaster (1966). This has partly stemmed from the development of the hedonic pricing method, also referred to in Chapters 2 and 3, as are two other areas of research that have emerged, discrete choice analysis and panel data, outlined in Chapter 3.

As shown in Chapter 6, the first two features, TSA and CGE, were very much in their infancy with regard to their application to tourism in the 1990s; their investigation has expanded rapidly in the first years of this century. Both represent significant advances in the contribution economics can make to understanding the magnitude and importance of tourism in both developed and developing countries. While their consideration in Chapter 6 includes the problems of deriving indicators of the size (TSA) and impact (CGE) of tourism, it is worth highlighting that their development is hampered by two factors mentioned in Chapter 1. There, the brief discussion raised the problems of the conceptualization of tourism in defining whether it is an industry or a market and the reliance on secondary data. The development of TSAs is dependent on the willingness of countries to delineate tourism and institute the collection of data specific to it. Likewise, for progress to be made in the relevance of CGE modelling, there are further difficulties in deciding on the variables to be included and the degree to which exogenous ones determine the results obtained. Another issue is the need to develop more dynamic models. There are also discrepancies in estimates of income and income multipliers at the aggregate level using CGE modelling compared with the established I–O and multiplier analyses. Moreover, it has recently been acknowledged that CGE only considers impacts, and therefore has less relevance to the enactment and implementation of policies, because it is being recognized that environmental and social effects should be taken account of. This has led researchers to consider the integration of CGE modelling with CBA. The problem is that the two methods have different purposes, methods and application. Therefore, the results from each are not directly comparable. In CBA, a change in the level of economic activity cannot be measured because it is a partial equilibrium method. The CGE method cannot show whether a project, say an event such as the Olympics, is worth undertaking as it does not capture many benefits and costs, specifically social ones; it is also much less detailed than CBA. A further problem with the integration of the two methods is a double-counting issue. The CBA and CGE methods are broadly complementary and consistent with each other; nevertheless, more research is required for integration to be possible in practice.

The examination, at some length, of tourism development and growth theories, particularly the latter, in Chapter 6 illustrates graphically the lack of a research connection within and between disciplines. The investigation of the spatial dimension of tourism in economics, such as that in relation to core–periphery issues, the competition between and within destinations, the growth of specific holiday locations and even the life-cycle of resorts, has been conducted with virtually no reference to the field of urban and regional economics and its growth and location theory. It is also curious that the elements of New Economic Geography (NEG) that emerged in tourism studies in the mid-1990s, considered in Chapter 6, echoes almost exactly the development from geographic origins of spatial economics.

The application of IT to tourism has accelerated at an incredible rate since the 1990s and attention to its emergence as an area of research has not developed appreciably beyond description of its impact on tourists' means of choosing holiday destinations, and making travel, accommodation and tourism services arrangements with the suppliers of these elements. The effect of the continuing development of IT on consumer behaviour has, however, influenced the structure of supply, particularly the intermediaries, determining their competitiveness, efficiency, innovation and productivity. These factors are very much within the purview of industrial economics and thus should attract researchers to consider their impact on tourism supply, examined in Chapters 4 and 5, with some reference to an environmental dimension in Chapters 8, 9 and 10. It is only recently that economists have become more interested in this aspect of tourism; the factors are strongly interrelated so will be considered together here. In economics the basis for ascertaining the effect of increased productivity, the attainment of greater efficiency and innovation and ultimately an enhanced competitive position and growth, is within production function analysis. A fundamental problem in deriving tourism production functions is the difficulty of establishing what constitutes the sector's inputs and outputs. This is not only because of the non-availability of appropriate data, but also the conceptual issue of what constitutes tourism outputs. Consequently, the measurement of outputs poses almost insurmountable problems. Estimating productivity, an emerging interest in tourism, is driven by investment, innovation, labour skills, enterprise and competition. These drivers are extremely difficult to identify. Efficiency, allocative and technical, to maximize output for given inputs, or to maximize revenue or profits, or minimize costs, is not easy to estimate in tourism. One issue is the significance of its reliance on un-priced public goods, especially natural resources, but the main difficulty is again the availability of data, particularly the unwillingness of tourism businesses to disclose their records. The increasing interest in innovation is in a similar position to the estimation of productivity and efficiency with respect to its conceptualization and the effect it has on competitiveness. With all these elements of tourism, research into their effect is still in its infancy; a fundamental requirement is for more empirical research to be undertaken.

What tourism governance comprises (the term was not in common use when the first edition of this book was published) has been delineated in Chapter 1 in the section identifying the scope and content of tourism economics. As perceived by Wanhill (2007), while it embodies what would be expected under the heading, such as taxation, policy and management issues, he also includes under it the consideration of public goods and the methods adopted for their valuation and management as non-priced resources. However, this last aspect is more within the purview of wider environmental issues and really stands alone in terms of management and policy intervention to safeguard the resources base of tourism, as shown in Chapters 9 and 10. The continued significance of tourism as an activity, because it raises many issues as to its effects, particularly on the grounds of welfare, warrants research of its governance. Wanhill's evidence shows that governance is not as strong an economics research interest as it should be; in his survey it represents less than 10 per cent of papers submitted to *Tourism Economics*.

Poverty alleviation that embraces pro-poor initiatives, the attainment of the triple bottom line of economic, environmental and social and value chain processes considered in Chapter 9, has entered quite strongly into the wider context of tourism studies. It is an issue of growing importance, coming within the purview of welfare economics. However, although it figures strongly in the ecological and environmental fields, as indicated below under Section 11.2.3, it is not an area that has appreciably attracted tourism economists to undertake research into it.

Globalization and its influence on tourism first surfaced as an issue warranting attention in the 1990s (Mowforth and Munt, 1998). Its widespread impact had been recognized earlier in mainstream economics, which is traced at some length in Chapter 7, before looking at its role in tourism. In that chapter the positive and negative effects on developing countries are identified, linking it to poverty alleviation, examined above. In Chapter 8 it is considered in relation to likely environmental impacts and its possible interaction with tourism. It is an area of study that would benefit from more empirical investigation in connection with tourism: it is not identified as a research topic in submissions to *Tourism Economics*.

11.2.3 Themes, issues and methods under-researched in economic tourism studies

- the application of CBA to the appraisal of tourism development projects;
- sustainability – conditions to attain and indicators;
- policy instruments to enhance tourism benefits and mitigate costs;
- crisis management;

- inter- and intragenerational equity;
- definition and conceptualization of tourism.

As shown in some detail in Chapter 9, in which an exposition and evaluation of CBA was given, it has not been applied to tourism to any great extent. A likely explanation is that it is an expensive exercise, both in terms of cost and time, for commercial entities to undertake project appraisals of tourism using such a wide-ranging method. In destinations, especially in developing countries, it would even stretch the resources of the public sector if large-scale tourism projects were to be envisaged as a means of securing economic development and growth. Another reason may well be that tourism is seen as an environmentally and socially benign activity, therefore obviating the need to consider the social benefits and costs of the development it engenders. A further explanation of its neglect might be that it has been subject to much criticism by a faction of mainstream economists. Furthermore, it is a method that has continuously been refined theoretically over many years, resulting in it being more complicated operationally. However, CBA has been accepted in environmental economics as being an appropriate way of appraising projects that would have an impact on environmental resources, particularly natural ones. As indicated above in discussing its relationship with CGE, there is a growing acknowledgement that CBA can contribute to identifying and evaluating elements of the impact of tourism that other approaches ignore or cannot capture. It is certainly an economic method that until now has been virtually absent from tourism studies and thus is very much an under-utilized approach to the appraisal of specific projects and the sector's impact generally.

Sustainability, examined in Chapter 8, is a concept that in recent years appears to have had more attention paid to it than any other aspect of tourism and has engendered debate over a long period without any consensus having been reached on its conceptualization, definition and operationalization. That it is a neglected area of the economic study of tourism is borne out by the very low number of submissions on the topic to the journal *Tourism Economics* (Wanhill, 2007); it constitutes less than 2 per cent. It is curious that even in the economic approaches to tourism studies that have been conducted, the subject has not drawn on developments in the field of environmental economics that have set down the bases of the conditions to prevail for sustainability to occur and indicators that it is on a path of being achieved. While it is recognized that there is a need for tourism development to be sustainable and to determine how this sustainability can be attained, the empirical investigation of it is in a most unsatisfactory state because of the lack of consensus referred to above. Moreover, at an operational level, as stated in Chapter 8, the term sustainability is misused, as it refers to commercial considerations of financial viability in the long run.

There is very little empirical research that can be truly considered as concerned with sustainable tourism in the sense of embracing the triple bottom line of environmental and social dimensions as well as the economic.

Research into tourism policy has tended to concentrate on central government intervention with respect to regional economic development, in which the promotion of tourism is undertaken to contribute to it, or the sector, if in its infancy, is given protected status to facilitate its growth. Taxation in the tourism economic field is mostly concerned with its impact on welfare and the competitive position of the sector. It is not a topic that has attracted much attention in economic tourism studies; it represents only 2 per cent of submissions to *Tourism Economics*. Research into the use of taxes and subsidies relevant to environmental protection and enhancement has barely entered the purview of economists working in the study of tourism. The kinds of policy instruments, a firmly established aspect of policy in environmental economics, and their potential role in the context of tourism are identified and explained in Chapter 10. They can be employed to enhance benefits and mitigate costs, both of a private and social nature. There was some consideration of their utilization in the 1990s, virtually confined to materials and energy use and control of waste generation in the hospitality sector, but interest in that area has waned, as has any application with regard to the wider environmental impact of tourism.

Crisis management, considered in Chapter 5 in relation to the supply side of tourism, but included under governance in Wanhill (2007), started to be a topic of concern in the 1990s but became of much greater importance this century with the advent of natural disasters and acts of terrorism. It has been given quite wide coverage in the more general tourism literature, but it has generated little research in tourism economics; it accounts for only 1 per cent of submissions to *Tourism Economics*. As the rationale of crisis management is founded in complexity and chaos theory it should be of considerable interest, given the debate on the conceptualization of tourism outlined in Chapter 1. However, more importantly, it has implications for the whole approach to the analysis of tourism, the characteristics of which, introduced in the same chapter, suggest that as a phenomenon it should be considered within the analytical framework of chaos theory.

As shown in Chapters 8 and 9, inter- and intragenerational equity has been a very prominent feature of the consideration of sustainability (it is embodied in definitions of it), reflecting its quite extensive examination as a characteristic of market failure in ecological and environmental economics. In the wider tourism literature it has received much attention, being referred to in the pro-poor and triple bottom line issues identified in the above two chapters, particularly in the review of responsible tourism in Chapter 9. Notwithstanding the fact that questions of intragenerational equity are referred to in Chapter 9, it is, however, an area of research that is largely neglected in the tourism economics field. Nevertheless, it is an issue that will become more important with the increasing pressure of tourism on communities in destinations and the incidence of globally significant natural disasters, which tend to be more frequent in developing countries. There are also implications for future generations in these countries of current trends in both globalization and tourism.

Unlike their widespread coverage in the wider arena of tourism studies (see Section 1.2.2 in Chapter 1) the definition and conceptualization of tourism have not commanded much attention in tourism economics; they are completely absent from submissions to *Tourism Economics*. The relatively brief glance at the two aspects in Chapter 1, Section 1.2.5, was primarily concerned with the need for clarification as to whether tourism is an industry or a market (Wilson, 1998) because of the effect it has on the selection of published data on the sector and the validity of the results obtained. However, there is a more fundamental issue on the economic approach to the analysis of tourism that was introduced in Chapter 1, Section 1.2.6. This is the possibility that the current economic analytical framework is too narrow; this issue is discussed below.

11.3 An overall appraisal of the current state of tourism economics research

The critical review of the development of tourism economics over the recent past in the previous section might suggest that its performance has been somewhat modest. This is certainly not the case. It should be emphasized that the strengths of the subject are founded in the long established features of research identified in Section 11.2.1 above that continue to break new ground. These research areas epitomize the contribution economics has made to the study of tourism that is acknowledged in the wider context of tourism research, for example the revival of interest in a political economy approach, originally suggested by de Kadt (1979), Britton (1982) and more recently by Bianchi, (2002) and Williams (2004), and the emergence of New Economic Geography, referred to in Chapter 6. Holden (2005) credits economics as being the only exception to the generally weak analysis of tourism in the social sciences.

The attributes of economics are its precision, rigorousness, objectivity, explanatory and predictive qualities, which demonstrate the impact of tourism activity and given policies on consumer behaviour, the distribution of income and wealth, allocation and optimal use of capital, land and labour resources and efficiency. The discipline also possesses methods, identified above, that can be directly applied to the operation and practice of tourism, such as those concerning the appraisal of capital projects, the generation of income and employment, the development and growth of tourism and the meeting of environmental targets. This is the foundation on which the economic analysis of tourism has developed over the last two decades.

In the literature search for this book, it quickly became apparent that economic publications, particularly in journals, have increasingly concentrated on the development of quantitative modelling outside the well-established ones of demand and forecasting, pricing, macroeconomic impact assessment and within international trade theory. Econometric analysis has been extended into accommodation, attractions, competitiveness, destination choices, employment, innovation, and operations of the intermediaries, seasonality, tourism development and growth. The increase in the use of quantitative methodologies and the application of econometrics has underlined the rigour of papers and their acceptance in mainstream economics.

It is worth emphasizing the geographical shift in the sources of papers, as indicated in the preface to this edition of the book. It is evident from the names and academic institutions of the authors of a significant number of papers that contributions from Asian countries are increasing. This is partly a reflection of the development of the participation internationally of scholars studying tourism in universities in countries such as China, Hong Kong SAR, Singapore and Taiwan. This contribution has also arisen because of the collaboration between Western universities and their counterparts in these Asian countries; for example, the Tourism and Travel Institute at Nottingham has established strong links with China.

Two further factors have raised the profile of the economic study of tourism. The first is the launching in 1995 of the subject's dedicated journal, *Tourism Economics*, in the UK, which quickly became an internationally reputable publication. Given the proliferation of tourism journals over the last twenty years, referred to below in Section 11.4, it is surprising that there have been no other journals introduced that focus primarily on the economics of the sector. Nevertheless, increasingly, well-respected journals, such as the *Annals of Tourism Research* and *Tourism Management*, long associated with the wider tourism literature, publish papers emanating from economists. The second factor was the inauguration of the International Association for Tourism Economics (IATE) at its first conference in Majorca, at the University of the Balearics in October 2007. This promises to be an important forum for the promotion of the economics of tourism. Very likely recognizing the increasing importance of the contributions on tourism from Asia, the second IATE conference in 2009 was scheduled to be in Thailand.

11.4 The wider context of the economic analysis of tourism

11.4.1 Introduction

Since the early 1990s there has been a transformation in tourism as an activity, with what almost amounts to a revolution in all aspects of it, that has raised innumerable new issues demanding attention. This has been echoed in the academic field, which has responded by expanding the volume of research in many new directions, several hitherto not anticipated. This expansion is exemplified by the explosion in the number of journals, many of which are concerned with quite specialized aspects of the sector. Currently (Srinivas, 2009) the number of journals concerned with tourism and related areas covering leisure, recreation, sport and the environment, has risen from 20 to 92 in less than twenty years. The list is by no means comprehensive, as Srinivas does not claim to have included every journal in the world, especially those not in the English language. In addition, there are articles that are tourism related in the journals of the applied disciplinary fields in economics; ecological and environmental economics are obvious examples. There are also contributions from other applied fields of the discipline; for example cultural, industrial and public sector economics. Furthermore, there are many articles on tourism in the conventional economic publications. However, tourism economics is not without its critics; it is still not fully recognized as a field worthy of study in the mainstream discipline.

11.4.2 Perspectives from the wider context of the study of tourism on its research in economics

The review so far conducted has only implicitly noted the issue of the future direction of economic tourism research. The questioning of the subject's single disciplinary approach, and indeed of other disciplines studying tourism that adhere to positivist scientific predictive, specialized deterministic, cause and effect method, has been raised in Chapter 1, in Section 1.2 on the conceptualization of tourism. The recognition and discussion of the difficulties of conceptualizing and analysing tourism are gathering momentum as indicated there and certainly hold implications for economic research. This issue is now reconsidered.

There is a danger that the economic quantitative studies that figure strongly in the submissions to the journals will isolate the subject, undermining its undoubted relevance to new research trends that are increasingly likely to transcend traditional disciplinary boundaries. Debate has occurred within economics because the system of thought and analytical method of much of mainstream thinking have emphasized the attainment of equilibrium outcomes. The subject have been criticized for its acceptance of the equilibrium condition and the associated restrictive assumptions that have narrowed the accepted orthodoxy. This is perceived as divorcing it from real world issues and problems. For instance, conventional consumer behaviour concepts have been challenged by economic psychology and the traditional theories of the firm and markets called in question by alternative views, such as those propounded by the Austrian school, which considers dynamic and disequilibrium circumstances. The range of different approaches in the discipline (reviewed by Greenaway et al. 1996) suggests that a pluralistic attitude is required, with cross-fertilization of concepts, theories and methods, both within and from outside the subject.

Contributions from other disciplines prominent in the tourism literature, such as geography and sociology, and to a lesser extent cultural studies, ecology, politics and psychology, seldom refer to economic analysis. Echoing the debate within economics, the criticisms of those from other disciplines regarding restrictive assumptions, the abstraction and simplification of real life circumstances and the heavy reliance on secondary data have some validity. It is possible that there is an element of cynicism by researchers in other disciplines about economic analysis, especially concerning the reliability of secondary data. This concern is partially justified, especially if the sources are tourism bodies. Mowforth and Munt (2003) and Wheeller (2004) are sceptical of the reliability of data emanating from the international organizations and some governments that have their own policy agendas to promote. Economists understandably retort that researchers from other disciplines do not fully comprehend the subject's methods. It is acknowledged that much quantitative economic research on tourism is very much at a micro level and therefore too specific to address what are basically macro issues. Despite many aspects of economic research being concerned with policy, does such specialization suggest that the subject is not actually connecting with that in the more general sphere of tourism enquiry and practice? Is it possible that economists are only communicating with each other and past everyone else in other disciplines? It is worrying that much relevant economic research is being largely ignored elsewhere.

Thus there is a need to convince those coming from other perspectives to acknowledge, and indeed accept, the significance of economics' contribution. A few examples from the literature research illustrate how this is the case. In examining the concept of carrying capacity, not one instance was found of any reference to the economic element. Another area of the paucity of an economic perspective is in publications on information technology (IT) in tourism. In a review by Sheldon (2006), of 78 references cited, she was only able to cite four that had any economic content, which is inexplicable, given that IT is essentially about such matters as competitiveness, efficiency, innovation diffusion, marketing and productivity, firmly within the purview of economics. In the relatively recent interest in crisis management, Ritchie (2004) makes no mention of the relevance of economics, referring only to business and environmental management, planning, geography and political science. To date little reference has been made, in the wider context, of the almost revolutionary conceptual and analytical developments in economics, for example the crucial empirical importance of the concept of market failure and welfare economics, the analytical methods of CBA in the environmental field with respect to the notion of TEV, the operationalization of the methods of valuation of non-traded resources and the rapid development of CGE modelling, which increasingly meets the criticisms levelled at economics referred to above.

11.5 The future direction of research on tourism

11.5.1 Introduction

It is clear from the literature review for this book that in the wider context the study of tourism is continuing to be turned upside down with regard to both the most appropriate analytical framework and the scope and content of research. The observations on the state of the economic study of tourism above suggest that economics needs to engage more with other disciplines to show its relevance and part of this must be to acknowledge and contribute to the debate on inter-and trans-disciplinarity. This issue has been anticipated in Sections 11.2.3 and 11.4.1 above. However, it was first raised in Chapter 1 concerning the attention given there to the definition and conceptualization of tourism. It was also included in the section on sustainability and tourism in Chapter 8 with regard to the academic dimension. Given the growing interest in, and indeed significance for, future research into tourism, the issue needs to be revisited. A brief rehearsal of the essence of the debate as to the future structure and direction is warranted.

11.5.2 Research issues in the wider context of tourism studies: agendas for the future?

The origins for an appraisal of the whole approach of research into tourism have come almost entirely from geographers and sociologists, for example Hardy (2002), Lew et al. (2004), Harrison (2007) and further references given in Chapters 1 and 8. There are several bases for the continuing and inconclusive debate, of which the most significant are the frustration with the intractability of deriving a theory of tourism, the tendency for research to be fragmented and ill-directed so that the desire to develop a tourism discipline is thwarted, the tendency for the many disciplinary studies of the sector to adhere to their own analytical frameworks and methodologies and the increasing specialization of papers in the subject's journals. Additionally, academic institutional structures and ethos with regard to tourism curricula are cited as issues (Airey and Johnson, 1999; Tribe, 2001). There are also fears that the development of tourism studies threatens the disintegration of disciplines, for instance as stated by Hall and Page (2006) with respect to geography. While the factors identified above have been cited as contributing to what is perceived by scholars as a hindrance to the development of tourism studies, there are two that are considered to be fundamental and need to be addressed; both were identified in Chapter 1.

The first is the recognition of tourism as a complex, dynamic, evolving, fragmented, unstable phenomenon, subject to much uncertainty and therefore not susceptible to accepted conventional analysis. This spawned the debate on the conceptualization of tourism that began in the 1990s and is ongoing; references on this aspect are given in Chapters 1 and 8. The second factor, in a sense emanating from the first, is that the characteristics of tourism are similar to events that can be explained by chaos theory, an explanation of which was given in Chapter 1. Effectively, chaos theory is the foundation on which a theoretical framework for the analysis of tourism can be erected. It is contrasted by both Faulkner and Russell (1997) and McKercher (1999) with the positivist, deterministic, linear, scientific paradigm, on which research in the social sciences, especially economics, has traditionally been based. Ritchie (2004), in an article on the application of chaos theory to crisis management, endorses the earlier contributions of Faulkner and Russell and McKercher in suggesting tourism is subject to volatility, exogenous events of considerable magnitude and sudden changes in consumer behaviour. He argues for a holistic, proactive, strategic approach to research that has implications for its operation.

Closely associated with the positing of the chaos theoretical approach is the argument that tourism analysis can no longer be encompassed within a single discipline. As indicated in Chapter 1, it is proposed (Tribe 1997, 2000, 2004; Lew et al. 2004; Wheeller, 2004; Coles, 2006; Weed 2006; Harrison 2007) that research should be at least multidisciplinary, ideally, recognizing the benefit of breaking down conventional disciplinary boundaries in 'post-disciplinary', 'knowledge-based' and 'research synthesis' frameworks.

The concentration on the fundamental issues of defining and conceptualizing tourism has tended to overshadow what may be considered as more prosaic and specific matters on future research strategies and agendas. There is a reluctance to identify particular aspects of concern with respect to policies on tourism, or its operation and management. In the context of wider tourism studies, there are pleas to reconsider the approach to research that are not quite as radical as abandoning the objective scientific method. Representative examples, drawn initially from the wider sphere of tourism studies of what has been published on the current state of research and its content, serve to show that they are given in rather general terms.

Pearce (2005) sets out in a rather abstract way what he sees as the way forward. He perceives the state of tourism research as being complex, diversified, fragmented and focusing on specific sectors or forms, such as demand for it and its impact, and studied by a wide range of disciplines. He considers that tourism research needs to be more explicit in how it contributes to the larger picture and to recognize that past studies are related to those in the future. Pearce also states that what is required is theorization, methodological rigour, sound techniques and operationalization. In summary, he calls for research to be more concerned with linking the literatures of different disciplines, to be comprehensive, structured and integrated in a systematic approach, with an emphasis on the interrelationship between and functioning of systems, and to have a clear sense of direction so that it moves from description to understanding.

Shaw and Williams (2002), in a book with the title *Critical Issues in Tourism*, do not specifically indicate what the future direction of research should be as they draw no conclusions. The content focuses on current issues relating to the broad categories of international tourism and globalization, access to tourism consumption, covering tourists' motivation and behaviour, the production of tourism services, including the leisure industries, entrepreneurship, employment and labour markets, and tourism environments, encompassing mass, urban, rural and sustainable tourism.

Lew et al. (2004), as editors in a collection of contributions from a largely geographic perspective, emphasize cultural, historic, tourist movement, place, space, environmental, planning, policies, social and governance dimensions. The editors, in an introductory chapter, identify three issues in tourism studies by a key word search of the CABI abstracts, for the last quarter of the twentieth century and beginning of the twenty-first. The three issues are those with a social science, geographic and economic orientation respectively, the first two reflecting the content of their book. However, the last is of interest even though it is not detailed enough to be of great value in tracing the changes in research interests in the economics of tourism. The broad categories cover tourism development, the sector's impact, policy, growth and three somewhat vague ones labelled respectively analysis, evaluation and economics. In 2002, by far and away the most prominent was the impact of tourism, with 82 articles, followed by development, with 38; growth was next with 11 and policy articles, numbering only three. The Lew et al. (2004) book has a concluding chapter in which future research themes and challenges are added to those outlined above; they are the need for more cultural approaches, greater attention to the scale of enquiries, such as those concerning globalization and the global–local nexus, placing tourism relationships in a framework to articulate tourist flows, space and time, consideration of environmental, especially sustainability issues, recognition of the economic and social contexts of tourism, more investigation of gender issues and the under-representation of the interest and needs of communities in destinations. Finally, Lew et al. (2004) identify the necessity of taking account of the external environment, particularly concerning the risk and uncertainty of events outside the control of tourists and tourism supply entities in a rapidly changing world; this is especially so in circumstances which are characterized as having economic, political and social inequities.

Hall and Page (2006), in their concluding chapter on the future of geographical research, assert that tourism is now seen as being an intrinsic part of human existence. Specific tourism issues that should be addressed are not identified in what is effectively an overview of the role of geography in the study of tourism and its ability to cross over into other disciplines. They do, however, reinforce the subject's core rationale of considering the dimensions in tourism of movement, people, places, space and time. Another feature of the discipline that they perceive as a contributory asset to tourism research is its two branches of human and physical geography, the latter focusing on the environmental aspects. They also posit that the attributes of geographers is their holistic problem-solving approach at an operational level, and their capabilities of engaging with stakeholders, from a position of recognizing their needs, to finding solutions relevant to communities and policy-makers.

11.5.3 Perceived future tourism research issues in economics

In the preparation of the new edition of this book, apart from the information provided by Wanhill (2007), little was found on what economists think are the big research issues facing their discipline and its place in the wider context of tourism studies. The firm impression gained is that economists remain largely unconcerned about reflecting at a general level on the state and future direction of their research into tourism, but they certainly do consider it in relation to their specific field of study (see for example, Dwyer and Forsyth 2006). To an extent scholars such as Nightingale (1978), Lieper (1979), and Wilson (1998), quoted in Chapter 1, with regard to the conceptualization of tourism and whether it is an industry or market, are exceptions.

In a fairly recent collection of papers by Matias et al. (2007), ostensibly on research advances in the economics of tourism, the introductory chapter suggests which issues require attention, consequent on the development of and fundamental structural changes in tourism. The editors briefly identify the in- and outbound macro- and microeconomic, environmental and socio-cultural drivers and impacts of tourism. Also the spatial and temporal effects are recognized, as is the significance of economic growth, the measurement of the size and importance of the tourism sector and adverse effects on the quality of life. Some emphasis is given to the acute problem of the paucity of statistics on tourism flows, nature and duration of holidays, transport modes patterns, infrastructure provision, tourism products (attractions, facilities and services), accommodation, tourists' choices of destinations and tourist expenditures. The contributions included in Matias et al. (2007) are claimed to be indicative of the new research challenges and trends covering new analytical frameworks, new operational tools and applied country and regional applications. This claim is not fully justified as several of the chapters consist of approaches and methods already in existence.

11.6 The future direction of economics tourism research

The appraisal of the state of research in tourism economics and other disciplines above in Sections 11.2, 11.3 and 11.4 does not reveal many prescriptions for its future direction except to make good shortcomings in current research interests. The reader may consider that the glance at the research challenges and issues identified in other disciplines concerned with the study of tourism in Section 11.5.2 is a distraction and of no consequence for economics, hence the question mark in the chapter heading. Moreover, even Section 11.5.3 on perceived future research issues in economics could be interpreted as simply continuing to dwell on existing ones. This understandable conclusion is questioned here.

The overriding issue in the wider context of the study of tourism that may well dictate future research is the analytical framework in which it is conducted. Tourism's rapid growth, until the global economic and financial crisis emerged in 2007, was phenomenal and generated fundamental changes in the pattern and nature of its demand and supply, giving rise to wide-ranging structural changes in the sector. While the advent of the crisis undoubtedly reduced the level of both domestic and international tourism, it has actually reinforced the profound changes in demand, supply and the sector's structure. A further factor is the occurrence of natural disasters and the increasing frequency of political unrest and terrorism. These events are adding to the recognition in the academic study of tourism of its complexity and its characteristics that tend to bear out its nature and thus its susceptibility to analysis within the chaos theory framework. This is in close association with the increasing misgivings over the efficacy of the hitherto unchallenged dominance of the positivist scientific method in the social sciences. Other factors, such as globalization, emergence of worldwide environmental issues, particularly the paramount goal of pursuing sustainability and the interrelationship of cultural, ecological, economic, political and social systems in which power structures are embedded, are compelling reasons for there to be a more holistic approach to tourism research. Effectively, all these factors call for inter- and trans-disciplinary research in which traditional disciplinary boundaries might be blurred, as was suggested in Chapter 1.

Do these possibilities hold implications for the future of economics research into tourism? They might well do if the subject is to contribute to resolving problems and determining policy initiatives at the operational level. Within the mainstream discipline there are signs of pressure to broaden its perspective coming from the newer fields of ecological and environmental economics. While it was shown in Chapter 9 that environmental economics largely conforms to conventional analytical approaches, the ecological research field has widened the scope of economics by acknowledging the relevance of and embracing the natural sciences, sociology, cultural, ethical and political studies and welfare economics, which recognizes the normative elements of the subject. Furthermore, to an extent, urban and regional economics, considered in Chapter 6, has stressed the spatial dimension of the discipline. The preliminary observations by Matias et al. (2007), given in Section 11.5.3 from an economic viewpoint, are of value in that they correctly identify ongoing issues and challenges that require resolution, particularly those relating to the lack of availability of data.

Looking outside economics at the contribution of other disciplines to the study of tourism, examined in Section 11.5.2, is instructive for it adds force to the need to broaden the perspective of the subject; this should determine the future of economics research. Attention is concentrated on the approach in geography, particularly the perception by Hall and Page (2006) as to the position of their subject. Their insights echo the observations made in the previous paragraph. Geography has the advantage of a tradition of a very wide approach to what it studies, covering the physical environment and the human and economic aspects of the world. In this respect it has a more holistic paradigm, embodying the place, movement and time dimensions of tourism that are the very essence of the phenomenon. The association between economics and geography is acknowledged in this book by indicating in Chapter 6 that the latter is the foundation of spatial economics and by giving an exposition of the New Economic Geography. Geographers have made a sound contribution to the definition and conceptualization of tourism and have undertaken research on aspects, identified below, that economics has not.

It is not an inappropriate observation to make that tourism economics research has not fully taken cognizance of the developments within its own subject's boundaries, nor within other disciplines. This field of economics is largely continuing to work within the traditional positivist paradigms of micro and macroeconomics, in which a number of elements of tourism are still being considered at a theoretical rather than at an empirical level. It is instructive to reiterate three examples of areas of study that are sketchy in terms of applied research; the environmental issues concerning the impact on the performance of tourism businesses of the technological improvements affecting the use of non-renewable resources, the relief of poverty and the pursuit of sustainable tourism.

It is only recently that study has been undertaken into the impact of investment in new technologies to reduce the use of non-renewable and energy resources and the generation of waste. The examination of the California effect and Porter hypothesis in Chapter 8, concerning the effect on efficiency, productivity and profitability, and therefore competitiveness (very much the kind of research that should be undertaken by economists), of the voluntary adoption of innovative technologies to reduce the environmental impact of businesses, revealed that very little research has been undertaken by the discipline in relation to tourism (Razumova, 2007). Poverty alleviation, and the securing of the triple bottom line in tourism destinations, considered to be a crucial ongoing issue globally, was investigated in Chapter 9. It is another neglected area of study in tourism economics. Most work has been conducted by geographers, who have investigated the role of responsible tourism, particularly its potential to relieve poverty. Advocacy and discussion about achieving sustainable tourism have rumbled on for more than thirty years and have yet to become an operational reality. Again, it is not an area of research that has commanded serious attention from tourism economists, mainly because of problems concerning its definition and conception. However, this is curious given that there have been significant developments in environmental economics, (see Chapter 8) in the identification of the conditions for sustainability to occur and indicators that it has been achieved, which could be the basis for applying these factors to sustainable tourism.

11.7 Postscript

The economics of tourism is in somewhat of a quandary. It quite rightly values its long-standing and established theoretical credentials and its rigorous quantitative methods of analysis with explanatory and predictive power way ahead of all the other disciplines involved in the study of tourism. However, the discussion throughout this book, and especially in the previous section, suggests that the way in which tourism is developing, and the global cultural, economic, financial, environmental, political and social situation, demands that academic research analytical approaches in the future will need to change radically. With regard to specific issues, high priority will need to be given to the impact on tourism of its rate and pattern of growth. Other notable challenges are: climate change and global warming; environmental and ecological degradation and biodiversity loss; air, land and water pollution; depletion of productive and energy resources; globalization of trade in goods and services, and intra- and intergenerational inequalities. Moreover, at a more specific level the emphasis will need to be on resolving the problems of an operational and practical nature in the sector. While the continued development of quantitative methods (the strength of economics) is desirable, there are very likely many instances which demand approaches that are pragmatic, qualitative and ad hoc.

REFERENCES

Aaronovich, S. and Sawyer, M. (1975) *Big Business*, London: Macmillan.

Abowd, J. and Card, D.A. (1989) 'On the covariance structure of earnings and hours changes', *Econometrica*, 57: 411–55.

ABTA (2008) *Annual Report and Financial Highlights for the Year Ended 30 June 2008*, London: ABTA.

Adams, B. and Parmenter, B. (1999) *General Equilibrium Models* in *Valuing Tourism: Methods and Techniques*, Canberra: Bureau of Tourism Research, 3–12.

Adams, P.D. and Parmenter, B.R. (1995) 'An applied general equilibrium analysis of the economic effects of tourism in a quite small, quite open economy', *Applied Economics*, 27: 985–94.

ADB (1995) Indonesia–Malaysia–Thailand Growth Triangle Development Project, Regional Technical Assistance 5550, vol. VI Tourism, Manila: Asian Development Bank.

Adkins, L. (1995) *Gendered work: sexuality family and the labour market*, Buckingham and Philadelphia: Open University Press.

Aghion, P. and Howitt, P. (1992) 'A model of growth through creative destruction', *Econometrica*, 60: 323–51.

—— (1998) *Endogenous Growth Theory*, Cambridge, Mass.: MIT Press.

Aghion, P., Harris, C., Howitt, P., Vickers, J. (2001) 'Competition, imitation, and growth with step-by-step innovation', *Review of Economic Studies*, 68: 467–92.

Agiomirgianakis, G., Vlassis, M. and Thompson, H. (2006) *International Trade – International Economic Relations* (in Greek), Athens: Rosili.

Airey, D. and Johnson, S. (1999) 'The content of tourism degree courses in the UK', *Tourism Management*, 20: 229–35.

Airline Business (2003) *IT Trends Survey 2003* (CD-ROM).

—— (2008) *Airline Rankings – Passenger*, August, 64–70.

Airports Council International (2007) *ACI Europe Position on Allocation of Slots*, Online. Available HTTP: <http://www.acieurope.org/upload/ACI%20EUROPE%20POSITION_Slot%20allocation%20June%2007.pdf> (accessed 11 February 2008).

Akerlof, A.G. (1970) 'The Market for Lemons: Quality Uncertainty and the Market Mechanism', *Quarterly Journal of Economics*, 84: 488–500.

Alam, A. (1995) 'The new trade theory and its relevance to the trade policies of developing countries', *The World Economy*, 23: 367–85.

Alavalapati, J.R.R. and Adamowicz, W.L. (2000) 'Tourism impact modelling for resource extraction regions', *Annals of Tourism Research*, 27: 188–202.

Alegre, J. and Juaneda, C. (2006) 'Destination loyalty: consumers' economic behaviour', *Annals of Tourism Research*, 33: 684–706.

Alegre, J. and Pou, L. (2006) 'The length of stay in the demand for tourism', *Tourism Management*, 27: 1343–55.

Alexandersson, G., Nash, C. and Preston, J. (2008) 'Risk and reward in rail contracting', *Research in Transportation Economics*, 22: 31–35.

Allinson, G., Ball, S., Cheshire, P.C., Evans, A.W. and Stabler, M.J. (1996) *The Value of Conservation?* London: English Heritage.

American Society of Travel Agents (2008) *ASTA Agency Profile*, Online. Available HTTP: <http://www.asta.org/about/content.cfm?Item Number=2853&navItemNumber=518> (accessed 26 April 2009).

Amin, S. (1976) *Unequal Development: An Essay on the Social Formations of Peripheral Capitalism*, New York: Monthly Review Press.

Anas, A. (1990) 'Taste heterogeneity and urban spatial structure: the logit model and monocentric city reconciled', *Journal of Urban Economics*, 28: 315–35.

Anastassopoulos, G., Filippaios, F. and Phillips, P. (2007) *An 'Eclectic' Investigation of Tourism Multinationals' Activities: Evidence from the Hotels and Hospitality Sector in Greece*, Hellenic Observatory Papers on Greece and Southeast Europe, GreeSE Paper No. 8, November, London: London School of Economics.

Anderson, D.M. (2003) Testimony: *Hearing on the Harmful Algal Bloom and Hypoxia Research Amendments Act of 2003*, Washington, DC: Subcommittee on Environment, Technology and Standards, US House of Representatives.

Anderson, L.G. (1995) 'Privatizing open access fisheries: individual transferable quotas', in Bromley, D.W. (ed.) *Handbook of Environmental Economics*, Oxford: Blackwell, 453–74.

Anderson, R.J. and Crocker, T.D. (1971) 'Air pollution and property values', *Urban Studies*, 8: 171–80.

Andronikou, A. (1987) 'Cyprus: management of the tourist sector', *Tourism Management*, 7: 127–29.

Archer, B.H. (1973) *The Impact of Domestic Tourism*, Occasional Papers in Economics, no. 2, Bangor: University of Wales Press.

—— (1976) *Demand Forecasting in Tourism*, Occasional Papers in Economics, no. 9, Bangor: University of Wales Press.

—— (1977a) *Tourism Multipliers: The State of the Art*, Occasional Papers in Economics, no. 11, Bangor: University of Wales Press.

—— (1977b) *Tourism in the Bahamas and Bermuda: Two Case Studies*, Occasional Papers in Economics, no. 10, Bangor: University of Wales Press.

—— (1989) 'Tourism and island economies: impact analyses', in C.P. Cooper (ed.) *Progress in Tourism, Recreation and Hospitality Management, vol. 1*, London: Belhaven, 125–134.

—— (1995) 'Importance of tourism for the economy of Bermuda', *Annals of Tourism Research*, 22: 918–30.

Archer, B.H. and Fletcher, J. (1996) 'The economic impact of tourism in the Seychelles', *Annals of Tourism Research*, 23: 32–47.

Archer, B.H., Cooper, C. and Ruhanen, L. (2005) 'The positive and negative impacts of tourism', in Theobold, W.F. (ed.) 3rd edn, *Global Tourism*, Oxford: Butterworth Heinemann, 79–102.

Armstrong, M. and Porter, R. (2007) *Handbook of Industrial Organization Vol. 3*, Oxford: Elsevier.

Armstrong, M., Cowan, S. and Vickers, J. (1994) *Regulatory Reform: Economic Analysis and British Experience*, Cambridge, Mass.: MIT Press.

Arrow, K.J. (1962) 'The economic implications of learning by doing', *Review of Economic Studies*, 29: 155–73.

—— (1975) 'Vertical integration and communication', *Bell Journal of Economics*, 6: 173–83.

—— (1985) 'The economics of agency', in Pratt, J. and Zeckhauser, R. (eds), *Principals and Agents*, Boston: Harvard Business School, 37–51.

Artus, J.R. (1972) 'An econometric analysis of international travel', *IMF Staff Papers*, 19: 579–614.

Arvanitis, P. and Zenelis, P. (2008) 'Africa', in Graham, A., Papatheodorou, A. and Forsyth, P. (eds) *Aviation and Tourism: Implications for Leisure Travel*, Aldershot: Ashgate, 303–12.

Asabere, P.K., Hachey, G. and Grubaugh, S. (1989) 'Architecture, historic zoning, and the value of homes', *Journal of Real Estate, Finance and Economics*, 2: 181–95.

Ashworth, G. and Stabler, M.J. (1988) 'Tourism development planning in Languedoc: Le mission impossible?', in Goodall, B. and Ashworth, G. (eds) *Marketing in the Tourism Industry: the Promotion of Destination Regions*, London: Croom Helm, 187–97.

Ayres, R. (1998) 'Ecothermodynamics and the Second Law', *Ecological Economics*, 26: 198–210.

Azariadis, C. and Drazen, A. (1990) 'Threshold externalities in economic development', *Quarterly Journal of Economics*, 105: 501–26.

Bagguley, P. (1990) 'Gender and labour flexibility in hotels and catering', *Service Industries Journal*, 10: 737–47.

Baier, S.L. and Glomm, G. (2001) 'Long-run growth and welfare effects of public policies with distortionary taxation', *Journal of Economic Dynamics and Control*, 25: 2007–42.

Bain, J.S. (1956) *Barriers to New Competition*, Cambridge, Mass.: Harvard University Press.

Baker, S., Kousis, M., Richardson, D. and Young, S. (1997) *The Politics of Sustainable Development: Theory, Policy, and Practice within the European Union*, London: Routledge.

Balchin, P.N., Kieve, J.L. and Bull, G.H. (1988) *Urban Land Economics and Public Policy*, 4th edn, London: Macmillan.

Barbier, E.B. (1992) 'Community-based development in Africa', in Swanson, T.M. and Barbier, E.B. (eds) *Economics for the Wilds*, London: Earthscan.

Baretje, R. (1982) 'Tourism's external account and the balance of payments', *Annals of Tourism Research*, 9: 57–67.

BarOn, R.R. (1979) 'Forecasting tourism—theory and practice', TTRA (Travel and Tourism Research Association) Tenth Annual Conference Proceedings, University of Utah.

—— (1983) 'Forecasting tourism by means of travel series over various time spans under specified scenarios', Third International Symposium on Forecasting, Philadelphia.

Barro, R.J. (1990) 'Government spending in a simple model of endogenous growth', *Journal of Political Economy*, 98: S103–S125.

—— (1991) 'Economic growth in a cross-section of countries', *Quarterly Journal of Economics*, 106: 409–43.

Barro, R.J. and Sala-i-Martin, X. (1992) 'Public finance in models of economic growth', *Review of Economic Studies*, 54: 646–61.

—— (2004) *Economic Growth*, Cambridge, Mass.: MIT Press.

Bastakis, C., Buhalis, D. and Butler, R. (2004) 'The perception of small and medium sized tourism accommodation providers on the impacts of the tour operators' power in Eastern Mediterranean', *Tourism Management*, 25: 151–70.

Basu, K. (1993) *Lectures in Industrial Organisation Theory*, Oxford: Basil Blackwell.

Bateman, I.J., Willis, K.G., Garrod, G.D., Doktor, P., Langford, I. and Turner, R.K. (1992) 'Recreation and environmental preservation value of the Norfolk Broads: a contingent valuation study', Unpublished Report, Environmental Appraisal Group, University of East Anglia.

Baum, T. and Mudambi, R. (1995) 'An empirical analysis of oligopolistic hotel pricing', *Annals of Tourism Research*, 22: 501–16.

Baumol, W.J. (1982) 'Contestable markets: an uprising in the theory of industry structure', *American Economic Review*, 72: 1–15.

Beioley, S. (1995) 'Green tourism—soft or sustainable?', *Insights*, May: B75–B89.

Ben-Akiva, M. and Lerman, S.R. (1985) *Discrete Choice Analysis – Theory and Application to Travel Demand*, Cambridge, Mass.: MIT Press.

Bennett, J. and Blamey, R. (2001) *The Choice Modelling Approach to Environmental Valuation*, Cheltenham: Edward Elgar.

Bennett, M.M. (1993) 'Information technology and travel agency: a customer service perspective', *Tourism Management*, 14: 259–66.

Benos, N. (2008) 'Education policy, growth and welfare', *Education Economics* (forthcoming), available online

Berkes, F. (ed.) (1989) *Common Property Resources: Ecology and Community-Based Sustainable Development*, London: Belhaven.

Best Western (2009) *About Best Western*, Online. Available HTTP: <http://www.bestwestern.com/aboutus/index.asp> (accessed 3 May 2009).

Bhagwati, J.N. (1968) 'Distortions and immiserizing growth: a generalization', *Review of Economic Studies*, 35: 481–85.

Bianchi, R. (2002) 'Towards a new political economy of global tourism' in Sharpley, R. and Telfer, D. (eds) *Tourism and Development: Concepts and Issues,* Clevedon: Channel View Publications, 265–99.

Blake, A.T. (2000) *The Economic Effects of Tourism in Spain*, Tourism and Travel Research Institute Discussion Paper, 2001/2, Nottingham University Business School.

—— (2005) *The Economic Impact of the London 2012 Olympics*, Report of the Christel DeHaan Tourism and Travel Research Institute, Nottingham: Nottingham University Business School.

Blake, A.T. and Sinclair, M.T. (2003) 'Tourism crisis management: US response to September 11', *Annals of Tourism Research*, 30: 813–32.

Blake, A.T., Gillham, J. and Sinclair, M.T. (2006) 'CGE Tourism Analysis and Policy Modelling', in Dwyer, L. and Forsyth, P. (eds) *International Handbook on the Economics of Tourism*, Cheltenham: Edward Elgar Publishing, 301–15.

Blake, A.T, Sinclair, M.T. and Sugiyarto, G. (2003) 'Quantifying the effect of foot and mouth disease on tourism and the UK economy', *Tourism Economics*, 9: 449–65.

Blake, A.T., Arbache, J.S., Sinclair, M.T. and Teles, V. (2008) 'Tourism and poverty relief', *Annals of Tourism Research*, 35: 107–26.

Blake, A.T., Durbarry, R., Sinclair, M.T. and Sugiyarto, G. (2000) *Modelling Tourism and Travel using Tourism Satellite Accounts and Tourism Policy and Forecasting Models*, Tourism and Travel Research Institute Discussion Paper, 2001/4, Nottingham University Business School.

Blinder, A.S. and Deaton, A.S. (1985) 'The time-series consumption revisited', *Brookings Papers on Economic Activity*, 465–521.

Blundell, R., Pashardes, P. and Weber, G. (1993) 'What do we learn about consumer demand patterns from micro data?', *American Economic Review*, 83: 570–97.

Board, J., Sinclair, M.T. and Sutcliffe, C.M.S. (1987) 'A portfolio approach to regional tourism', *Built Environment*, 13: 124–37.

Bockstael, N.E. (1995) 'Travel cost methods,' in Bromley, D.W. (ed.) *The Handbook of Environmental Economics*, Oxford: Blackwell, 655–671.

Boettke, P.J. and Leeson, P.T. (2003) '28A. 'Postwar heterodox economics, the Austrian School of Economics, 1950–2000', in Samuels, W. Biddle, J.E. and Davis, J.B. (eds) *A Companion to the History of Economic Thought*. Blackwell Publishing, 446–52.

Bote Gómez, V. (1988) *Turismo en espacio rural: rehabilitatión del patrimonio sociocultural y de la economía local*, Madrid: Editorial Popular.

—— (1990) *Planificación económica del turismo: de una estrategia masiva a una artesanal*, Mexico: Editorial Trillas.

—— (1993) 'La necesaria revalorización de la actividad turística española en una economía terciarizada e integrada en la CEE', *Estudios Turísticos*, 118: 5–26.

—— (1996) 'Tourism demand and supply in Spain', in Barke, M. Newton, M. and Towner, J. (eds) *Tourism in Spain: Critical Perspectives*, Wallingford: C.A.B. International.

Bote Gómez, V., Sinclair, M.T., Sutcliffe, C.M.S. and Valenzuela, M. (1989) 'Vertical integration in the British/Spanish tourism industry', in *Leisure, Labour and Lifestyles: International Comparisons, Tourism and Leisure. Models and Theories*, Proceedings of the Leisure Studies Association Second International Conference, Brighton, Conference Papers no. 1, 8, 39.

Boulding, K.E. (1966) 'The economics of the coming spaceship earth', in Jarrett, H. (ed.) *Environmental Quality in a Growing Economy*, Baltimore: Johns Hopkins University Press.

Bowes, G. (2004) 'Thais target women to shed sleazy image,' *The Observer* 17th October, 10.

Bowers, J. (1997) *Sustainability and Environmental Economics*: *An Alternative Approach*, Harlow: Addison Wesley Longman.

Boyfield, K., Starkie, D., Bass, T. and Humphreys, B. (2003) *A Market in Airport Slots*, London: Institute of Economic Affairs.

Bramwell, B. (2007) 'Opening up new spaces in the sustainable tourism debate', *Tourism Recreation Research*, 32: 1–9.

Bramwell, B. and Lane, B. (2000) *Tourism Collaboration and Partnerships, Politics, Practice and Sustainability*, Clevedon: Channel View.

Bramwell, B., Henry, I.P., Jackson, G., Goytia Prat, A., Richards, G. and van der Straaten, J. (eds) (1996) *Sustainable Tourism Management Principles and Practice*, Tilburg: Tilburg University Press.

Brander, J.A. (1981) 'Intra-industry trade in identical commodities', *Journal of International Economics*, 11: 1–14.

Brander, J.A. and Krugman, P.R. (1983) 'A reciprocal dumping model of international trade', *Journal of International Economics*, 15: 313–21.

Brau, R., Lanza, A. and Pigliaru, F. (2003) 'How fast are the tourism countries growing? The cross-country evidence', Fondazione Eni Enrico Mattei. *Nota di Lavoro*, No. 85.

Braun, P.A., Constantinides, G.M. and Ferson, W.E. (1993) 'Time non-separability in aggregate consumption: international evidence', *European Economic Review*, 37: 897–920.

Breathnach, P., Henry, M., Drea, S. and O'Flaherty, M. (1994) 'Gender in Irish tourism employment', in Kinnaird, V. and Hall, D. (eds) *Gender: A Tourism Analysis*, Chichester: John Wiley.

Brierton, U.A. (1991) 'Tourism and the environment', *Contours*, 5: 18–19.

Briguglio, L., Archer, B., Jafari, J. and Wall, G. (1996) (eds) *Sustainable Tourism in Islands and Small States*: *Issues and Policies*, London: Pinter.

Britton, S.G. (1982) 'The political economy of tourism in the third world', *Annals of Tourism Research*, 9: 331–58.

—— (1991) 'Towards a critical geography of tourism', *Environment and Planning D. Society and Space*, 9: 451–78.

Brockhoff, K. (1975) 'The performance of forecasting groups in computer dialogue and face to face discussion', in Limestone, H.A. and Turoff, M. (eds) *The Delphi Method Techniques and Applications*, Reading, Mass.: Addison-Wesley.

Brooke, A., Kendrick, D. and Meeraus, A. (1988) *GAMS: A User's Guide*, San Francisco: The Scientific Press.

Brookshire, D., Eubanks, L. and Randall, A. (1983) 'Estimating option price and existence values for wildlife resources', *Land Economics*, 59: 1–15.

Brown, G. Jr, and Henry, W. (1989) *The Economic Value of Elephants*, London Environmental Economics Centre Paper 89–12, University College London.

Brown, G. and Mendelsohn, R. (1984) 'The hedonic travel cost method', *Review of Economics and Statistics*, 66: 427–33.

Brozen, Y. (1971) 'Bain's concentration and rates of return revisited', *Journal of Law and Economics*, 14: 351–69.

Bryden, J.M. (1973) *Tourism and Development: A Case Study of the Commonwealth Caribbean*, Cambridge: Cambridge University Press.

Buchanan, J.M. (1968) *Demand and Supply of Public Goods*, Chicago, Ill.: Rand McNally.

Buckley, R.C. (2002) 'A global triple-bottom-line evaluation for ecotourism', Paper delivered at the World Ecotourism Summit, Quebec.

—— (2003a) *Case Studies in Ecotourism*, Wallingford: CABI.

—— (2003b) 'Environmental inputs and outputs in ecotourism; geotourism with a positive triple bottom line?' *Journal of Ecotourism*, 2: 76–82.

Buhalis, D. (2006) 'The impact of information technology on tourism competition', in Papatheodorou, A. (ed.) *Corporate Rivalry and Market Power: Competition Issues in the Tourism Industry*, London: I.B. Tauris, 143–71.

Bull, A. (1999) 2nd edn, *The Economics of Travel and Tourism*, Melbourne: Addison Wesley Longman.

Burgan, B.J. and Mules, T. (2001) 'Reconciling cost benefit and economic impact assessment for event tourism', *Tourism Economics*, 7: 321–30.

Burns, A.C. and Ortinau, D.J. (1979) 'Underlying perceptual patterns in husband and wife purchase decision influence assessments', *Advances in Consumer Research*, 6: 372–76.

Burns, P. and Cleverdon, R. (1995) 'Destination on the edge? The case of the Cook Islands', in Conlin, M.V. and Baum, T. (eds) *Island Tourism*, Chichester: John Wiley.

Burns, P. and Holden, A. (1995) *Tourism: A New Perspective*, London: Prentice Hall.

Butler, R.W. (1980) 'The concept of a tourist area cycle of evolution: implications for management of resources', *Canadian Geographer*, 14: 5–12.

—— (1999) 'Sustainable tourism: a state of the art review', *Tourism Geographies*, 1: 7–25.

—— (2005) 'Developing the destination: difficulties in achieving sustainability', in Alejziak, W. and Winiarski, R. (eds) *Tourism in Scientific Research*, Krakow: University of Information Technology and Management, Rzeszow, 33–46.

CAA (2000) *The 'Single Till' and the 'Dual Till' Approach to the Price Regulation of Airports*, London: CAA.

—— (2005) *Financial Protection for Air Travellers and Package Holidaymakers in the Future*, CAP 759, London: CAA.

—— (2006) *Reforming Airport Slots Allocation in Europe: Making the most of a Valuable Resource*, London: CAA.

—— (2007) *ATOL Business 29*, London: CAA.

—— (2008) *Air Travel Trust: Report and Accounts*, London: CAA.

Caballero, R.J. (1993) 'Durable goods: an explanation for their slow adjustment', *Journal of Political Economy*, 101: 351–84.

Cai, J., Leung, P. and Mak, J. (2006) 'Tourism's forward and backward linkages', *Journal of Travel Research*, 45: 36–52.

Campbell, J.Y. and Mankiw, N.G. (1991) 'The response of consumption to income: a cross-country investigation', *European Economic Review*, 35: 715–21.

Cannon, E.S. (2000) 'Economies of scale and constant returns to capital: a neglected early contribution to the theory of economic growth', *American Economic Review*, 90: 292–95.

Capó-Parrilla, J., Riera Font, A. and Rosselló-Nadal, J. (2007) 'Tourism and long term growth: a Spanish perspective', *Annals of Tourism Research*, 34: 709–26.

Carlton, D.W. (1979) 'Vertical integration in competitive markets under uncertainty', *Journal of Industrial Economics*, 27: 189–209.

Carson, R. (1962) *Silent Spring*, Boston: Houghton Mifflin.

Casagrandi, R. and Rinaldi, S. (2002) 'A theoretical approach to tourism sustainability', *Conservation Ecology*, 6: 13–27.

Castelberg-Koulma, M. (1991) 'Greek women and tourism: women's co-operatives as an alternative form of organization', in Redclift, N. and Sinclair, M.T. (eds) *Working Women: International Perspectives on Labour and Gender Ideology*, London and New York: Routledge.

Cater, C. and Cater, E. (2007) *Marine Ecotourism: Between the Devil and the Deep Blue Sea*, Wallingford: CABI.

Cater, E. and Goodall, B. (1992) 'Must tourism destroy its resource base?', in Mannion, A.M. and Bowlby, S.R. (eds) *Environmental Issues in the 1990s*, Chichester: John Wiley, 307–313.

Cater, E. and Lowman, G. (eds) (1994) *Ecotourism: a sustainable option?* Chichester: Wiley.

Cazes, G. (1972) 'Le Rôle du tourisme dans la croissance économique: rèflexions à pártir de trois examples antillais', *The Tourist Review*, 27: 93–98 and 144–48.

Center Parcs (2009) *Discover Center Parcs*, Online. Available HTTP: <http://www.centerparcs.co.uk> (accessed 5 May 2009).

Chadha, B. (1991) 'Wages, profitability and growth in a small open economy', *IMF Staff Papers*, 38: 59–82.

Chakwin, N. and Hamid, N. (1996) 'The economic environment in Asia for investment', in Fry, C. and Oman, C. (eds) *Investing in Asia*, Paris: OECD.

Chamberlin, E.H. (1933) *The Theory of Monopolistic Competition*, Cambridge, Mass.: Harvard University Press.

Chant, S. (1997) 'Gender and tourism employment in Mexico and the Philippines', in Sinclair, M.T. (ed.) *Gender, Work and Tourism*, London and New York: Routledge.

Chapman, D. (1999) *Environmental Economics: Theory, Application and Policy*, Harlow: Addison Wesley.

Chen, B.L. (2006) 'Economic growth with an optimal public spending composition', *Oxford Economic Papers*, 58: 123–36.

Chen, J.J. and Dimou, I. (2005) 'Expansion strategy of international hotel firms', *Journal of Business Research*, 58: 1730–40.

Cheshire, P.C. and Sheppard, S. (1995) 'On the price of land and the value of amenities', *Economica*, 62: 247–68.

Cheshire, P.C. and Stabler, M.J. (1976) 'Joint consumption benefits in recreational site "surplus": an empirical estimate', *Regional Studies*, 10: 343–51.

Chevas, J.P, Stoll, J. and Sellar, C. (1989) 'On the commodity value of travel time in recreational activities', *Applied Economics*, 21: 711–22.

Christaller, W. (1933) *Die zentrale Orte in Süddeutschland*, Jena: Gustav Fischer Verlag.

—— (1963) 'Some considerations of tourism location in Europe: the peripheral regions – underdeveloped countries – recreation areas,' *Regional Science Association Papers*, 12: 95–105.

—— (1966) *Central Places in Southern Germany*, Trans. C.W. Baskin, Englewood Cliffs: Prentice Hall.

Christou, E. and Eaton, J. (2000) 'Management competencies for graduate trainees of hospitality and tourism programs', *Annals of Tourism Research*, 27: 1058–61.

Clapp, J.M. (1986) 'Interdependent behaviour with relocation costs – the comparative statics of spatial history', *Journal of Regional Science*, 26: 33–46.

Clarke, C.D. (1981) 'An analysis of the determinants of the demand for tourism in Barbados', PhD thesis, Fordham University, USA.

Clarke, J. (1997), A framework of approaches to sustainable tourism', *Journal of Sustainable Tourism*, 5: 224–33.

Clarke, R. and Davies, S.W. (1982) 'Market structure and price-cost margins', *Economica*, 49: 277–88.

Clawson, M. and Knetsch, J.L. (1966) *Economics of Outdoor Recreation*, Baltimore: Johns Hopkins University Press.

Clewer, A., Pack, A. and Sinclair, M.T. (1990) 'Forecasting models for tourism demand in city dominated and coastal areas', *European Papers of the Regional Science Association*, 69: 31–42.

—— (1992) 'Price competitiveness and inclusive tour holidays in European cities', in Johnson, P. and Thomas, B. (eds) *Choice and Demand in Tourism*, London: Mansell, 123–43.

Coase, R.H. (1960) 'The problem of social cost', *Journal of Law and Economics*, 3: 1–44.

Coccossis, H. (1996), Tourism and sustainability: perspectives and implications, in Priestley, G.K., Edwards, J.E., Coccossis, H. (eds) *Sustainable Tourism? European Experiences*, Wallingford: CABI, 1–21.

Coles, T., Hall, C.M. and Duval, D.T. (2006) 'Tourism and post-disciplinary enquiry', *Current Issues in Tourism*, 9: 293–319.

Combs, J.P., Kirkpatrick, R.C., Shogren, J.F. and Herriges, J.A. (1993) 'Matching grants and public goods: a closed-ended contingent valuation experiment', *Public Finance Quarterly*, 21: 178–95.

Commission of the European Communities (1994) *Report by the Commission to the Council, to the European Parliament and the Economic and Social Committee on Community Measures affecting Tourism, Council Decision 92/421/EEC*, Brussels: Commission of the European Communities.

—— (1995a) *The Role of the Union in the Field of Tourism, Commission Green Paper*, COM (95) 97, Brussels: Commission of the European Communities.

—— (1995b) *Consultation on the basis of the Green Paper: A Step further towards Recognition of Community Action to Assist Tourism, Forum on European Tourism*, Brussels: Commission of the European Communities.

—— (1996) *Proposal for a Council Decision on a First Multiannual Programme to Assist European Tourism 'Philoxenia' (1997–2000)*, COM (96) 168, Brussels: Commission of the European Communities.

Common, M. (1973) 'A note on the use of the Clawson method', *Regional Studies*, 7: 401–6.

Common, M. and Stagl, S. (2005) *Ecological Economics: An Introduction*, Cambridge: Cambridge University Press.

Competition Commission (2002) *P&O Princess Cruises plc and Royal Caribbean Cruises Ltd – A Report on the Proposed Merger*, London: Competition Commission.

Conlin, M.V. and Baum, T. (1995) *Island Tourism*, Chichester: John Wiley.

Connor, S. (2008a) 'The world's oceans at risk from rising acidity', *Independent*, 23rd May, 11.

—— (2008b) 'Revealed: polluting impact of humans on the oceans', *Independent*, 15th February, 10–11.

—— (2008c) 'Tiny snail crucial to Antarctic life may be wiped out', *Independent*, 19th February, 11.

Conrad, J.M. (1995) 'Bioeconomic models of the fishery', in Bromley, D.W. (ed.) *Handbook of Environmental Economics*, Oxford: Blackwell, 405–432.

Contractor, J.F. and Kundu, K.S. (1998) 'Modal choice in the world of alliances: analyzing organizational forms in the international hotel sector', *Journal of International Business Studies*, 29: 325–58.

Cooper, C.P. (1992) 'The life cycle concept and tourism', in Johnson, P. and Thomas, B. (eds) *Choice and Demand in Tourism*, London: Mansell, 145–160.

Cooper, C.P. and Wanhill, S. (1997) (eds) *Tourism Development: Environmental and Community Issues*, Chichester: Wiley.

Cooper, C.P., Fletcher, J., Gilbert, D., Fyal, A. and Wanhill, S. (2004) *Tourism: Principles and Practice*. 3rd edn, Harlow: Financial Times/ Prentice Hall.

—— (2008) *Tourism: Principles and Practice*, 4th edn, Harlow: Financial Times/Prentice Hall.

Copeland, B.R. (1991) 'Tourism, welfare and de-industrialization in a small open economy', *Economica*, 58: 515–29.

Copeland, B.R. and Taylor, M.S. (2004) 'Trade, growth and the environment', *Journal of Economic Literature*, 42: 7–71.

Cornelissen, S. (2005) *The Global Tourism System*. Aldershot: Ashgate.

Costantino, C. and Tudini, A. (2005) 'How to develop an accounting framework for ecologically sustainable tourism', in Lanza, A., Markandya, A. and Pigliari, F. (eds) *The Economics of Tourism and Sustainable Development*, Cheltenham: Edward Elgar, 104–72.

Costanza, R. (1984) 'Natural resource valuation and management: towards an ecological economics', in Jansson, A.M. (ed.) *Integration of Economy and Ecology: An Outlook for the Eighties*, Stockholm: University of Stockholm Press, 7–18.

—— (1989) 'What is ecological economics?', *Ecological Economics*, 1: 1–17.

—— (1998) 'The value of ecosystems services: putting the issues into perspective', *Ecological Economics*, 25: 67–72.

Costanza, R., d'Arge, R., de Groot, R., Farber, S., Grasso, M., Hannon, B., Naeem, S., Limburg, K., Paruelo, J., O'Neill, R.V., Raskiri, R.,

Sutton, P. and van den Bett, M. (1997) 'The value in the world's ecosystem services and natural capital', *Nature*, 387 (6630): 253–60.

Cowling, K. and Waterson, M. (1976) 'Price-cost margins and market structure', *Economica*, 43: 267–74.

Crompton, J.L. (1979) 'Motivations for pleasure vacation', *Annals of Tourism Research*, 6: 408–24.

Crouch, G.I. and Ritchie, J.R.B. (2006) 'Destination competitiveness', in Dwyer, L. and Forsyth, P. (eds) *International Handbook on the Economics of Tourism*, Cheltenham: Edward Elgar Publishing, 419–33.

Cruise Lines International Association (2008) *2008 CLIA Cruise Market Overview*, Online. Available HTTP: <http://www.cruising.org/Press/overview2008> (accessed 10 May 2009).

Curry, S. (1994) 'Cost-benefit analysis', in Witt, S.F. and Moutinho, L. (eds) 2nd edn, *Tourism marketing and management,* New York: Prentice Hall, 504–9.

Curry, S. and Morvaridi, B. (1992) 'Sustainable tourism: illustrations from Kenya, Nepal and Jamaica', in Cooper, C.P. and Lockwood, A. (eds) *Progress in Tourism, Recreation and Hospitality Management, vol. 4*, London: Belhaven, 131–139.

Cyert, R.M. and March, J.G. (1963) *A Behavioural Theory of the Firm*, Englewood Cliffs, NJ: Prentice Hall.

Cyert, R.M. and Simon, H.A. (1983) 'The behavioural approach: with emphasis on economies', *Behavioural Science*, 28: 95–108.

D'Amore, L.J. (1992) 'Promoting sustainable tourism: the Canadian approach', *Tourism Management*, 13: 258–62.

Dalkey, N. and Helmer, O. (1963) 'An experimental application of the Delphi method of the use of experts', *Management Science*, 9: 458–67.

Daly, H. (1977) *Steady State Economics*, San Francisco: Freeman.

—— (1991) 2nd edn, *Steady State Economics*, Washington, DC.: Island Press.

Daly, H. and Cobb, J. Jr. (1989) *For the Common Good: Towards Community, the Environment, and a Sustainable Future*, Boston: Beacon Press.

Daly, H. and Farley, J. (2004) *Ecological Economics: Principles and Applications*, Washington, DC: Island Press.

Daneshkhu, S. (1996) 'Inter-Continental gains four hotels in Malaysia', *Financial Times*, 9 September.

Dann, G.M.S. (1981) 'Tourist motivation: an appraisal', *Annals of Tourism Research*, 8: 187–219.

Dardin, R. (1980) 'The value of life: new evidence from the marketplace', *American Economic Review*, 70: 1077–82.

Darnell, A., Johnson, P. and Thomas, B. (1992) 'Modelling visitor flows at the Beamish Museum', in Johnson, P. and Thomas, B. (eds) *Choice and Demand in Tourism*, London: Mansell, 161–174.

Dasgupta, A.K. (1972) *Cost-benefit analysis: Theory and Practice*, Basingstoke: Macmillan.

Dasgupta, P. (1993) *An Inquiry into Well-Being and Destitution*, Oxford: Oxford University Press.

Dasgupta, S.A., Mody, S.R. and Wheeler, D. (1995) *Environmental Regulation and Development: A Cross-Country Empirical Analysis,* World Bank Policy Research Department Working Paper 1448.

Dauvergne, P. (2005) *Handbook of Global Environmental Politics,* Cheltenham: Edward Elgar.

Davidson, J.E., Hendry, D.F., Srba, F. and Yeo, S. (1978) 'Econometric modelling of the aggregate time-series relationship between consumers' expenditure and income in the United Kingdom', *Economic Journal,* 88: 661–92.

Davies, B. and Downward, P. (1998) 'Competition and Contestability in the UK Package Tour Industry: Some Empirical Observations', *Tourism Economics,* 4: 241–51.

—— (2001) Industrial organization and competition in the UK tour operator/travel agency business 1989–93: an econometric investigation, *Journal of Travel Research* 39: 411–25.

—— (2006) 'Structure, conduct, performance and industrial organization in tourism", in Dwyer, L. and Forsyth, P. (eds) *International Handbook on the Economics of Tourism,* Cheltenham: Edward Elgar, 117–37.

Davies, S. (1989) 'Concentration', in Davies, S., Lyons, B. with Dixon, H. and Gerowski, P. *Economics of Industrial Organisation,* London and New York: Longman. 73–126.

Davies, S., Lyons, B. with Dixon, H. and Gerowski, P. (1989) *Economics of Industrial Organisation,* London and New York: Longman.

Davis, H.L. (1970) 'Dimensions of marital roles in consumer decision-making', *Journal of Marketing Research,* 7: 168–77.

Davis, O.A. and Whinston, A.B. (1961) 'The economics of urban renewal', in Wilson, J.Q. (ed.) *Urban Renewal: The Record and the Controversy,* Cambridge, Mass.: MIT Press.

Davis, R.K. (1964) 'The value of big game hunting in a forest', *Natural Resources Journal,* 3: 238–49.

Deaton, A.S. (1992) *Understanding Consumption,* Oxford: Clarendon Press.

Deaton, A.S. and Muellbauer, J. (1980a) 'An almost ideal demand system', *American Economic Review,* 70: 312–26.

—— (1980b) *Economics and Consumer Behaviour,* Cambridge: Cambridge University Press.

Debbage, K.G. (1990) 'Oligopoly and the resort cycle in the Bahamas', *Annals of Tourism Research,* 17: 513–27.

de Blust, M. (2008) *Travel Agents & Tour Operators in the EU in the Context of Sustainable Management and Development of Tourism,* Presentation at the Lithuanian Tourism Association, 19th November.

de Jong, H.W. and Shepherd W.G. (eds) (1986) *Mainstreams in Industrial Organization,* Boston, Mass.: Kluwer.

de Kadt, E. (1979) *Tourism: Passport to Development,* Oxford: Oxford University Press.

de la Fuente, A. (2002) 'On the sources of convergence: a close look at the Spanish regions', *European Economic Review,* 46: 569–99.

Deloitte (2006) *Hospitality 2010: A Five Year Wake-Up Call,* London: Deloitte.

Delos Santos, J.S., Ortiz, E.M., Huang, E. and Secretario, F. (1983) 'Philippines', in Pye, E.A. and Lin, T.-B. (eds) *Tourism in Asia: The Economic Impact*, Singapore: Singapore University Press.

de Mello Jr, L.R. and Sinclair, M.T. (1995) *Foreign Direct Investment, Joint Ventures and Endogenous Growth*, Studies in Economics no. 95/13, University of Kent at Canterbury.

de Mello, M.M. and Fortuna, N. (2005), Testing alternative dynamic systems for modelling tourism demand', *Tourism Economics*, 11: 517–37.

de Mello, M.M. and Nell, K.S. (2005) 'The forecasting ability of a cointegrated VAR system of the UK tourism demand for France Spain and Portugal', *Empirical Economics*, 30: 277–308.

de Mello, M.M., Pack, A. and Sinclair, M.T. (2002) 'A system of equations model of UK tourism demand in neighbouring countries. *Applied Economics*, 34: 509–21.

Demsetz, H. (1969) 'Information and efficiency: another viewpoint', *Journal of Law and Economics*, 12: 1–22.

—— (1974) 'Two systems of belief about monopoly', in Goldschmid, H.J., Mann, H.M. and Weston, J.F. (eds) *Industrial Concentration: The New Learning*, Boston: Little Brown, 164–84.

Desvousges, W.S., Smith, V.K. and McGivney, M.P. (1983) *Comparison of Alternative Approaches for Estimating Recreation and Related Benefits of Water Quality Improvements*, Washington, DC: US Environmental Protection Agency, EPA 230–05–83–001.

Deutsche Bahn (2009) Services, Online. Available HTTP: <http://www.bahn.de/p/view/service/service_uebersicht.shtml> (accessed 3 April 2009).

Dharmaratne, G.S. (1995) 'Forecasting tourist arrivals in Barbados', *Annals of Tourism Research*, 22: 804–18.

Diakomihalis, M. and Lagos, D. (2008) 'Economic impacts: estimation of yachting in Greece via the tourism satellite account', *Tourism Economics*, 14: 871–87.

Diamond, D. and Richardson, R. (1996) *The Economic Significance of the British Countryside*, London: The Countryside Business Group.

Diamond, J. (1974) 'International tourism and the developing countries: a case study in failure', *Economica Internazionale*, 27: 601–15.

—— (1977) 'Tourism's role in economic development: the case re-examined', *Economic Development and Cultural Change*, 25: 539–53.

di Benedetto, C.A. and Bojanic, D.C. (1993) 'Tourism area life cycle extensions', *Annals of Tourism Research*, 20: 557–70.

Dicken, P. and Lloyd, P. (1991) *Location in Space: Theoretical Perspective in Economic Geography*, Harlow: Prentice Hall.

Dickens, P. (2003) 'Changing our environment, changing ourselves: critical realism and trans-disciplinary research', *Interdisciplinary Science Reviews*, 28: 95–105.

Dickie, M. and Gerking, S. (1991) 'Willingness to pay for ozone control: inferences from the demand for medical care', *Journal of Environmental Economics and Management*, 21: 1–16.

Dickie, M., Delorme Jr, C.D. and Humphreys, J.M. (1997) 'Hedonic prices, goods-specific effects and functional form: inferences from cross section – time series data', *Applied Economics*, 29: 239–49.

Dietrich, M. (1994) *Transaction Cost Economics and Beyond*: *Towards a New Economics of the Firm*, Routledge: London.

Dimou, I. (2004) *Expansion Strategies of International Hotel Firms*: *A Transaction Cost Economics and Agency Theory Approach*, PhD Thesis, Guildford: University of Surrey.

Dingle, P.A.J.M. (1995) 'Practical green business', *Insights*, March: C35–C45.

Divisekera, S. (1995) 'An econometric model of international visitor flows to Australia', *Australian Economic Papers*, 34: 291–308.

—— (2003) 'A model of demand for international tourism', *Annals of Tourism Research*, 30: 31–49.

Divisekera, S. and Kulendran, N. (2006) 'Economic effects of advertising on tourism demand: a case study', *Tourism Economics*, 12: 187–205.

Dixit, A.K. (1982) 'Recent developments in oligopoly theory', *American Economic Review Papers and Proceedings*, 72: 12–17.

Dixit, A.K. and Stiglitz, J.E. (1977) 'Monopolistic competition and optimum product diversity', *American Economic Review*, 67: 297–308.

Dixon, J.A., Scura, L.F. and van't Hof, T. (1993) 'Meeting ecological and economic goals: marine parks in the Caribbean', *Ambio,* 22: 117–25.

Dixon, P.B., Parmenter, B.R., Powell, A.A. and Wilcoxen, P.J. (1992) *Notes and Problems in Applied General Equilibrium Economics*, Amsterdam: North-Holland.

Doan, T.M. (2000) 'The effects of ecotourism in developing nations; an analysis of case studies', *Journal of Sustainable Development*, 8: 288–304.

Dodgson, J. and Topham, N. (1998) *Bus Deregulation and Privatisation*, Aldershot: Avebury.

Doganis, R. (2002) *Flying off Course*: *the Economics of International Airlines*, 3rd edn, London: Routledge.

—— (2005) *The Airline Business,* 2nd edn, London: Routledge.

Dolnicar, S., Crouch, G.I., Devinney, T., Huybers, T., Louviere, J.J. and Oppewal, H. (2008) 'Tourism and discretionary income allocation: heterogeneity among households', *Tourism Management*, 29: 44–52.

Domar, E.D. (1946) 'Capital expansion, rate of growth and employment', *Econometrica*, 14: 137–47.

—— (1947) 'Expansion and employment', *American Economic Review*, 37: 34–55.

Domberger, S. (1986) 'Economic regulation through franchise contracts', in Kay, J., Mayer, C. Thompson, D. (eds) *Privatisation and Regulation*: *The UK Experience*, Ch. 14, Oxford: Clarendon Press.

Douglas, N. (1997) 'Applying the life cycle model to Melanesia', *Annals of Tourism Research*, 24: 1–22.

Dowling, R.K. (ed.) (2006a) *Cruise Ship Tourism*, Wallingford: CABI.

—— (2006b) 'Looking ahead: the future of cruising', in R.K. Dowling (ed.) *Cruise Ship Tourism*, Wallingford: CABI, 414–34.

—— (2006c) 'The cruising industry', in Dowling, R.K. (ed) *Cruise Ship Tourism*, Wallingford: CABI, 3–17.

Drobny, A. and Hall, S.G. (1989) 'An investigation of the long run properties of aggregate non-durable consumers' expenditure in the United Kingdom', *Economic Journal*, 99: 454–60.

DTLR (Department for Transport, Local Government and the Regions) (2001) *Multi Criteria Analysis: A Manual*, Online. Available HTTP: <http://www.dtlr.gov.uk/about/multicriteria/07.htm> (accessed 6 March 2009).

Duesenberry, J.S. (1949) *Income, Saving and the Theory of Consumer Behaviour*, Cambridge, Mass.: Harvard University Press.

Dunning, J.H. (1981) *International Production and the Multinational Enterprise*, London: George Allen and Unwin.

Dunning, J.H. and Kundu, S.K. (1995) 'The internationalisation of the hotel industry – some new findings from a field ftudy', *Management International Review*, 35: 101–33.

Dunning, J.H. and McQueen, M. (1982a) *Transnational Corporations in International Tourism*, New York: United Nations Centre for Transnational Corporations.

—— (1982b) 'Multinational corporations in the international hotel industry', *Annals of Tourism Research*, 9: 69–90.

Dupeyras, A. (2002) 'OECD including the employment dimension in APEC', in Cockerell, N. and Spurr, R. (eds) *Best Practice in Tourism Satellite Account Development in APEC Member Economies*, Singapore: Asia Pacific Economic Cooperation (APEC) Secretariat Publication No 202-TR-01-1, 35–40.

Durbarry, R. and Sinclair, M.T. (2003) 'Market shares analysis: the case of French tourism demand', *Annals of Tourism Research*, 30: 927–41.

Dwyer, L. and Forsyth, P. (1994) 'Foreign tourism investment: motivation and impact', *Annals of Tourism Research*, 21: 512–37.

—— (eds) (2006) *International Handbook on the Economics of Tourism*, Cheltenham: Edward Elgar.

Dwyer, L., Forsyth, P. and Spurr, R. (2004) 'Evaluating tourism's economic effects: new and old approaches', *Tourism Management*, 25: 307–17.

—— (2006) 'Economic evaluation of special events', in Dwyer, L. and Forsyth, P. (eds) *International Handbook on the Economics of Tourism*, Cheltenham: Edward Elgar, 316–55.

—— (2007) 'State investment in major events: integrating economic impact modelling with cost-benefit analysis', Paper presented at the first Conference of the International Association for Tourism Economics, Palma, Majorca 25th to 27th October.

Dwyer, L., Forsyth, P., Spurr, R. and Van Ho, T. (2003) 'Tourism's contribution to a state economy: a multi-regional general equilibrium analysis', *Tourism Economics*, 9: 431–48.

Eadington, W.R. and Redman, M. (1991) 'Economics and tourism', *Annals of Tourism Research*, 18: 41–56.

Eber, S. (ed.) (1992) *Beyond the Green Horizon: Principles for Sustainable Tourism*, Godalming: World Wide Fund for Nature.

EC (2000) *Towards Quality Urban Tourism*, Brussels: EC.

—— (2004) *Video on Passenger Rights*, Brussels: EC.

—— (2005) *Community Guidelines on Financing of Airports and Start-up Aid to Airlines Departing from Regional Airports*, 2005/C 312/01, Brussels: EC.

—— (2007a) *EU–US First Stage Air Transport Agreement*, Presentation made by the Directorate-General Energy and Transport/Air Transport Directorate, Brussels: EC.

—— (2007b) *Proposal for a Regulation of the European Parliament and of the Council on a Code of Conduct for Computerised Reservation Systems: Impact Assessment COM(2007) 709, SEC(2007) 1497,* Brussels: EC.

—— (2007c) *Case No COMP/M.4601 – KarstadtQuelle/MyTravel Notification of 26 March 2007 pursuant to Article 4 of Council Regulation No 139/2004,* Brussels: EC.

—— (2007d) *Case No COMP/M.4600 – TUI/ First Choice Notification of 4 April 2007 pursuant to Article 4 of Council Regulation No 139/2004,* Brussels: EC.

Economist (1996) 'Go to dreamland, forget the mosques', *Economist,* 17 August.

Eggertson, T. (1990) *Economic Behaviour and Institutions,* New York: Cambridge University Press.

Eisenhardt, M. (1989) 'Agency theory: an assessment and review', *Academy of Management Review,* 14: 57–74.

Endo, K. (2006) 'Foreign direct investment in tourism – flows and volumes', *Tourism Management,* 27: 600–614.

Englin, J. and Mendelsohn, R. (1991) 'A hedonic travel cost analysis for valuation of multiple components of site quality: the recreational value of forest management', *Journal of Environmental Economics and Management,* 21: 275–90.

Esty, D.C. and Porter, M.E. (2002) 'Ranking national environmental regulation and performance: a leading indicator of future competitiveness', in *The Global Competitive Report 2001–2002,* New York: Oxford University Press.

E-Tid (2003) *Court Rules Ryanair Strasbourg Funding Illegal,* e-tid.com, 24th July.

Eurocontrol (2008) *Eurocontrol and ACI Europe join Forces to Combat Airport Congestion and Reduce Fuel Burn and Emissions,* Press Release 28th October, Brussels: Eurocontrol.

Evans, A.W. (1982) 'Externalities, rent seeking and town planning', *Discussion Papers in Urban and Regional Economics* no. 10, University of Reading: Department of Economics.

—— (1985) *Urban Economics: An Introduction* Oxford: Blackwell.

—— (2004) *Economics and Land Use Planning,* Oxford: Blackwell.

Evans, N. and Stabler, M.J. (1995) 'A future for the package tour operator in the 21st century?', *Tourism Economics,* 1: 245–63.

Fairbairn-Dunlop, P. (1994) 'Gender, culture and tourism development in Western Samoa', in Kinnaird, V. and Hall, D. (eds) *Tourism: A Gender Analysis,* Chichester: John Wiley.

Fair Trade Federation (2008) About Fair Trade, Online. Available HTTP: <http: www.fairtradefederation.org> (accessed 5 December 2008).

Falk, M. (2008) 'A hedonic price model for ski lift tickets', *Tourism Management* 29: 1172–1184.

Farrell, B.H. and Twining-Ward, L. (2004) 'Reconceptualising tourism', *Annals of Tourism Research,* 31: 274–95.

—— (2005) 'Seven steps towards sustainability: tourism in the context of new knowledge,' *Journal of Sustainable Tourism,* 13: 109–22.

Faulkner, B. (2001) 'Towards a framework for tourism disaster management', *Tourism Management,* 22: 135–47.

Faulkner, B and Russell, R. (1997) 'Chaos and complexity in tourism: in search of a new perspective', *Pacific Tourism Review*, 1: 93–102.

Farver, J.A.M. (1984) 'Tourism and employment in the Gambia', *Annals of Tourism Research*, 1: 249–65.

Fayed, H. and Fletcher, J. (2002) 'Globalisation of economic activity issues for tourism', *Tourism Economics*, 8: 207–30.

Feichtinger, G., Hartly, R.F., Kort, P.M. and Veliov, V.M. (2005) 'Environmental policy, the Porter hypothesis and the composition of capital: effects of learning and technological progress', *Journal of Environmental Economics and Management*, 50: 434–46.

Fennell, D.A. (2007) *Ecotourism*, 3rd edn, London: Routledge.

Fennell, D.A. and Dowling, R.K. (2003) (eds) *Ecotourism Policy and Planning*, Wallingford: CABI.

Ffrench, R. (1972) 'The effect of devaluation on the foreign travel balance of Jamaica', *Social and Economic Studies*, 20: 443–59.

Field, B.C. and Field, M.K. (1998) 3rd edn, *Environmental Economics: An Introduction*, London: McGraw-Hill.

Field, D. and Pilling, M. (2003) 'Privatised airports blasted', *Airline Business*, 19: 10.

Filiatrault, P. and Ritchie, J.R.B. (1980) 'Joint purchasing decisions: a comparison of influence structure in family and couple decision-making unit', *Journal of Consumer Research*, 7: 131–40.

Fink, S. (1986) *Crisis Management: Planning for the Inevitable*, New York: American Association of Management.

Fishbein, M. (1963) 'An investigation of the relationships between beliefs about an object and the attitude toward that object', *Human Relationships*, 16: 233–40.

Fitch, A. (1987) 'Tour operators in the UK: survey of the industry, its markets and product diversification', *Travel and Tourism Analyst*, March: 29–43.

Flavin, M. (1981) 'The adjustment of consumption to changing expectations about future income', *Journal of Political Economy*, 89: 974–1009.

Fleischer, A. and Felsenstein, D. (2000) 'Support for rural tourism: does it make a difference?', *Annals of Tourism Research*, 27: 1007–24.

Fletcher, J.E. (1986a) *The Economic Impact of International Tourism on the National Economy of the Republic of Palau*, Madrid: WTO/UNDP.

—— (1986b) *The Economic Impact of Tourism on Western Samoa*, Madrid: WTO/UNDP.

—— (1987) *The Economic Impact of International Tourism on the National Economy of the Solomon Islands*, Madrid: WTO/UNDP.

Fletcher, J.E. and Archer, B.H. (1991) 'The development and application of multiplier analysis', in Cooper, C.P. (ed.) *Progress in Tourism, Recreation and Hospitality*, *Management*, Vol. *1*, London: Belhaven, 28–47.

Fletcher, J. and Westlake, J. (2006) 'Globalisation', in Dwyer, L. and Forsyth, P. (eds) *International Handbook on the Economics of Tourism*, Cheltenham: Edward Elgar, 464–80.

Flogenfeldt, T. (1999) 'Traveler geographic origin and market segmentation: the multi trips destination case', *Journal of Travel and Tourism Marketing*, 8: 111–18.

Florida Tax Watch (2000) *The Benefits and Costs of Tourism to Florida*, Florida: FTW, August.

Ford, D.A. (1989) 'The effect of historic district designation on single-family home prices', *AREUEA Journal*, 17: 353–62.

Forsyth, T. (1995) 'Business attitudes to sustainable tourism: responsibility and self regulation in the UK outgoing tourism industry', Paper presented at the Sustainable Tourism World Conference, Lanzarote, Spain.

Foster, J. (1997) (ed.) *Valuing Nature? Ethics, Economics and the Environment*, London: Routledge.

Frank, A.G. (1966) 'The development of underdevelopment', *Monthly Review*, 18.

Franses, P.H. and McAleer, M. (1998) 'Cointegration analysis of seasonal data', *Journal of Economic Surveys*, 12: 651–78.

Frechtling, D.C. (2006) 'An assessment of visitor expenditure methods and models', *Journal of Travel Research*, 45: 26–35.

Freeman, A.M. (1979) *The Benefits of Environmental Improvement*, Washington, DC: Resources for the Future.

Friedman, M. (1957) *A Theory of the Consumption Function*, Princeton, NJ: University Press.

—— (1968) 'The role of monetary policy', *American Economic Review*, 38: 1–17.

Fritz, R.G. (1987) 'An empirical comparison of systems of demand equations: an application to visitor expenditure in resort destinations', *Philippine Review of Business and Economics*, 24: 79–102.

Fritz, R.G., Brandon, C. and Xander, J. (1984) 'Combining the time-series and econometric forecast of tourism activity', *Annals of Tourism Research*, 11: 219–29.

FTO (2009a) *Pricing and Profit*, Online. Available HTTP: < http://www.fto.co.uk/operators-factfile/pricing-and-profit> (accessed 25 April 2009).

FTO (2009b) *Tour Operators and Travel Agents*, Online. Available HTTP: <http://www.fto.co.uk/operators-factfile/tour-operators/> (accessed 25 April 2009).

Fujita, M. and Krugman, P. (1995) 'When is the economy monocentric? Von Thünen and Chamberlin unified', *Regional Science and Urban Economics*, 25: 505–28.

Fujita, M. and Mori, T. (1996) 'The role of ports in the making of major cities: self-organisation and hub-effects', *Journal of Development Economics*, 49: 93–120.

—— (2005) *Frontiers of New Economic Geography*, Institute of Developing Economies, Discussion Paper 27, Chiba, Japan.

Fujita, M. and Thisse, J.F. (1997) 'Économie géographique, problèmes anciens et nouvelles perspectives', *Annales d'Économie et de Statistique*, 45: 38–87.

Fujita M., Krugman P. and Mori, T. (1999) 'On the evolution of hierarchical urban systems', *European Economic Review*, 43: 209–51.

Gabrynowicz, S. (2003) *The Relationship between Environmental Management and Economic Performance*, South Australia: Department for Environment and Heritage, 6.

Gago, A., Labandeira, X., Picos, F. and Rodriguez, M. (2009) 'Specific and general taxation of tourism activities: evidence from Spain', *Tourism Management*, 30: 381–392.

Gale, D. (2008) 'Hotels 325', *Hotels Magazine*, Online. Available HTTP:<http://www.hotelsmag.com/article/CA6575623.html>(accessed 25 February 2009).

Gallagher, K.P. (2004) *Free Trade and the Environment. Mexico, NAFTA and Beyond*. Palo Alto, Calif. Stanford University Press.

GAMS - General Algebraic Modelling System (2008) *An Introduction to GAMS*, Online. Available HTTP: <http://www.gams.com> (accessed 14 May 2009).

Garin-Muñoz, T. and Amaral, T.P. (2000) 'An econometric model for international tourism flows to Spain', *Applied Economics Letters*, 7: 525–29.

Garrod, B. and Wilson, J.C. (2003) (eds) *Marine Ecotourism Issues and Experiences*, Clevedon: Channel View.

Garrod, G. (1991a) 'The environmental economic impact of woodland: a two-stage hedonic price model of the amenity value of forestry in Britain', *Applied Economics*, 24: 715–28.

—— (1991b) 'Some empirical estimates of forest amenity value', Countryside Change Unit Working Paper, no. 13. Newcastle-upon-Tyne: University of Newcastle.

—— (1991c) 'The hedonic price method and the valuation of the countryside characteristics', Countryside Change Unit Working Paper, no. 14. Newcastle-upon-Tyne: University of Newcastle.

Garrod, G. and Willis, K. (1990) 'Contingent valuation techniques: a review of their unbiasedness, efficiency and consistency', Countryside Change Unit Working Paper, no. 10, Newcastle-upon-Tyne: University of Newcastle.

Garrod, G., Pickering, A. and Willis, K. (1991) 'An economic estimation of the recreational benefit of four botanic gardens', Countryside Change Unit Working Paper, no. 25, Newcastle-upon-Tyne: University of Newcastle.

Garrod, G., Willis, K. and Saunders, C.M. (1994) 'The benefits and costs of the Somerset levels and moors ESA', *Journal of Rural Studies*, 10: 131–45.

GEMPACK - General Equilibrium Modelling PACKage (2009) *GEMPACK Home Page*, Online. Available HTTP: <http://www.monash.edu.au/policy/gempack.htm> (accessed 14 May 2009).

Georgescu-Roegen, N. (1971) *The Entropy Law and the Economic Process* Cambridge, Mass.: Harvard University Press.

—— (1992) 'Tourism planning and destination life cycle', *Annals of Tourism Research*, 19: 752–70.

Geurts, M.D. (1982) 'Forecasting the Hawaiian tourist market', *Journal of Travel Research*, 11: 18–21.

Geurts, M.D. and Ibrahim, I.B. (1975) 'Comparing the Box-Jenkins approach with the exponentially smoothed forecasting model: application to Hawaii tourists', *Journal of Marketing Research*, 12: 182–88.

Gidwitz, T. (2005) *The Deadly Tides*, Online. Available HTTP: <http://www.Tomgidwitz.com/main/tides.htm> (accessed 7 January 2009)

Gilpin, A. (2000) *Environmental Economics: A Critical Overview*, Chichester: Wiley.

Ginsburgh, V. and Keyzer, M. (2002) *The Structure of Applied General Equilibrium Models*. Cambridge, Mass.: MIT Press.

Glaesser, D. (2006) *Crisis Management in the Tourism Industry*, 2nd edn, Oxford: Butterworth-Heinemann.

Go, F. and Pine, R. (1995) *Globalization Strategy in the Hotel Industry*, London: Routledge.

Godbey, G. (1988) 'Play as a model for the study of tourism', Paper presented at the Leisure Studies Association 2nd International Conference, Leisure, Labour and Lifestyles: International Comparisons, June, Brighton: University of Sussex.

Goelder, C.R., Ritchie, J.R.B. and McIntosh, R.W. (2005) *Tourism: Principles, Practices, Philosophies*, 8th edn, Chichester: Wiley.

Goldman, G., Nakazawa, A. and Taylor, D. (1994) *Cost-Benefit Analysis of Local Tourism Development*: Western Rural Development Centre: Oregon State University.

Gonzalez, P. and Moral, P. (1996) 'Analysis of tourism trends in Spain', *Annals of Tourism Research*, 23: 739–54.

Goodall, B (1992) 'Environmental auditing for tourism', in Cooper, C.P. and Lockwood, A. (eds) *Progress in Tourism, Recreation and Hospitality Management, vol. 4*, London: Belhaven, 60–74.

Goodall, B. and Stabler, M.J. (1994) 'Tourism–environment issues and approaches to their solution', in Voogd, H. (ed.) *Issues in Environmental Planning*, London: Pion, 78–99.

Goodstein, E.S. (1999) 2nd edn, *Economics and the Environment*, New York: Wiley.

Gordon, I.R. (1994) 'Crowding, competition and externalities in tourism development: a model of resort life cycles', *Geographical Systems*, 1: 289–308.

Gordon, I.R. and Goodall, B. (1992) 'Resort cycles and development processes', *Built Environment*, 18: 41–56.

Gorman, W.M. (1980) 'A possible procedure for analysing quality differentials in the egg market', *Review of Economic Studies* 47: 843–56.

Gössling, S. and Hall, C.M. (eds) (2006) *Tourism and Environmental Change: Ecological, Social Economic and Political Relationships*. London: Routledge.

Graham, A. (2008a) 'Trends and characteristics of leisure travel demand', in Graham, A., Papatheodorou, A. and Forsyth, P. (eds) *Aviation and Tourism: Implications for Leisure Travel*, Aldershot: Ashgate, 21–33.

—— (2008b) *Managing Airports: An International Perspective*, 3rd edn, Oxford: Butterworth-Heinemann.

Gray, H.P. (1970) *International Travel—International Trade*, Lexington, Mass.: D.C. Heath.

—— (1966) 'The demand for international travel by the United States and Canada', *International Economic Review*, 7: 83–92.

Greaker, M. (2006) 'Spillovers in the development of new pollution abatement technology: a new look at the Porter hypothesis', *Journal of Environmental Economics and Management*, 52: 411–20.

Green, H. and Hunter, C. (1992) 'The environmental impact assessment of tourism development', in Johnson, P. and Thomas, B. (eds) *Perspectives on Tourism Policy*, London: Mansell.

Green, H. and Hunter, C. and Morre, B. (1990a) 'Application of the Delphi technique in tourism', *Annals of Tourism Research*, 17: 270–79.

—— (1990b) 'Assessing the environmental impact of tourism development: use of the Delphi technique', *Tourism Management*, 11: 111–20.

Greenaway, D., Bleaney, M. and Stewart, I. (1996) *Guide to Modern Economics*, London and New York: Routledge.

Greenwich Village Society for Historic Preservation (2009) *About Us – Mission*, Online. Available HTTP: <http://www.gvshp.org/_gvshp/about/mission.htm> (accessed 28 April 2009).

Grossman, G.M. and Helpman, E. (1990a) 'Comparative advantage and long run growth', *American Economic Review*, 80: 796–815.

—— (1990b) 'Trade, innovation, and growth', *American Economic Review*, 80: 86–91.

—— (1991) *Innovation and Growth in the Global Economy*, Cambridge, Mass.: MIT Press.

—— (1994) 'Endogenous innovations in the theory of growth', *Journal of Economic Perspectives*, 8: 23–44.

Grossman, G.M. and Krueger, A.B. (1995) 'Economic growth and the environment', *Quarterly Journal of Economics*, 112: 353–77.

Guerrier, Y. (1986) 'Hotel manager: an unsuitable job for a woman?', *The Service Industries Journal*, 6: 227–40.

Guiver, J., Lumsdon, L., Weston, R. and Ferguson, M. (2007) 'Do buses help meet tourism objectives? The contribution and potential of scheduled buses in rural destination areas', *Journal of Transport Policy*, 14: 275–82.

Gunadhi, H. and Boey, C.K. (1986) 'Demand elasticities of tourism in Singapore', *Tourism Management*, 7: 239–53.

Gunn, C.A. (1987) 'A perspective on the purpose and nature of tourism research methods', in Ritchie, J.R.B. and Goeldner, C.R. (eds) 2nd edn, *Travel, Tourism and Hospitality Research*: *A Handbook for Managers and Researchers,* New York: Wiley, 3–11.

Gurung, C.P. and de Coursey, M. (1994) 'The Annapurna conservation area project: a pioneering example of sustainable tourism' in Cater, E. and Lowman, G. (eds) *Ecotourism*: *A Sustainable Option?*, Chichester: Wiley 177–94.

Haas, P.M. (ed.) (2003) *Environment in the new Global Economy*, Cheltenham: Edward Elgar.

Halfpenny, E.A. (2002) *Marine Ecotourism*: *Impacts, International Guidelines and Best Practice Case Studies*, Burlington, Vt: The International Ecotourism Society.

Hall, C.M. and Boyd, S. (eds) (2005) *Nature Based Tourism in Peripheral Areas*, Clevedon: Channel View Publications.

Hall, C.M. and Highman, J. (eds) (2005) *Tourism, Recreation and Climate Change*, Clevedon: Channel View.

Hall, C.M. and Lew, A.A. (eds) (1998) *Sustainable Tourism*, Harlow: Addison Wesley Longman.

Hall, C.M. and Page, S.J. (2006) *The Geography of Tourism and Recreation*: *Environment, Place and Space*, London: Routledge.

Hall, D. and Richards, G. (eds) (2000) *Tourism and Sustainable Community Development*, London: Routledge.

Hamilton, K. (2000) *Genuine Savings as a Sustainability Indicator*, Working Paper 77, Washington, DC: World Bank.

Hamilton, K. and Hartwick, J. (2005) 'Investing exhaustible resource rents and the path of consumption', *Canadian Journal of Economics*, 58: 615–21.

Hampton, M.P. and Christensen, J. (2007) 'Competing industries in islands: a new tourism approach', *Annals of Tourism Research*, 34: 998–1020.

Hanley, N. (1988) 'Using contingent valuation to value environmental improvements', *Applied Economics*, 20: 541–49.

—— (1989) 'Problems in valuing environmental improvements resulting from agricultural policy changes: the case of nitrate pollution', Discussion Paper no. 89/1, University of Stirling: Economic Department.

Hanley, N. and Ruffell, R. (1992) 'The valuation of forest characteristics', Queen's Discussion Paper, 849.

Hanley, N. and Spash, C. (1993) *Cost-Benefit Analysis and the Environment*, Aldershot: Edward Elgar.

Hanley, N., Mourato, S. and Wright, R.E. (2001) 'Choice modelling approaches: a superior alternative for environmental valuation', *Journal of Economic Surveys*, 15: 435–62.

Hanley, N., Shogren, J.F. and White, B. (2001) *Introduction to Environmental Economics*, Oxford: Oxford University Press.

Hannah, L. and Kay, J.A. (1977) *Concentration in Modern Industry*: *Theory, Measurement and the UK Experience*, London: Macmillan.

Hansen, L.T. and Hallam, A. (1991) 'National estimates of the recreational value of streamflow', *Water Resources Research*, 27: 167–75.

Hardin, G. (1968) 'The tragedy of the commons', *Science*, 162: 1243–48.

Hardy, A., Beeton, R.J.S. and Pearson, L. (2002) 'Sustainable tourism: an overview of the concept and its position in relation to conceptualisations of tourism', *Journal of Sustainable Tourism*, 10: 475–96.

Harrigan, K.R. (1985) 'Exit barriers and vertical integration', *Academy of Management Journal*, 28: 686–97.

Harris, C. (1954) 'The market as a factor in the localisation of industry in the United States', *Annals of the Association of American Geographers*, 64: 315–48.

Harrison, A.J.M. and Stabler, M.J. (1981) 'An analysis of journeys for canal-based recreation', *Regional Studies*, 15: 345–58.

Harrison, D. (1992) 'International tourism and the less developed countries: the background', in Harrison, D. (ed.) *Tourism and the Less Developed Countries*, London: Belhaven, 1–18.

—— (1996) 'Sustainability and tourism: reflections from a muddy pool', in Briguglio, L., Archer, B., Jafari, J. and Wall, G. (eds) *Sustainable Tourism in Islands and Small States*; *Issues and Policies*, London: Pinter 69–89.

—— (2007) 'Towards developing a framework for analysing tourism phenomena: a discussion', *Current Issues in Tourism*, 10: 61–86.

Harrod, R.F. (1939) 'An essay in dynamic theory', *Economic Journal*, 49: 14–33.

Hartwick, J. (1977) 'Intergenerational equity and the investment of rents from exhaustible resources', *American Economic Review*, 67: 972–74.

Hartwick, J.M. and Olewiler, N.D. ((1998) 2nd edn, *The Economics of Natural Resource Use*, New York: Harper Row.

Harvey, A.C. (1993) *Time Series Models*, Hemel Hempstead: Harvester Wheatsheaf.

Hawkins, C.J. and Pearce, D.W. (1971) *Capital Investment Appraisal*, London: Macmillan

Hawkins, D. and Mann, S. (2007) 'The World Bank's role in tourism development', *Annals of Tourism Research*, 34: 348–63.

Hay, D.A. and Morris, D.J. (1991) *Industrial Economics and Organization*: *Theory and Evidence*, Oxford: Oxford University Press.

Hayashi, F. (1985) 'The effect of liquidity constraints on consumption: a cross-sectional analysis', *Quarterly Journal of Economics*, 100: 183–206.

Hayek, F.A. (1949) *Individualism and Economic Order*, London: Routledge & Kegan Paul.

Hazari, B.R. and Sgro, P.M. (2004) *Tourism, Trade and National Welfare*, Boston, Mass.: Elsevier.

Heberlein, T. and Bishop, R. (1986) 'Assessing the validity of contingent valuation: three experiments', *Science of the Total Environment*, 56: 99–107.

Heckman, J.J. (1974) 'Life-cycle consumption and labor supply: an exploration of the relationship between income and consumption over the life cycle', *American Economic Review*, 64: 188–94.

Hellenic Association of Tourist and Travel Agencies (2006) 'Greece as an international cruising hub in the Eastern Mediterranean', *Cruise in Greece Conference*, September.

Helpman, E. and Krugman, P.R. (1993) *Market Structures and Foreign Trade*, Cambridge, Mass.: MIT Press.

Helu Thaman, K. (1992) 'Beyond Hula, hotels and handicrafts', *In Focus*, 4, Summer: 8–9.

Hendry, D.F. (1983) 'Econometric modelling: the "consumption function" in retrospect', *Scottish Journal of Political Economy*, 30: 193–220.

—— (1995) *Dynamic Econometrics*, Oxford: Oxford University Press.

Hendry, D.F. and Mizon, G.E. (1978) 'Serial correlation as a convenient simplification, not a nuisance', *Economic Journal*, 88: 549–63.

Heng, T.M. and Low, L. (1990) 'The economic impact of tourism in Singapore', *Annals of Tourism Research*, 17: 246–69.

Hennessy, S. (1994) 'Female employment in tourism development in south-west England', in Kinnaird, V. and Hall, D. (eds) *Tourism*: *A Gender Analysis*, Chichester: John Wiley.

Henscher, D. and Brewer, A. (2001) *Transport*: *An Economics and Management Perspective*, Oxford: Oxford University Press.

Hickman, L. (2007) *The Final Call*, London: Transworld Publishers.

Hicks, L. (1990) 'Excluded women: how can this happen in the hotel world?', *The Service Industries Journal*, 10: 348–63.

Hill, J., Marshall, I. and Priddey, C. (1994) *Benefiting Business and the Environment: Case Studies of Cost Savings and New Opportunities from Environmental Initiatives*, London: Institute of Business Ethics.

Hirschmann, A.O. (1958) *The Strategy of Economic Development*, New Haven: Yale University Press.

—— (1964) 'The paternity of an index', *American Economic Review*, 54: 761–62.

Hirshleifer, J. (1982) *Research in Law and Economics, vol. 4, Evolutionary Models in Economics and Law*, London: JAI Press.

Hjalager, A.M. (1996) 'Tourism and the environment: the innovation connection', *Journal of Sustainable Devlopment*, 4: 201–18.

—— (1998) 'Environmental regulation of tourism: impact on business innovation', *Progress in Tourism and Hospitality Research*, 4: 17–30.

—— (2007) 'Stages in the economic globalization of tourism', *Annals of Tourism Research*, 34: 437–57.

HKIA (2003) *Various Press Releases*, Online. Available HTTP: <http://www.hongkongairport.com/eng/media/press-releases/ex_423.html> (accessed 26 February 2009).

Hohl, A. and Tisdell, C.A. (1993) 'How useful are environmental safety standards in economics? The example of safe minimum standards for protection of species', *Biodiversity and Conservation*, 2: 168–81.

Holden, A. (2005) *Tourism Studies and the Social Sciences*, London: Routledge.

—— (2007) 2nd edn, *Environment and Tourism*. London: Routledge.

Holecek, D., Forsberg, P. and Myers, J. (1994), *Travel and Tourism: Issues in Tax Policy*, Brussels: WTTC.

Holland, S.M., Ditton, R.B. and Graefe, A.R. (1998) 'An ecotourism perspective on billfish fisheries', *Journal of Sustainable Tourism*, 6: 97–115.

Holloway, J.C. and Taylor, N. (2006) *The Business of Tourism*, 7th edn, Harlow: Financial Times/Prentice Hall.

Honey, M. (1999) *Ecotourism and Sustainable Development: Who Owns Paradise?*, Washington, DC: Island Press.

Hoover, E.M. and Giarratani, F. (1999) *An Introduction to Regional Economics*, 3rd edn, Morgantown: West Virginia University. Online. Available HTTP: <http://www.rri.wvu.edu/WebBook/Giarratani/main.htm> (accessed 7 April 2009).

Hotels Magazine (2008) *Corporate 300 Ranking*, Online. Available HTTP: <http://www.hotelsmag.com> (accessed 25 February 2009).

Hotelling, H. (1931) 'The economics of exhaustible resources', *Journal of Political Economy*, 39: 137–75.

—— (1949) *The Economics of Public Recreation*, The Prewitt Report, Land and Recreation Planning Division, National Park Service, US Department of the Interior, Washington, DC.

Hough, D.E. and Kratz, C.G. (1983) 'Can "good" architecture meet the market test?', *Journal of Urban Economics*, 14: 40–54.

Hughes, H. (1981) 'A tourism tax: the cases for and against", *International Journal of Tourism Management*, 2: 196–206.

Hunter, C. (1995) 'On the need to re-conceptualise sustainable tourism development', *Journal of Sustainable Tourism*, 3: 155–65.

—— (1997) 'Sustainable tourism as an adaptive paradigm', *Annals of Tourism Research*, 24: 850–67.

Hunter, C. and Green, H. (1995) *Tourism and the Environment: A Sustainable Relationship?* London: Routledge.

Huybers, T. and Bennett, J. (2003) 'Environmental management and the competitiveness of nature-based tourism destinations', *Environmental and Resources Economics*, 24: 213–33.

Hymer, S.H. (1976) *The International Operations of National Firms: A Study of Direct Investment*, Cambridge, Mass.: MIT Press.

Iatrou, K. and Oretti, M. (2007) *Airline Choices for the Future: From Alliances to Mergers*, Aldershot: Ashgate.

Iatrou, K. and Tsitsiragou, E. (2008) 'Leisure travel, network carriers and alliances', in Graham, A., Papatheodorou, A. and Forsyth, P. (eds) *Aviation and Tourism: Implications for Leisure Travel*, Aldershot: Ashgate, 137–46.

Ioannides, D. and Debbage, K.G. (1998) 'Neo-Fordism and flexible specialisation in the travel industry: dissecting the polyglot', in Ioannides, D. and Debbage, K.G. (eds) *The Economic Geography of the Tourist Industry: A Supply-Side Analysis*, London: Routledge, 99–122.

Ioannides, D. and Petridou-Daugthrey, E. (2006) 'Competition in the travel distribution system: the US travel retail sector', in Papatheodorou, A. (ed.) *Corporate Rivalry and Market Power: Competition Issues in the Tourism Industry*, London: I.B. Tauris, 124–42.

Institute of Business Ethics (IBE) (1994) *Benefiting Business and the Environment*, London: IBE.

Inter-Continental Hotels and Resorts (2008) *Our History*, Online. Available HTTP: <http://www.interconti.com> (accessed 9 December 2008).

International Hotels Environment Initiative (IHEI) (1993) *Environmental Management for Hotels: The Industry Guide to Best Practice*, Oxford: Butterworth Heinemann.

Iso-Ahola, S.E. (1982) 'Towards a social psychological theory of tourism motivation: a rejoinder', *Annals of Tourism Research*, 9: 256–61.

Jaffe, A.B., Peterson, S.R., Portney, P.R. and Stavins, R.N. (1995) 'Environmental regulation and the competitiveness of U.S. manufacturing: what does the evidence tell us?', *Journal of Economic Literature*, 33: 132–63.

JAL (2009) About the *JAL World Hotels Program*, Online. Available HTTP: <http://www.jal.co.jp/jwh/information_e.html> (accessed 12 May 2009).

Jappelli, T. and Pagano, M. (1988) 'Liquidity constrained households in an Italian cross-section', Centre for Economic Policy Research Discussion Paper no. 257.

Jennings, G.R. (2007) 'Advances in tourism research: theoretical paradigms and accountability', in Matias, A., Nijkamp, P. and Neto, P. (eds) *Advances in Modern Tourism Research: Economic Perspectives*, Heidelberg: Physica-Verlag, 9–35.

Jensen, C.M. and Meckling, H.W. (1976) "Theory of the firm: managerial behavior, agency costs and ownership structure", *Journal of Financial Economics*, 3: 305–60.

Jensen, T.C. and Wanhill, S. (2002) 'Tourism's taxing times: value added tax in Europe and Denmark', *Tourism Management*, 23: 67–79.

Johnson, C. and Vanetti, M. (2005) 'Locational strategies of international hotel chains', *Annals of Tourism Research*, 32: 1077–99.

Johnson, P. and Ashworth, J. (1990) 'Modelling tourism demand: a summary review', *Leisure Studies*, 9: 145–60.

Johnson, P. and Thomas, B. (1990) 'Measuring the local employment impact of a tourist attraction: an empirical study', *Regional Studies*, 24: 395–403.

Johnston, R.J. and Tyrrell, T.J. (2005) 'A dynamic model of sustainable tourism', *Journal of Travel Research*, 44: 124–34.

—— (2008) 'Tourism sustainability, resiliency and dynamics: towards a more comprehensive perspective', *Tourism and Hospitality Research*, 8: 14–24.

Jones, C.I. (1999) 'Growth: with or without scale effects', *American Economic Review*, 89: 139–44.

Jones, P. and Pizam, A. (1993) *The International Hospitality Industry – Organizational and Operational Issues*, London: Pitman.

Jorgensen, F. and Solvoll, G. (1996) 'Demand models for inclusive tour charter: the Norwegian case', *Tourism Management*, 17: 17–24.

Jundin, S. (1983) 'Barns uppfattning om konsumtion, sparande och arbete (Children's conceptions about consumption, saving and work)', Stockholm: The Stockholm School of Economics, EFI (doctoral dissertation, English summary).

Kahneman, D., Slovic, P. and Tversky, A. (1982) *Judgment under Uncertainty: Heuristics and Biases*, Cambridge: Cambridge University Press.

Kalecki, M. (1939) *Essays in the Theory of Economic Fluctuations*, London: Allen & Unwin.

Keane, M.J. (1997) 'Quality and pricing in tourism destinations', *Annals of Tourism Research*, 24: 117–30.

Keogh, B. (1990) 'Public participation in community tourism planning', *Annals of Tourism Research*, 17: 449–65.

Kent, P. (1990) 'People, places and priorities: opportunity sets and consumers' holiday choice', in Ashworth, G. and Goodall, B. (eds) *Marketing Tourism Places*, London: Routledge, 42–62.

—— (1991) 'Understanding holiday choices', in Sinclair, M.T. and Stabler, M.J. (eds) *The Tourism Industry: An International Analysis*, Wallingford: CAB International, 165–183.

Khan, H., Seng, C.F. and Cheong, W.K. (1990) 'Tourism multiplier effects in Singapore', *Annals of Tourism Research*, 17: 408–18.

Kinnaird, V., Kothari, U. and Hall, D. (1994) 'Tourism: gender perspectives', in Kinnaird, V. and Hall, D. (eds) *Tourism: A Gender Analysis*, Chichester: John Wiley.

Kirchler, E. (1988) 'Household economic decision-making', in van Raaij, W.F., van Veldhoven, G.M. and Wärneryd, K-E. (eds) *Handbook of Economic Psychology*, Dordrecht: Kluwer.

Kirzner, I.M. (1973) *Competition and Entrepreneurship*, Chicago, Ill.: Chicago University Press.

Klein, R. (2005) *Playing Off the Ports: BC and the Cruise Tourism Industry*, Vancouver: Canadian Centre for Policy Alternatives.

Klein, B., Crawford, R.G. and Alchian, A.A. (1978) 'Vertical integration, appropriable rents and the competitive contracting process', *Journal of Law and Economics*, 21: 297–326.

Kliman, M.L. (1981) 'A quantitative analysis of Canadian overseas tourism', *Transportation Research*, A,15: 487–97.

Kling, C.L. and Crocker, J.R. (1999) 'Recreation demand models for environmental valuation', in van de Bergh, J.C.J.M. (ed.) *Handbook of Environmental and Resource Economics*, Cheltenham: Edward Elgar.

Knowles, T., Diamantis, D. and El-Mourhabi, J. (2001) *The Globalization of Tourism and Hospitality: A Strategic Perspective*, London: Continuum.

Knudsen, O. and Parnes, A. (1975) *Trade Instability and Economic Development*, Lexington, Mass.: D.C. Heath.

Korca, P. (1991) *Assessment of the Environmental Impacts of Tourism*, proceedings of an International Symposium on the Architecture of Tourism in the Mediterranean, Istanbul, Turkey: Yildiz University Press.

Kostis, C., Papatheodorou, A. and Parthenis, S. (2008) *Are Community Airports Mature Enough for the Introduction of Market Mechanisms Concerning the Airport Slot Allocation? A Survey in the Greek Industry* 12th Annual Conference of the Air Transport Research Society, Athens, Greece.

Kotler, P. and Keller, K. (2008) *Marketing Management*, Harlow: Pearson Education.

Krippendorf, J. (1987) *The Holiday Makers*, London: Heinemann.

Krugman, P.R. (1980) 'Scale economies, product differentiation and the pattern of trade', *American Economic Review*, 70: 950–59.

—— (1989a) 'Industrial organization and international trade', in Schmalensee, R. and Willig, R.D. (eds) *Handbook of Industrial Organization*, Amsterdam: North Holland.

—— (1989b) 'New trade theory and the less developed countries', in Calvo, G. and World Institute for Development Economics Research (eds) *Debt, Stabilization and Development*, Oxford: Basil Blackwell.

—— (1991a) *Geography and Trade*, Cambridge Mass.: MIT Press.

—— (1991b) 'Increasing returns and economic geography', *Journal of Political Economy*, 99: 483–99.

—— (1995) *Development, Geography and Economic Theory*, Cambridge, Mass.: MIT Press.

Krugman, P.R. and Obstfeld, M. (2009) *International Economics: Theory and Policy*, 8th edn, Boston: Pearson Education.

Krugman, P.R. and Venables, A. (1995) 'Globalisation and the inequality of nations', *Quarterly Journal of Economics*, 110: 857–80.

Kulendran, N. and King, M. (1997) 'Forecasting international quarterly tourist flows using error-correction and time-series models', *International Journal of Forecasting*, 13: 319–27.

Kuniyal, J.C. (2005) 'Solid waste management in the Himalayan trails and expedition summits', *Journal of Sustainable Tourism*, 13: 391–410.

Lakatos, I. and Musgrave, A. (eds) (1970) *Criticism and the Growth of Knowledge*, Cambridge: Cambridge University Press.

Lamminmaki, D. (2007) 'Outsourcing in Australian hotels: a transaction cost economics perspective', *Journal of Hospitality & Tourism Research*, 31: 73–110.

Lancaster, K.J. (1966) 'A new approach to consumer theory', *Journal of Political Economy*, 84: 132–57.

—— (1971) *Consumer Demand: A New Approach*, New York: Columbia University Press.

Lanza, A., Markandya, A. and Pigliaru, F. (2005) *The Economics of Tourism and Sustainable Development*, Cheltenham: Edward Elgar.

Lawson, F. (1995) *Hotels and Resorts – Planning, Design and Refurbishment*, London: Architectural Press.

Lawton, C. (2002) *Cleared for Take-off: Structure and Strategy in Low Fare Airline Business*, Aldershot: Ashgate.

Layard, P.R.G. and Glaister, S. (eds) (1994) 2nd edn, *Cost-benefit analysis* Cambridge: Cambridge University Press.

Lea, J.P. (1993) 'Tourism development ethics in the third world', *Annals of Tourism Research*, 20: 701–15.

LECG (1999) *Quantitative Techniques in Competition Analysis*, Research Paper 17, London: Office of Fair Trading.

Ledesma-Rodriguez, F.J., Navarro-Ibáñez, M. and Pérez-Rodríguez, J.V. (2001) 'Panel data and tourism: a case study of Tenerife', *Tourism Economics*, 7: 75–88.

Lee C-K and Han S-Y (2002) 'Estimating the use and preservation values of national parks tourism resources using a contingent valuation method', *Tourism Management*, 23: 531–40.

Lee, K, Var, T. and Blaine, T.W. (1996) 'Determinants of inbound tourist expenditures', *Annals of Tourism Research*, 23: 527–42.

Lee, G. (1987) 'Tourism as a factor in development cooperation', *Tourism Management*, 8: 2–19.

Lee, W. (1991) 'Prostitution and tourism in South-East Asia', in Redclift, N. and Sinclair, M.T. (eds) *Working Women: International Perspectives on Labour and Gender Ideology*, London and New York: Routledge.

Leiper, N. (1979) 'The framework of tourism: towards a definition of tourism, tourist and the tourist industry', *Annals of Tourism Research*, 6: 390–407.

—— (1990) *Tourism Systems*, Palmerston North: Massey University.

Leontidou, L. (1994) 'Gender dimensions of tourism in Greece: employment, subcultures and restructuring', in Kinnaird, V. and Hall, D. (eds) *Tourism: A Gender Analysis*, Chichester: John Wiley.

Lerner, A.P. (1934) 'The concept of monopoly and the measurement of monopoly power', *Review of Economic Studies*, 1: 157–75.

Levine, M.E. (1987) 'Airline competition in deregulated markets: theory, firm strategy and public policy', *Yale Journal on Regulation*, 4: 393–494.

Levinson, A. (1996) 'Environmental regulations and manufacturers' location choice: evidence from the Census of Manufacturers', *Journal of Public Economy*, 62: 5–29.

Lew, A.A. and McKercher, B. (2006) 'Modelling tourist movements: a local destination analysis', *Annals of Tourism Research*, 33: 403–23.

Lew, A.A., Hall, C.M. and Williams, A.M. (eds) (2004) *A Companion to Tourism*, Oxford: Blackwell.

Lichfield, N. (1988) *Economics of Urban Conservation*, Cambridge: Cambridge University Press.

Liebenstein, H. (1950) 'Bandwagon, snob and Veblen effects in the theory of consumers' demand', *Quarterly Journal of Economics*, 64: 183–201.

Liebermann, M.B. and Montgomery, D.B. (1988) 'First-mover advantages', *Strategic Management Journal*, 9: 41–58.

Li, G. (2004) *Modelling and Forecasting UK Tourism Demand in Western Europe: Illustrations of TVP-LAIDS Models' Superiority over Other Econometric Approaches*, unpublished PhD Thesis, Guildford: University of Surrey, UK.

Li, G., Song, H. and Witt, S.F. (2004) 'Modelling tourism demand: a dynamic linear AIDS approach', *Journal of Travel Research*, 43: 141–50.

—— (2006) 'Time varying parameter and fixed parameter linear AIDS: an application to tourism demand forecasting', *International Journal of Forecasting*, 22: 57–71.

Lim, C. (2006) 'A survey of tourism demand modelling practice: issues and implications', in Dwyer, L. and Forsyth, P. (eds) *International Handbook on the Economics of Tourism*, Cheltenham: Edward Elgar, 45–72.

Lin, T.-B. and Sung, Y.-W. (1983) 'Hong Kong', in Pye, E.A. and Lin, T-B. (eds) *Tourism in Asia: The Economic Impact*, Singapore: Singapore University Press.

Lindberg, K. and Huber, R.M. (1993) 'Economics issues in ecotourism management', in Lindberg, K. and Hawkins, D.E. (eds) *Ecotourism: A Guide for Planners and Managers*, North Bennington, Vt: The Ecotourism Society, 82–115.

Lindberg, K. and Johnson, R.L. (1997) 'The economic values of tourism's social impacts', *Annals of Tourism Research*, 24: 90–116.

Linder, S.B. (1961) *An Essay on Trade and Transformation* London: John Wiley.

Little, I.M.D. and Mirlees, J.A. (1974) *Project Appraisal and Planning for Developing Countries*, London: Heinemann.

Little, J.S. (1980) 'International travel in the UK balance of payments', *New England Economic Review*, May: 42–55.

Littlechild, S.C. (1986) *The Fallacy of the Mixed Economy*, 2nd edn, London: Institute of Economic Affairs.

Liu, Z. (2003) 'Sustainable tourism development: a critique', *Journal of Sustainable Tourism*, 11: 459–75.

Lockwood, M., Loomis, J. and DeLacy, T. (1993) 'A contingent valuation survey and benefit-cost analysis of forest preservation in East Gippsland, Australia', *Journal of Environmental Management*, 38: 233–43.

Loeb, P.D. (1982) 'International travel to the United States: an econometric evaluation', *Annals of Tourism Research*, 9: 7–20.

Loewenstein, G. (1987) 'Anticipation and the value of delayed consumption', *Economic Journal*, 97: 666–84.

Lombardi, P. and Sirchia, G. (1990) 'Il quarterre 16 IACF di Torino', in R. Roscelli (ed.) *Misurare Nell'Incertezza*, Turin: Celid.

Long, V.H. (1991) 'Government–industry–community interaction in tourism development in Mexico', in Sinclair, M.T. and Stabler, M.J. (eds) *The Tourism Industry: An International Analysis*, Wallingford: CAB International.

Loomis, J.B., Creel, M. and Park, T. (1991) 'Comparing benefit estimates from travel cost and contingent valuation using confidence intervals from Hicksian welfare measures', *Applied Economics*, 23: 1725–31.

Lorenz, E. (1963) 'Deterministic non-period flow', *Journal of Atmospheric Science*, 20: 1340–141.

Lösch, A. (1940) *Die raumliche Ordnung der Wirtschaft*, Jena: Gustav Fischer Verlag.

Lovelock, J. (1979) *Gaia: A New Look at Life on Earth*, Oxford: Oxford University Press.

Lucas, Jr, R.E. (1972) 'Expectations and the neutrality of money', *Journal of Economic Theory*, 90: 103–24.

—— (1977) 'Understanding business cycles', in *Stabilization of the Domestic and International Economy*, Carnegie-Rochester Series on Public Policy, vol. 5: 7–30.

—— (1988) 'On the mechanics of economic growth', *Journal of Monetary Economics*, 22: 3–42.

Luck, M. (2007) *Encylopaedia of Tourism and Recreation in Marine Environments*, Wallingford: CABI.

Lumsden, L.M. and Swift, J.S. (1998) 'Ecotourism at the crossroad: the case of Costa Rica', *Journal of Sustainable Tourism*, 6: 155–72.

Lundberg, D.E. (1989) *The tourism business*, New York: Van Nostrand Reinhold.

Lundberg, D.E., Krishnamoorthy, M. and Stavenga, M.H. (1995) *Tourism Economics*, New York: John Wiley.

Lyons, B. (1989) 'Barriers to entry', in Davies, S., Lyons, B. with Dixon, H. and Gerowski, P. *Economics of Industrial Organisation*, London: Longman. 26–72.

Maarten, C.W.J. (1993) *Microfoundations: A Critical Enquiry*, London: Routledge.

MacCurdy, T.E. (1982) 'The use of time-series processes to model the error structure of earnings in longitudinal data analysis', *Journal of Econometrics*, 18: 83–114.

McIntosh, R.W. and Goeldner, C.R. (1990) *Tourism Principles, Practices, Philosophy*, 6th edn, New York: John Wiley.

McKercher, B. (1999) 'A chaos approach to tourism', *Tourism Management*, 20: 425–34.

Mackie, P.J. and Preston, J.M. (1996) *The Local Bus Market: A Case Study of Regulatory Change*, Aldershot: Avebury.

McKinsey (2003) 'Off-shoring and beyond', *The McKinsey Quarterly*, Special Edition 4.

—— (2007) 'How businesses are using Web 2.0: a McKinsey global survey', *The McKinsey Quarterly*, March.

McQueen, M. (1983) 'Appropriate policies towards multinational hotel corporations in developing countries', *World Development*, 11: 141–52.

Maddison, D. (2001) *The Amenity Value of the Global Climate*, London: Earthscan.

Makridakis, S. (1986) 'The art and science of forecasting: an assessment and future directions', *International Journal of Forecasting*, 2: 15–39.

Mangion, M.L., Durbarry, R. and Sinclair, M.T. (2005) 'Tourism competitiveness: price and quality tourism competitiveness: price and quality', *Tourism Economics*, 11: 45–68.

Manta (2009a) *Tour Operators Companies in the United States*, Online. Available HTTP: <http://www.manta.com/mb_35_B72D5000_000/tour_operators> (accessed 25 April 2009).

Manta (2009b) *Travel Agencies in the United States*, Online. Available HTTP: <http://www.manta.com/mb_35_B72D4000_000/travel_agencies?search=travel%20agencies> (accessed 25 April 2009).

March, J.G. (1962) 'The business firm as a political coalition', *Journal of Politics*, 24: 662–78.

March, J.G. and Simon, H.A. (1958) *Organizations*, New York: John Wiley.

Marks, K. and Howden, D. (2008) 'Vast and growing fast, a garbage tip that stretches from Hawaii to Japan', *Independent*, February 5th, 2.

Marriott (2009) *Marriott Rewards*, Online. Available HTTP: <http://www.marriott.com/rewards/rewards-program.mi> (accessed 6 May 2009).

Martin, C.A. and Witt, S.F. (1987) 'Tourism demand forecasting models: choice of appropriate variable to represent tourists' cost of living', *Tourism Management*, 8: 233–46.

—— (1988) 'Substitute prices in models of tourism demand', *Annals of Tourism Research*, 15: 255–68.

—— (1989) 'Forecasting tourism demand: a comparison of the accuracy of several quantitative methods', *International Journal of Forecasting*, 5: 1–13.

Martin, S. (1993) *Advanced Industrial Economics*, Cambridge, Mass.: Basil Blackwell.

Marx, K. (1967) *Capital (Centennial Edition of Das Kapital, 1867)*, New York: International Publishers.

Mason, E.S. (1957) *Economic Concentration and the Monopoly Problem*, Cambridge, Mass.: Harvard University Press.

Mathieson, A. and Wall, G. (1982) *Tourism: Economic, Physical, and Social Impacts*. London: Longman.

Mathisen, T.A. and Solvoll, G. (2007) 'Competitive tendering and structural changes: an example from the bus industry', *Journal of Transport Policy*, 15: 1–11

Matias, A., Nijkamp, P. and Neto, P. (eds) (2007) *Advances in Modern Tourism Research: Economic Perspectives*, Heidleberg: Physica-Verlag.

Means, G. and Avila, R. (1986) 'Econometric analysis and forecasts of US international travel: using the new TRAM model', *World Travel Overview*, 1986/87: 90–107.

—— (1987) 'An econometric analysis and forecast of US travel and the 1987 TRAM model update', *World Travel Overview*, 1987/88: 102–23.

Melville, J.A. (1995) 'Some empirical results for the airline and air transport markets of a small developing country', PhD thesis, University of Kent at Canterbury.

Middleton, V. (2002) 'A national strategy for visitor attractions', in Fyall, A., Garrod, B. and Laersk, A. (eds) *Managing Visitor Attractions: New Directions*, Oxford: Butterworth Heinemann, 5–15.

Middleton, V.T.C. and Hawkins, R. (1993) 'Practical environmental policies in travel and tourism', *Travel and Tourism Analyst*, 6: 63–76, London: Economist Intelligence Unit.

Middleton, V.T.C. and Jackie, C. (2001) *Marketing in Travel and Tourism*, Oxford: Butterworth Heinemann.

Miller, G. and Twining-Ward, L. (2005) *Monitoring for a Sustainable Tourism Transition: The Challenge of Developing and Using Indicators*, London: CABI.

Miller, R. (2002) 'A private sector perspective in APEC', in Cockerell, N. and Spurr, R. (eds) *Best Practice in Tourism Satellite Account Development in APEC Member Economies*, Singapore: Asia Pacific Economic Cooperation (APEC) Secretariat Publication No 202-TR-01-1, 41–52.

Milne, S.S. (1998) 'Tourism and sustainable development: exploring the global-local nexus', in Hall C.M. and Lew A.A. (eds) *Sustainable Tourism: A Geographical Perspective*, Harlow: Longman, 35–48.

Mintel (2007) 'European cruises', *Travel and Tourism Analyst*, August.

Mitchell, F. (1970) 'The value of tourism in East Africa', *East Africa Economic Review*, 2: 1–21.

Mitchell, R.C. and Carson, R.T. (1989) *Using Surveys to Value Public Goods: The Contingent Valuation Method*, Washington, DC: Resources for the Future.

Moeller, G.H. and Shafer, E.L. (1987) 'The Delphi technique: a tool for long-range tourism and travel planning', in Ritchie, J.R.B. and Goeldner, C.R. (eds) *Travel Tourism and Hospitality Research*, New York: Wiley.

Momsen, J.H. (1994) 'Tourism, gender and development in the Caribbean', in Kinnaird, V. and Hall, D. (eds) *Tourism: A Gender Analysis*, Chichester: Wiley.

Moorhouse, J.C. and Smith, M.S. (1994) 'The market for residential architecture: 19th century row houses in Boston's South End', *Journal of Urban Economics*, 35: 267–77.

Morley, C.L. (1992) 'A microeconomic theory of international tourism demand', *Annals of Tourism Research*, 19: 250–67.

—— (1994) 'Experimental destination choice analysis', *Annals of Tourism Research*, 21: 780–91.

Morris, D. (2000) *Competition Policy and Regulation in the UK: A New Era*, Lecture given in Lancaster University on 29th November.

Moscardo, G. (2008) 'Sustainable tourism innovation: challenging basic assumptions', *Tourism and Hospitality Research*, 8: 4–13.

Mott MacDonald (2006) *Study on the Impact of the Introduction of Secondary Trading at Community Airports: Volume I*, Technical Report prepared for the European Commission (DG TREN) by Mott MacDonald's Aviation Group in association with Oxera, Hugh O'Donovan and Keith Boyfield Associates.

Mowforth, M. and Munt, I. (1998) *Tourism and Sustainability: New Tourism in the Third World*, London: Routledge.

—— (2003) 2nd. edn, *Tourism and Sustainability*, London: Routledge.

Mudambi, R. (1994) 'A Ricardian excursion to Bermuda: an estimation of mixed strategy equilibrium', *Applied Economics*, 26: 927–36.

Muroi, H. and Sasaki, N. (1997) 'Tourism and prostitution in Japan', in Sinclair, M.T. (ed.) *Gender, Work and Tourism*, London and New York: Routledge, 180–219.

Murphy, P.E. (1985) *Tourism: A Community Approach*, New York: Methuen.

Naude, W.A. and Saayman, A. (2005) 'Determinants of tourist arrivals in Africa: a panel data regression analysis', *Tourism Economics*, 11: 365–91.

Nelson, C.R. and Plosser, C.I. (1982) 'Trends and random walks in macroeconomic time series', *Journal of Monetary Economics*, 10: 139–62.

Nelson, P.H. (1974) 'Advertising as information', *Journal of Political Economy*, 81: 729–45.

Nelson, R.R. and Winter, S.G. (1982) *An Evolutionary Theory of Economic Change*, Cambridge, Mass.: Harvard University Press.

NERA (1999) *Merger Appraisal in Oligopolistic Markets*, Research Paper 19, London: Office of Fair Trading.

—— (2001) *The Role of Market Definition in Monopoly and Dominance Inquiries*, Economic Discussion Paper 2, London: Office of Fair Trading.

Neves Sequiera, T. and Campos, T. (2005) 'International tourism and economic growth: a panel data approach', Fondazione Eni Enrico Mattei. *Nota di Lavoro* No 141.

Newbould, G. (1970) *Management and Merger Activity*, Liverpool: Guthstead.

Newsome, D., Moore, S.A. and Dowling, R.K. (2001) *Natural Area Tourism: Ecology, Impacts and Management*, Clevedon. Channel View Publications.

Niels, G. and van Dijk, R. (2006) 'Market definition in the tourism industry', in Papatheodorou, A. (ed.) *Corporate Rivalry and Market Power: Competition Issues in the Tourism Industry*, London: I.B. Tauris, 187–200.

Nightingale, J. (1978) 'On the definition of "industry" and "market"', *Journal of Industrial Economics*, 27: 31–40.

Nijkamp, P. (1975) 'A multicriteria analysis for project evaluation: economic–ecological evaluation of a land reclamation project', *Papers of the Regional Science Association*, 35: 87–111.

—— (1988) 'Culture and region: a multidimensional evaluation of movements', *Environment' and Planning B: Planning and Design*, 15: 5–14.

NOAA (1993) *Report of the NOAA Panel on Contingent Valuation*, 12th March, Washington, DC: National Oceanic and Atmospheric Administration.

Noreen, E. (1988) 'The economics of ethics: a new perspective on agency theory", *Accounting Organization and Society*, 13: 359–69.

North, D.C. (1990) *Institutional Change and Economic Performance*, Cambridge: Cambridge University Press.

Norton, G.A. (1984) *Resource Economics*, London: Edward Arnold.

Nowak, J.J. and Sahli, M. (2007) 'Coastal tourism and Dutch disease in a small island economy', *Tourism Economics,* 13: 49–65.

Nowotny, H. (2003) *The Potential of Transdisciplinarity*, Paris: Rethinking Interdisciplinarity.

Oates, W.E. (1999) *The RFF Reader in Environmental and Resource Management*, Washington DC: Resources for the Future.

Obstfeld, M. (1990) 'Intertemporal dependence, impatience, and dynamics', *Journal of Monetary Economics*, 26: 45–75.

ODI (2001) *Pro-Poor Tourism Strategies*: *Making Tourism Work for the Poor*, *A Review of Experience*, Nottingham: Russell Press.

—— (2007a) *Assessing How Tourism Revenues Reach the Poor*, Briefing paper 21, June, London: ODI.

—— (2007b) *Can Tourism offer Pro-poor Pathways to Prosperity?*, Briefing paper 22, June, London: ODI.

Office of Fair Trading (2007) 'British Airways to pay record £121.5m penalty in price fixing investigation', Press Release August 1st, London: Office of Fair Trading.

Official Journal of the European Union (2009) *Regulation (EC) No 80/2009 of the European Parliament and of the Council of 14 January 2009 on a Code of Conduct for Computerised Reservation Systems and Repealing Council Regulation (EEC) No 2299/89*, Luxembourg: European Union.

O'Hagan, J.W. and Harrison, M.J. (1984) 'Market shares of US tourist expenditure in Europe: an econometric analysis', *Applied Economics*, 16: 919–31.

Oi, W.Y. and Hurter, A.P. (1965) *Economics of Private Truck Transportation*, Dubuque, Iowa: William C. Brown.

Oppermann, M. (1995) 'Travel life cycle', *Annals of Tourism Research*, 22: 535–52.

Oxera (2009) 'Overruled: the state aid case against Ryanair and Charleroi Airport', *Agenda*, January.

Pack, A. and Sinclair, M.T. (1995a) *Tourism, Conservation and Sustainable Development, Indonesia*, Report for the Overseas Development Administration, London.

—— (1995b) *Tourism, Conservation and Sustain-able Development, India*, Report for the Overseas Development Administration, London.

Paelinck, J.H.P. (1976) 'Qualitative multiple criteria analysis, environmental protection and multiregional development', *Papers of the Regional Science Association*, 36: 59–74.

Page, S.J. and Dowling, R.K. (2002) *Ecotourism*, Harlow: Prentice Hall; Pearson Education imprint.

Palmquist, R.B. (1999) 'Hedonic models', in van der Bergh, J.C.J.M. (ed.) *Handbook of Environmental and Resource Economics*, Cheltenham: Edward Elgar.

—— (2005) 'Property value models', in Maler, K.G. and Vincent, J. (eds) *Handbook of Environmental Economics Vol. 2*, London: Elsevier, 763–819.

Pambudi, D., McCaughey, N. and Smyth, R. (2009) 'Computable general equilibrium estimates of the impact of the Bali bombing on the Indonesian economy", *Tourism Management*, 30: 232–39.

Papadopoulos, S.I. (1987) 'Strategic marketing techniques in international tourism', *International Marketing Review*, 71–84.

Papatheodorou, A. (1999) 'The demand for international tourism in the Mediterranean region', *Applied Economics*, 31: 619–30.

—— (2000) *Evolutionary Patterns in Tourism*: *A Spatial Industrial Organisation Approach*, DPhil Thesis, Oxford: University of Oxford.

—— (2001a) 'Tourism, transport geography and industrial economics: a synthesis in the context of Mediterranean islands'. *Anatolia*, 12: 23–34.

—— (2001b) 'Why people travel to different places?', *Annals of Tourism Research*, 28: 164–79.

—— (2002a) 'Civil aviation regimes and leisure tourism in Europe', *Journal of Air Transport Management*, 8(6): 381–88.

—— (2002b) 'Exploring competitiveness in Mediterranean resorts', *Tourism Economics*, 8: 133–50.

—— (2003a) 'Corporate strategies of British tour operators in the Mediterranean region: an economic geography approach', *Tourism Geographies*, 5: 280–304.

—— (2003b) 'Do we need airport regulation?', *Utilities Journal*, 6: 35–37.

—— (2003c) 'Exploring the determination of student performance in university modules and streams', *Applied Economics*, 35: 1859–64.

—— (2003d) 'Modelling tourism development – a synthetic approach,' *Tourism Economics*, 9: 407–30.

—— (2004) 'Exploring the evolution of tourist resorts', *Annals of Tourism Research*, 31: 219–37.

—— (2006a) 'Conclusion: the need for constructive policymaking', in Papatheodorou, A. (ed.) *Corporate Rivalry and Market Power: Competition Issues in the Tourism Industry*, London: I.B. Tauris, 201–5.

—— (2006b) 'Corporate rivalry, market power and competition issues in tourism: an introduction', in Papatheodorou, A. (ed.) *Corporate Rivalry and Market Power: Competition Issues in the Tourism Industry*, London: I.B. Tauris, 1–19.

—— (2006c) 'Liberalisation and deregulation for tourism', in Buhalis, D. and Costa, C. (eds) *Tourism Management Dynamics: Trends, Management and Tools*, Oxford: Butterworth-Heinemann, 68–77.

—— (2006d) 'The cruise industry – an industrial organisation perspective', in Dowling, R. (ed.) *Cruise Ship Tourism*, Wallingford: CABI, 31–40.

—— (2008) 'The impact of civil aviation regimes on leisure travel', in Graham, A., Papatheodorou, A. and Forsyth, P. (eds) *Aviation and Tourism: Implications for Leisure Travel*, Aldershot: Ashgate, 49–57.

Papatheodorou, A. and Arvanitis, P. (2009) 'Spatial evolution of airport traffic and air transport liberalization: the case of Greece', *Journal of Transport Geography* 17: 402–412.

Papatheodorou, A. and Iatrou, K. (2008) 'Leisure travel: implications for airline alliances', *International Review of Aerospace Engineering* 1: 332–42.

Papatheodorou, A. and Lei, Z. (2006) 'Leisure travel in Europe and airline business models: a study of regional airports in Great Britain', *Journal of Air Transport Management*, 12: 47–52.

Papatheodorou, A. and Platis, N. (2007) 'Airline deregulation, competitive environment and safety', *Rivista di Politica Economica*, I-II 2007: 221–42.

Papatheodorou, A. and Song, H. (2005) 'International tourism forecasts: a time series analysis of world and regional data', *Tourism Economics*, 11: 11–24.

Park, J.R., Stabler, M.J., Mortimer, S.R., Jones, P.J., Ansell, D.J. and Parker, G.P.D. (2004) 'The use of multiple criteria decision analysis to evaluate the effectiveness of landscape and habitat enhancement mechanisms: an example for the South Downs', *Journal of Environmental Planning and Management*, 47: 773–93.

Parry, B. and Drost, R. (1995) 'Is chaos good for your profits?', *International Journal of Contemporary Hospitality Management*, 7: 1–3.

Pawson, I.G., Stanford, D.D., Adams, V.A. and Nurbu, M. (1984) 'Growth of tourism in Nepal's Everest region: impact on the physical environment and structure of human settlements', *Mountain Research and Development*, 4: 237–46.

Pearce, D.G. (1987) *Tourism Today: A Geographical Analysis*, Harlow: Longman.

—— (1989) 2nd edn, *Tourism Development*, Harlow: Longman.

—— (2005) 'Advancing tourism research: issues and responses', in Alejziak, W. and Winiarski, R. (eds) *Tourism in Scientific Research*, Krakow-Rzeszow: Academy of Physical Education, University of Information Technology and Management in Rzeszow, 7–20.

Pearce, D.G. and Butler, R.W. (eds) (1993) *Tourism Research: Critiques and Challenges*, London: Routledge.

—— (1999) *Contemporary Issues in Tourism Development*, London: Routledge.

Pearce, D.W. (1983) 2nd edn, *Cost Benefit Analysis*, Basingstoke: Macmillan.

—— (1991) *Blueprint 2: Greening the World Economy*, London: Earthscan.

Pearce, D.W. and Nash, C.A. (1981) *The Social Appraisal of Projects: A Text in Cost-Benefit Analysis*, London: Macmillan.

Pearce, D.W. and Turner, R.K. (1990) *Economics of Natural Resources and the Environment*, London: Harvester Wheatsheaf.

Pearce, D.W., Markandya, A. and Barbier, E.B. (1989) *Blueprint for a Green Economy*, London: Earthscan.

Pearson. L.J., Tisdell, C.A. and Lisle, A.J. (2002) 'The impact of Noosa national park on surrounding property values: an application of the hedonic price method', *Economic Analysis and Policy*, 21: 158–71.

Peltzman, S. (1977) 'The gains and losses from industrial concentration', *Journal of Law and Economics*, 20: 229–63.

Pendleton, L. (1999) 'Reconsidering the hedonic versus RUM debate in the valuation of recreational environmental amenities', *Resources and Energy Economics*, 21: 167–89.

Perman, R., Ma, Y., McGilvray, J. and Common, M. (2003) *Natural Resource and Environmental Economics*, Harlow: Pearson.

Perrett, K.E., Hopkins, J.M., Pickett, M.W. and Walmsley, D.A. (1989) *The Effects of Bus Deregulation in the Metropolitan Areas*, TRRL Research Report 210, London: Department of Transport.

Philips, R. (2005) *Pricing and Revenue Optimization*, Stanford: Stanford Business Books.

Pischke, J.-S. (1991) 'Individual income, incomplete information and aggregate consumption', Industrial Relations Section working paper no. 289, Princeton University, NJ.

Plog, S.C. (1973) 'Why destination areas rise and fall in popularity', *Cornell HRA Quarterly*, November, 13–16.

Pollard, H.J. (1976) 'Antigua, West Indies: an example of the operation of the multiplier process arising from tourism', *Revue de Tourisme*, 3: 30–34.

Poon, A. (1988) 'Innovation and the future of Caribbean tourism', *Tourism Management*, 9: 213–20.

—— (1990) 'Flexible specialization and small size – the case of caribbean tourism', *World Development* 18:109–23.

Porter, M.E. (1979) 'How competitive forces shape strategy', *Harvard Business Review*, March/April.

—— (1980) *Competitive Strategy: Techniques for Analysing Industries and Competitors*, New York: Free Press.

—— (1990) *The Competitive Advantage of Nations*, New York: Free Press.

—— (1998) *Competitive Strategy: Techniques for Analysing Industries and Competitors: With a New Introduction*, New York: Free press.

Porter, M.E. and van der Linde (1995) *Green and Competitive: Ending the Stalemate*, Cambridge Mass.: Harvard Business Review.

Porter, R.C. (1982) 'The new approach to wilderness preservation through cost-benefit analysis', *Journal of Environmental Economics and Management*, 9: 59–80.

Posner, M.W. (1961) 'International trade and technical change', *Oxford Economic Papers*, 13: 323–41.

Prais, S.J. (1976) *The Evolution of Giant Firms in Britain*, National Institute of Economic and Social Research, Economic and Social Studies, 30, Cambridge: Cambridge University Press.

Pred, A. (1966) *The Spatial Dynamics of U.S. Urban Industrial Growth*, Cambridge, Mass.: MIT Press.

Preston, J. (1991) 'Explaining competition practices in the bus industry: the British experience', *Transport Planning and Technology*, 15: 277–94.

Pritchard, A. and Morgan, N. (2000) 'Privileging the male gaze: gendered tourism landscapes', *Annals of Tourism Research*, 27: 884–905.

Pro-Poor Tourism (2008) *Pro-Poor Tourism*, Online. Available HTTP: <http://www.propoortourism.org.uk> (accessed 4 November 2008).

Przeclawski, K. (1993) 'Tourism as the subject of interdisplinary research,' in Pearce, D.G. and Butler, R.W. (eds) *Tourism Research: Critiques and Challenges*, London: Routledge, 9–19.

Puga, D. and Venables, A. (1996) 'The spread of industry: spatial agglomeration in economic development', *Journal of the Japanese and International Economies*, 10: 440–64.

Puhipan (1994) 'Boycott paradise', *In Focus*, 12, summer: 10–11.

Purcell, K. (1997) 'Women's employment in UK tourism: gender roles and labour markets', in Sinclair, M.T. (ed.) *Gender, Work and Tourism*, London and New York: Routledge, 35–59.

Pyo, S.S., Uysal, M. and McLellan, R.W. (1991) 'A linear expenditure model for tourism demand', *Annals of Tourism Research*, 18: 443–54.

Qualls, W.J. (1982) 'Changing sex roles: its impact upon family decision making', *Advances in Consumer Research*, 9: 151–62.

Quayson, J. and Var, J. (1982) 'A tourism demand function for the Okanagan, BC', *Tourism Management*, 3: 108–15.

Radisson Blu (2009) *About Us, Online*. Available HTTP: <http://www.radissonblu.com/about-us/blu> (accessed 2 May 2009).

Raguraman, K. (1997) 'Airlines as instruments for nation building and national identity: case study of Malaysia and Singapore', *Journal of Transport Geography*, 5: 239–56.

Randall, A. (1993) 'The problem of market failure', in Dorfman, R. and Dorfman, N.S. (eds) *Economics of the Environment*, 3rd edn, New York: Norton.

Razumova, M., Rey-Maquiera, J. and Lozano, J. (2007) *Environment-Competitiveness Relationship in Tourism: Application of the Porter Hypothesis*, paper presented at the first conference of the International Association for Tourism Economics (IATE) conference. Majorca: Palma, 25th-27th October.

Redefining Progress (1995) *The Genuine Progress Indicator: Summary of Data and Methodology*, San Francisco: Redefining Progress.

Reekie, W.D. (1984) *Markets, Entrepreneurs and Liberty: An Austrian View of Capitalism*, Brighton: Wheatsheaf.

Research Institute of Urban Environment and Human Resources (2008) *HERMES Project*, Online. Available HTTP: <http://www.uehr.gr/> (accessed 1 November 2008).

Retter, C. (2008) 'TUI Sells Hapag-Lloyd Unit in 4.45 Billion Euro Deal', *Bloomberg*, 12th October. Online. Available HTTP: <http://www.bloomberg.com/apps/news?pid=20601100&sid=aOE3s3wv2_z8&refer=germany> (accessed 8 May 2009).

Richardson, H.W. (1969) *Elements of Regional Economics*, Harmondsworth: Penguin.

—— (1972) *Input-Output and Regional Economics*, London: Weidenfeld & Nicolson.

—— (1979) *Regional and Urban Economics*, London; Pitman.

Rindfleisch, A. and Heide, B.I. (1997) 'Transaction cost analysis: past, resent and future applications', *Journal of Marketing*, 61: 30–54.

Ritchie, B.W. (2004) 'Chaos, crisis and disasters: a strategic approach to crisis management in the tourism industry', *Tourism Management*, 25: 669–83.

Ritchie, B.W., Burns, P. and Palmer, C. (2005) (eds) *Tourism Research Methods: Integrating Theory with Practice*, Wallingford: CABI.

Ritchie, J.R.B. and Crouch, G.I. (2003) *The Competitive Destination: A Sustainable Tourism Perspective*, Wallingford: CABI.

Rivera-Batiz, L.A. and Romer, P.M. (1991) 'Economic integration and endogenous growth', *Quarterly Journal of Economics*, 106: 531–55.

Robertson, R. (1992) *Globalization, Social Theory and Global Culture*, London: Sage Publications.

Robinson, J. (2001) 'Socio-cultural dimension of sustainable tourism development: achieving the vision', in Varma, H. (ed.) *Island Tourism in Asia and the Pacific*, Madrid: UNWTO, 78–86.

Roden, A. (2004) 'First challenges GNER as East Coast franchise race restarts', *Rail*, February 18–March 2.

Rodrik, D. (1995) 'Getting intervention right: how South Korea and Taiwan grew rich', *Economic Policy*, 20: 53–107.

Rodriquez, A. (2002) 'Determining the entry choice for international expansion: the case of the Spanish hotel industry', *Tourism Management*, 23: 597–607.

Romer, P.M. (1986) 'Increasing returns and long-run growth', *Journal of Political Economy*, 94: 1002–37.

—— (1990) 'Endogenous technological change', *Journal of Political Economy*, 98: S71–S102.

—— (1994) 'The origins of endogenous growth', *Journal of Economic Perspectives*, 8: 3–22.

Romeril, M. (1989a) 'Tourism and the environment: accord or discord', *Tourism Management*, 10: 204–8.

—— (1989b) 'Tourism: the environmental dimension', in Cooper C.P. (ed.) *Progress in Tourism, Recreation and Hospitality Management, Vol. 1*, London: Belhaven 103–13.

Roscelli, R. and Zorzi, F. (1990) 'Valutazione di progetti di riqualificazione urbana', in Roscelli, R. (ed.) *Misurare Nell'Incertezza*, Turin: Celid.

Rosen, S. (1974) 'Hedonic prices and implicit markets: production differentiation in pure competition', *Journal of Political Economy*, 82: 34–55.

Rosenberg, J. (2006) *The Follies of Growth Theory*, London: Verso.

Rosenberg, M. (1956) 'Cognitive structure and attitudinal effect', *Journal of Abnormal and Social Psychology*, 53: 367–72.

Ross, S. and Wall, G. (1999) 'Ecotourism: towards a congruence between theory and practice', *Tourism Management*. 20: 123–32.

Rubinstein, A. (1979) 'Equilibrium in supergames with the overtaking criterion', *Journal of Economic Theory*, 21: 1–9.

Rugg, D. (1973) 'The choice of journey destination: a theoretical and empirical analysis', *Review of Economics and Statistics*, 55: 64–72.

Saarinen, J. (2006) 'Traditions of sustainability in tourism studies', *Annals of Tourism Research,* 33: 1121–40.

Saaty, R.W. (1987) 'The analytic hierarchy process: what it is and how it is used', *Mathematical Modelling*, 9: 161–76.

Sahli, M. and Nowak, J.J. (2007) 'Does inbound tourism benefit developing countries? A trade theoretic approach', *Journal of Travel Research*, 45: 426–34.

Sakai, M.Y. (1988) 'A micro-analysis of business travel demand', *Applied Economics*, 20: 1481–96.

Salop, S.C. (1979a) 'Monopolistic competition with outside goods', *Bell Journal of Economics*, 10: 141–56.

—— (1979b) 'Strategic entry deterrence', *American Economic Review Papers and Proceedings*, 69: 335–38.

Samples, K.C. and Bishop, R.C. (1981) *The Lake Michigan Angler: A Wiscon-sonp Profile*, Madison, Wisc.: University of Wisconson Sea Grant Institute.

Samuelson, P. (1948) 'International trade and the equalization of factor prices', *Economic Journal*, 58: 163–84.

—— (1949) 'International factor price equalization once again', *Economic Journal*, 59: 181–97.

Sargent, T. and Wallace, N. (1976) 'Rational expectations and the theory of economic policy', *Journal of Monetary Economics*, 2: 169–83.

Savage, I. (1984) *The Deregulation of Bus Services*, Aldershot: Gower.

Scarf, H. (2008) 'Computation of general equilibria', *The New Palgrave Dictionary of Economics*, 2nd edn, Basingstoke: Palgrave Macmillan.

Schaeffer, P.V. and Millerick, C.A. (1991) 'The impact of historic district designation on property values: an empirical study', *Economic Development Quarterly*, 5: 301–12.

Scherer, F.M. (1967) 'Research and development resource allocation under rivalry', *Quarterly Journal of Economics*, 81: 359–94.

Scherer, F.M. and Ross, D. (1990) *Industrial Market Structure and Economic Performance*, Boston: Houghton Mifflin Company.

Scheyvens, R. (2002) *Tourism and Development: Empowering Communities:* Harlow: Pearson Education.

Schmalensee, R. (1972) *The Economics of Advertising*, Amsterdam: North Holland.

Schmalensee, R. and Willig, R.D. (eds) (1989) *Handbook of Industrial Organization, Vols 1 and 2*, Amsterdam: North Holland.

Schmoll, G.A. (1977) *Tourism Promotion*, London: Tourism International Press.

Schumacher, E.F. (1973) *Small is Beautiful: A Study of Economics as if People Mattered*, London: Blond Briggs.

Schumpeter, J.A. (1996) *Capitalism, Socialism and Democracy*, London: Routledge.

Schwaninger, M. (1989) 'Trends in leisure and tourism for 2000 to 2010: scenario with consequences for planners', in Witt, S.F. and Moutinho, L. (eds) *Tourism Marketing and Management Handbook*, Hemel Hempstead: Prentice Hall.

Scitovsky, T. (1954) 'Two concepts of external economies', *Journal of Political Economy*, 62: 143–51.

Scott, J. (1997) 'Chances and choices: women and tourism in Northern Cyprus', in Sinclair, M.T. (ed.) *Gender, Work and Tourism*, London and New York: Routledge, 60–90.

Scott, J. (2001) 'Gender and sustainability in Mediterranean island tourism', in Ioannides, Y., Apostolopoulos, E. and Sönmez, E. (eds) *Mediterranean Islands and Sustainable Tourism Development: Practices, Management and Policy*, London: Continuum, 87–107.

Scottish Tourist Board (1993) *Going Green: Guideline for the Scottish Tourism Industry*, Edinburgh: Scottish Tourist Board.

Seely, R.L., Iglarsh, H.J. and Edgell, D.L. (1980) 'Utilizing the Delphi technique at international conferences: a method for forecasting international tourism conditions', *Travel Research Journal*, 1: 30–35.

Seetaram, N. (2008) 'Mauritius', in Graham, A., Papatheodorou, A. and Forsyth, P. (eds) *Aviation and Tourism: Implications for Leisure Travel*, Aldershot: Ashgate, 313–21.

Sen, A. (1979) 'Rational fools', in Hahn, F. and Hollis, M. (eds) *Philosophy and Economic Theory*, Oxford: Oxford University Press.

Sessa, A. (1984) 'Comments on Peter Gray's "The contribution of economics to tourism"', *Annals of Tourism Research*, 11: 283–302.

Shafer, E.L., Moeller, G.H. and Getty, R.E. (1974) 'Future leisure environment', Forest Research Paper NE-301, USDA Forest Experiment Station, Pennsylvania.

Sharpley, R. and Telfer, D.J. (2002) *Tourism and Development*, Clevedon: Channel View.

Shaw, G. and Williams, A.M. (2002) 2nd edn, *Critical Issues in Tourism*, Oxford: Blackwell.

Sheldon, P.J. (1986) 'The tour operator industry: an analysis', *Annals of Tourism Research*, 13: 349–65.

—— (1990) 'A review of tourism expenditure research', in Cooper, C.P. (ed.) *Progress in Tourism, Recreation and Hospitality Management*, Vol. 2, London: Belhaven, 399–403.

—— (1994) 'Tour operators', in Witt, S.F. and Moutinh, L. (eds) *Tourism Management and Marketing Handbook*, 2nd edn, Hemel Hempstead: Prentice Hall.

—— (1997) *Tourism Information Technology*, Wallingford: CABI Publishing.

—— (2006) 'Tourism information technology', in Dwyer, L. and Forsyth, P. (eds) *International Handbook on the Economics of Tourism*, Cheltenham: Edward Elgar, 399–418.

Sigala, M. (2003) 'The information and communication technologies productivity impact on the UK hotel sector', *International Journal of Operations and Production Management*, 23: 1224–45.

—— (2004) 'Collaborative supply chain management in the airline sector: the role of global distribution systems (GDS)', *Advances in Hospitality and Leisure*, 1: 103–21.

Sigala, M. and Christou, E. (2003) 'Enhancing and complementing the instruction of tourism and hospitality courses through the use of on-line educational tools', *Journal of Hospitality & Tourism Education*, 15: 6–16.

Silverman, D. (2004) *Qualitative Research: Theory, Method and Practice*. Thousand Oaks, Calif.: Sage Publications.

Simon, H.A. (1955) 'A behavioural model of rational choice', *Quarterly Journal of Economics*, 69: 99–118.

—— (1957) *Models of Man*, New York: Wiley.

—— (1961) *Administrative Behavior*, 2nd edn, New York: Macmillan.

Sinclair, M.T. (1990) *Tourism Development in Kenya*, Washington, DC: World Bank.

—— (1991a) 'The economics of tourism', in Cooper C.P. (ed.) *Progress in Tourism, Recreation and Hospitality Management*, vol. 3, London: Belhaven, 1–27.

—— (1991b) 'The tourism industry and foreign exchange leakages in a developing country', in Sinclair, M.T. and Stabler, M.J. (eds) *The Tourism Industry: An International Analysis*, Wallingford: CABI, 185–204.

—— (1991c) 'Women, work and skill: economic theories and feminist perspectives', in Redclift, N. and Sinclair, M.T. (eds) *Working Women: International Perspectives on Labour and Gender Ideology*, London and New York: Routledge.

—— (1992) 'Tour operators and tourism development policies in Kenya', *Annals of Tourism Research*, 19: 555–58.

—— (ed.) (1997a) *Gender, Work and Tourism*, London and New York: Routledge.

—— (1997b) 'Issues and theories of gender and work in tourism', in Sinclair, M.T. (ed.) *Gender, Work and Tourism*, London: Routledge, 1–15.

Sinclair, M.T. and Bote Gómez, V. (1996) 'Tourism, the Spanish economy and the balance of payments', in Barke, M., Newton, M. and Towne, J. (eds) *Tourism in Spain: Critical Perspectives*, Wallingford: CABI.

Sinclair, M.T. and Stabler, M.J. (eds) (1991) *The Tourism Industry: An International Analysis*, Wallingford: CABI.

Sinclair, M.T. and Sutcliffe, C.M.S. (1978) 'The first round of the Keynesian income multiplier', *Scottish Journal of Political Economy*, 25: 177–86.

—— (1988a) 'The estimation of Keynesian income multipliers at the sub-national level', *Applied Economics*, 20: 1435–44.

—— (1988b) 'Negative multipliers: a case for disaggregated estimation', *Tijdschrift Voor Economische en Sociale Geografie*, 79: 104–7.

—— (1989) 'Truncated income multipliers and local income generation over time', *Applied Economics*, 21: 1621–30.

Sinclair, M.T. and Tsegaye, A. (1990) 'International tourism and export instability', *Journal of Development Studies*, 26: 487–504.

Sinclair, M.T., Alizadeh, P. and Atieno Adero Onunga, E. (1992) 'The structure of international tourism and tourism development in Kenya', in Harrison, D. (ed.) *Tourism and the Less Developed Countries*, London: Belhaven.

Sinclair, T., Clewer, A. and Pack, A. (1990) 'Hedonic prices and the marketing of package holidays: the case of tourism resorts in Malaga', in Ashworth, G. and Goodall, B. (eds) *Marketing Tourism Places*, Routledge, London, 85–103.

Sindiyo, D.M. and Pertet, F.N. (1984) 'Tourism and its impact on wildlife in Kenya', *Industry and Environment*, 7: 14–19.

Slovic, P, Fischoff, B. and Lichtenstein, S. (1977) 'Behavioural decision theory', *Annual Review of Psychology*, 28: 1–39.

Small Luxury Hotels of the World (2009) *About Us*, Online. Available HTTP: <http://www.slh.com/aboutus.html> (accessed 10 May 2009)

Smeral, E. (1988) 'Tourism demand, economic theory and econometrics: an integrated approach', *Journal of Travel Research*, 26: 38–43.

—— (1998) 'The impact of globalization on small and medium enterprises: new challenges for tourism policies in European countries', *Tourism Management*, 19: 371–80.

—— (2006) 'Tourism satellite accounts: a critical assessment', *Journal of Travel Research*, 45: 92–98.

Smeral, E. and Witt, S.F. (1996) 'Econometric forecasts of tourism demand to 2005', *Annals of Tourism Research*, 23: 891–907.

Smith, A. (1994) 'Imperfect competition and international trade', in Greenaway, D. and Winters, L.A. (eds) *Surveys in International Trade*, Oxford: Basil Blackwell.

Smith, C. and Jenner, P. (1984) 'Tourism and the environment', *Travel and Tourism Analyst*, 5: 68–86.

Smith, S. (1998) 'Tourism as an industry – debate and concepts', in Ioannides, D. and Debbage, K.G. (eds) *The Economic Geography of the Tourism Industry – A Supply Side Analysis*, London: Routledge, 31–52.

Smith, S. and Wilton, D. (1997) 'TSAs and the WTTC/WEFA methodology: different satellites or different planets?', *Tourism Economics*, 3: 249–63.

Smith, S. and Wilton, D. (1989) *Tourism Analysis: A Handbook*, Harlow: Longman.

Smith, V.K. and Desvouges, W. (1986) *Measuring Water Quality Benefits*, Boston, Mass.: Kluwer.

Smith, V.K., Palmquist, R.B. and Jakus, P. (1991) 'Combining Farrel frontier and hedonic travel cost models for valuing estuarine quality', *Review of Economics and Statistics*, 63: 694–99.

Sol Meliá Cuba (2008) *Cuba hotels and resorts*, Online. Available HTTP: <http://www.solmeliacuba.com/cuba-hotels-resorts> (accessed 6 November 2008).

Solow, R.M. (1956) 'A contribution to the theory of economic growth', *Quarterly Journal of Economics*, 70: 65–94.

Song, B.-N. and Ahn, C.-Y. (1983) 'Korea', in Pye, E.A. and Lin,T.-B. (eds) *Tourism in Asia: The Economic Impact*, Singapore: Singapore University Press.

Song, H. and Witt, S.F. (2000). *Tourism Demand Modelling and Forecasting: Modern Econometric Approaches*, Cambridge: Pergamon.

Song, H. and Turner, L. (2006) 'Tourism demand forecasting', in Dwyer, L. and Forsyth, P. (eds) *International Handbook on the Economics of Touris*, Cheltenham: Edward Elgar, 89–114.

Song, H. and Li, G. (2008) 'Tourism demand modelling and forecasting – a review of recent research', *Tourism Management*, 29: 203–20.

Sönmez, S.F. (1998) 'Tourism, terrorism, and political instability', *Annals of Tourism Research*, 25: 416–56.

Southgate, C. and Sharpley, R. (2002) 'Tourism development and the environment, in Sharpley, R. and Telfer, D.J. (eds) *Tourism and Development*, Clevedon: Channel View, 231–62.

Spence, A.M. (1977) 'Entry, capacity, investment and oligopolistic pricing', *Bell Journal of Economics*, 8: 534–44.

Spurr, R. (2006) 'Tourism satellite accounts', in Dwyer, L. and Forsyth, P. (eds) *International Handbook on the Economics of Tourism*, Cheltenham: Edward Elgar Publishing, 283–300.

Srinivas, H. (2009) *Tourism Related Journals*, Online. Available HTTP: <http://www.gdrc.org/uem/eco-four/journals.html> (accessed 5 March 2009).

Stabler, M.J. (1995) 'Research in progress on the economic and social value of conservation', in Burman, P., Pickard, R. and Taylor, S. (eds) *The Economics of Architectural Conservation*, York: Institute of Advanced Architectural Studies, University of York.

—— (1996a) 'The emerging new world of leisure quality: does it matter and can it be measured?', in Collins, M. (ed.) *Leisure in Different Words, vol. 2, Leisure in Industrial and Post-Industrial Societies*, Eastbourne: Leisure Studies Association.

—— (1996b) 'Managing the leisure natural resource base: utter confusion or evolving consensus?', Paper presented at the World Leisure and Recreation Association Fourth World Congress, Free Time and the Quality of Life for the 21st Century, Cardiff, July.

—— (1997) (ed.) *Tourism and Sustainability: Principles to Practice*, Wallingford: CABI.

—— (1998) 'The economic evaluation of the role of conservation and tourism in the regeneration of historic urban destinations', in Laws, E., Faulkner, B., Moscardo, G. (eds) *Embracing and Managing Change in Tourism*, London: Routledge, 235–63.

—— (1999) 'Environmental aspects of tourism: applications of cost-benefit analysis', in Baum, T. and Mudambi, R. (eds) *Economic and Management Methods for Tourism and Hospitality Research*, Chichester: Wiley, 233–67.

Stabler, M.J. and Goodall, B. (1996) 'Environmental auditing in planning for sustainable island tourism', in Briguglio, L., Archer, B., Jafari, J. and Wall, G. (eds) *Sustainable Tourism in Islands and Small States: Issues and Policies*, London: Pinter (Cassell), 170–196.

—— (1997) 'Environmental awareness, action and performance in the tourism industry: a case study of the hospitality sector in Guernsey', *Tourism Management*, 18: 19–33.

Stacey, B.G. (1982) 'Economic socialization in the pre-adult years', *British Journal of Social Psychology*, 21: 159–73.

Stavrinoudis, T. (2006a) 'Timeshare in Greece: an investigation of the causes for its unsatisfactory development', *Tourism Today*, 6: 171–177.

—— (2006b) 'Advantages, opportunities and policy guidelines concerning the development of timeshare combined with cruises in Greece and Turkey', *Tourism in Marine Environments*, 3: 25–34.

—— (2008) 'Applying S.W.O.T. analysis methodology in the formulation of propositions aiming at a more effective operational application of timeshare in Greece', *Tourismos*, 3: Autumn, 113–138.

Stem, C.J., Lassole, J.P., Lee, D.R. and Deshler, D.J. (2003) 'How "eco" is ecotourism? A comparative case study of ecotourism in Costa Rica', *Journal of Sustainable Tourism*, 11: 322–347.

Stergiou, D., Airey, D. and Riley, M. (2008) 'Making sense of tourism teaching', *Annals of Tourism Research*, 35: 631–49.

Stern Report (2006) *Stern Review on the Economics of Climate Change*, London: HM Treasury Office.

Stiglitz, J.E. (1989) 'Imperfect information in the product market', in Schmalensee, R. and Willig, R.D. (eds) *Handbook of Industrial Organization, vol. 1*, Amsterdam: North Holland.

Strasbourg and Bas-Rhin Chamber of Commerce and Industry (2003) *Communiqué*, 26th August.

Stronge, W.B. and Redman, M. (1982) 'US tourism in Mexico: an empirical analysis', *Annals of Tourism Research*, 9: 21–35.

Sugiyarto, G., Blake, A. and Sinclair, M.T. (2003) 'Tourism and globalization: economic impact in Indonesia', *Annals of Tourism Research*, 30: 683–701.

Sutcliffe, C.M.S. and Sinclair, M.T. (1980) 'The measurement of seasonality within the tourist industry: an application of tourist arrivals in Spain', *Applied Economics*, 12: 429–41.

Svoronou, E. and Holden, A. (2004) 'Ecotourism as a tool for nature conservation: the role of WWF Greece in the Dadia-Lefkimi-Soufli forest reserve in Greece', *Journal of Sustainable Tourism*, 12: 456–67.

Swain, M.B. (ed.) (1995) 'Gender in tourism', special issue, *Annals of Tourism Research*, 22, 2.

Sweeting, J.E.N. and Wayne, S.L. (2006) 'A shifting tide: environmental challenges and cruise industry responses', in Dowling, R.K. (ed.) *Cruise Ship Tourism*, Wallingford: CABI, 327–337.

Syriopoulos, T. (1995) 'A dynamic model of demand for Mediterranean tourism', *International Review of Applied Economics*, 9: 318–36.

Syriopoulos, T. and Sinclair, M.T. (1993) 'An econometric study of tourism demand: the AIDS model of US and European tourism in Mediterranean countries', *Applied Economics*, 25, 12: 1541–52.

Talberth, J., Cobb, C. and Slattery, N. (2006) *The Genuine Progress Indicator: A Tool for Sustainable Development*, Oakland, Calif.: Redefining Progress.

Tamoepeau, S. (2008) 'South Pacific', in Graham, A., Papatheodorou, A. and Forsyth, P. (eds) *Aviation and Tourism: Implications for Leisure Travel*, Aldershot: Ashgate, 323–31.

Tan, L. (1992) 'A Heckscher-Ohlin approach to changing comparative advantage in Singapore's manufacturing sector', *Weltwirtschaftliches Archiv*, 128: 288–309.

Taylor, P. (1995), 'Measuring changes in the relative competitiveness of package tour destinations', *Tourism Economics* 1: 169–82.

—— (1998) 'Mixed strategy pricing behaviour in the UK package tour industry', *International Journal of the Economics of Business*, 5: 29–46.

Telfer, D.J. and Wall, G. (1996) 'Linkages between tourism and food production', *Annals of Tourism Research*, 23: 635–53.

Teo, P. (2002) 'Striking a balance for sustainable tourism: implications for the discourse on globalisation', *Journal of Sustainable Tourism*, 10: 459–74.

TEPRO (1985) *The Tourism Sector in the Community – A Study of Concentration, Competition and Competitiveness*, Luxembourg: Commission of the European Communities.

Tietenberg, T. (1997) *The Economics of Global Warming*, Cheltenham: Edward Elgar.

—— (2006) 7th edn, *Environmental and Natural Resource Economics*, London: Pearson.

Tirole, J. (1988) *The Theory of Industrial Organization*, Cambridge, Mass.: MIT Press.

Tobey, J.A. (1990) 'The effects of domestic environmental policies on world trade: an empirical test', *Kyklos*, 43: 191–209.

Toms, M. (2001) *Thoughts on Airport Regulation*, University of Westminster/Cranfield University Airport Economics and Finance Symposium, March.

Toner, J.P. (2001) *The London Bus Tendering Regime: Principles and Practice*. 7th International Conference on Competition and Ownership in Land Passenger Transport, Molde, Norway.

Tongzon, J. (2001) 'Efficiency measurement of selected Australian and other international ports using data envelopment analysis', *Transportation Research Part A*, 35: 113–28.

Tooman, L.A. (1997) 'Application of the life-cycle model in tourism', *Annals of Tourism Research*, 21: 214–34.

Toulantas, G. (2001) 'Valuing tour operators during volatile times', *Deloitte & Touche Leisure Review*, 4: 5–11.

Tourism Concern (1995) 'Our holidays, their homes', special issue on people displaced by tourism, *In Focus*, 15, spring: 3–13.

TravelMole (2003a) 'Ryanair Boss defends Business Model', *Travelmole*, 25th September. Online. Available HTTP: <http://www.travelmole.com> (accessed 25 September 2003).

—— (2003b) 'Princess Deal is Sealed', *Travelmole*, 22nd April. Online. Available HTTP: <http://www.travelmole.com> (accessed 22 April 2003).

Tremblay, P. (1989) 'Pooling international tourism in Western Europe', *Annals of Tourism Research*, 16: 477–91.

—— (1998) 'The economic organization of tourism', *Annals of Tourism Research*, 25: 837–59.

Tribe, J. (1997) 'The indiscipline of tourism', *Annals of Tourism Research*, 24: 638–57.

—— (1999) 2nd edn, *The Economics of Leisure and Tourism*, Oxford: Butterworth Heinemann.

—— (2000) 'Indisciplined and unsubstantiated', *Annals of Tourism Research*, 27: 809–13.

—— (2001) 'Research paradigms and the tourism curriculum', *Journal of Travel Research*, 39: 442–48.

—— (2004) 'Knowing about tourism: epistemological issues', in Phillimore, J. and Goodson, L. (eds) *Qualitative Research in Tourism: Ontologies, Epistemologies and Methodologies.*

Triplett, J.E. (1975), 'Consumer demand and characteristics of consumption goods', in Terlckyj, N. (ed.) *Household Production and Consumption*, Washington, DC: National Bureau of Economic Research, 305–23.

Troyer, W. (1992) *The Green Partnership Guide*, Toronto: Canadian Pacific Hotels and Resorts.

Tsartas, P. (1992) 'Social and economic impacts of tourist development on the islands of Serifos and Ios, Greece', *Annals of Tourism Research*, 19: 516–33.

Tsartas, P. and Lagos, D. (2004) 'Critical appraisal of regional tourism development and policy in Greece', *Journal of Applied Economics and Management*, 2: 15–31.

TUI (2008) *Company Presentation*, Online. Available HTTP: <http://www.tui-group.com> (accessed 26 November 2008).

Turner, L.W. and Witt, S.F. (2001) 'Factors influencing demand for international tourism: tourism demand analysis using structural equation modelling revisited', *Tourism Economics*, 7, 21–38.

Turner, R.K. (ed.) (1988) *Sustainable Environmental Management Principles and Practice*, London: Belhaven.

Turner, R.K., Pearce, D.W. and Bateman, I. (1994) *Environmental Economics: An Elementary Introduction*, London: Harvester Wheatsheaf.

Twining-Ward, L. and Farrell, B.H. (2005) 'Sustainable tourism', in Miller, G. and Twining-Ward, L. (eds) *Monitoring for a Sustainable Tourism Transition: The Challenge of Developing and Using Indicators*, Wallingford: CABI.

Ungson, G.R., Braunstein, D.N. and Hall, P.D. (1981) 'Managerial information processing: a research review', *Administrative Science Quarterly*, 26: 116–34.

United Kingdom National Accounts (2007) *The Blue Book*, London: The Office for National Statistics.

UNCTAD (1988) *Trade and Development Report*, Geneva: UNCTAD.

United Nations Statistics Division (2008a) 'The SNA as a System', available from: http://unstats.un.org/unsd/sna1993/tocLev8.asp?L1 = 1&L2 = 1 (accessed on 07/11/2008).

—— (2008b) 'Detailed Structure and Explanatory Notes', available from: http://unstats.un.org/unsd/cr/registryr/egcst.asp?Cl = 27&Lg = 1 (accessed on 07/11/2008).

UNWTO (1988) *The Problems of Protectionism and Measures to Reduce Obstacles to International Trade in Tourism Services*, Madrid: UNWTO.

—— (1995) *GATS Implications for Tourism: The General Agreement on Trade in Services and Tourism*, Madrid: UNWTO.

—— (1999a) *Tourism Satellite Accounts (TSA): The Conceptual Framework*, Madrid: UNWTO.

—— (1999b) *Tourism Taxation: Striking a Fair Deal*, Madrid: UNWTO.

—— (2000) *Measuring Total Tourism Demand*, Madrid: UNWTO.

—— (2002) *Tourism in the Age of Alliances, Mergers and Acquisitions*, Madrid: UNWTO.

—— (2003) *Worldwide Cruise Ship Activity*, Madrid: UNWTO.

—— (2006) *Tourism and the Least Developed Countries: A Sustainable Opportunity to Reduce Poverty*, Madrid: UNWTO.

—— (2007) *Sustainable Tourism Eliminating Poverty*, Madrid: UNWTO.

—— (2008a) *Task Force for the Protection of Children in Tourism*, Online. Available HTTP: <http://www.unwto.org/protect_children/index.php> (accessed 7 November 2008).

—— (2008b) *Tourism Highlights 2008 Edition*, Madrid: UNWTO.

US Census Bureau (2005) *Establishment and Firm Size: 2002 Economic Census – Accommodation and Food Services*, Washington: US Census Bureau.

US Department of Transportation (2004) *Computer Reservation Systems (CRS) Regulation: Final Rule*, Online. Available HTTP: <http://www.dot.gov/affairs/Computer%20Reservations%20System.htm> (accessed 6 March 2009).

USTOA (2009) *USTOA Fact Sheet*, Online. Available HTTP: < http://www.ustoa.com/fastfacts.cfm> (accessed 4 May 2009).

Utell (2009) *About Us*, Online. Available HTTP: <http://www.utell.com/about.html> (accessed 11 May 2009).

Uysal, M. and Crompton, J.L. (1984) 'Determinants of demand for international tourist flows in Turkey', *Tourism Management*, 5: 288–97.

Van der Ploeg, F. and Tang, P. (1994) 'Growth, deficits and research and development in the global economy', in van der Ploeg, F. (ed.) *The Handbook of International Macroeconomics*, Oxford: Basil Blackwell.

Van Doorn, J.W.M. (1984) 'Tourism forecasting and the policymaker: criteria of usefulness', *Tourism Management*, 5: 24–39.

Van Doren, C.S., Koh, Y.K. and McCahill, A. (1994) 'Tourism research: a state of the art citation analysis', in Seaton, A.V. (ed.) *Tourism: The State of the Art*, Chichester: John Wiley.

Varian, H. (2006) *Intermediate Microeconomics: A Modern Approach* London: Norton.

Varley, R.C.G. (1978) 'Tourism in Fiji: some economic and social problems', Occasional Papers in Economics, no. 12, Bangor: University of Wales Press.

Veblen, T. (1899) *The Theory of the Leisure Class*, New York: Mentor.

Venables, A. (1996) 'Equilibrium locations of vertically linked industries', *International Economic Review*, 37: 341–59.

Ventura, J. (1997) 'Growth and interdependence', *Quarterly Journal of Economics*, 112: 57–84.

Vernon, R. (1966) 'International investment and international trade in the product cycle', *Quarterly Journal of Economics*, 80: 190–207.

Vickers, J. and Yarrow, G. (1988) *Privatisation: An Economic Analysis*, Cambridge, Mass.: MIT Press.

Vidal, J. (2007) 'CO_2 output from shipping twice as much as airlines', *Guardian*, Saturday, March 3rd.

Vlamis, P. (2007) 'Default risk of corporate entities: a critical assessment', *International Corporate Rescue*, 4: 99–105.

Vogel, D. (1995) *Trading Up: Consumers and Environmental Regulations in the Global Economy*, Cambridge Mass.: Harvard University Press.

Voogd, H. (1988) 'Multicriteria evaluation: measures, manipulation and meaning: a reply', *Environment and Planning B: Planning and Design*, 15: 65–72.

Wackernagel, M. and Rees, W. (1996) *Our Ecological Footprint: Reducing Human Impact on the Earth*, Gabriola Island, BC: New Society Publishers.

Wahab, S. and Pigram, J.J. (1997) *Tourism Development and Growth: The Challenge of Sustainability*, London: Routledge.

Wahab, S. and Cooper, C. (2001) *Tourism in the Age of Globalisation*, London: Routledge.

Wall, G. (1994) 'Ecotourism: old wine in new bottles', *Trends*, 31: 4–9.

—— (1997) 'Is ecotourism sustainable?', *Environmental Management*, 21: 483–91.

Wandner, S.A. and Van Erden, J.D. (1980) 'Estimating the demand for international tourism using time series analysis', in Hawkins, D.E. Shafer, E.L. and Rovelstad, J.M. (eds) *Tourism Planning and Development Issues*, Washington, DC: George Washington University.

Wanhill, S.R.C. (1980) 'Charging for congestion at tourist attractions', *International Journal of Tourism Management*, 1: 168–74.

—— (1982) 'Evaluating the resource costs of tourism', *Tourism Management*, 3: 208–11.

—— (1988) 'Tourism multipliers under capacity constraints', *Service Industries Journal*, 8: 136–42.

—— (2006) 'Competition in visitor attractions', in Papatheodorou, A. (ed.) *Corporate Rivalry and Market Power: Competition Issues in the Tourism Industry*, London: I.B. Tauris, 172–86.

—— (2007) 'What tourism economists do (the distribution of submissions to the journal *Tourism Economics*)' paper delivered at the 6th DeHaan Tourism Management Conference, Tourism and Travel Research Institute, 18th December.

Wattanakuljarus, A. and Coxhead, I. (2008) 'Is tourism-based development good for the poor?: a general equilibrium analysis for Thailand', *Journal of Policy Modelling*, 30: 929–55.

Wearing, S. and Neil, J. (1999) *Ecotourism: Impacts, Potentials and Possibilities*, Oxford: Butterworth Heinemann.

Weaver, D.B. (1998) *Ecotourism in the Less Developed World*, Wallingford: CABI.

Weaver, D.B. (ed.) (2001) *The Encyclopaedia of Ecotourism*, Wallingford: CABI.

Weber, A. (2002) 'Concerning location of industries, part one: the pure theory of location', *Géographie, Economie, Société*, 4: 363–76.

Webber, A.G. (2001) 'Exchange rate volatility and co-integration in tourism demand', *Journal of Travel Research*, 39: 398–405.

Weed, M. (2006) 'Undiscovered public knowledge: the potential of research synthesis approaches in tourism research', *Current Issues in Tourism*, 9: 256–68.

Wheat, S. (1999) 'To go or not to go to Indonesia', *Tourism in Focus*, 33.

Wheeler, M. (1994) 'The emergence of ethics in tourism and hospitality', *Progress in Tourism, Recreation and Hospitality Management*, 6: 647–654.

Wheeller, B. (1994) 'Egotourism, sustainable tourism and the environment: a symbiotic, symbolic or shambolic relationship', in Seaton, A.V. (ed.) *Tourism: The State of the Art*, Chichester: Wiley, 647–54.

—— (2004) 'The truth? The hole truth. Everything but the truth. Tourism and knowledge: a septic sceptic's perspective', *Current Issues in Tourism*, 7: 467–77.

White, K.J. (1982) 'The demand for international travel: a system-wide analysis for US travel to Western Europe', Discussion Paper no. 82–28, University of British Columbia, Canada.

White, P.R. (2008) *Public Transport: Its Planning, Management and Operation*, 5th edn, London: Routledge.

White, P.R. and Farrington, J. (1998) 'Bus and coach deregulation and privatization in Great Britain with particular reference to Scotland', *Journal of Transport Geography*, 6: 135–41.

White, P.R. and Tough, S. (1995) 'Alternative tendering systems and deregulation in Britain', *Journal of Transport Economics and Policy*, 29: 275–90.

Williams, A.M. (2004) 'Toward a political economy of tourism', in Lew, A.A., Hall. C.M. and Williams, A.M. (eds) *A Companion to Tourism*, Oxford: Blackwell, 61–73.

Williams, A.M., King, R. and Warnes, T. (2004) 'British second homes in Southern Europe: shifting nodes in the scapes and flows of migration and tourism', in Hall, M. and Muller, D.K. (eds) *Tourism, Mobility and Second Homes*, Clevedon: Channel View 97–112.

Williams, P. and Hobson, J.S.P. (1995) 'Virtual reality and tourism: fact or fantasy', *Tourism Management*, 16: 423–27.

Williams, S. (2004) *Critical Concepts in the Social Sciences, Vol. 3: Tourism Development and Sustainability*, London: Routledge.

Williamson, O.E. (1985) *The Economic Institutions of Capitalism*, New York: Free Press.

—— (1989) 'Transaction cost economies', in Schmalensee, R. and Willig, R.D. (eds) *Handbook of Industrial Organization, vol. 1*, Amsterdam: North Holland.

—— (1986) *Economic Organization: Firms, Markets and Policy Control*, Brighton: Wheatsheaf.

—— (1996) *The Mechanisms of Governance*, New York: Oxford University Press.

Willis, K.G. (1989) 'Option value and non-user benefits of wildlife conservation', *Journal of Rural Studies*, 5: 245–56.

Willis, K.G. and Garrod, G. (1991a) 'An individual travel cost method of evaluating forest recreation', *Journal of Agricultural Economics*, 42: 33–42.

—— (1991b) 'Valuing open access recreation on inland waterways: on-site recreational surveys and selection effects', *Regional Studies*, 25: 511–24.

—— (1991c) 'Landscape values: a contingent valuation approach and case study of the Yorkshire Dales National Park', Countryside Change Unit Working Paper 21, Newcastle-upon-Tyne: University of Newcastle.

—— (1993a) 'The value of waterside properties: estimating the impact of waterways and canals on property values through hedonic price models and contingent valuation methods', Countryside Change Unit Working Paper 44, Newcastle-upon-Tyne: University of Newcastle.

—— (1993b) 'Valuing wildlife: the benefits of wildlife trusts', Countryside Change Unit Working Paper 46, Newcastle upon Tyne: University of Newcastle.

Willis, K.G., Garrod, G. and Dobbs, I.M. (1990) 'The value of canals as a public good: the case of the Montgomery and Lancaster Canals', Countryside Change Unit Working Paper 5, Newcastle-upon-Tyne: University of Newcastle.

Willis, K.G., Garrod, G, Saunders, C. and Whitby, M. (1993) 'Assessing methodologies to value the benefits of environmentally sensitive areas', Countryside Change Unit Working Paper 39, Newcastle upon Tyne: University of Newcastle.

Wilson, K. (1998) 'Market/industry confusion in tourism economic analysis', *Annals of Tourism Research* 25: 803–17.

Wilson, P. (1994) 'Tourism earnings instability in Singapore, 1972–88', *Journal of Economic Studies*, 21: 41–51.

Witt, C.A., Witt, S.F. and Wilson, N. (1994) 'Forecasting international tourist flows', *Annals of Tourism Research*, 21: 612–28.

Witt, S.F. (1980) 'An econometric comparison of UK and German foreign holiday behaviour', *Managerial and Decision Economics*, 1: 123–31.

Witt, S.F. and Martin, C.A. (1987) 'Econometric models for forecasting international tourism demand', *Journal of Travel Research*, 25: 23–30.

—— (1989) 'Demand forecasting in tourism and recreation', in Cooper, C.P. (ed.) *Progress in Tourism, Recreation and Hospitality Management, vol. 1*, London: Belhaven, 4–32.

Wolf, H. (2004) 'Airport privatisation and regulation – getting the institutions right', in Forsyth, P., Gillen, D. Niemeier, H.M. and Starkie, D. (eds) *The Economic Regulation of Airports: Recent Developments in Australasia, North America and Europe*, Aldershot: Ashgate, 201–11.

Wolf, M. (2004) *Why Globalization Works*, New Haven: Yale University Press.

World Commission on Environmental Development (WCED) (1987) *Our Common Future* Oxford: Oxford University Press.

WTTC (1994) *Green Globe: An Invitation to Join*, London: WTTC.

—— (2007) *Progress and Priorities, 2007*, London: WTTC.

WTO (1999) *World Trade Report*, Geneva: WTO.

—— (2000) *Conclusions and Recommendations of the Expert Meeting on Tourism organized by UNCTAD*, S/C/W/149, 23rd May, Geneva: WTO.

—— (2006) *The General Agreement on Trade in Services: An Introduction*, Online. Available HTTP: <http://www.wto.org> (accessed 30 March 2009).

—— (2007) *Report of the Second Session of the Review Mandated under Paragraph 5 of the Annex on Air Transport Services held on 2nd October 2007*, S/C/M/89, 19th November, Geneva: WTO.

Xie, J. (1996) *Environmental Policy Analysis: A General Equilibrium Approach*, Aldershot: Avebury.

Young, A. (1991) 'Learning by doing and the dynamic effects of international trade', *Quarterly Journal of Economics*, 106: 369–405.

Young, E.H. (2003) 'Balancing conservation with development in marine-dependent communities', in Zimmerer, K.S. and Bassett, T.J. (eds) *Political Ecology: An Integrative Approach to Geography and Environment-Development Studies*, London: Guilford Press, 29–49.

Zacharatos, G. (1986) *Tourism Consumption: Estimation Methods and their Usefulness in the Study of Tourism Impacts on the National Economy* (in Greek), Athens: Centre for Planning and Economic Research (KEPE).

Zahedi, F. (1986) 'The analytic hierarchy process: a survey of the method and its application', *Interfaces*, 16: 96–108.

Zeldes, S.P. (1989) 'Consumption and liquidity constraints: an empirical investigation', *Journal of Political Economy*, 97: 305–46.

Zenelis, P. and Papatheodorou, A. (2008) *Low Cost Carriers' Penetration: A Comparative Case Study of Greece & Spain*, 12th Annual Conference of the Air Transport Research Society, Athens, Greece.

Zhang, J. and Jensen, C. (2007) 'Comparative advantage: explaining tourism flows', *Annals of Tourism Research* 34: 223–43.

Zhou, D., Yanagida, J.F., Chakravorty, V. and Leung, P. (1997) 'Estimating economic impacts from tourism', *Annals of Tourism Research*, 24: 76–89.

INDEX